AF411428

Multiscale Impacts of Anthropogenic and Climate Changes on Tropical and Mediterranean Hydrology

Multiscale Impacts of Anthropogenic and Climate Changes on Tropical and Mediterranean Hydrology

Editors

Luc Descroix
Gil Mahé
Alain Laraque
Olivier Ribolzi
Guillaume Lacombe

MDPI • Basel • Beijing • Wuhan • Barcelona • Belgrade • Manchester • Tokyo • Cluj • Tianjin

Editors

Luc Descroix
IRD, UMR PALOC (Patrimoines
Locaux, Environnement et
Globalisation)
IRD/MNHN, Paris
France

Gil Mahé
IRD, UMR HSM IRD/
Univ Montpellier,
France

Alain Laraque
IRD, UMR GET
IRD/CNRS/CNES/UPS
France

Olivier Ribolzi
IRD, UMR GET
IRD/CNRS/CNES/UPS
France

Guillaume Lacombe
CIRAD, UMR G-EAU
Institut Agronomique et
Vétérinaire Hassan II
Morocco

Editorial Office
MDPI
St. Alban-Anlage 66
4052 Basel, Switzerland

This is a reprint of articles from the Special Issue published online in the open access journal *Water* (ISSN 2073-4441) (available at: https://www.mdpi.com/journal/water/special_issues/Tropical_Mediterranean_Hydrology).

For citation purposes, cite each article independently as indicated on the article page online and as indicated below:

LastName, A.A.; LastName, B.B.; LastName, C.C. Article Title. *Journal Name* **Year**, *Volume Number*, Page Range.

ISBN 978-3-0365-1236-5 (Hbk)
ISBN 978-3-0365-1237-2 (PDF)

Cover image courtesy of Alain Laraque.
It depicts the Canal Casiquiare in Venezuela, South America.

Contents

About the Editors

Luc Descroix is a hydrologist of IRD (Institut de Recherche pour le Développement, the French Research Institute for Sustainable Development, Marseille, France). His lab PALOC (Patrimoines Locaux, Environnement et Globalisation) is located in the Muséum National d'Histoire Naturelle, in Paris, France. L Descroix defended his PhD in geography at Lyon (France) Université Lumière, and he holds his HDR (the degree to become senior researcher in Europe) in hydrology at Grenoble, France, Université Joseph Fourier. His research focuses on the impact of land use and climate changes on the water cycle, through the evolution of water resources (streamflows, groundwater, ponds and lakes, wetlands, soil water, etc.), soil water holding capacity and river recession coefficients. He is also very interested in the role of trees in the field, in the cities, in the societies and their impact on local and regional climates.

Gil Mahé is a research director at the Institute of Research for Development. He has worked on the hydrology and climatology of Africa for more than three decades and spent many years working in cooperation with national and international institutions in many African countries. He has studied hydroclimatic variability and its impact on river regimes and water resources, and the impact of human activities on the environment. He also studied erosion and solid transport and the alteration of the sediment transport by dams. He leads the FRIEND-Water program of UNESCO/IHP and he is the president of the International Commission of Surface Waters of the International Association of Hydrological Sciences.

Alain Laraque is a potamologist at the French National Research Institute for Sustainable Development. For more than three decades, he has been studying the main hydrosystems (Amazon, Congo, Orinoco) that converge in the Intertropical Atlantic Ocean, with the study of their hydroclimatic, hydrosedimentological, geochemical and geodynamic functioning, as well as the anthropic impacts (dams, etc.). He also has experience in hydro-saline processes within hill reservoirs in semi-arid environments.

Olivier Ribolzi is a research director at the Institute of Research for Sustainable Development (IRD) in the field of hydro-geochemistry, having gained over 20 years of experience in conducting and coordinating interdisciplinary research in IRD, mainly in Southeast Asia, West Africa and the Mediterranean region. His main research interests include the effect of land use/land use change on the hydro-biogeochemical response of headwater catchments, soil erosion and sediment yield, groundwater–stream water interactions, and the occurrence/spread of contaminants. His publishing credits include more than 70 articles in refereed journals.

Guillaume Lacombe is a researcher in hydrology and agricultural water management at CIRAD (French agricultural research and international cooperation organization). He has authored more than 25 peer-reviewed journal articles, assessing how human activities alter the spatial and temporal variability of water resources over multiple scales, in order to improve the efficiency and sustainability of water uses. Geographic focuses include developing countries in Asia and Africa, where he has been collaborating with research for development organizations and universities.

Editorial

Special Issue "Multiscale Impacts of Anthropogenic and Climate Changes on Tropical and Mediterranean Hydrology"

Gil Mahé [1], Luc Descroix [2,3,*], Alain Laraque [4], Olivier Ribolzi [5] and Guillaume Lacombe [6]

[1] IRD HSM, Case MSE, Université Montpellier, 300 avenue Emile Jeanbrau, 34090 Montpellier, France; gilmahe@hotmail.com

[2] UMR PALOC IRD/MNHN/Sorbonne Université, Museum National d'Histoire Naturelle, 75231 Paris, France

[3] LMI PATEO, UGB, St. Louis 46024, Senegal

[4] Géosciences Environnement Toulouse, Université de Toulouse, CNES, CNRS, IRD, UPS, 31400 Toulouse, France; alain.laraque@ird.fr

[5] Géosciences Environnement Toulouse, Université de Toulouse, CNES, CNRS, IRD, UPS, 31400 Toulouse, France; olivier.ribolzi@ird.fr

[6] CIRAD, UMR G-EAU, Université Montpellier, AgroParisTech, CIRAD, INRAE, Institut Agro, IRD, 34398 Montpellier, France; guillaume.lacombe@cirad.fr

* Correspondence: luc.descroix@ird.fr

Citation: Mahé, G.; Descroix, L.; Laraque, A.; Ribolzi, O.; Lacombe, G. Special Issue "Multiscale Impacts of Anthropogenic and Climate Changes on Tropical and Mediterranean Hydrology". *Water* **2021**, *13*, 491. https://doi.org/10.3390/w13040491

Received: 14 January 2021
Accepted: 8 February 2021
Published: 14 February 2021

Publisher's Note: MDPI stays neutral with regard to jurisdictional claims in published maps and institutional affiliations.

1. Introduction

Most tropical and Mediterranean areas, landscapes, soils, and territories are experiencing new vulnerabilities, facing global warming [1] and accelerating changes in land-use in all kinds of ecosystems [2]. Long droughts, dry spells, rainfall intensification, and an increase in the number of storms and cyclones make agriculture as well as land management and water and sediment control more difficult. In many regions, the population increase is too strong to allow cropping and rural activities to easily reach Boserupian behavior. The intensification of the climatic cycle induced by global warming commonly leads to an acceleration of the hydrological cycle, increasing the occurrence of flooding, inundation, as well as droughts and water shortages [3,4]. Human actions and overall rural activity can strongly modify water runoff and infiltration, then water balance, by increasing infiltration and buffering the water cycle, or on the contrary, by reducing the soil permeability, increasing runoff, and accelerating the water cycle [5–7]. Rural practices are commonly suspected to decrease the soil water-holding capacity. This could lead to a rise in flooding occurrence and intensity downstream, and could cause an edaphic drought even in areas where no climatic drought is observed.

These kinds of changes influence hydrology and erosion/sedimentation processes, with implications for various risks: food insecurity, natural disasters (landslides, flood damage), water shortages, and pathogenic contaminations. The accelerating pace of these environmental changes leaves limited time for adaptation. While the need for "climate change adaptation" is recognized, there is also a need for land-use change adaptation and a better assessment of contamination risks [8]. To date, there is limited understanding of the processes linking land-use management to these risks.

There is a need to understand causal chains of processes, from land-use, vegetation structure, soil, erosion, hydrological connectivity (surface and subsurface) infiltration, and their related risks in order to better manage these risks and reduce them [9]. Multi-scale approaches are relevant because there are multi-scale impacts; micro-plots and hillslopes are relevant to understanding processes controlling soil detachment and the partition of rainfall into surface and subsurface flow in relation with interacting land-use and soil surface dynamics. At the watershed level, local dynamics translate into risks of hydrological extremes, erosion, and contamination, threatening downstream ecosystems, infrastructures, and populations [10].

1

2. Spatial and Temporal Scales

This Special Issue presents results of studies from many different parts of the world, mainly from Africa and Europe, including the Mediterranean area, Asia [11,12], and South America [13]. The papers showcase studies under equatorial, tropical humid, semi-arid, and Mediterranean climate. All spatial scales are investigated, from the very local scale of gullies in Spain [14] to the largest rivers of the world, the Amazon and Orinoco rivers in South America [13] and Congo River in Central Africa [15]. The whole African continent is also the topic of the paper of Dieulin et al. [16] (2020), presenting a new continental monthly rainfall dataset at the half-degree square scale covering the years 1940 to 1999, computed only from observations from ground-based stations. Many time scales are also investigated, from daily events [17,18] to interannual variability [15,16].

The topics studied cover a large panel of hydrological questions. Many of them deal with the study of time series variability, rainfall [16], discharges [19,20], sediment transport [11–13,21], recession coefficient [22,23], climate change impacts [17,24], or anthropogenic impacts [2,11,20,21]. Drought is also a topic of notable interest in Mediterranean and tropical areas [23–25].

This Special Issue includes several papers dealing with topics at the continental or subcontinental scale. The quality of rainfall data is, for instance, of very high importance to improve the efficiency of hydrological modeling through calibration/validation experiments. The HydroSciences Montpellier Laboratory (HSM) has a long experience in collecting and managing hydro-climatological data. Dieulin et al. [16] present the results of several years of studies to elaborate a reference dataset in order to build monthly rainfall grids over the African continent over a period of 60 years (1940–1999). The large quantity of data collected (about 7000 measurement points) allowed for interpolation using only ground-based observed data, with no statistical use of a reference period. The mean annual grid was compared to the Climate Researh Unit (CRU) grid with HSM_SIEREM data (database HydroSciences Montpellier_Système d'Information Environnementales pour les Ressources en Eau et leur Modélisation) being closer to the observed rain gauge values. The statistical tests, computed on the observed and gridded data, detected a rupture in the rainfall regime around 1979/1980, at the scale of the whole continent, which seems in line with some results in different parts of Africa (North Africa [21]; Congo basin in Central Africa [15]). At the monthly time scale, the most widely observed signal over the period of 1940–1999 was a significant decrease of the austral rainy season between March and May, which has not been previously well documented.

3. Hydrosedimentary Regimes of the Three Largest Rivers of the World

Laraque et al. [13,15] also deal with very large scales, as they cover the three largest rivers of the world by discharge value, all located in equatorial-tropical areas, the Amazon and Orinoco rivers in South America and Congo River in Central Africa. These papers highlight low-documented areas. In South America, the study is focused on the very curious and much unknown Casiquiare Canal, linking both the giant Orinoco and Amazon rivers, investigated during a dedicated international expedition in 2000 [13]. The results presented for the Congo River cover more than one century of measurements and represent the most recently documented synthesis on the Congo River water, with an investigation of dissolved and suspended matters fluxes [15]. In the flat basin of the Congo River, the absence of mountain chains and the extent of its coverage by dense rainforest explains how chemical weathering (10.6 t km^{-2} year^{-1} of total dissolved solids) slightly predominates physical erosion (9.3 t km^{-2} year^{-1} of total suspended solids), followed by organic production (4.5 t km^{-2} year^{-1} of dissolved organic carbon). As the interannual mean discharges are similar, it can be assumed that these interannual averages of material fluxes, calculated over the longest period (2006–2017) of monthly monitoring of its sedimentology and bio-physical-chemistry, are therefore representative of the flow record available since 1902 (with the exception of the wet decade of 1960). Without neglecting the indispensable in situ work, the contributions of remote sensing and numerical modelling should be increasingly used to try to circumvent the dramatic lack of field data that persists in this basin.

Regarding the South American rivers, the aim is to present a review and synthesis of the hydrological and sedimentological knowledge of the world's longest natural defluence, the Casiquiare channel, including its first hydrosedimentary, which was conducted 9–12 September 2000 at the bifurcation and mouth during the expedition 'Humboldt-Amazonia 2000'. This fluvial system connects the two largest river of South America and the main conclusions of this study explain why the Casiquiare can be qualified of a 'cameleon' channel which is taking a significant proportion of flow (20% to 30%) from the Upper Orinoco basin to the Amazon basin.

4. Climatic or Anthropogenic Impacts on River Regimes

In several studies, authors search for the possible origin, climatic or anthropogenic, of the interannual variability observed in the times series of rainfall and discharges. In Central Africa, Ebodé et al. [20] study the impact of anthropization and climate change on two forested watersheds (of Ntem and Nyong) in the south of the country, under equatorial and sub-equatorial climates. The results of this study show that in Central Africa, annual discharges have decreased significantly since the 1970s, and yet the decline in annual rainfall does not become significant until the 2000s. The discharges of the rainy seasons (spring and autumn) recorded the most important changes, following variations in the rainfall patterns of the dry seasons (winter and summer) that preceded them. Winters experienced a significant decrease in precipitation between the 1970s and 1990s, which caused a drop in spring flows. Their rise, which began in the 2000s, is also accompanied by an increase in spring flows, which nevertheless seems to have been very moderated in the case of the Nyong. Conversely, between the 1970s and 1990s, there was a joint increase in summer rainfall and autumn flows. A decrease in summer rainfall was noted since the 2000s, and is also noticeable in autumn flows. Maximum flows have remained constant on the Nyong despite the slight drop in rainfall. This seems to be a consequence of changes in land-use patterns (diminution of forest and increases in the amount of impervious areas). The decrease in maximum flows noted on the Ntem could be linked to the slight drop in precipitation during the rainy season that generates it. Natural factors, such as the general decrease in precipitation during the winter and the reduction in the area occupied by water bodies, could justify the decrease in minimum flows observed in the two watersheds.

Amoussou et al. [17] study the impact of climate change from regional climate model outputs on extreme rainfall events for the Mono River, between Benin and Togo. This research characterizes the future changes in extreme rainfall and air temperature from observational datasets and regional climate models (RCMs) that were used to analyze climatic variations in space and time and fit a generalized extreme value (GEV) model to investigate the extreme rainfalls and their return periods. The authors show that the annual maxima of daily precipitation is expected to increase by 2050 and could be of benefit to the ecosystem services and socioeconomic activities in the Mono River basin, but could also be a threat due to unpredictable consequences of more extreme events.

Bodian et al. [18] investigate the recent trend of hydroclimatic variables in the Senegal River basin based on 36 rain gauge stations and three hydrometric stations not influenced by hydraulic structures. They concentrate on the description of insight views regarding the consequences of the changes in rainfall regimes on the hydrological characteristics of several stations of the Senegal River over several decades. The Mann–Kendall and Pettitt's tests were applied for the annual rainfall time series from 1940 to 2013 to detect the shift and the general trend of the annual rainfall. In addition, trends of the average annual flow rate (AAFR), maximum daily flow (MADF), and low flow rate (LFR) were evaluated before and after annual rainfall shift.

5. Changes in Hydrological Cycle and Water Quality

Descroix et al. [24] explore the evolution of the salinity in the inverse estuaries of two Senegalese rivers in West Africa since the end of the 1968–1993 drought. The results show that in both estuaries, the mean water salinity values have markedly decreased since the end of the drought. However, the Saloum estuary remains a totally inverse estuary, while for the Casamance River, the estuarine turbidity maximum (ETM) is the location of the salinity maximum, and it moves according to the seasons. The West African mangroves are among the few in the world that have been markedly increasing since the beginning of the 1990s and the end of the peak of the dry period in West Africa, as mangrove growth is favored by the relative salinity reduction in this area.

Going further inside the hydrological cycle, Descroix et al. [23] discuss the recent dynamics of the Fouta Djallon's hydrological functioning. The evolution of the runoff and depletion coefficients are analyzed as well as their correlations with the rainfall and vegetation cover. The latter is described at three different space scales and with different methods. Twenty-five years after the end of the 1968–1993 major drought, annual discharges continue to slowly increase, nearly reaching the long-term average, as natural reservoirs which were emptied to sustain streamflows during the drought have been replenishing since the 1990s, explaining the slow increase in discharges. However, another important trend has been detected since the beginning of the drought, namely the increase in the depletion coefficient of some of the Fouta Djallon upper basins, as a consequence of the reduction in the soil water-holding capacity. After observing that this increase and subsequent decrease in the depletion coefficient is localized in parts of the Upper Niger basin and the Konkouré basin, this paper identifies the factors possibly linked with the basins' storage capacity trends. Two pieces of "good news" are highlighted: first, the densely populated areas of the summit plateau are also shown to be the ones where vegetation cover is not threatened and where the ecological intensification of rural activities is ancient; second, most of the river basins providing the Sahel with freshwater have recovered the discharge and soil water-holding capacity levels that they had before the drought.

Bader et al. [22] develops a methodology to compare the evolution of the runoff and depletion coefficients in the Senegal Niger and Gambia rivers. The seasonal flow recession observed at 54 gauging stations in these basins from 1950 to 2016 is represented by empirical and usual conceptual models that express the daily depletion factor (K). Compared to conventional conceptual models, an empirical model representing K as a polynomial of the decimal logarithm of discharge (Q) is shown to provide better representations of K and better discharge forecasting at horizons from 1 to 120 days for most stations. The relationship between specific discharge (Qs) and K, not monotonous, is highly homogeneous in some sub-basins but differs significantly between the Senegal and Gambia basins on the one hand and the Niger basin on the other.

Concerning the Mediterranean, three studies deal with complementary topics: rainfall, runoff, and erosion.

Cavus and Aksoy [25] study the spatial drought characterization for the Seyhan River in Turkey, investigating the spatial and temporal characteristics of drought using standardized precipitation index and different spatialization techniques. They primarily use the standardized precipitation index (SPI) from monthly precipitation data at a 12-month time scale for 19 meteorological stations scattered over the river basin. The drought with the highest severity in each year is defined as the critical drought of the year. Frequency analysis was applied on the critical drought to determine the best-fit probability distribution function by using the total probability theorem. The sole frequency analysis is insufficient in drought studies unless it is numerically related to other factors, such as the severity, duration, and intensity. Yilmaz et al. [19] use statistical methods for daily streamflow estimation at ungauged basins in Turkey. The single donor station drainage area ratio (DAR) method, the multiple-donor stations drainage area ratio (MDAR) method, the inverse similarity weighted (ISW) method, and its variations with three different power parameters (1, 2, and 3) are applied to the two main sub-basins of the Euphrates basin in Turkey to estimate daily streamflow data. Each station in each basin is considered in turn as the target station where there are no streamflow data. Their results show that MDAR and ISW give improved performance compared to DAR for estimating the daily streamflow for seven out of eight target stations in the Middle Euphrates basin and for four out of seven target stations in the Upper Euphrates basin. Higher Nash-Sutcliffe Efficiency (NSE) values for both MDAR and ISW are mostly obtained with the three most physically similar donor stations in the Middle Euphrates basin and with the two most physically similar donor stations in the Upper Euphrates basin. Hout et al. [14] use lidar data to survey the evolution of the morphology of gullies on the slopes of a lake in semi-arid southern Spain. Due to the nature of the geological formations at the lakeshore level, gully erosion is very high and leads to silting up of the dam. This study was conducted to estimate the sediment input from the bank gullies. The combination of two configurations of nadir and oblique photography allowed the authors to obtain a complete high-resolution modeling of complex bank gullies with overhangs. They develop a complex methodology to allow erosion to be quantified between two dates at the gully scale. The results reveal significant lakeshore erosion activity by bank gullies, since the annual inflow from the banks is estimated at 39 t ha^{-1} year^{-1}.

6. Erosion, Land Cover, and Land Use

Three more papers investigate erosion surveys and their relationships with land cover and land use. Two of these studies use spatial data to retrieve land-cover characteristics.

In northern Laos (Song et al. [12]), the replacement of traditional crops by tree plantations, such as teak trees, has led to a dramatic increase in floods and soil loss and to the degradation of basic soil ecosystem services, such as water filtration by soil, fertility maintenance, etc. The authors [12] hypothesized that conserving understory under teak trees would protect soil, limit surface runoff, and help reduce soil erosion. It is shown that teak planting does not have an anti-erosive power at all; this land cover shows the highest runoff coefficient and soil loss, associated with the highest crusting rate; this is due to the kinetic energy of rain drops falling from the broad leaves of the tall teak trees down to bare soil, devoid of plant residues, thus leading to severe soil surface crusting and soil detachment. Hence, teak tree plantation owners could divide soil loss by 14 by keeping understory, such as broom grass, within teak tree plantations. Overall, promoting understory, such as broom grass, in teak tree plantations would (1) limit surface runoff and improve soil infiltrability, thus increasing the soil water stock available for both root absorption and groundwater recharge, and (2) mitigate soil loss while favoring soil fertility conservation.

Ma et al. [11] study the hydrological impacts of expanding rubber plantations on a watershed of 100 km^2 in the Xishuangbanna prefecture, Yunnan Province, China. The influence of land-cover change on streamflow recorded since 1992 was isolated from that of rainfall variability using cross-simulation matrices produced with the monthly lumped conceptual water balance model GR2M. Their results indicate a statistically significant reduction in wet and dry season streamflow from 1992 to 2002, followed by an insignificant increase until 2006. Analysis of satellite images from 1992, 2002, 2007, and 2010 shows a gradual increase in the areal percentage of rubber tree plantations at the watershed scale. However, there were marked heterogeneities in land conversions (between forest, farmland, grassland, and rubber tree plantations), and in their distribution across elevations and slopes, among the studied periods. Possible effects of this heterogeneity on hydrological processes, controlled mainly by infiltration and evapotranspiration, are discussed in light of the hydrological changes observed over the study period. The authors suggest pathways to improve the eco-hydrological functionalities of rubber tree plantations, particularly those enhancing dry-season base flow, and recommend how to monitor them.

In Morocco, North Africa, Ezzaouini et al. [21] assess the sediment inputs into the dam's lake on the Bouregreg River. They apply the Modified Universal Soil Loss Equation (MUSLE) model to predict solid transport, calibrated by two years of solid transport measurements from four main gauging stations at the entrance of the Sidi Mohamed Ben Abdellah dam. The application of the MUSLE on the basin demonstrated relatively small differences between the measured values and those expected for the calibrated version; these differences are, for the non-calibrated version, +5% and +102% for the years 2016/2017 and 2017/2018, respectively, and between −33% and +34% for the calibrated version. Furthermore, the measured and modeled volumes that do not exceed 1.78×10^6 m^3/year^{-1} remain well below the dam's siltation rate of 9.49×10^6 m^3/year^{-1}, obtained from a continuous bathymetry survey, which means that only 18% of the dam's sediment comes from upstream. This seems very low because it is calculated from only two years. The main hypothesis that is formulated is that the majority of the sediments of the dam most probably come from the erosion of its banks.

7. Concluding Remarks

This Special Issue shows the great interest of the scientific community in investigating how to characterize the environmental impact of anthropogenic and climatic changes, and also determining which is the main source of changes depending on the regional situations and thematic issues. This first issue might lead to following issues on more specific topics, according to, for instance, the 23 unsolved problems in hydrology [26], or future issues that are focused on different regional areas. Future issues on this topic might also focus more on water resources challenges in urban tropical and Mediterranean areas, which constitute increasing shares of the populations in these areas. These densely populated areas are also often localized near the coast, which brings up specific concerns about the coastal damages caused by human interventions (such as dams and irrigations) and sea level rising on the inland water cycle [27]. The impact of global warming and the increasing occurrence of extreme events on runoff generation and discharge dynamics must also be addressed in the future. Changes in the water-holding capacity of soil (at the point scale) and basins (at the regional scale) will impact river regimes through the evolution of runoff and recession coefficients as indexes of these major possible dynamics.

Conflicts of Interest: The authors declare no conflict of interest.

References

1. Tramblay, Y.; Koutroulis, A.; Samaniego, L.; Vicente-Serrano, S.M.; Volaire, F.; Boone, A.; Le Page, M.; Carmen Llasat, M.; Albergel, C.; Burak, S.; et al. Challenges for drought assessment in the mediterranean region under future climate scenarios. *Earth Sci. Rev.* **2020**, *210*, 103348. [CrossRef]
2. Song, X.P.; Hansen, M.C.; Stehman, S.V.; Potapov, P.V.; Tyukania, A.; Vermote, E.F.; Townshend, J.R. Global land change from 1982 to 2016. *Nature* **2018**, *560*, 639–643. [CrossRef]
3. Eccles, R.; Zhang, H.; Hamilton, D. A review of the effects of climate change on riverine flooding in subtropical and tropical regions. *J. Water Clim. Chang.* **2019**, *10*, 687–707. [CrossRef]
4. Tramblay, Y.; Llasat, M.C.; Randin, C.; Coppola, E. Climate change impacts on water resources in the Mediterranean. *Reg. Environ. Chang.* **2020**, *20*, 83. [CrossRef]
5. Mahé, G.; Dray, A.; Paturel, J.E.; Crès, A.; Koné, F.; Manga, M.; Crès, F.N.; Djoukam, J.; Maïga, A.H.; Ouédraogo, M.; et al. Climatic and anthropogenic impacts on the flow regime of the Nakambe River in Burkina. In *Bridging the Gap between Research and Practice, Proceedings of the Fourth International FRIEND Conference, Cape Town, South Africa, 18–22 March 2002*; Van Lannen, H., Demuth, S., Eds.; IAHS Publ. 274; IAHS Publications: Oxfordshire, UK, 2002; pp. 69–76. Available online: https://iahs.info/uploads/dms/12264.14-Mahe-et-al.--F28--pp.-69-76-revised.pdf (accessed on 22 March 2020).
6. Descroix, L.; Guichard, F.; Grippa, M.; Lambert, L.; Panthou, J.; Mahe, G.; Gal, L.; Dardel, C.; Quantin, G.; Kergoat, L.; et al. Surface hydrology evolution in the Sahelo-Sudanian stripe: An updated review. *Water* **2018**, *10*, 748. [CrossRef]
7. Ribolzi, O.; Evrard, O.; Huon, S.; Rouw, A.; De Silvera, N.; Latsachack, O.; Soulileuth, B.; Lefèvre, I.; Pierret, A. From shifting cultivation to teak plantation: Effect on overland flow and sediment yield in a montane tropical catchment. *Sci. Rep.* **2017**, *12*, 1–12. [CrossRef]
8. Toteu, S.F.; Mahé, G.; Moritz, R.; Sracek, O.; Davies, T.C.; Ramasamy, J. Environmental, health and social legacies of mining activities in Sub-Saharan Africa. *J. Geochem. Explor.* **2020**, *209*. [CrossRef]
9. Lacombe, G.; Valentin, C.; Sounyafong, P.; Rouw, A.; De Soulileuth, B.; Silvera, N.; Pierret, A.; Sengtaheuanghoung, O.; Ribolzi, O. 2018. Linking crop structure, throughfall, soil surface conditions, runoff and soil detachment: 10 land uses analyzed in Northern Laos. *Sci. Total Environ.* **2017**, *616–617*, 1330–1338. [CrossRef]

10. Morán-Ordóñez, A.; Duane, A.; Gil-Tena, A.; De Cáceres, M.; Aquilué, N.; Guerra, C.; Geijzendorffer, I.R.; Fortin, M.-J.; Brotons, L. Future impact of climate extremes in the Mediterranean: Soil erosion projections when fire and extreme rainfall meet. *Land Degrad. Dev.* **2020**, *31*, 3040–3054. [CrossRef]
11. Ma, X.; Lacombe, G.; Harrison, R.; Xu, J.; van Noordwijk, M. Expanding Rubber Plantations in Southern China: Evidence for Hydrological Impacts. *Water* **2019**, *11*, 651. [CrossRef]
12. Song, L.; Boithias, L.; Sengtaheuanghoung, O.; Oeurng, C.; Valentin, C.; Souksavath, B.; Sounyafong, P.; de Rouw, A.; Soulileuth, B.; Silvera, N.; et al. Understory Limits Surface Runoff and Soil Loss in Teak Tree Plantations of Northern Lao PDR. *Water* **2020**, *12*, 2327. [CrossRef]
13. Laraque, A.; Lopez, J.; Yepez, S.; Georgescu, P. Water and Sediment Budget of Casiquiare Channel Linking Orinoco and Amazon Catchments, Venezuela. *Water* **2019**, *11*, 2068. [CrossRef]
14. Hout, R.; Maleval, V.; Mahe, G.; Rouvellac, E.; Crouzevialle, R.; Cerbelaud, F. UAV and LiDAR Data in the Service of Bank Gully Erosion Measurement in Rambla de Algeciras Lakeshore. *Water* **2020**, *12*, 2748. [CrossRef]
15. Laraque, A.; Moukandi N'kaya, G.; Orange, D.; Tshimanga, R.; Tshitenge, J.; Mahé, G.; Nguimalet, C.; Trigg, M.; Yepez, S.; Gulemvuga, G. Recent Budget of Hydroclimatology and Hydrosedimentology of the Congo River in Central Africa. *Water* **2020**, *12*, 2613. [CrossRef]
16. Dieulin, C.; Mahé, G.; Paturel, J.; Ejjiyar, S.; Tramblay, Y.; Rouché, N.; EL Mansouri, B. A New 60-Year 1940/1999 Monthly-Gridded Rainfall Data Set for Africa. *Water* **2019**, *11*, 387. [CrossRef]
17. Amoussou, E.; Awoye, H.; Totin Vodounon, H.; Obahoundje, S.; Camberlin, P.; Diedhiou, A.; Kouadio, K.; Mahé, G.; Houndénou, C.; Boko, M. Climate and Extreme Rainfall Events in the Mono River Basin (West Africa): Investigating Future Changes with Regional Climate Models. *Water* **2020**, *12*, 833. [CrossRef]
18. Bodian, A.; Diop, L.; Panthou, G.; Dacosta, H.; Deme, A.; Dezetter, A.; Ndiaye, P.; Diouf, I.; Vischel, T. Recent Trend in Hydroclimatic Conditions in the Senegal River Basin. *Water* **2020**, *12*, 436. [CrossRef]
19. Yilmaz, M.; Onoz, B. A Comparative Study of Statistical Methods for Daily Streamflow Estimation at Ungauged Basins in Turkey. *Water* **2020**, *12*, 459. [CrossRef]
20. Ebodé, V.; Mahé, G.; Dzana, J.; Amougou, J. Anthropization and Climate Change: Impact on the Discharges of Forest Watersheds in Central Africa. *Water* **2020**, *12*, 2718. [CrossRef]
21. Ezzaouini, M.; Mahé, G.; Kacimi, I.; Zerouali, A. Comparison of the MUSLE Model and Two Years of Solid Transport Measurement, in the Bouregreg Basin, and Impact on the Sedimentation in the Sidi Mohamed Ben Abdellah Reservoir, Morocco. *Water* **2020**, *12*, 1882. [CrossRef]
22. Bader, J.; Dacosta, H.; Pouget, J. Seasonal Variations of the Depletion Factor during Recession Periods in the Senegal, Gambia and Niger Watersheds. *Water* **2020**, *12*, 2520. [CrossRef]
23. Descroix, L.; Faty, B.; Manga, S.; Diedhiou, A.; Lambert, L.A.; Soumaré, S.; Andrieu, J.; Ogilvie, A.; Fall, A.; Mahé, G.; et al. Are the Fouta Djallon Highlands Still the Water Tower of West Africa? *Water* **2020**, *12*, 2968. [CrossRef]
24. Descroix, L.; Sané, Y.; Thior, M.; Manga, S.; Ba, B.; Mingou, J.; Mendy, V.; Coly, S.; Dièye, A.; Badiane, A.; et al. Inverse Estuaries in West Africa: Evidence of the Rainfall Recovery? *Water* **2020**, *12*, 647. [CrossRef]
25. Cavus, Y.; Aksoy, H. Spatial Drought Characterization for Seyhan River Basin in the Mediterranean Region of Turkey. *Water* **2019**, *11*, 1331. [CrossRef]
26. Blöschl, G.; Bierkens, M.F.; Chambel, A.; Cudennec, C.; Destouni, G.; Fiori, A.; Kirchner, J.W.; McDonnell, J.J.; Savenije, H.H.; Sivapalan, M.; et al. Twenty-three unsolved problems in hydrology (UPH)—A community perspective. *Hydrol. Sci. J.* **2019**, *64*, 1141–1158. [CrossRef]
27. Hzami, A.; Heggy, E.; Amrouni, O.; Mahe, G.; Maanan, M.; Abdeljaouad, S. Alarming coastal vulnerability of the deltaic and sandy beaches of North Africa. *Nat. Sci. Rep.* **2021**, *11*, 2320. [CrossRef] [PubMed]

 *water*

Article

Seasonal Variations of the Depletion Factor during Recession Periods in the Senegal, Gambia and Niger Watersheds

Jean-Claude Bader [1,*], Honoré Dacosta [2] and Jean-Christophe Pouget [1]

[1]   G-EAU, IRD (AgroParisTech, Cirad, IRSTEA, MontpellierSupAgro, Univ Montpellier), BP5095, 34196 Montpellier, France; jean-christophe.pouget@ird.fr
[2]   Département de Géographie, FLSH-UCAD, BP 5005 Dakar-Fann, Senegal; honore.dacosta@ucad.edu.sn
*   Correspondence: jean-claude.bader@ird.fr

Received: 17 July 2020; Accepted: 29 August 2020; Published: 9 September 2020

Abstract: The daily depletion factor K describes the discharge decrease of rivers only fed by groundwater in the absence of rainfall. In the Senegal, Gambia and Niger river basins in West Africa, the flow recession can exceed 6 months and the precise knowledge of K thus allows discharge forecasts to be made over several months, and is hence potentially interesting for hydraulic structure managers. Seasonal flow recession observed at 54 gauging stations in these basins from 1950 to 2016 is represented by empirical and usual conceptual models that express K. Compared to conventional conceptual models, an empirical model representing K as a polynomial of the decimal logarithm of discharge Q gives better representations of K and better discharge forecasting at horizons from 1 to 120 days for most stations. The relationship between specific discharge Qs and K, not monotonous, is highly homogeneous in some sub-basins but differs significantly between the Senegal and Gambia basins on the one hand and the Niger basin on the other. The relationship $K(Q)$ evolves slightly between three successive periods, with values of K generally lower (meaning faster discharge decrease) in the intermediate period centered on the years 1970–1980. These climate-related interannual variations are much smaller than the seasonal variations of K.

Keywords: flow recession model; discharge forecast; Senegal River; Gambia River; Niger River

---

1. Introduction

1.1. Background, Context

According to [1], "the decrease of discharge corresponding to the emptying of groundwater without any precipitation is called the flow recession of a river". This stream flow regime has been the subject of numerous studies, including reviews [2,3].

According to these authors, the flow recession concerns "all natural storages feeding the stream, including unsaturated grounds and lakes". Among the oldest works, [4] established the differential equation governing the discharge of an aquifer in a homogeneous ground. This author derives two conceptual models that give the discharge rate of the groundwater into a river under certain boundary conditions. The first, which is the most widely used recession model [1,5], is described as linear [2]. Assuming a thick aquifer whose free surface is close to the horizontal, it expresses the discharge Q at time T as a function of discharge Q_0 at time T_0, with a constant α characteristic of the ground and the dimensions of the aquifer and dependent on the unit of time (but independent of T_0), or with a constant K equal to $e^{-\alpha}$, called "depletion factor" [6]:

$$Q(T) = Q_0 e^{-\alpha(T-T_0)} = Q_0 K^{(T-T_0)}. \tag{1}$$

The second model proposed by [4], described as non-linear [2], assumes a groundwater table limited by an impermeable horizontal bottom and vertical wall, draining into the watercourse at its base. It is expressed in this way, with a constant σ_0 characteristic of the soil and the dimensions of the aquifer and depending on both the unit of time T and the initial time T_0:

$$Q(T) = Q_0/(1 + \sigma_0(T - T_0))^2. \tag{2}$$

According to [5], the relationship (1) first applied by [7] is generally referred to as the "Maillet formula", while the relationship (2), used by [8] for African rivers, is generally referred to as the "Tison formula".

Reference [9] shows that these two formulas are particular cases of a more general conceptual model ($Q = ZV^m$) that represents the emptying of a reservoir under the assumption of a discharge Q proportional to a certain m power of the remaining water volume V, with a proportionality constant Z. This author finds the Maillet equation (1) with m equal to 1, his model then representing the emptying of a vertical cylinder by a porous plug at its base [1]. When m is different from 1, he obtains the following result where V_0 is the volume remaining at time T_0:

$$Q(T) = Q_0/(1 + \sigma_0(T - T_0))^n, \tag{3}$$

$$n = m/(m - 1), \tag{4}$$

$$\sigma_0 = (m - 1)Q_0/V_0. \tag{5}$$

with $m = 2$, the "Coutagne equation" (3) gives the Tison equation (2) and represents the emptying of a vertical cylindrical reservoir by a parabolic spillway with a vertical axis [10].

Some authors propose the empirical addition of a constant term W to Maillet's equation [11–15], to Tison's equation (Maillet, in [4]) or to that of Coutagne [9,16]. The relationships thus obtained are called here "generalized Maillet equation" (6) and "generalized Coutagne equation" (7):

$$Q(T) = W + (Q_0 - W)e^{-\alpha(T-T_0)}, \tag{6}$$

$$Q(T) = W + (Q_0 - W)/(1 + \sigma_0(T - T_0))^n. \tag{7}$$

in these relationships, the parameters α and σ_0 depend on the time unit and σ_0 also depends on the initial time T_0, which can be set arbitrarily. With the relationship (6), the recession can be represented indifferently from the flow rates Q_A and Q_B at the initial times T_A and T_B if the relationship (8) is respected. With the relationship (7), Q_A and Q_B at the initial times T_A and T_B, associated with σ_A and σ_B, must verify the relationship (9).

$$Q_B = W + (Q_A - W)e^{-\alpha(T_B-T_A)}, \tag{8}$$

$$\sigma_B = \sigma_A/(1 + \sigma_A(T_B - T_A)) = \sigma_A((Q_B - W)/(Q_A - W))^{1/n}. \tag{9}$$

Relationship (10) is a particular case of the generalized Coutagne equation obtained after changing the variable by translation over time. It is used in different forms with the parameters λ, μ and r by some authors [13,17–21]:

$$Q(T) = \mu + \lambda/T^r. \tag{10}$$

Reference [22] proposes an empirical adaptation of the Maillet formula, by a change of variable with a power function over time. So the "Horton formula", sometimes referred to as the double exponential [2], is expressed as follows with the parameters β and p:

$$Q(T) = Q_0 e^{-\beta(T-T_0)^p}. \tag{11}$$

Finally, recession is often represented by the sum of Maillet formulas [4,6,23–26] or Maillet formulas and Tison formulas [27].

The most commonly used conceptual models to represent recession, listed by [3,28], can all be represented by the generalized Maillet (6), generalized Coutagne (7) or Horton (11) formulas, or by combinations of these formulas. The calibration of these conceptual or empirical models is usually based on the analysis of certain relationships in observed data, using either a graphical or least squares method. These relationships, sometimes interpreted as envelope curves to distinguish the origin of the flows (slow variable feeding by the aquifers and more rapid variable feeding for the other origins), are as follows:

- Relationship between $Q(T)$ and $Q(T + dT)$: correlation method. With $dT = 1$ day, [29] uses this mean relationship as an empirical recession curve, and [30] represents the upper envelope curve by a line passing through the origin, whose slope corresponds to the depletion factor K of the Maillet equation (1). With $dT = 5$ days, [31] represents the mean relationship by a line passing through the origin, whose slope corresponds to K^5.
- Relationships between T and $\text{Log}(Q)$ established for different recession sequences. Reference [32] uses these relationships and interprets their slope in each point as $-\alpha$ (Maillet's formula). He derives an empirical model giving α as a function of Q and T. Ref. [33] set the Maillet formula by representing the relationships obtained for the different sequences by linear relationships, whose common slope corresponds to $-\alpha$.
- Relationship between $\text{Log}(Q)$ and $\text{Log}(-dQ/dT)$. Refs. [34,35] represent the low envelope curve of this relationship by sections of linear functions, whose slope and ordinate at the origin correspond respectively to $(n + 1)/n$ and $\text{Log}(n\sigma_0 Q_0^{-1/n})$, where n and σ_0 are the parameters of the Coutagne equation (3). The 3/2 slope obtained for low flows gives $n = 2$, according to the Tison equation (2). Reference [36] obtains an average relationship with a slope close to 1, according to the Maillet equation (1).
- Relationship between Q and $-dQ/dT$. For different periods varying in distance from the beginning of recession, [37] represent the upper and lower envelopes of this relationship by lines passing through the origin, whose slope corresponds to the α parameter of the Maillet equation (1). These authors find that α is not very variable for low envelopes associated with baseflow from groundwater, while it decreases over time for high envelopes.

Each of the above methods, therefore, consists in representing, generally by one or more lines, a relationship between a primary variable ($Q(T)$, T, $\text{Log}(Q)$ or Q) and a secondary variable ($Q(T + dT)$, $\text{Log}(Q)$, $\text{Log}(-dQ/dT)$ or $-dQ/dT$). This can be problematic for rivers in West Africa, where Q and $-dQ/dT$ can decrease by several orders of magnitude during recession sequences spread over several months. Any elementary function adjusted to Q or $-dQ/dT$ by the least squares method can then lead to very large relative errors for the lowest values of $Q(T + dT)$ or $-dQ/dT$. Conversely, adjusting a function on logarithms of Q or $-dQ/dT$ can be unfavorable and allow large relative errors for the highest values of Q or $-dQ/dT$.

The method used by [5,38], derived from the correlation method, does not have these disadvantages. From the mean Q_m of observed discharge, these authors calibrate the Maillet equation by calculating the ratio $Q_m(T + dT)/Q_m(T)$, interpreted as K^{dT}. They deduce K from the resulting ratios calculated for different dT values.

To study the flow recession on the Senegal River basin in West Africa, reference [39] uses a quite similar method to the previous one. Instead of $Q_m(T + dT)/Q_m(T)$, they calculate for $dT = 1$ day the daily values of $Q(T + dT)/Q(T)$ that corresponds to the daily depletion factor K of the Maillet formula. Despite their wide dispersion between 0.5 and 1, the values obtained show an average seasonal evolution that can be represented by an empirical function of time T, which is counted from an origin set at the same date each year. This function represents K with the same accuracy for very different discharges, which is an advantage over the methods mentioned above. For most of the stations analyzed, K does not vary monotonously with T: from a low value at the beginning of recession, K increases to a maximum value and then decreases rapidly until the end of recession. Among the commonly used conceptual models, only the generalized Coutagne model (7), used with a negative W, makes it possible to represent such a variation of K (for a demonstration see text S1).

1.2. Objectives of the Project

This study, which is an extension of the work of [39], focuses on the depletion factor K in the Senegal, Gambia and Niger river basins in West Africa. Precise knowledge of K, always situated between 0 and 1 by definition, is of interest for two main reasons:

- The knowledge of groundwater, since K describes the recharge of the river flow by the groundwater in the absence of rainfall. A high K coefficient, close to 1, corresponds to a slow decrease in the discharge, thanks to important contributions from the water tables. Conversely, a low K coefficient corresponds to a rapid decrease in the flow of rivers, with little discharge supply from groundwater. K thus characterizes the capacity of groundwater to supply rivers. This capacity depends both on geological and pedological characteristics (transmissivity, storage, permeability) and on the abundance of previous rainfall inputs to the basin.
- The forecasting of discharge during recession periods, which can extend more than 6 months in the region. Flow forecasts can, therefore, be made over very long horizons of several months, which can be of great interest to managers of hydraulic structures such as dams.

For 54 stations located in the Senegal, Gambia and Niger river basins, we model the flow recession of rivers from observed values of the daily depletion factor K over the period 1950–2016. In addition to the most frequently used Maillet model (K constant), other models are tested to describe seasonal variations of K as a function of time T or discharge Q, using formulas whose form is empirical or imposed by the generalized Coutagne equation (7). We compare the performance of the different models in representing K and in forecasting discharge Q at all possible forecast horizons during the recession sequences enabled by the models. Finally, we compare seasonal variations and interannual variations of K.

The following sections present the method, data, results and conclusions of the study. All dates are expressed in day/month/year.

2. Materials and Methods

2.1. Methods

2.1.1. Observed Values of the Daily Depletion Factor K

According to equation (1), the daily depletion factor $K(T)$ is equal to $Q(T + 1)/Q(T)$ with T expressed in days. But to limit their dispersion, its values are calculated over 3 consecutive days from the flows observed at times $T - 1$ and $T + 2$:

$$K(T) = (Q(T+2)/Q(T-1))^{1/3}. \tag{12}$$

Only the values $K(T)$ based on flows $Q(T - 1)$ and $Q(T + 2)$ belonging to the same recession sequence are retained, with T verifying the following 3 conditions:

- C1: times $T - 2$ to $T + 2$ fall within the usual recession period, which starts and ends on invariable dates $T1$ and $T2$, expressed in days/months;
- C2: between times $T - 2$ and $T + 2$, the discharge is known at a daily time step with no missing data and decreases in the broad sense, meaning that $Q(T + k) \le Q(T + k - 1)$ for all integer numbers k between −1 and 2;
- C3: the discharge $Q(T + 2)$ is greater than a minimum threshold $Qthr$ arbitrarily set at 0.1 $\text{m}^3 \cdot \text{s}^{-1}$.

2.1.2. Analyses Performed

We tested 5 models to represent the depletion factor K:

- The Maillet's conceptual model (model 0). This model was chosen because it is the most often used model of recession.
- An empirical model representing K as a function of time (model 1).
- An empirical model representing K as a function of discharge (model 2).
- The generalized Coutagne conceptual model. This model was chosen because it is the only usual conceptual model capable of representing the non-monotonic evolution of K observed in the Senegal River basin during the recession period (see above). The generalized Coutagne model is used to represent K as a function of time (model 3) or as a function of discharge (model 4).

For each station, we represent the seasonal variations of $K(T)$ with these five models calibrated using the least squares method:

- model 0: average Km of the observed values of $K(T)$, constant value compatible with the Maillet formula;
- model 1: empirical polynomial function f of the delay $D(T)$ elapsed between the initial time T_0 (invariable, expressed in day/month) and T, defined with parameters A_1 to A_6 by:

$$K = f(D) = \sum_{i=0}^{i=6} \left(A_i \times D^i \right);\tag{13}$$

- model 2: empirical function g of the discharge $Q(T)$, polynomial of the decimal logarithm of Q defined with parameters $B1$ to $B6$ by:

$$K = g(Q) = \sum_{i=0}^{i=6} \left(B_i \times (\log Q)^i \right);\tag{14}$$

- model 3: function f_b of the delay $D(T)$, verifying with the parameters Q_0, W, σ_0 and n the relationship (15) imposed by the generalized Coutagne formula:

$$K = f_b(D) = \frac{W + (Q_0 - W)(1 + \sigma_0 + \sigma_0 D)^{-n}}{W + (Q_0 - W)(1 + \sigma_0 D)^{-n}};\tag{15}$$

- model 4: function g_b of the discharge $Q(T)$, verifying with the parameters Q_0, W, σ_0 and n the relationship (16) imposed by the generalized Coutagne formula:

$$K = g_b(Q) = \frac{W + (Q_0 - W)\left(((Q_0 - W)/(Q - W))^{1/n} + \sigma_0 \right)^{-n}}{Q}.\tag{16}$$

Calibrated on all the N observed values of K, each of these models makes it possible to represent hydrographs $Q(T)$ that differ between different recession periods by homothety for models 0, 1 and 3 and by time translation for models 0, 2 and 4 (for a demonstration see Text S2). The models are classified according to the Nash and Sutcliffe efficiency coefficient C_{NSE0} relative to the N modeled K values ($C_{NSE0} = 0$ for model 0 based on the Maillet formula), with R_a rank ranging from 1 for maximum C_{NSE0} to 5 for minimum C_{NSE0}. The models' ability to forecast recession flows is then evaluated for all entire H forecast horizons between 1 and 120 days.

Let a sequence of uninterrupted recession be limited by the times $T_b - 2$ and $T_b + N_p + 1$, during which conditions C1 to C3 are checked for any T between T_b and $T_b + N_p - 1$. These conditions make it possible to calculate N_p observed values of K with equation (12) between the times Tb and $T_b + N - 1$, and also to calculate $N_p \times (N_p + 1)/2$ forecasted discharges over the sequence with each recession model: N_p values are forecast on a first sub sequence at times $T_b + 1$ to $T_b + N_p$ from the discharge observed at time T_b, for horizons H from 1 to N_p days; $N_p - 1$ values are forecast on a second sub sequence at times $T_b + 2$ to $T_b + N_p$ from the discharge observed at time $T_b + 1$, for horizons H from 1 to $N_p - 1$ days; a value is predicted on a N_pth sub sequence at time $T_b + N_p$ from the discharge observed at time $T_b + N_p - 1$, for a horizon of 1 day. For each sub sequence, each forecast discharge is calculated with the recession model applied to the previous day's discharge (observed for the first of the sub sequence and forecast for the following ones).

The accuracy of the forecast discharges is evaluated as follows: for each horizon H, using the relative standard error R_{rmse} (ratio between the standard error S_e and the average Q_m of the corresponding observed flows) and the Nash and Sutcliffe efficiency coefficient C_{NSE1} for the N_e values forecast at horizon H over all the recession periods defined above; using the C_{NSE2} for the N_t values predicted at the different horizons (with $N_t = \sum_{H=1}^{H=120} N_e$).

2.2. Data

The data used are the average daily discharges observed under natural conditions between 1950 and 2016 at 54 gauging stations (Tables 1–3 and Figure 1) located in the upper basins of the Senegal River (up to Kaédi), the Gambia River (in Senegal, upstream of Gouloumbo) and the Niger River (in Mali, up to Ansongo). These data come from the former ORSTOM/IRD databases developed and enriched through various projects [39–43], often carried out in partnership with data-producing or -managing organizations (*Direction Nationale de l'Hydraulique du Mali, Organisation pour la Mise en Valeur du fleuve Gambie, Organisation pour la Mise en Valeur du fleuve Sénégal, Service de Gestion et de Planification des Ressources en Eau du Sénégal*).

Table 1. Senegal basin-list of stations analyzed, with their catchment area and mean specific discharge Q_{sm} over the period 1970–1979.

Sub-Basin	River	Station	Rank	Long. (° W)	Lat. (° N)	Area (km²)	Q_{sm} l·s⁻¹·km⁻¹
High Bafing	Bafing	Pont km17	1	12°09′	10°29′	18	17.45
High Bafing	Samenta	Doureko	2	11°42′	11°18′	225	17.38
High Bafing	Kioma	Teliko	3	11°53′	11°22′	360	17.34
High Bafing	Kioma	Salouma	4	11°42′	11°17′	775	15.96
High Bafing	Téné	Bebele	5	11°49′	11°01′	3470	17.59
High Bafing	Bafing	Balabori	6	11°22′	11°18′	11,600	13.29
Bafing	Bafing	Boureya	7	10°44′	11°45′	14,800	12.77
Bafing	Bafing	Daka Saidou	8	10°37′	11°57′	15,700	12.50
Bafing	Bafing	Makana	9	10°17′	12°33′	22,000	10.44
Bafing	Bafing	Soukoutali	10	10°25′	13°12′	27,800	9.11
Bafing	Bafing	Dibia	11	10°48′	13°14′	33,500	7.74
Faleme	Faleme	Moussala	12	11°18′	12°31′	3103	10.04
Faleme	Faleme	Fadougou	13	11°23′	12°31	9300	5.75
Faleme	Faleme	Gourbassy	14	11°38′	13°24′	17,100	5.01
Faleme	Faleme	Kidira	15	12°13′	14°27′	28,900	3.61
Bakoye	Bakoye	Toukoto	16	09°53′	13°27′	16,500	2.94
Bakoye	Baoule	Siramakana	17	09°53′	13°35′	59,500	0.56
Bakoye	Bakoye	Oualia	18	10°23′	13°36′	84,700	1.05
Bakoye	Bakoye	Kalé	19	10°39′	13°43′	85,600	1.05
Senegal	Senegal	Galougo	20	11°03′	13°51′	128,400	2.79
Senegal	Senegal	Gouina	21	11°06′	14°00′	128,600	2.83
Senegal	Senegal	Kayes	22	11°27′	14°27′	157,400	2.32
Senegal	Senegal	Bakel	23	12°27′	14°54′	218,000	2.36
Senegal	Senegal	Matam	24	13°15′	15°39′	230,000	1.99
Senegal	Senegal	Kaédi	25	13°30′	16°08′	253,000	1.53

Table 2. Gambia basin—list of stations analyzed, with their catchment area and mean specific discharge Q_{sm} over the period 1970–1979.

Sub-Basin	River	Station	Rank	Long. (° W)	Lat. (° N)	Area (km^2)	Q_{sm} l·s^{-1}km^{-1}
tributaries	Sili	pont routier	26	12°16′	13°32′	90	9.12
tributaries	Diarha	pont routier	27	12°37′	12°36′	760	10.00
tributaries	Tiokoye	pont routier	28	12°32′	12°34′	950	9.41
tributaries	Diaguery	pont routier	29	12°05′	12°38′	1010	7.52
tributaries	Niokolokoba	pont routier	30	12°44′	13°04′	3000	2.65
tributaries	Koulountou	Gué (parc)	31	13°29′	12°47	5350	4.88
tributaries	Koulountou	Missira Gounasse	32	13°37′	13°12′	6200	5.13
Gambia	Gambia	Kedougou	33	12°11′	12°33′	7550	10.69
Gambia	Gambia	Mako	34	12°21′	12°52′	10,450	8.86
Gambia	Gambia	Simenti	35	13°18′	13°02′	20,500	6.42
Gambia	Gambia	Wassadou amont	36	13°22′	13°21′	21,200	6.20
Gambia	Gambia	Wassadou aval	37	13°23′	13°21′	33,500	3.96

Table 3. Niger basin—list of stations analyzed, with their catchment area and mean specific discharge Q_{sm} over the period 1970–1979.

Sub-Basin	River	Station	Rank	Long. (° W)	Lat. (° N)	Area (km^2)	Q_{sm} l·s^{-1}km^{-1}
High Niger	Sankarani	Selingue	38	08°13′	11°38′	34,200	9.51
High Niger	Niger	Banankoro	39	08°40′	11°41′	70,740	12.72
High Niger	Niger	Koulikouro	40	07°33′	12°51′	120,000	10.60
High Niger	Niger	Ke Macina	41	05°21′	13°58′	141,000	8.41
Bani	Degou	Manankoro	42	07°27′	10°27′	1550	5.52
Bani	Banifing	Kolondieba	43	06°51′	11°03′	3050	3.36
Bani	Baoulé	Madina Diassa	44	07°40′	10°47′	7880	6.60
Bani	Baoulé	Dioila	45	06°48′	12°31	32,500	3.25
Bani	Bani	Douna	46	05°54′	13°12′	101,600	2.73
Bani	Bani	Beneny Kegny	47	04°54′	13°23′	116,000	2.47
Bani	Bani	Sofara	48	04°05′	14°05′	125,400	1.96
Niger	Niger	Mopti	49	04°12′	14°30′	281,600	3.00
Niger	Niger	Akka	50	04°14′	15°24′	-	-
Niger	Niger	Diré	51	03°23′	16°16′	-	-
Niger	Niger	Koryoumé	52	03°01′	16°40′	-	-
Niger	Niger	Tossaye	53	00°35′	16°57′	-	-
Niger	Niger	Ansongo	54	−00°30′	15°40′	-	-

Among the discharges obtained by translating the levels observed at the stations, all the reconstructed values are excluded from the analyses of the recession regime. Data influenced by large reservoir dams upstream (at Soukoutali and downstream from 1987 on the Senegal basin (Manantali dam); at Sélingué and downstream from 1981 on the Niger basin (Sélingué dam)) are also excluded.

The stations analyzed control basins whose area ranges from 18 to more than 280,000 km^2. Their mean interannual specific discharges Q_{sm} (calculated by integrating certain reconstituted values) vary between 0.56 and 17.59 l·s^{-1}·km^{-2} (Tables 1–3). The study, therefore, concerns very diverse stations, for which the data also cover the study period in a very variable way (varying numbers of missing observations).

The annual peak flood in the region occurs generally in mid September, except in the Inner Niger Delta (DIN) and downstream, where the flood mainly formed upstream spreads very slowly (Figure 2). The general decrease in flow occurs then and persists until the arrival of flows caused by the following monsoon, usually in May or June or even in July downstream of the DIN. The origin T_0 of the times is therefore set to the previous 15 September for all stations and the T_1 and T_2 start and end dates of the usual recession period are set to 15 September and 31 May, respectively, except for the following stations:

- T_1 = 15 October and T_2 = 30 June for Akka on Niger (rank 50), due to the slower spread of flows as they pass the DIN;
- T_1 = 15 November and T_2 = 31 July for for Diré, Koryoumé, Tossaye and Ansongo on Niger (ranks 51 to 54), for the same reasons as Akka;
- T_1 = 15 September and T_2 = 30 April for Moussala and Siramakana in the Senegal watershed (ranks 12 and 17), Diaguery and Niokolokoba in the Gambia watershed (ranks 29 and 30) and Manankoro in the Niger watershed (rank 42), where the discharge ceases each year before the end of April.

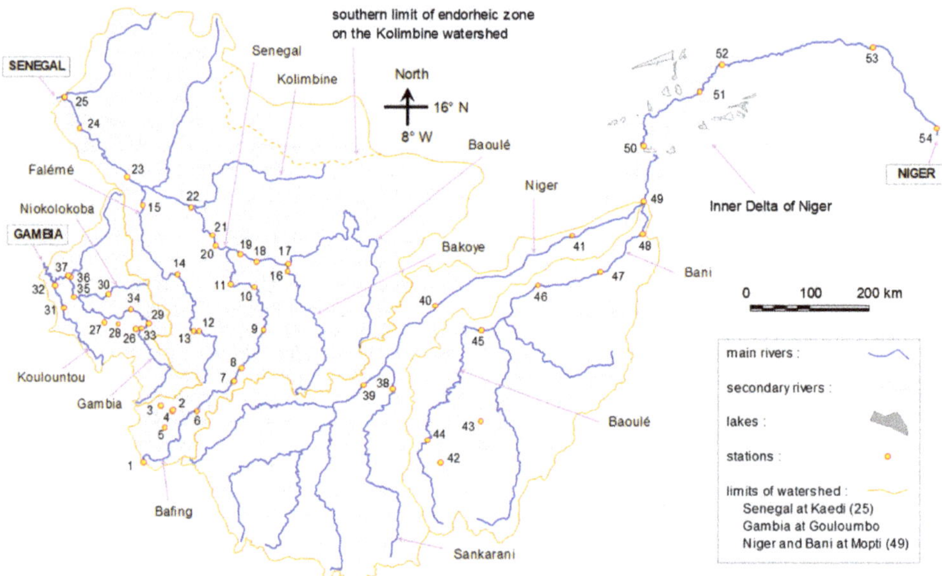

Figure 1. Hydrographic networks of the Senegal River upstream of Kaédi, Gambia upstream of Gouloumbo and Niger upstream of Ansongo, based on the maps of [40] for Senegal, [41] for Niger and [42] for Gambia.

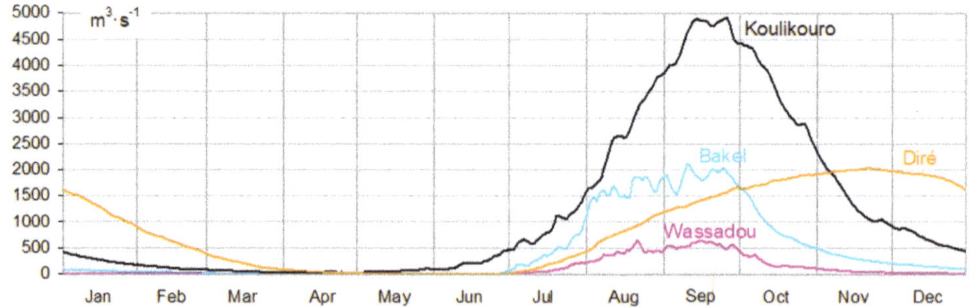

Figure 2. Median annual hydrographs for Senegal at Bakel (rank 23), Gambia at Wassadou upstream (rank 36), Niger at Koulikouro (rank 40) and Niger at Diré (rank 51) over the period 1970–1979.

3. Results

3.1. Calibration of Models Representing K

3.1.1. Results for Daka Saidou Station

This station has the largest number (N = 10,966) of observed values of K (mean: K_m = 0.9714). The average evolution of K as a function of D (growth from September to the beginning of February and then decline until the end of May, Figure 3) is better represented by model 1 than by model 3 (C_{NSE0} = 0.347 and 0.296 respectively). Similarly, the mean evolution of K as a function of Q is better represented by model 2 than by model 4 (C_{NSE0} = 0.392 and 0.335 respectively): K increases strongly as a function of Q under low flows; under Q greater than about 25 $m^3 \cdot s^{-1}$, K decreases as a function of Q until the highest flows are reached (Figure 4).

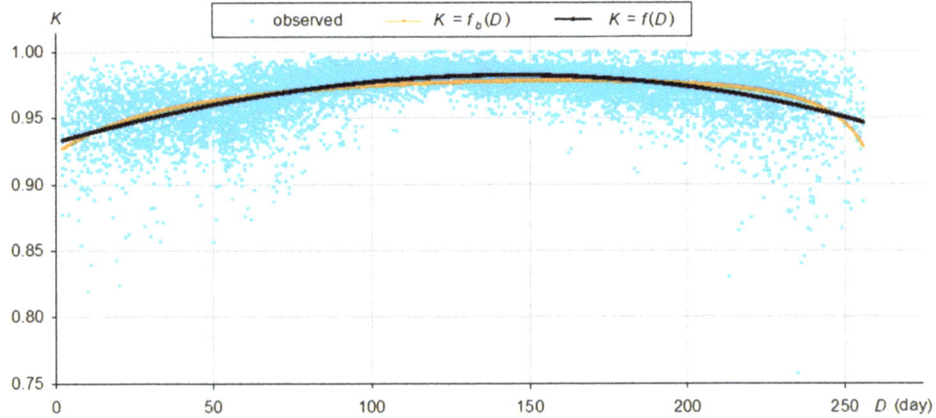

Figure 3. Relationship between D (time elapsed since 15 September) and K at Daka Saidou on the Bafing (row 8; N = 10,966 observed values from 5 october 1952 to 20 october 2016; C_{NSE0} = 0.347 for model 1 with $K = f(D)$, C_{NSE0} = 0.296 for model 3 with $K = fb(D)$).

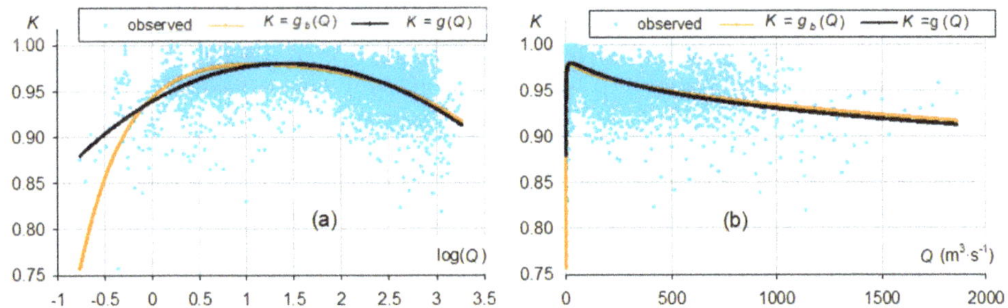

Figure 4. Relationship between discharge Q and K at Daka Saidou on the Bafing (row 8; 10,966 points observed from 5 october 1952 to 20 october 2016; C_{NSE0} =0.392 for model 2 with $K = g(Q)$; C_{NSE0} = 0.335 for model 4 with $K = gb(Q)$), with abscissa in logarithmic values on the left (**a**) and natural values on the right (**b**).

3.1.2. Results for All Stations

For the calibration of functions f and g of models 1 and 2, the degree of the polynomial of D or $\log(Q)$ is generally limited to 3. However, a higher degree (between 4 and 6) is used to improve the calibration for 30% of the stations for f and 41% for g, mainly for the Niger basin. To set the functions

f_b and g_b of models 3 and 4, a superior limit of 10 is imposed on parameter n and Q_0 is arbitrarily set to the highest discharge Q associated with observed K (with no effect on the quality of the model, see above: relation (9)).

The parameters of the different models calibrated for each station are given in the Supplementary Materials (Tables S1–S3 for models 0 to 2 and Table S4 for models 3 and 4), with the C_{NSE0} of calibration, the R_a classification rows and the extremities of the calibration domains (D_{min} and D_{max} for f and f_b and Q_{min} and Q_{max} for g and g_b). Figure 5 compares model performance in describing the observed values of K. For the average of their rank R_a over all stations, indicated in brackets, the models rank from best to worst as follows: model 2 (1.2), model 1 (2.3), model 4 (2.9), model 3 (3.9) and model 0 (4.6). The best results are obtained with model 2 for 81% of the stations, with model 1 for 13% of the stations, with model 3 for 4% of the stations and with model 4 for 2% of the stations. The results of the two best models are detailed below.

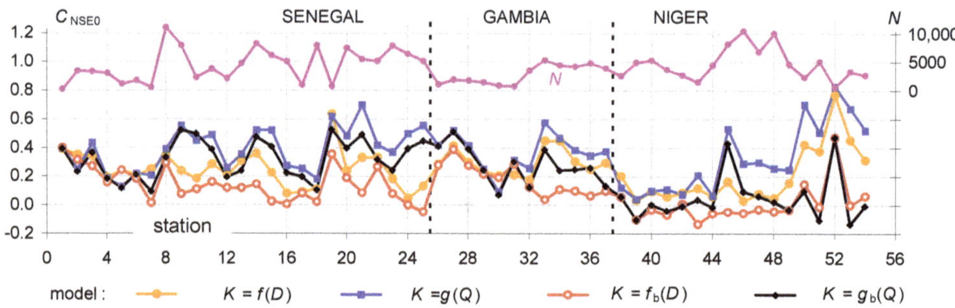

Figure 5. Efficiency coefficient C_{NSE0} and number N of calibration of models 1 to 4 describing K as a function of D or Q, for each station analyzed.

3.1.3. Results for Model 1, Describing K as an Empirical Function of D

For most stations in the Senegal and Gambia basins (Figure 6a–g), the evolution of K as a function of time D elapsed since 15 September is large and very close to that observed at Daka Saidou: K increases as a function of D for 3 to 5 months from mid September on, then decreases until the end of May. Most downstream stations (ranks 23 to 25 for Senegal; 32 and 35 to 37 for Gambia) are characterized by smaller amplitude of K variations and some differences at the beginning and end of the recession.

Compared to the above stations, those in the Niger Basin have very different seasonal variations of K, with smaller amplitude (Figure 6h–j): K decreases with D for about 2 months from a very high value at the beginning of the recession period, and then changes relatively little for most stations.

3.1.4. Results for Model 2, Describing K as an Empirical Function of Q

For most stations in the Senegal and Gambia (Figure S1) basins, the functions $K = g(Q)$ are similar to those of Daka Saidou, as shown in Figures 7a–e and 8a,b, where discharge Q on the abscissa is mostly replaced by specific discharge Q_s (flow divided by catchment area) to facilitate readability. Under low flows, K increases rapidly with Q_s. Under higher flows, K gradually decreases as a function of Q_s with curves often quite close to each other at most stations in the same sub basin such as Bafing, Falémé and Senegal (Figure 7b,c,e). As an exception to these general trends, there is an additional inflection and growth of the function $g(Q)$ under the highest flow rates, at the following stations: Siramakana and Niokolokoba (ranges 17 and 30) that drain the northernmost and, therefore, least watered, sub basins among the stations analyzed in the Senegal and Gambia basins respectively; Bakel, Matam and Kaedi (ranges 23 to 25) in the Senegal basin and Missira Gonasse, Simenti, Wassadou upstream and Wassadou downstream (ranges 32 and 35 to 37) in the Gambia basin, the most downstream stations in these basins.

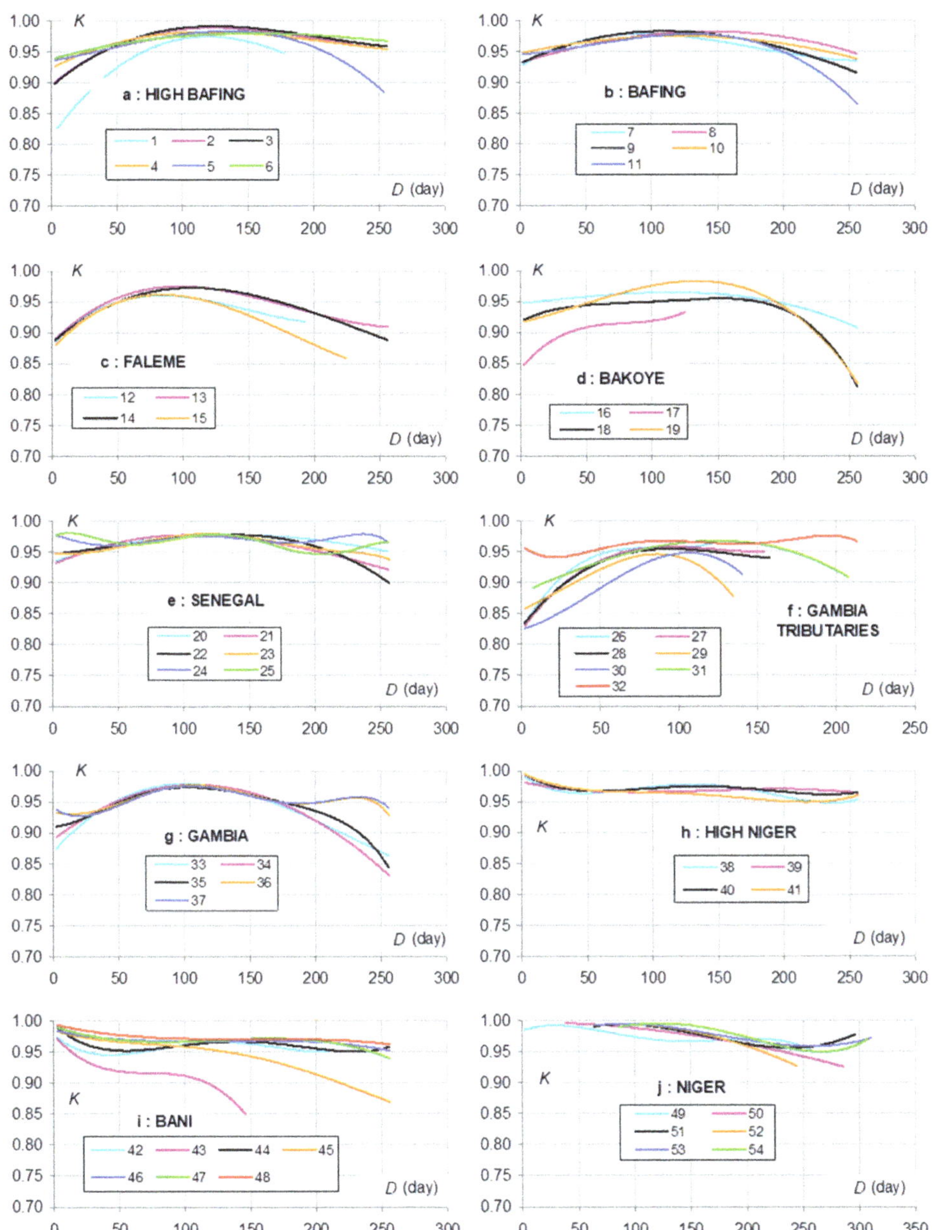

Figure 6. Model 1 representing K as a function of the time D elapsed since the previous 15 September ($K = f(D)$) at the 54 stations located in the Senegal, Gambia and Niger basins, by sub-basins: High Bafing (**a**); Bafing (**b**); Faleme (**c**); Bakoye (**d**); Senegal (**e**); Gambia tributaries (**f**); Gambia (**g**); High Niger (**h**); Bani (**i**); Niger (**j**).

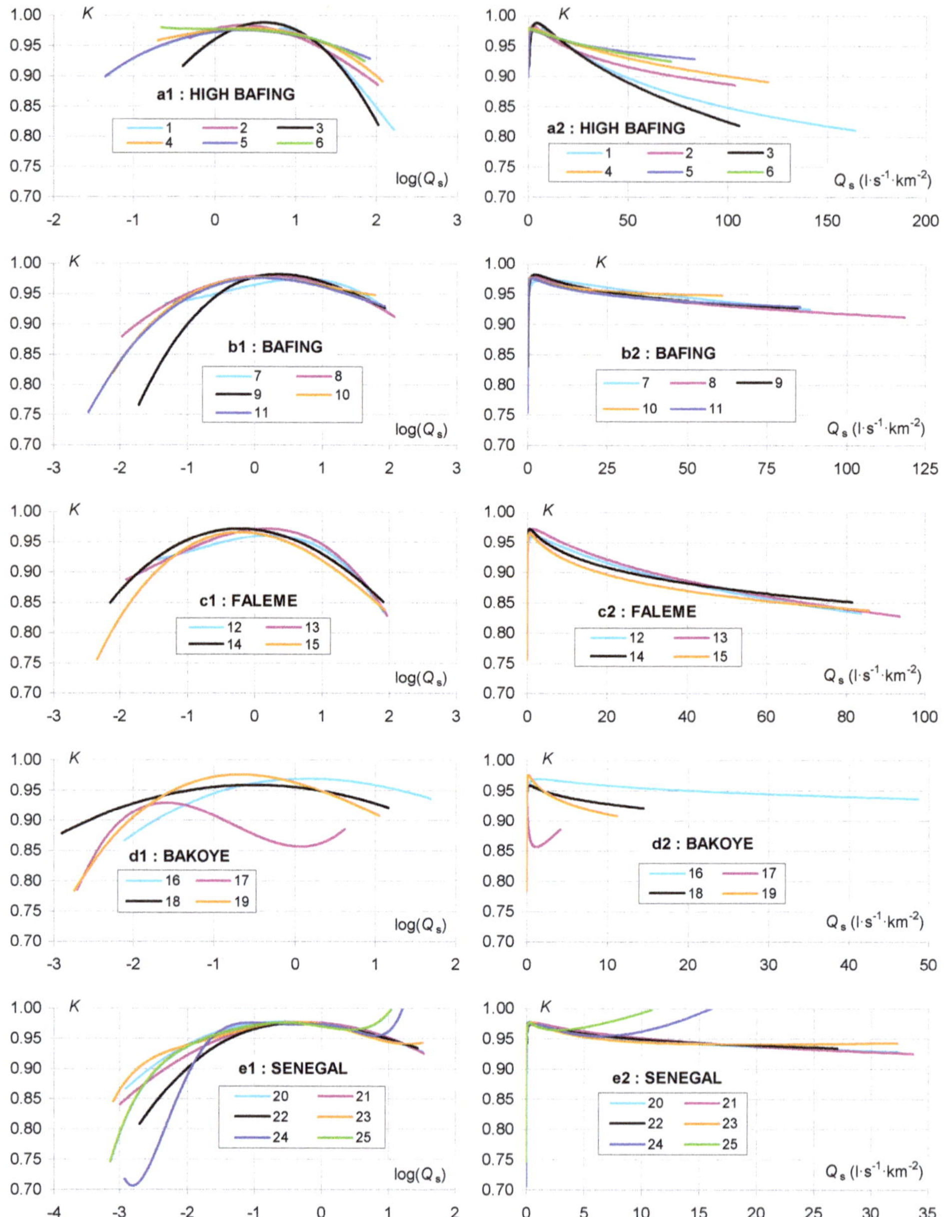

Figure 7. Model 2 representing K as a function of the specific discharge Q_s at the 25 stations located in the Senegal basin (relations $K = g(Q)$), with abscissa in logarithmic values on the left (index 1) and natural values on the right (index 2), by sub-basins: High Bafing (**a1,a2**), Bafing (**b1,b2**), Faleme (**c1,c2**), Bakoye (**d1,d2**), Senegal (**e1,e2**).

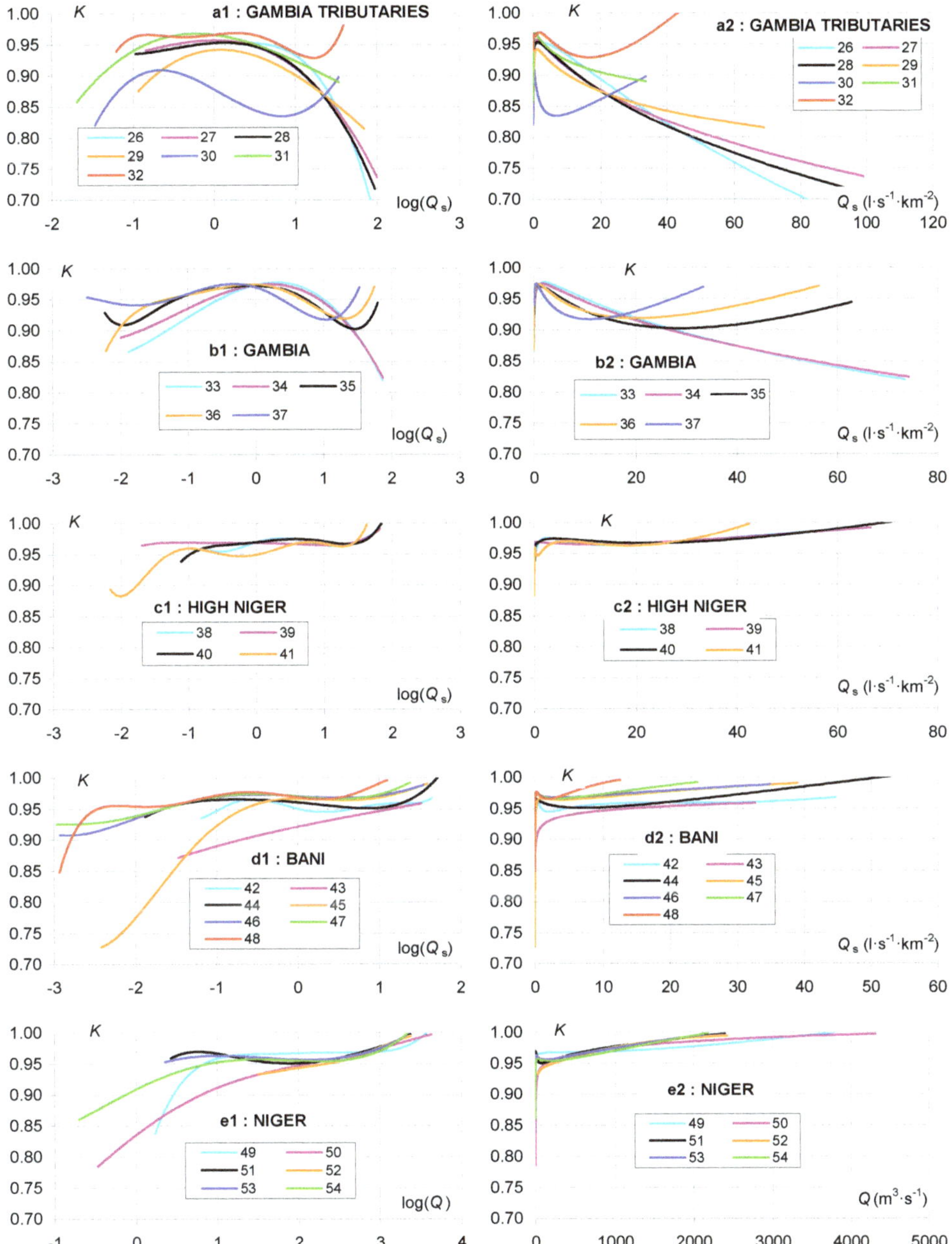

Figure 8. Model 2 representing K as a function of the specific discharge Q_s or the discharge Q at stations 26 to 54 in the Gambia and Niger basins (relations $K = g(Q)$), with abscissa in logarithmic values on the left (index 1) and natural values on the right (index 2), by sub-basins: Gambia tributaries (**a1,a2**), Gambia (**b1,b2**), High Niger (**c1,c2**), Bani (**d1,d2**), Niger (**e1,e2**).

For stations in the Niger Basin (Figure S2 and Figure 8c–e), the $K = g(Q)$ functions differ significantly from previous ones with a smaller amplitude of variation and a reverse direction of variation (except under low flows). For four stations (rows 43, 49, 50 and 52), the K rate increases as a function of discharge over the entire tidal range. For the other stations, K generally increases as a function of discharge at low water, slightly decreases at medium water and increases at high water.

3.2. Performance of the Recession Models in Forecasting Discharge

3.2.1. Calculation Details

For each station, the discharge chronicle used to calculate the observed K rates is also used to evaluate the discharge forecasts made with the different recession models, at all entire H horizons between 1 and 120 days. For model 0 characterized by a constant K rate characteristic of each station, we use the value K_0 which maximizes the overall C_{NSE2} of the N_t forecasts calculated at the different horizons H. This optimal rate K_0 differs slightly from the average K_m of the observed K rates (Figure 9; Table S1). But the differences observed between basins are identical with K_0 and K_m, with a distribution function that is low for the Gambia, intermediate for Senegal and high for Niger. The K_0 value is particularly low for Siramakana and Niokolokoba stations (ranges 17 and 30), mentioned above for their relations $K = g(Q)$.

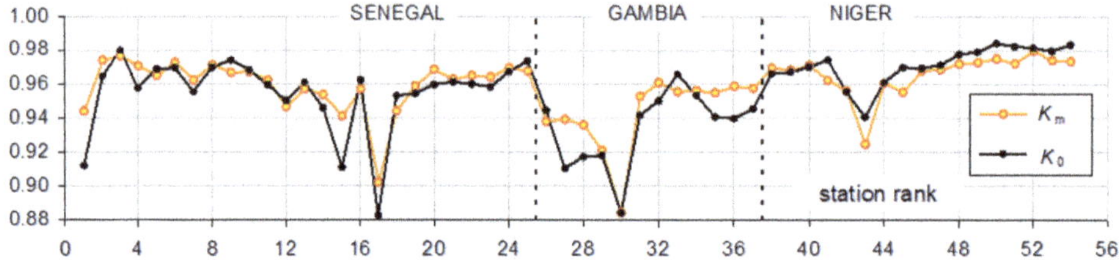

Figure 9. Mean K_m of K observed and optimal value K_0 maximizing the efficiency coefficient C_{NSE2} of the N_t discharge forecasts made with model 0 (Maillet) at horizons H between 1 and 120 days.

The discharge forecasts calculated with models 1 to 4 are performed for D and Q values within their calibration ranges. For some stations (mainly in the Niger basin), however, the function g tends to overestimate K under the highest flows, with values very close to or slightly higher than 1. The application range of g under high water for these stations is therefore reduced, with a limit Q_{lim} lower than Q_{max} (Table S3). In this case, model 2 is applied with a value $g(Q_{lim})$ for all flows above Q_{lim}.

3.2.2. Results for All Forecasting Horizons

Unlike parameter K_0 of model 0, the parameters of models 1 to 4 are not optimized to maximize the overall C_{NSE2} from forecasts to different forecast horizons, but simply based on observed values of K. Notwithstanding, these models perform better (higher C_{NSE2}) than model 0 for most stations, except models 3 and 4 in the Niger basin (Figure 10). The number of stations for which a model gives the best results (maximum C_{NSE2}) is 3, 10, 40, 0 and 1 for models 0 to 5 respectively, whose average rankings based on C_{NSE2} are 3.8, 2.4, 1.4, 4.3 and 3.1. Overall, model 2 is, therefore, the best for flow forecasts at different horizons H between 1 and 120 days.

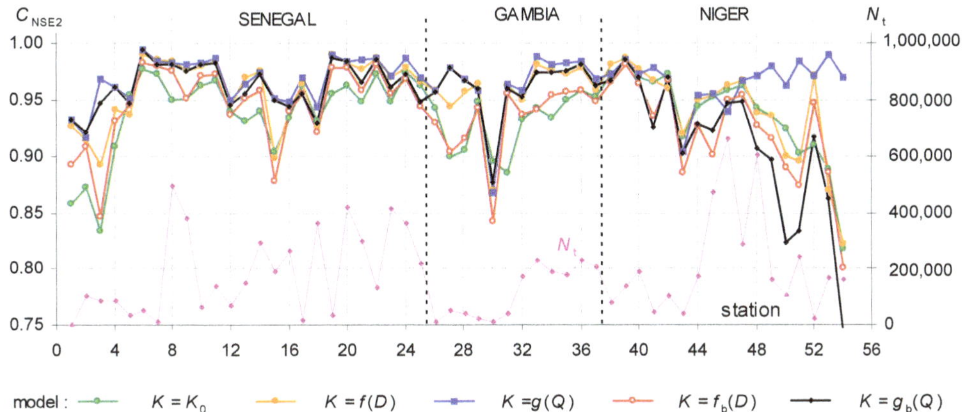

Figure 10. C_{NSE2} efficiency coefficient and number N_t of the discharge forecasts made with the different K models (optimal K_0 value or D or Q functions) at all H horizons between 1 and 120 days.

3.2.3. Results for Each Forecasting Horizon

For all models, the accuracy of flow forecasts, expressed by C_{NSE1}, tends to decrease with time horizon H, as shown in Figure 11 for Daka Saidou station and Figures S3 and S4 for Kedougou and Mopti. For Daka Saidou, the superiority of models 1 and 2 over the others is verified for each horizon.

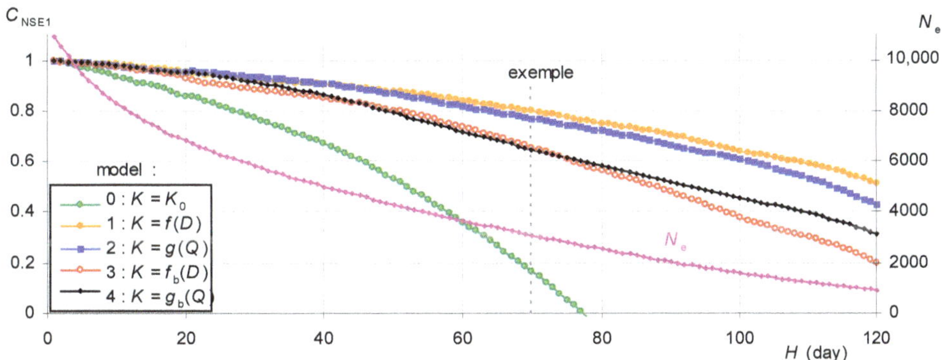

Figure 11. Performance of models in forecasting discharge during recession periods at Daka Saidou on the Bafing, according to the forecast horizon.

The following result shows that model 2 provides the best overall forecasts for all stations for each forecast horizon: for each horizon H between 1 and 120 days, the average for the different stations of the relative standard error R_{rmse} is always lower with model 2 than with the other models (Figure 12). The advantage of model 2 increases with the forecast horizon, since compared to model 0, it gives a lower average R_{rmse} value of 15% to 30% for each horizon between 1 and 30 days, 30% to 36% for each horizon between 31 and 60 days and 36% to 41% for each horizon between 61 and 120 days.

The values of the Nash and Sutcliffe coefficient C_{NSE1} and the relative standard error R_{rmse} for flow forecasting at each station are given for all H horizon multiples of 5 days in Supplementary Materials (Tables S5 and S6).

Example: Forecasting of N_e = 3082 discharge values at horizon H = 70 days gives C_{NSE1} equal to 0.165, 0.798, 0.769, 0.654 and 0.647 with models 0, 1, 2, 3 and 4, respectively

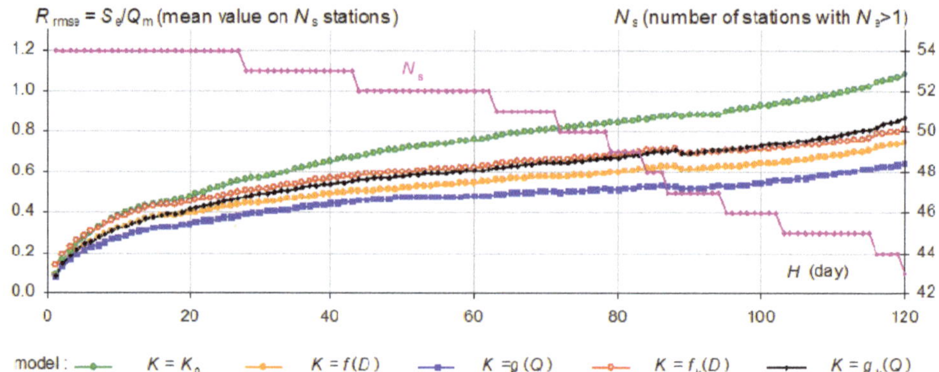

Figure 12. Average model performance in forecasting discharge at all stations, according to the forecast horizon H (average of the relative standard error R_{rmse}, calculated on the N_s stations for which the number N_e of discharge forecasts at horizon H is greater than 1).

3.3. Interannual Evolution of the Recession Regime

3.3.1. Results for Daka Saidou Station

For each station, model 2, based on all available data from 1950 to 2016, represents the observed values of K with some errors. For most stations, the chronological accumulation C_e of these errors does not vary randomly according to the chronological rank R_c. On the contrary, C_e shows fairly clear overall slope breaks under low flows, high flows or all flows as the case may be. However, the slope dC_e/dR_c of C_e is equal to the error of the model. When fairly regular over a certain period, this slope then corresponds to a systematic error of the model, which justifies a particular calibration of the model in this period. For the Daka Saidou station, for example (Figure 13), the overall slope of $C_e(R_c)$ under high flows, initially negative (K is generally underestimated by the model), increases significantly and even becomes positive from about March 1976 (K is strongly overestimated), then decreases while remaining positive from about December 1994 (K is overestimated). The model can, therefore, be calibrated over three periods 1 to 3 (October 1952 to February 1976; March 1976 to November 1994; December 1994 to October 2016) to better represent K. The three curves $K = g(Q)$ obtained differ quite clearly under high flows, with the lowest values obtained in period 2 and the highest in period 1 (Figure 14). However, their general shape remains identical to that of the average model based on all the data.

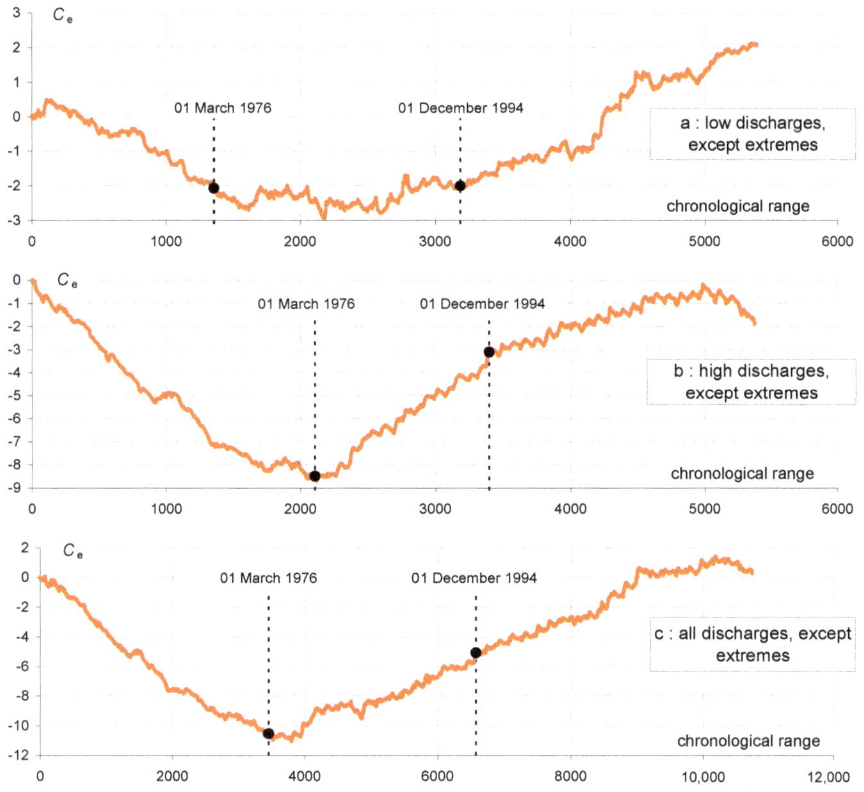

Figure 13. Chronological evolution of the cumulative error C_e of model 2 with respect to the $N = 10,966$ observed values of K on the Bafing at Daka Saidou, for low flows with frequency of exceedance between 0.5 and 0.99 (**a**), high flows with frequency between 0.01 and 0.5 (**b**), and all flows except extreme flows, with frequency between 0.01 and 0.99 (**c**).

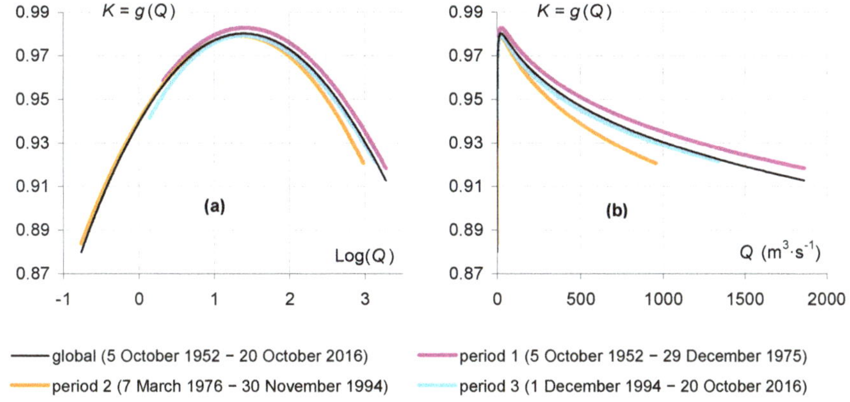

Figure 14. Model 2 representing K as a function of discharge Q, calibrated on successive periods 1 to 3 and on the overall period, for the Bafing at Daka Saidou (station 8), with abscissa in logarithmic values on the left (**a**) and natural values on the right (**b**).

3.3.2. Results for All Stations

By reproducing this analysis of the cumulative errors $C_e(R_c)$ for each station, we obtain the following results (Figure 15):

- for 44 stations, a transition between period 1 and 2 can be detected by $C_e(R_c)$ overall change in slope under high flows (38 stations) or low flows (6 stations). Except for the station of rank 1, for which data are lacking, this transition can be dated between March 1968 and March 1983, depending on the station. For 39 of these 44 stations, the transition from period 1 to period 2 corresponds to an increase in errors in the average model and therefore a decrease in K observed for the same Q, like for Daka Saidou;

- for 26 stations, a transition between period 2 and 3 can be detected by a change in $C_e(R_c)$ slope under high flows (25 stations) or under low flows (1 station). Except for stations 29 to 31, for which data are lacking, this transition can be dated between March 1988 and December 1994, depending on the station. For these 26 stations, the transition from period 2 to period 3 corresponds to a decrease in the errors of the average model and therefore to an increase in K observed for the same Q, like for Daka Saidou;

- the following results are observed for the average K_m of the observed values of K (Table S1): K_m decreases by an average of 1.2% between periods 1 and 2 and increases by an average of 0.8% between periods 2 and 3. This evolution of K_m is observed at all stations with rare exceptions. With $\alpha = -Log(K)$ according to equation (1), these average changes in K over all the stations correspond to an average increase of 0.0120 d^{-1} between periods 1 and 2 and a decrease of 0.0080 d^{-1} between periods 2 and 3 for the parameter α generally used with the Maillet formula. This evolution of α between periods 1 and 2 is consistent with the results presented by [44,45] for several basins of West and Central Africa, [46] for the Niger River and [47] for the Bani.

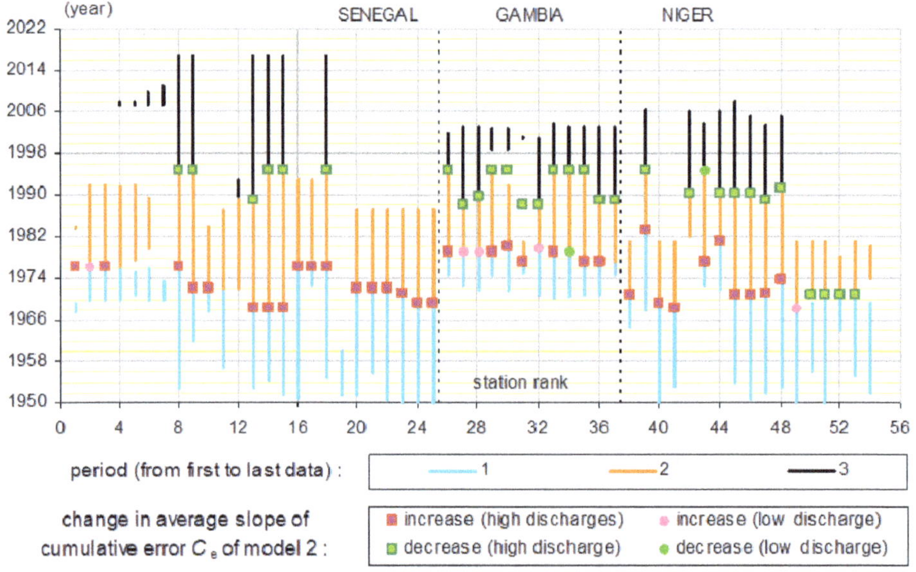

Figure 15. Division of the data series into successive periods for each station, based on the dates of change in the overall slope of the time accumulation C_e of the model 2 error.

For each station, model 2 is finally calibrated successively on the three periods 1, 2 and 3, defined from the transition dates determined above for the stations concerned. For the other stations, these dates are arbitrarily fixed (or specified for stations 1 and 29 to 31) from the dates determined at neighboring stations: the transition between periods 1 and 2 for 8 stations (ranks 1, 4 to 7, 11, 37 and 54); the transition between periods 2 and 3 for 9 stations (ranks 4 to 7, 12, 27 and 29 to 31).

For each station, as for Daka Saidou (Figure 14), the general shape of the $g(Q)$ curves set over successive periods is identical to that of the average model set over all the data (see Figures S5 and S6 for examples). The main differences already noted between basins on the $g(Q)$ curves set on all the data are, therefore, found over each period 1 to 3, in particular for the $g(Q)$ slope on most of the upper part of the tidal range: mainly negative in the Senegal and Gambia basins and positive in the Niger basin.

For all stations, the amplitude of variation of the average function $g(Q)$ over its field of application is much greater than the offsets observed between functions $g(Q)$ set over successive periods. On average for all stations, this amplitude $\max(g(Q))-\min(g(Q))$ represents 13% of the average observed rate K_m over all the data. It is much higher than the differences observed between the average K_m values calculated over periods 1 and 2 (Figure 16). The seasonal variations of K, therefore, appear to be much greater than its interannual variations

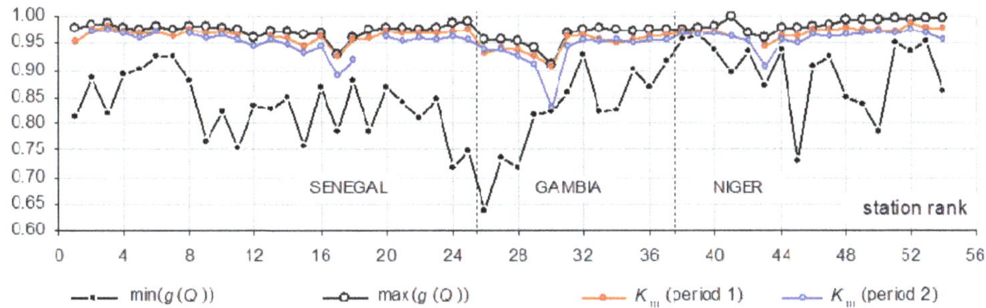

Figure 16. Mean K_m of K observed over each period 1 and 2 and minimum and maximum values of K modeled as a function of discharge (model $K = g(Q)$ calibrated on all observations).

4. Discussion

For the representation of the daily depletion factor K and the discharge forecasting at horizons 1 to 120 days during recession regimes over the Senegal, Gambia and Niger basins, empirical models 1 ($K = f(D)$) and 2 ($K = g(Q)$) give better overall results than conceptual models based on the Maillet (model 0) and generalized Coutagne (models 3 and 4) formula. The best (model 2) represents K by a polynomial of the decimal logarithm of the discharge Q, with a Nash and Sutcliffe C_{NSE0} coefficient varying between 0.04 and 0.83 (mean 0.37, median 0.37) compared to the observed values of K over the 1950–2016 period. Compared to the most frequently used Maillet formula, it improves flow forecasting with on average a 30% to 41% decrease of relative standard error R_{rmse}, for each forecast horizon H between 31 and 120 days.

Model 2 shows a fairly homogeneous relationship over some areas or sub-basins (Bafing, Falémé, Senegal, Upper Niger) between the specific discharge Q_s and K. On average over each basin, according to this model, K evolves during the recession period as follows:

- In the Senegal and Gambia basins, K is generally quite low at the beginning of the recession (between 0.80 and 0.95 under the highest discharges Q). It then increases over time ($K(Q)$ decreasing) to reach a fairly high value (between 0.96 and 0.98), for a specific flow rate Q_s between 0.3 and $2\ l{\cdot}s^{-1}{\cdot}km^{-2}$ depending on the sub-basin concerned, before decreasing ($K(Q)$ strongly increasing) until the end of the recession. The downstream stations are characterized by a decrease of K at the beginning of the recession ($K(Q)$ increasing), from very high values, close to 1 under the highest flows;
- In the Niger basin, K is generally very high at the beginning of the recession (higher than 0.975 under the highest flows). It then decreases ($K(Q)$ slightly increases) to a value greater than 0.95,

reached for a specific discharge Q_s between 7 and 17 $l \cdot s^{-1} \cdot km^{-2}$ upstream of the Inner Delta or for a discharge Q of about 133 $m^3 \cdot s^{-1}$ for stations downstream. Then, K evolves relatively little according to the flow and in a variable way according to the station, until the end of the recession.

The basins also differ in the K_0 value of K which optimizes discharge forecasts at horizons 1 to 120 days with the model 0 (Maillet formula). Overall, K_0 is quite low in the Gambia basin (average = 0.934; median = 0.941), medium in the Senegal basin (average = 0.955; median = 0.960) and quite high in the Niger basin (average = 0.972; median = 0.972). These differences may be related to the geological and soil characteristics of the three basins, but also to the rainfall regime. Indeed, the Gambia basin, which has the least rainfall in its upstream part according to [48], has the fastest drying up, while the Niger basin, which has the most rainfall according to [48], has the slowest drying up.

Finally, for most stations, the analysis of the chronological accumulation of errors on K with model 2 shows successive periods that differ slightly for model calibration. According to the available data, for each station we identified two or three periods with transition dates ranging from 1968 to 1983 between periods 1 and 2 and from 1988 to 1994 between periods 2 and 3. Generally, the average K_m of K observed decreases between periods 1 and 2 and increases between periods 2 and 3. The faster depletion during period 2 is probably related to a decrease in groundwater reserves, as a result of reduced rainfall in West Africa during this period [48,49]. Over each period, the function $K = g(Q)$ differs slightly from the average function based on all the data. However, the differences between periods are much smaller than the amplitude of variation of $g(Q)$ over its field of application. On average for all stations, seasonal variations of K represent 13% of the average observed K_m rate (15% for Senegal; 15% for Gambia; 9% for Niger), while interannual variations of K_m are only about 1% between successive periods 1 and 2. For each basin, the seasonal variations of the daily depletion factor K are, therefore, much greater than its climate-related interannual variations.

Supplementary Materials: The following are available online at http://www.mdpi.com/2073-4441/12/9/2520/s1: mathematical demonstrations (Texts S1 and S2) and detailed results (Tables S1–S6 and Figures S1–S6) for interested readers. However, this material is not essential for the understanding of the article. A list of variables and parameters used is given in the Abbreviations.

Author Contributions: conceptualization, J.-C.B.; data curation, J.-C.B.; formal analysis, J.-C.B.; investigation, J.-C.B., H.D., J.-C.P.; methodology, J.-C.B.; project administration, J.-C.B.; resources, J.-C.B.; software, J.-C.B.; supervision, J.-C.B.; validation, J.-C.B., H.D., J.-C.P.; visualization, J.-C.B., H.D., J.-C.P.; writing—original draft, J.-C.B.; writing—review & editing, J.-C.B., H.D., J.-C.P. All authors have read and agreed to the published version of the manuscript.

Funding: This research received no external funding.

Conflicts of Interest: the authors declare no conflict of interest.

Abbreviations

List of Variables and Parameters

A_i (d^{-i})	with i integer between 0 and 6: model 1 parameters ($K = f(D)$), polynomial coefficients of D
B_i (-)	with i integer between 0 and 6: model 2 parameters ($K = g(Q)$), polynomial coefficients of $\log(Q)$
C_e (-)	chronological accumulation of model 2 errors in relation to observed values of K
C_{NSE0} (-)	Nash and Sutcliffe model efficiency coefficient, equal to 1-((standard error)/(standard deviation))^2, for the modeling of K
C_{NSE1} (-)	Nash and Sutcliffe model efficiency coefficient, for the forecast of Q at a given horizon H
C_{NSE2} (-)	Nash and Sutcliff model efficiency coefficient, for the forecast of Q at all horizons H
D (d)	period elapsed since the previous 15 September
D_{max} (d)	upper limit of the calibration range of the function $f(D)$
D_{min} (d)	lower limit of the calibration range of the function $f(D)$
f (-)	relationship giving K as a function of D (model 1, empirical)
f_b (-)	relationship giving K as a function of D (model 3, based on the generalized Coutagne formula)

g (-)	relationship giving K as a function of Q (model 2, empirical)
g_b (-)	relationship giving K as a function of Q (model 4, based on the generalized Coutagne formula)
H (d)	horizon of discharge forecast
i (-)	rank of a recession sequence
j (-)	rank of a recession sequence
K (-)	daily discharge depletion factor, between 0 (end of flow) and 1 (flow constancy)
K_0 (-)	optimal constant value of K, maximizing the C_{NSE2} of the N_t forecast discharges with the recession model 0 at all H horizons between 1 and 120 days
K_m (-)	average of the N observed values of K
Log (-)	neperian logarithm
log (-)	decimal logarithm
m (-)	positive exponent of volume V, used in the Coutagne theory
n (-)	positive exponent of time in the Coutagne formula
N (-)	number of observed values of K
N_e (-)	number of discharge values forecast by recession model, for a given forecast horizon H
N_s (-)	number of stations for which N_e is greater than 1
N_t (-)	number of discharge values forecast per recession model, for all forecast horizons H between 1 and 120 days
p (-)	positive exponent of time used in the Horton formula
P (-)	percentage of stations, between 0 (no stations) and 100 (all stations)
Q (m$^3 \cdot$s^{-1})	discharge
Q_A (m$^3 \cdot$s^{-1})	discharge at time T_A
Q_B (m$^3 \cdot$s^{-1})	discharge at time T_B
Q_{lim} (m$^3 \cdot$s^{-1})	upper limit of the application range of function $g(Q)$
Q_m (m$^3 \cdot$s^{-1})	average of observed discharges
Q_{max} (m$^3 \cdot$s^{-1})	upper limit of the calibration range of function $g(Q)$
Q_{min} (m$^3 \cdot$s^{-1})	lower limit of the calibration range of function $g(Q)$
Q_0 (m$^3 \cdot$s^{-1})	discharge at the beginning of a recession sequence
Q_s (l$\cdot$s$^{-1} \cdot$km^{-2})	specific discharge (discharge per unit of drained area)
Q_{sm} (l$\cdot$s$^{-1} \cdot$km^{-2})	mean interannual specific discharge
Q_{thr} (m$^3 \cdot$s^{-1})	minimum discharge required to account for the observed values of K
r (-)	exponent of time used in the Otnes model
R_a	rank of a recession model for a given station, between 1 for the best (maximum C_{NSE}) and 5 for the worst (minimum C_{NSE})
R_c	chronological rank of K modeled
R_{rmse}	relative standard error of the discharges forecast per recession model, for a given time horizon
S (-)	homothety slope between hydrographs of distinct recession sequences
S_e (m$^3 \cdot$s^{-1})	standard error of the discharge forecasts calculated by a recession model, for a given horizon H
T (d)	time
T_A (d)	initial time
T_B (d)	initial time
T_b (d)	date of the oldest K observed during an uninterrupted sequence of recession
T_r (d)	time translation duration
T_0 (d)	initial time, used in the Maillet and Tison formulas
T_1 (d)	usual start date of the recession period, expressed in "dd/mm" format
T_2 (d)	usual end date of the recession period, expressed in "dd/mm" format
V (m^3)	remaining water volume in a reservoir used in the Coutagne theory
W (m$^3 \cdot$s^{-1})	constant used in the generalized Maillet and generalized Coutagne formulas. If positive, it is the theoretical value of the discharge reached after an infinite time during a recession phase
Z (m$^{3(1-m)} \cdot$s^{-1})	proportionality constant between Q and V^m in the Coutagne theory
α (d^{-1})	positive recession coefficient, used in the Maillet formula (exponential form)

β (d^{-p})   positive recession coefficient, used in the Horton formula
λ (m$^3 \cdot$s$^{-1} \cdot$d^r)   parameter used in the Otnes model
μ (m$^3 \cdot$s^{-1})   parameter used in the Otnes model
σ_0 (d^{-1})   positive recession coefficient, used in the Tison and Coutagne formulas

References

1. Roche, M. *Hydrologie de Surface*; Gauthier-Villars/ORSTOM: Paris, France, 1963; p. 430.
2. Hall, F.R. Base Flow Recessions—A review. *Water Resour. Res.* **1968**, *4*, 973–983. [CrossRef]
3. Tallaksen, L.M. A Review of Baseflow Recession Analysis. *J. Hydrol.* **1995**, *165*, 349–370. [CrossRef]
4. Boussinesq, J. Recherches théoriques sur l'écoulement des nappes d'eau infiltrées dans le sol et sur le débit des sources. *J. Math. Pure Appl.* **1904**, *10*, 5–78.
5. Lang, C. *Etiages et Tarissements: Vers Quelles Modélisations ? L'approche Conceptuelle et L'analyse Statistique en Réponse à la Diversité Spatiale des Ecoulements en Etiage des Cours d'eau de l'Est Français*; Thèse de l'université Paul Verlaine de Metz: Metz, France, 2007. Available online: https://tel.archives-ouvertes.fr/tel-00534656/document (accessed on 3 January 2014).
6. Barnes, B.S. The structure of discharge-recession curves. *Trans. Am. Geophys Union* **1939**, *20*, 721–725. [CrossRef]
7. Maillet, E. *Essai d'hydraulique Souterraine et Fluviale*; Herman, A., Ed.; Libraire Sci. Hall: Paris, France, 1905.
8. Tison, G. *Courbe de Tarissement, Coefficient D'écoulement et Perméabilité du Bassin*; Mém. A.I.H.S.: Helsinki, Finland, 1960; pp. 229–243.
9. Coutagne, A. *Meteorologie et Hydrologie. Etude Générale des Variations de Débit en Fonction des Facteurs qui les Conditionnent. 2me Partie. Les Variations de Débit en Période non Influencée par les Précipitations. Le Débit D'infiltration (Corrélations Fluviales Internes)*; La Houille Blanche: Grenoble, France, 1948; pp. 416–436. Available online: https://www.shf-lhb.org/articles/lhb/pdf/1948/07/lhb1948053.pdf (accessed on 24 June 2019).
10. Carlier, M. *Hydraulique Générale et Appliquee*; Eyrolles: Paris, France, 1972; p. 565.
11. Thomson, S.H. Hydrologic conditions in the chalk at Compton, West-Sussex. *Trans. Inst. Water Eng.* **1921**, *26*, 228–261.
12. Wicht, C.L. Determination of the effects of watershed-management on mountain streams. *Trans. Am. Geophys. Union* **1943**, *24*, 594–606. [CrossRef]
13. Toebes, C.; Strang, D.D. On recession curves, 1-Recession equations. *J. Hydrol. NZ* **1964**, *3*, 2–15.
14. Radczuk, L.; Szarska, O. Use of the flow recession curve for the estimation of conditions of river supply by underground water. *IAHS Publ.* **1989**, *187*, 67–74.
15. Clausen, B. Modelling streamflow recession in two Danish streams. *Nord. Hydrol.* **1992**, *23*, 73–88. [CrossRef]
16. Padilla, A.; Pulido-Bosh, A.; Mangin, A. Relative importance of baseflow and quickflow from hydrographs of karst spring. *Ground Water* **1994**, *32*, 267–277. [CrossRef]
17. Otnes, J. Uregulerte elvers vassføring i tørrværsperioder. *Nor. Geogr. Tidsskr.* **1953**, *14*, 210–218. [CrossRef]
18. Otnes, J. Tørrværskurven. In *Hydrologi i Praksis*; Otnes, J., Ræstad, E., Eds.; Ingeniørforlaget: Oslo, Norway, 1978; pp. 227–233.
19. Gjørsvik, O.G. *Grosetbekken. En Vurdering av Vannbalansen*; NVE Rep. 2, Part I, 1970; Norwegian Water and Energy Admin: Oslo, Norway, 1970.
20. Andersen, T. En Undersøkelse av Grunnvannsmagasinet i et Representativt Høyfjellsområde. Master's Thesis, University of Oslo, Oslo, Norway, 1972.
21. Tjomsland, T.; Ruud, E.; Nordseth, K. The physiographic influence on recession runoff in small Norwegian rivers. *Nord. Hydrol.* **1978**, *9*, 17–30. [CrossRef]
22. Horton, R.E. The role of infiltration in the hydrologic cycle. Trans. *Am. Geophys. Union* **1933**, *14*, 446–460. [CrossRef]
23. Werner, P.W.; Sundquist, K.J. On the groundwater recession curve for large watersheds. *IAHS Publ.* **1951**, *33*, 202–212.
24. Pereira, L.S.; Keller, H.M. Recession characteristics of small mountain basins, derivation of master recession curves and optimization of recession parameters. *IAHS Publ.* **1982**, *138*, 243–255.
25. Nutbrown, D.A. Normal mode analysis of the linear equation of groundwater flow. *Water Resour. Res.* **1975**, *11*, 979–987. [CrossRef]

26. Petras, I. An approach to the mathematical expression of recession curves. *Water S. Afr.* **1986**, *12*, 145–150.

27. Ishihara, T.; Takagi, F. A study of the variation of low flow. Disaster Prev. *Res. Inst. Kyoto Univ. Bull.* **1965**, *15*, 75–98.

28. Dewandel, B.; Lachassagne, P.; Bakalowicz, M.; Weng, P.; Al-Malki, A. Evaluation of aquifer thickness by analysing recession hydrographs. Application to the Oman ophiolite hard-rock aquifer. *J. Hydrol.* **2003**, *274*, 248–269. [CrossRef]

29. Langbein, W.B. Some channel-storage studies and their application to the determination of infiltration. *Trans. Am. Geophys. Union* **1938**, *19*, 435–445. [CrossRef]

30. Knisel, W.G. Baseflow recession analysis for comparison of drainage basin and geology. *J. Geophys. Res.* **1963**, *68*, 3649–3653. [CrossRef]

31. Lebaut, S. *L'apport de L'analyse et de la Modélisation Hydrologiques de Bassins Versants Dans la Connaissance du Fonctionnement d'un Aquifère: Les Grès d'Ardenne-Luxembourg*; Thèse de l'université de Metz: Metz, France, 2000. Available online: http://docnum.univ-lorraine.fr/public/UPV-M/Theses/2000/Lebaut_Sebastien.LMZ0010.pdf (accessed on 2 October 2019).

32. Federer, C.A. Forest transpiration greatly speeds streamflow recession. *Water Resour. Res.* **1973**, *9*, 1599–1604. [CrossRef]

33. Bako, M.D.; Hunt, D.N. Derivation of baseflow recession constant using computer and numerical analysis. *Hydrol. Sci. J.* **1988**, *33*, 357–367. [CrossRef]

34. Brutsaert, W.; Nieber, J.L. Regionalized drought flow hydrographs from a mature glaciated plateau. *Water Resour. Res.* **1977**, *13*, 637–643. [CrossRef]

35. Troch, P.A.; De Troch, P.; Brutsaert, W. Effective water table depth to describe initial conditions prior to storm rainfall in humid regions. *Water Resour. Res.* **1993**, *29*, 427–434. [CrossRef]

36. Vogel, R.M.; Kroll, C.N. Regional geohydrologic-geomorphic relationships for the estimation of low flow statistics. *Water Resour. Res.* **1992**, *28*, 2451–2458. [CrossRef]

37. Zecharias, Y.B.; Brutsaert, W. Recession characteristics of groundwater outflow and base flow from mountainous watersheds. *Water Resour. Res.* **1988**, *24*, 1651–1658. [CrossRef]

38. Lang, C.; Gilles, E. Une méthode d'analyse du tarissement des cours d'eau pour la prévision des débits d'étiage. *Norois* **2006**, *201*, 31–43. [CrossRef]

39. Bader, J.C.; Cauchy, S.; Duffar, L.; Saura, P. *Monographie Hydrologique du Fleuve Sénégal: De L'origine des Mesures Jusqu'en 2011*; Bader, J.C., Ed.; IRD: Marseille, France, 2015; pp. 79–920, ISBN 978-2-7099-1885-5.

40. Rochette, C.; Camus, H.; Danuc, R.; Pereira-Barreto, S. *Le Bassin du Fleuve Senegal. Monographies Hydrologiques, 1*; ORSTOM: Paris, France, 1974; p. 440, ISBN 2-7099-0344-X.

41. Brunet-Moret, Y.; Chaperon, P.; Lamagat, J.P.; Molinier, M. *Monographie Hydrologique du Fleuve Niger*; Monographies hydrologiques, 8; ORSTOM: Paris, France, 1986; pp. 402–510.

42. Lamagat, J.P.; Albergel, J.; Bouchez, J.M.; Descroix, L. *Monographie Hydrologique du Fleuve Gambie*; ORSTOM, OMVG: Dakar, Sénégal, 1990; p. 247.

43. Vauchel, P.; Guiguen, N.; Gomis, D.; Konaté, L. *Appui Institutionnel Aux Brigades Régionales du Ministère de L'hydraulique: Rapports D'avancement de Septembre 1998 à Septembre 2000*; IRD, SGPRE: Dakar, Sénégal, 2000; p. 88.

44. Olivry, J.C. *Etudes Régionales Sur Les Basses Eaux, Les Effets Durables du Déficit des Précipitations Sur Les Etiages et Les Tarissements en Afrique de L'ouest et du Centre*; XIIèmes journées hydrologiques de l'ORSTOM: Montpellier, France, 1996.

45. Bricquet, J.P.; Bamba, F.; Mahé, G.; Toure, M.; Olivry, J.C. Evolution récente des ressources en eau de l'Afrique atlantique. *Revue des Sciences de l'eau* **1997**, *3*, 321–337. [CrossRef]

46. Bricquet, J.P.; Mahé, G.; Bamba, F.; Olivry, J.C. Changements climatiques récents et modification du régime hydrologique du fleuve Niger à Koulikouro (Mali). *L'hydrologie Tropicale: Géoscience et Outil Pour le Développement*. 1995, Volume 238. Available online: https://books.google.fr/books?hl=fr&lr=&id=azUCAebIN-YC&oi=fnd&pg=PA157&dq=46.%09Bricquet,+J.P.%3B+Mah%C3%A9,+G.%3B+Bamba,+F.%3B+Olivry,+J.C.+Changements+climatiques+r%C3%A9cents+et+modification+du+r%C3%A9gime+hydrologique+du+fleuve+Niger+%C3%A0+Koulikouro+(Mali)&ots=MXLgArm6vz&sig=QUNj1kfsu8aHq4b-ivKMTMR0eII#v=onepage&q&f=false (accessed on 8 June 2020).

47. Mahé, G.; Olivry, J.C.; Dessouassi, R.; Orange, D.; Bamba, F.; Servat, E. Relations eaux de surface–eaux souterraines d'une rivière tropicale au Mali. *C.R. Acad. Sci.* **2000**, *330*, 689–692. [CrossRef]
48. L'Hôte, Y.; Mahé, G. *Afrique de L'ouest et Centrale: Précicpitations Moyennes Annuelles (Période 1951–1989)*; ORSTOM: Paris, France, 1996. Available online: https://horizon.documentation.ird.fr/exl-doc/pleins_textes/divers16-08/010058020.pdf (accessed on 24 August 2020).
49. Descroix, L.; Diongue, N.A.; Panthou, G.; Bodian, A.; Sane, Y.; Dacosta, H.; Malam, A.M.; Vandervaere, J.-P.; Quantin, G. Evolution récente de la pluviométrie en Afrique de l'ouest à travers deux régions: La Sénégambie et le bassin du Niger moyen. *Climatologie* **2015**, *12*, 25–43. [CrossRef]

Article

Climate and Extreme Rainfall Events in the Mono River Basin (West Africa): Investigating Future Changes with Regional Climate Models

Ernest Amoussou [1,2,3,4,*], Hervé Awoye [5,6], Henri S. Totin Vodounon [1,2], Salomon Obahoundje [3], Pierre Camberlin [4], Arona Diedhiou [3,7], Kouakou Kouadio [3,8,9], Gil Mahé [10], Constant Houndénou [2] and Michel Boko [2]

[1]   Département de Géographie et Aménagement du Territoire, Université de Parakou, Parakou BP 123, Benin; sourouhenri@yahoo.fr
[2]   Laboratoire Pierre PAGNEY, Climat, Eau, Ecosystème et Développement (LACEEDE), Université d'Abomey-Calavi, Cotonou 03 BP1122, Benin; constant500@yahoo.fr (C.H.); bokomichel@gmail.com (M.B.)
[3]   African Centre of Excellence on Climate Change, Biodiversity and Sustainable Development, Université Félix Houphouët Boigny (UFHB), 22 B.P. 582 Abidjan 22, Cote D'Ivoire; obahoundjes@yahoo.com (S.O.); arona.diedhiou@ird.fr (A.D.); kk.kouadio@yahoo.fr (K.K.)
[4]   Centre de Recherches de Climatologie (CRC), Biogéosciences, Université de Bourgogne, 6 boulevard Gabriel, F-21000 Dijon, France; camber@u-bourgogne.fr
[5]   Laboratoire d'Hydraulique et de Maîtrise de l'Eau (LHME), Université d'Abomey-Calavi, Cotonou 01 BP 526, Benin; herve.awoye@gmail.com
[6]   Department of Geography, University of Calgary, Calgary, AB T2N 1N4, Canada
[7]   Institute of Research for Development IRD, CNRS, Grenoble INP, IGE, University Grenoble Alpes, F-38000 Grenoble, France
[8]   Laboratory of Atmosphere Physics and Fluid Mechanics, UFHB, 22 BP 582 Abidjan 22, Cote D'Ivoire
[9]   Geophysical Station of Lamto (GSL), N'Douci BP 31, Cote D'Ivoire
[10] IRD, Laboratoire HydroSciences de Montpellier, Université de Montpellier 2, Case courrier MSE, Place Eugène Bataillon, 34095 Montpellier CEDEX 5, France; gil.mahe@ird.fr
*   Correspondence: ernestamoussou@gmail.com; Tel.: +229-95064746

Received: 28 November 2019; Accepted: 21 January 2020; Published: 16 March 2020

Abstract: This study characterizes the future changes in extreme rainfall and air temperature in the Mono river basin where the main economic activity is weather dependent and local populations are highly vulnerable to natural hazards, including flood inundations. Daily precipitation and temperature from observational datasets and Regional Climate Models (RCMs) output from REMO, RegCM, HadRM3, and RCA were used to analyze climatic variations in space and time, and fit a GEV model to investigate the extreme rainfalls and their return periods. The results indicate that the realism of the simulated climate in this domain is mainly controlled by the choice of the RCMs. These RCMs projected a 1 to 1.5 °C temperature increase by 2050 while the projected trends for cumulated precipitation are null or very moderate and diverge among models. Contrasting results were obtained for the intense rainfall events, with RegCM and HadRM3 pointing to a significant increase in the intensity of extreme rainfall events. The GEV model is well suited for the prediction of heavy rainfall events although there are uncertainties beyond the 90th percentile. The annual maxima of daily precipitation will also increase by 2050 and could be of benefit to the ecosystem services and socioeconomic activities in the Mono river basin but could also be a threat.

Keywords: Mono basin; extreme rainfall events; ENSEMBLE; regional climate models

---

1. Introduction

Climate change represents the latest in a series of environmental drivers of human conflict that have been identified in recent decades, following others, including drought, desertification, land degradation, failing water supplies, deforestation, fisheries depletion, and even ozone depletion [1]. Extreme hydroclimatic events, such as heavy rainfall, floods, and droughts, have detrimental effects on local populations and their socioeconomic environment. With the projected increase in the atmospheric greenhouse gases concentration, it is important to investigate the future evolution of the frequency and intensity of these extreme events [2–4]. In this respect, sub-Saharan Africa is greatly concerned because the populations are highly vulnerable and hydroclimatic phenomena are responsible for the majority of the reported victims of the natural hazards the region has experienced [5]. In the Gulf of Guinea, the Mono river basin, which extends over Benin and Togo, is influenced by two types of climate: A subequatorial climate in the south and a tropical wet climate over the rest of the watershed to the north. This climatic pattern is constrained by the seasonal shift of the Intertropical Convergence zone and associated lower and upper tropospheric flows over West Africa [2]. Interannual variations of these features as well as short-term rainfall and temperature variations result in a succession of drought and flood events. With ongoing global warming, we may experience an increasein rainfall intensity in the watershed [6]. Indeed, previous studies reported more frequent extreme rainfall events in many regions of the world and predicted increases in the future flood frequency in the West African Sahel [4,7].

The climate degradation could generate conflicts of water resource usage mainly in the transboundary basins. As most West African basins are trans-boundaries, this could be a source of conflict among different sectors of water users [8,9]. Brown and Crawford [10] proved that climate change has increased the conflict risk in trans-boundaries river basins like Volta shared by Burkina and Ghana. Additionally, as West African sub-Saharan agriculture systems are mainly rainfed agriculture systems, then any change or variation in rainfall could affect the production, and consequently lead to food insecurity. Moreover, the increase in temperature could accelerate the evaporative loss, and thus increase water competition [11–13].

Climate change impacts were previously assessed using raw simulations from global climate models (GCMs), which operate at coarse spatial resolutions ranging from 150 to 400 km [14–17]. Thus, the information generated by these models does not always match ground observations because GCMs do not explicitly consider the spatial heterogeneity of land surface conditions [18]. In addition, these models impose limits on the representation of certain atmospheric phenomena (e.g., mesoscale processes, localized orographic ascendance, etc.) [18,19]. The expected changes derived from GCM simulations should be considered with caution for climate impact and adaptation studies [3,20]. A more convenient way of assessing climate change impact on water resources, ecosystems, and agriculture is to use global climate model projections downscaled at a finer scale [21–24]. Many statistical approaches are used to downscale the coarse resolution GCM simulations and better account for the local or regional effects of topography, land use, and other forcings [3,17,25,26]. An alternative to statistical downscaling is dynamical downscaling with regional climate models (RCMs). With their higher spatial resolution as compared to GCMs, RCMs account for greater topographic diversity and more localized atmospheric dynamics [27]. Their ability to reproduce the main characteristics of hydro-climatological dynamics during extreme rainfall events has been demonstrated [28]. This good performance is mainly due to the higher spatial resolution of RCMs that better match the spatial scales at which mesoscale convective processes occur [29]. Despite these improvements, we still need to validate the RCM simulations before using them for climate change impact studies because they still feature systematic biases, which vary substantially from one model to another.

The aims of this paper were to assess the historical and future changes in key climatic variables (e.g., near-surface air temperature and precipitation, extreme rainfall) in the Mono river basin and derive conclusions regarding their impact on flood dynamics in the watershed upstream of the hydroelectric dam of Nangbéto. The RCM simulations over West Africa developed in the frame of the European project ENSEMBLES, CMIP5, and CORDEX-Africa. We assessed the performance of four RCMs to reproduce the mean air temperature, precipitation, and extreme rainfall events. An analysis of the changes in these climatic variables over 2028–2050 as projected by each RCM is presented, with an emphasis on the frequency and intensity of daily rainfall events. The selected RCMs were driven by two GCMs in order to assess the robustness of the simulations and the role of large and regional scales in the model differences. However, for this typical work, it was applied to a small watershed scale, like the Mono river basin.

This work is structured as follows: The study area is described followed by the used data; the statistical methods used for data analysis are presented as well; next, the results focus on the relevance of climatic fields and regimes simulated by the selected RCMs, the expected changes in mean near-surface air temperature, mean precipitation, the frequency and return periods of extreme rainfall events, and their potential impacts on the catchment hydrology and; finally, the results are discussed and some perspectives are highlighted.

2. Study Area

The Mono river basin is in West Africa and extends over 560 km from the north to the south [2]. This transboundary watershed covers 15,680 km^2 and is shared by Benin and Togo (Figure 1). This watershed is home to the Nangbéto hydroelectric dam that has been providing electricity to Benin and Togo since September 1987. As this hydroelectric dam is modifying the hydrology of the river, the main part of this study will focus on the Mono river basin upstream of the dam.

This watershed is patterned in the south by floodplains and plateaus, and higher landforms in the north and north-west, e.g., the Atakora Mountains with a height of 800 m and their southern extensions that are the Togo mountains.

Precipitation in Togo and Benin is constrained by the organization of the atmospheric circulation over West Africa as a whole, i.e., both that of the lower layers (humid monsoon flow from the southwest, and dry Harmattan flow from the northeast) and the flow of air in the lower troposphere and upper atmosphere (respectively African Easterly Jet (AEJ) and Tropical Easterly Jet (TEJ)) [2]. At the scale of the Mono river basin, this circulation is strongly linked to the energy gradients between the coastal plain and the Gulf of Guinea. The precipitations are also modulated by the interaction of the southwesterly humid monsoon flow and the landforms in the northwestern part of the watershed. The average annual precipitation ranges from 900 mm in the southeast, i.e., in the plains along the dry diagonal in the south of Togo, to 1200 mm in the northwestern uplands. There are two climatic domains in the basin: The subequatorial domain in the southern part of the basin with two rainy and two dry seasons and the tropical domain in the northern part with one rainy season and one dry season.

The hydrological regime is of a humid tropical type and consistent with the tropical climate that governs it. The uneven distribution of precipitation combined with heavy rainfall events over very short periods leads to recurring flooding events. These phenomena are aggravated by the increase in surface runoff due to land cover degradation [30]. As a result, the local population or riparian and their properties and livelihoods are more vulnerable than the population living upstream of the basin as evidenced by the severe floods of the past 12 years in the lower valley of the Mono river basin.

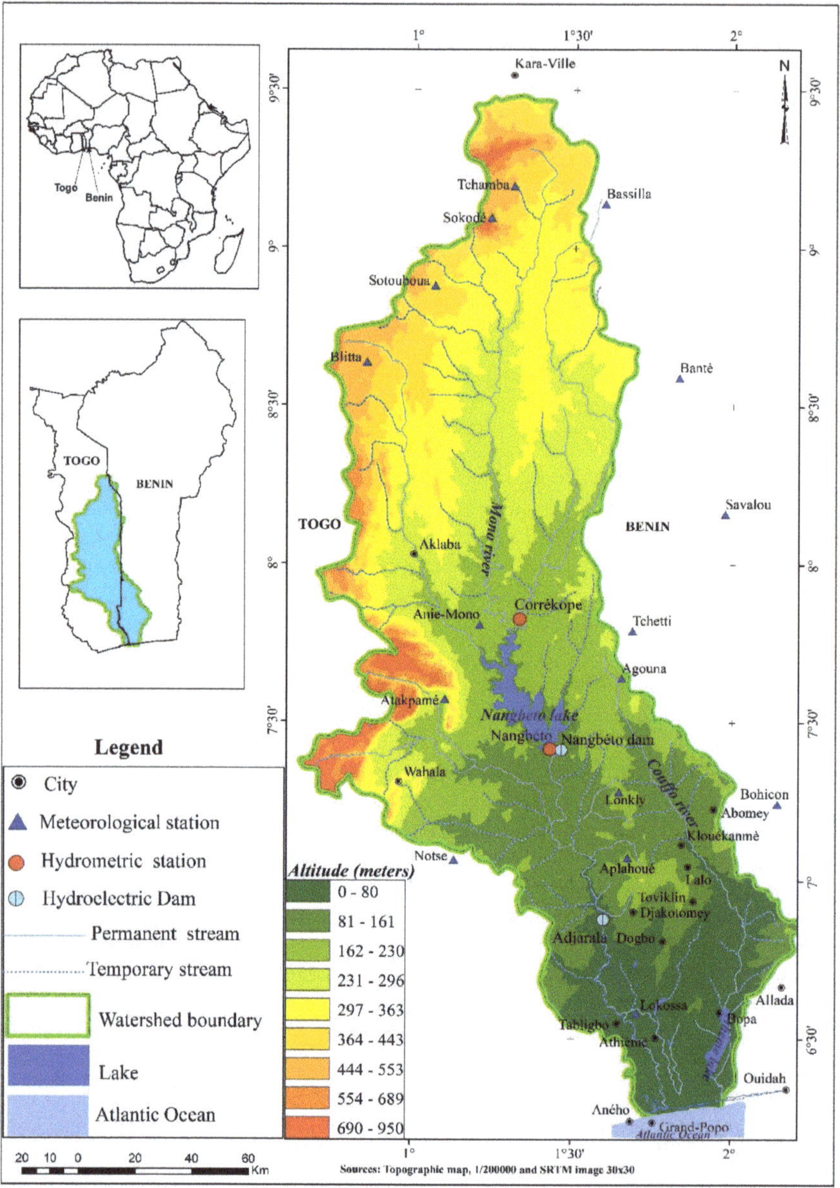

Figure 1. Location of the rainfall (blue triangles) and hydrometric stations (red dots) in the Mono river basin between Benin and Togo.

3. Materials and Methods

3.1. Observational Climate Data

We used daily rainfall data from 14 rain gauge stations located in the Mono river basin and its close vicinity, and daily near-surface air temperature from two meteorological stations, namely the Bohicon and Nangbéto stations. Those observed data were used to select the closest model to the observation (reality) for projection. Thus, the model closest to the reality was used. The data cover a 50-year period (1961–2010) and are provided by the Agency for the Air Navigation Safety in Africa and Madagascar (ASECNA) in Cotonou—Benin, the Direction de la Météorologie Nationale (DMN) of Togo, the Climatology Research Centre (CRC) of the University of Bourgogne in France, and the Global Historical Climate Network (GHCN, [31]).

We compared the gauged rainfall data with the gridded data from the RCM models. The rain-gauge data were spatially interpolated to a 0.5° resolution upstream of the Nangbéto hydro-electric dam using an ordinary kriging block (Figure 1). A climatological variogram was considered and an adjustment was made to a spherical variogram [22]. The range parameter of the adjusted model is 65 km and indicates the average distance of precipitation decorrelation between two stations. In contrast, the arithmetic mean of the two observed air temperature stations data was made and then used to compare to the model dataset.

In order to describe the climates that prevail in West Africa and compare them with the RCM simulations, gridded monthly data of precipitation and near-surface air temperature from 3 additional datasets were extracted over the region extending over 5–14° N and 1.2–3° E. These sets of data covering the years from 1988 to 2009 at a 0.5° spatial resolution are the GPCP Global Precipitation Climatology Project data [32], the near-surface air temperature data from CRU (Climatic Research Unit, [33]). The CRU data is chosen based on its multidiscipline application, namely in applied climatology, biogeochemical modelling, hydrology, and agricultural meteorology domains [33] and its performance [34].

3.2. Data from Regional Climate Models

Precipitation and near-surface air temperature data from the selected four RCMs developed in the framework of the ENSEMBLES European project [35] are listed in Table 1. They are extracted from the AMMA-ENSEMBLES database [17]. These RCMs simulations are centered on West Africa (19.8° S–35.2° N and 35.2° W–31.2° E), have a horizontal resolution of 50 km, and cover the period 1951–2050 (1980–2050 for the ICTP model). These RCMs simulations are driven by two CMIP3 global climate models (GCMs), namely ECHAM5 for MPI and ICTP models and HadCM3 for HC (METO) and SMHI models, following the IPCC A1B scenario. We used these RCM simulations at a daily time step for a region extending over 5–14° N and 1.2° W–3° E. METO and SMH simulations follow a 360-day calendar. The 18-year period spanning 1988–2005 was used for the comparison of the precipitation, and air temperature RCMs data with observations (GPCP for precipitation and CRU for temperature) were considered for the air temperature.

Table 1. Selected models.

Model Number	Regional Climate Model (RCM)		Driving Global Climate Model (GCM)	
	RCM	**Modeling Agency**	**GCM**	**Modeling Agency**
	CORDEX			
(1)	REMO	GERICS	CM5A-LR	IPSL
(2)	RegCM4-3	ICTP	MPI-ESM-MR	MPI-M
(3)	REMO	MPI-CSC	MPI-ESM-LR	MPI-M
(4)	SMHI	RCA4	CNRM-CERFACS	CNRM-CM5
(5)	SMHI	RCA4	CM5A-MR	IPSL
(6)	SMHI	RCA4	MPI-ESM-LR	MPI-M
	CMIP5			
(7)			CM5	CNRM
(8)			CM5A-LR	IPSL
(9)			CM5A-MR	IPSL
(10)			ESM-LR	MPI
(11)			ESM-MR	MPI

Table 1. *Cont.*

Model Number	Regional Climate Model (RCM)		Driving Global Climate Model (GCM)	
	RCM	Modeling Agency	GCM	Modeling Agency
	AMMA-ENSEMBLES			Scenario
(12)	RegCM3	ICTP	ECHAM5	A1B
(13)	HIRHAM	METNO	HadCM3	A1B
(14)	REMO	MPI-M	ECHAM	A1B
(15)	RCA	SMHI	HadCM3	A1B

NB: MPI-M—Max Planck Institute of Meteorology; MPI-CSC—Climate Service Center, Max Planck Institute for Meteorology; ICTP—International Centre for Theoretical Physics; SMHI—Swedish Meteorological and Hydrological Institute; CNRM—Centre National de Recherches Météorologiques; REMO—Regional Climate Model; RegCM—Regional Climate Model; RCA—Rossby Centre regional atmospheric model; GERICS—Climate Service Center Germany; HadCM3—Hadley Centre Coupled Model.

It is important to note that for the next step, International Centre for Theoretical Physics will be called (ICT); Swedish Meteorological and Hydrological Institute by SMH; Max Planck Institute of Meteorology by MPI, and Hadley Centre Coupled Model by METO.

We extracted the CORDEX, CMIP5, and AMMA-ENSEMBLES outputs over the Mono river basin at the outlet of Nangbéto, computed the spatial ensemble averages of each data type, and compared them with the interpolated (observed) data. Since the streamflow gauge data recorded are only available over 1988–2010, we used the period 1988 to 2010 (23 years) as the reference period and the period 2028 to 2050 (23 years) as the scenario period.

3.3. Statistical Methods

Observed and simulated climate data were analyzed for the sub-region and the Mono river basin itself. At the sub-regional scale, seasonal average maps were used to compare simulated precipitation and mean air temperature fields with the observational data over 1988–2010, and to assess the projected future changes for 2028–2050. The RCM whose outputs in the reference period is closer to observations can be considered for future projection.

At the catchment scale, the observed and simulated climate data were also spatially averaged and statistically processed. First, monthly averages of rainfall and air temperature were computed to describe the ability of the regional climate models to reproduce the present-day climatic regimes and to assess the projected future changes (2028–2050). The average frequency of the number of rainy days (i.e., a day with a rainfall amount higher than 1 mm) in each grid point in the Mono river basin was also computed. The contribution of the different classes of daily rainfall intensity to the total precipitation recorded on average in the watershed was determined from the calculation of deciles. The 99th percentile is usually used to characterize the extreme events [36–39]. This is the reason why the 99th percentile is used to assess the importance of heavy rains in the basin.

Observed hydroclimatic series do not always meet the assumption of data stationarity [40,41]. This is often the case for extreme rainfall and flood discharge. In the literature, some scholars [42,43] have suggested describing the distribution of extreme values with three different types of probability distribution for extreme order statistics while others [44–47] suggested using a generalized extreme value (GEV) distribution, which combines the three types of Fisher–Tippett extreme value distributions [48]. The GEV parameters use the generalized maximum likelihood (GML) method, which includes an additional constraint on the shape parameter to eliminate potentially invalid values of this parameter based on Tramblay et al.'s [4] study.

In this study, we used a GEV distribution to characterize extreme events. In the case of non-stationarity, the GEV parameters are not constant but are dependent on time or other covariates. To avoid having a value of zero for the scale parameter, αt, we considered the parameterization as $\phi t = \log(\alpha t)$ and thus restricted our analysis to linear dependencies [49]. However, many other types of dependencies can be linearized [41,44,48,50].

The return periods of a GEV model are defined by the cumulative distribution function (CDF), F(x), or the probability density function (PDF), f(x), hat are given by the formula:

$$F(x) = Pr[X \leq x] \text{ and } f(x) = \frac{dF(x)}{dx},$$ (1)

where Pr is the probability of the occurrence of an event and X stands for the random variable, which in this case is a series of annual daily maxima of rainfall. The quantile value, Q_T, is a magnitude of the extreme event that has a probability of $1/T$ and can be exceeded by a single event. A return period, T, associated with a quantile value, Q_T, was adapted from Stedinger et al.'s [51] equation and can be expressed as:

$$T = \frac{1}{[1 - F(Q_T)]}$$ (2)

The selection of the best GEV model adjustment was made by using the deviance test, which is based on the log-likelihood [20].

4. Results and Discussion

4.1. Spatial Distribution of the Annual Rainfall and Air Temperature for Three Datasets

Figures 2 and 3 present the ensemble average annual rainfall and mean air temperature for the historical period 1988–2005 (first column) and change between the reanalysis (GPCP) and models in (second column) the historical period as well as the projected changes for 2028–2050 (third column).

The spatial distribution of the observed rainfall (1988–2005 annual rainfall average, GPCP data) in the Mono river basin and surrounding areas is presented in Figure 2d. The observed rainfall pattern shows a diagonal axis of a high rainfall amount (over 1320 mm per year) from south-west to north-east along with the Togo and Atakora mountain ranges. A strong decrease in rainfall is also shown in the north towards the Sahelian zone (<600 mm/year). The GPCp reanalysis data captures the observed precipitation gradient over West Africa well [52] and this finding agrees with the study of Ntajal et al. [49] in Mono basin.

The ensemble averages were computed for CMIP5 and RCP8.5 for CORDEX. The CORDEX (Figure 2c) and CMIP5 (Figure 2b) display a greater rainfall estimation in the historical period compared to the AMMA-Ensemble (Figure 2a) and GPCP (Figure 2d), especially at the southern part. The validation made (second column of Figure 2) exhibits that CMIP5 and CORDEX overestimate the rainfall, which is about 340 mm/year. Nevertheless, the AMMA-Ensemble gives an estimation closer to the reanalysis (±20 mm/year).

There is a projected general decrease varying by 16 to 32 mm in the annual total precipitation. For instance, the CMIP5 model presents a decrease of about 24 mm and less than 24 mm, respectively, in the central and southern parts of the basin while the CORDEX estimates a decrease between 24 and 32 mm in the central and southern part of the basin. Nevertheless, in the northern part of the basin, an increase in the annual total precipitation of around 16 mm and in the range of 16 to 24 mm for the CMIP5 and CORDEX models, respectively, is projected. Finally, the AMMA-ENSEMBLES displays a decrease in the projection change ranging from 0 to 32 mm/year.

The temperature gradient was well shown by all the datasets, which increase in the northern direction (first column of Figure 3). The observed mean air temperature extracted from Climate Research Unit (CRU) varies slightly, with a maximum of 29.5 °C in the far north (the Sahel in Niger) and a minimum of 26 °C in altitude on the Togolese ridge (Figure 3d). The AMMA-Ensemble and CORDEX data give an underestimation of the temperature ranging from 0 to 1 °C while the CMIP5 presents an underestimation between 1 and 2 °C (second column of Figure 3) and the lowest rise is given by AMMA-ENSEMBLES. Unlike the rainfall, there is a very good agreement in the projected changes (increase) of air temperature from the AMMA-Ensemble, CMIP5, and CORDEX dataset.

The warming is expected to vary between 1.4 °C in the south to 1.8 °C towards the north. However, AMMA-ENSEMBLES gives the lowest projected changes among all the models (third column).

The performance of CORDEX RCMs in simulating the spatial distribution of the main precipitation and temperature features as well as the occurrence of other physical phenomena over West Africa was demonstrated by Gbobaniyi et al. [53]. Diallo et al. [54] also showed the importance of the ensemble mean of multiple models. They proved that the RCMs-GCMs alone are biased and these biases are reduced by averaging all RCM simulations, suggesting that multi-model RCM ensembles based on different driving GCMs help to compensate systematic errors from both the nested and the driving models [54]. This justified the use of multiple models. According to IPCC assessment reports, the temperature is projected to increase over West Africa for the end of the 21st century from the global climate simulation range between 3 and 6 °C above the late 20th century baseline depending on the emission scenario [55]. However, they also revealed that the onset and length of precipitation could also change.

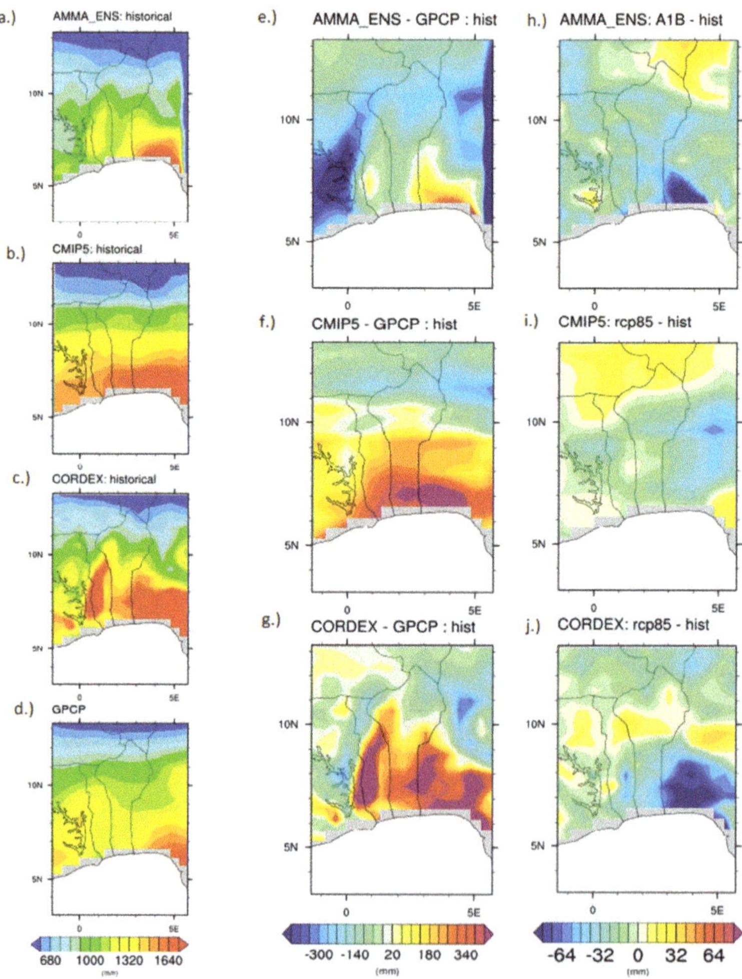

Figure 2. Ensemble average annual mean rainfall for 1988–2005 (first column), validation (second column over 1988–2005 period), and the projected changes with respect to the period 2028–2050 (third column).

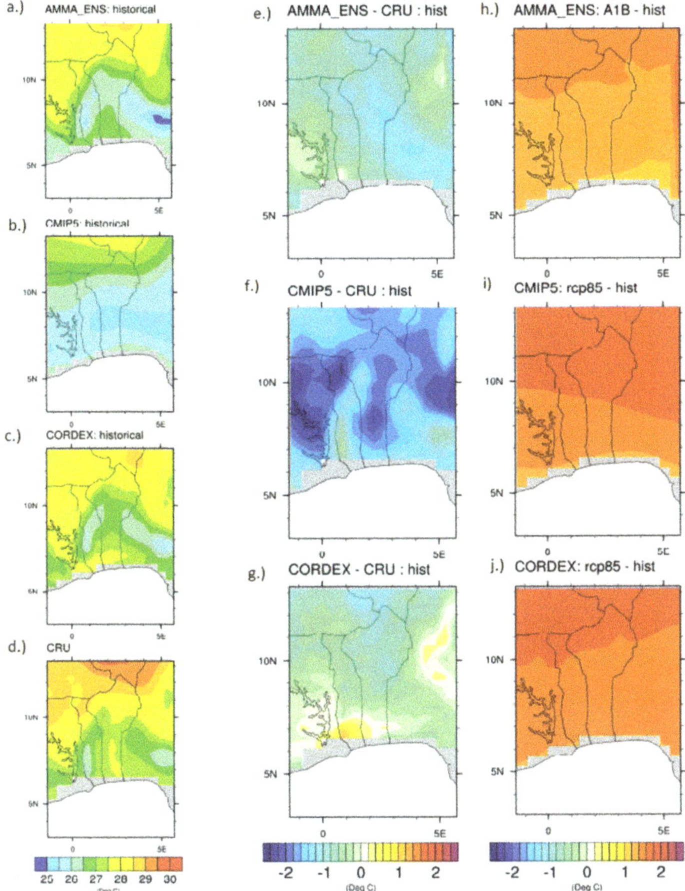

Figure 3. Ensemble average annual mean temperature for 1988–2005 (first column), validation (second column over 1988–2005 period), and the projected changes with respect to the period 2028–2050 (third column).

AMMA-ENSEMBLES data permits a better understanding of the processes underlying the West-African Monsoon (WAM) [56]. It provides and advances an understanding of the atmospheric processes over West Africa [57]. AMMA data were used in climate change assessment in the region. The finding of the current study is in accordance with Angelina et al.'s [58] study on the river Niger basin West Africa. They confirmed that all RCMs of AMMA-ENSEMBLES experiment displays an increase in air temperature while there was no consensus among them in precipitation variables.

4.2. Spatial Distribution of the Averaged Annual Precipitation and Air Temperature for AMMA-ENSEMBLES Models

An analysis of Figure 4 shows the simulated AMMA-ENSEMBLES model's precipitation dataset. The total annual precipitation data for the reference period (1988–2010 in the first raw) and projected period (2028–2050 in the second raw) and the changes in projection in the third raw are presented in Figure 4 for the MPI, METO, SMH, and ICT models. The total annual precipitation over the basin varies from 1200 to 1800 mm/year whatever the model. The total annual precipitation simulated in the reference period (1988–2010) ranges from 600 to 1800 mm while the observed annual precipitation is between 680 and 1640 mm over the chosen window. The spatial distribution of precipitation is almost similar in the reference period and the scenario period (2028–2050). In the south, a diagonal band of relatively low rainfall (<1000 mm/year) follows the Ghanaian-Togolese coast in relationto the diffluence of the monsoon winds and the seasonal rise of coastal cold water. However, we noted minor discrepancies, especially in the number of rainy days, as shown in Table 2, for METO and SMH. MPI and ICT simulate the same number of rainy days in the reference period and the scenario or projection period. By comparing it to the reference period (1988–2010) data, the models METO and SMH project a reduction in rainfall amounts during the projection period (2028–2050) by 2% and 4%, respectively [55]. In contrast, the fifth IPCC assessment reports a slight increase in precipitation over West Africa. Nevertheless, they could not be able to show any trend in precipitation in this zone.

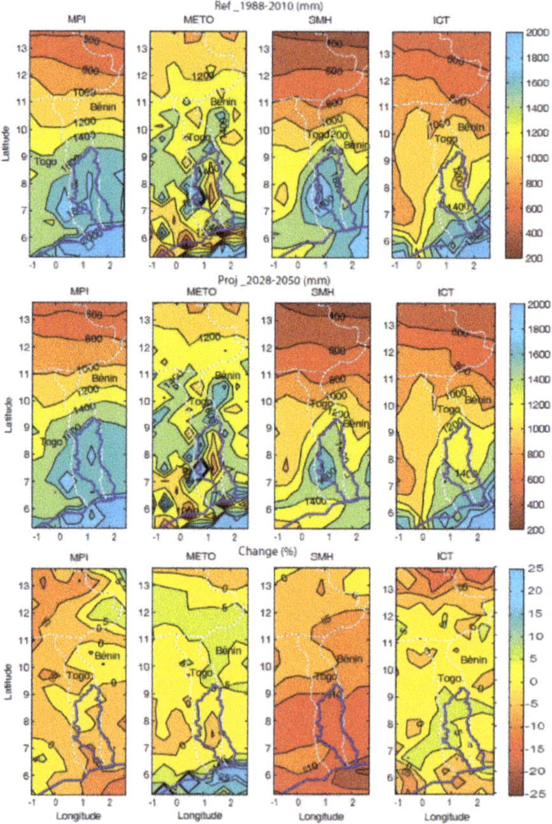

Figure 4. Spatial variation of the mean annual precipitation in the Mono river basin and surrounding areas as modeled by the four Regional Climate Models (RCMs). Upper panels: 1988–2010 reference period; Middle panels: 2028–2050 projection period; Bottom panels: variation between the reference and the scenario periods (%).

Table 2. Percentage (%) of the number of rainy days (>1 mm).

RCMs	Ref_1988–2010	Proj_2028–2050
MPI	42	42
METO	63	62
SMH	47	45
ICT	48	48

An analysis of the variations between the two periods shows moderate but contrasting changes from one model to the others, and across the domain. METO and ICT have projected a slight increase in rainfall ranging between 2% and 5% by 2050. Meanwhile, MPI projection shows no rainfall change whereas SMH has projected a strong rainfall deficit amounting up to 15% to 20% by 2050 in the south of the Mono river basin. These precipitation changes are associated with an increase in air temperature. This result is in accordance with the IPCC report [55] over the West Africa region. The same conclusion was made by Sylla et al. [59].

Figure 5 shows for the Mono river basin and surrounding areas, the spatial distribution of the simulated mean air temperature. It came out that the spatial distribution of the mean air temperature is well reproduced by the RCMs although most of these RCMs, especially SMH, underestimate the values. MPI simulates temperature variations closer to the observations and a much better spatial distribution. However, MPI still overestimates the average minimum air temperature (e.g., 27.5 °C on average for the whole catchment area) compared to the observed data (26.5 °C).

An analysis of Figure 5 for air temperature changes between the projection and reference periods confirms warming of a 1 °C magnitude order whatever the model. This fairly corroborates the 1.22 °C temperature increase obtained with the historical data and is also in good agreement with the SRES-A1B scenario. METO is the only RCM that projected a warming of more than 1.5 °C by 2050 all over the region, except on the coast of the Gulf of Guinea. Thus, the control of the large-scale driver is key to reproducing an accurate simulation of these dry zone features whose origin is mainly related to ocean–atmosphere interactions along the Ghana-Togo coast [60]. This finding agrees with some previous studies in West Africa. For instance, Sylla et al. [59] reported a gradual warming spatial variable reaching 0.5 °C per decade in recent years over West Africa, with a recovery from drought conditions. However, the amount of total annual precipitation of this recent decade is still lower compared with the decade before the 1970s' and 1980s' drought episodes [61,62]. Moreover, they also confirmed the increasing trend in the projected air temperature, and this could be in the range of 1.5 to 6.5 °C depending on the emission scenario. Furthermore, the projection of precipitation is with a lot of uncertainties. The West Africa region could experience change in the range of ±30% in the future associated with more intense extremes in the future climate but to a lesser extent [55,59].

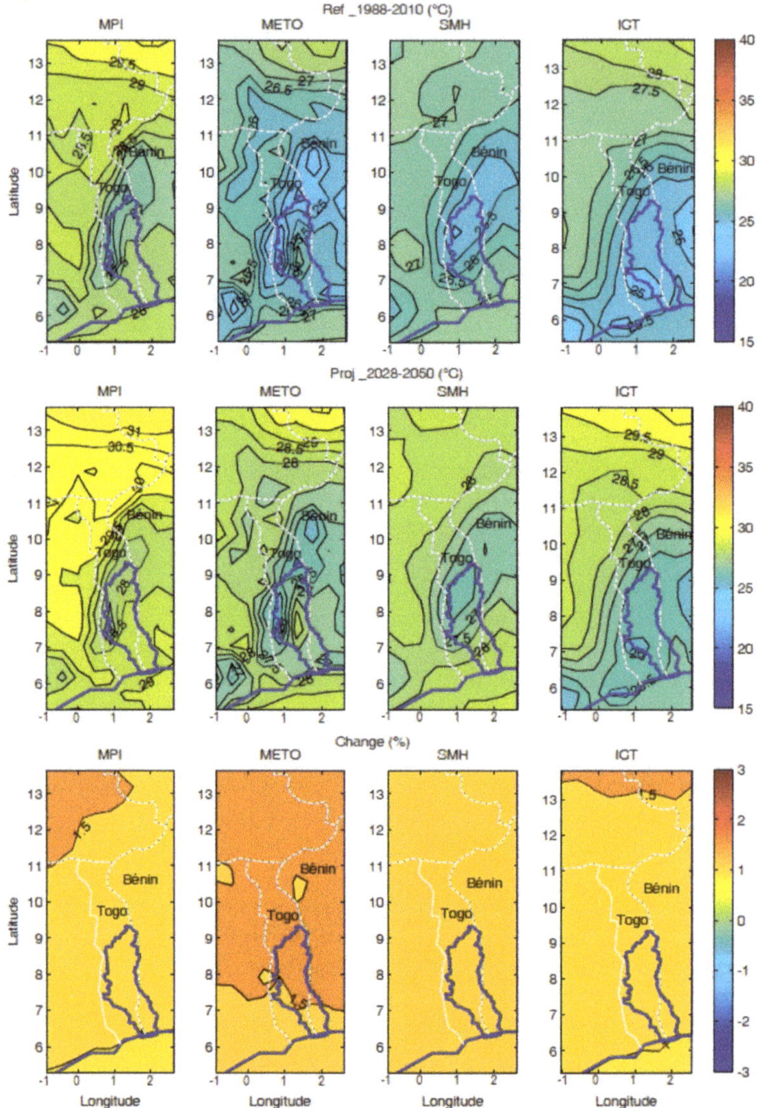

Figure 5. Spatial variation of the annual mean air temperature in the Mono river basin and surrounding areas as simulated by the four RCMs. Up: 1988–2010 reference period (in °C); Middle: 2028–2050 projection period (in °C); Down: variation between the reference and projection periods.

4.3. *Average Rainfall and Temperature Regimes in the Mono River Basin at the Nangbéto Outlet*

Figure 6 shows the average rainfall and air temperature regimes in the Mono river basin upstream of Nangbéto as obseved and modeled by the four RCMs.

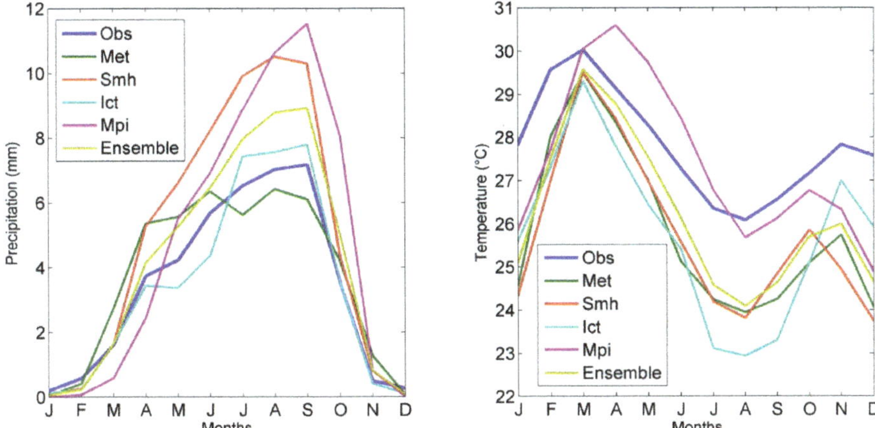

Figure 6. Comparison of the observed and simulated rainfall (mm/day) and air temperature (°C) regimes in the Mono river basin at the Nangbéto outlet (1988–2010 average).

As shown in Figure 6, the four RCMs reproduce the seasonal cycle of precipitation well. This has been proven earlier by some previous works in the region [54,57,63]. In the middle of the rainy season, MPI, SMH, and ICT overestimated the rainfall amounts while they also underestimated the rainfall amounts in January–February, which corresponds to the dry season. The smallest biases were obtained with ICT, thus making ICT the model that best replicated the rainfall regime of this watershed.

In most cases, the RCMs satisfactorily reproduce the temperature seasonality in the watershed. The simulated temperature values are underestimated, except MPI, which overestimates these values from April to July. These biases can significantly influence hydrological impact simulations as they will lead to an under(over)estimation of the high streamflow because of an under(over)estimation of the evaporative loss.

Figure 7 reveals the annual cycle of precipitation simulated by the RCMs for the reference period (1988–2010) and the projection period (2028–2050). For SMH, a moderate decrease in the rainfall regime is shown, especially from March to September. SMH fails to reproduce the observed rainfall regime. The other RCMs rarely show a significant decrease, except METO from June to August. A Fisher's test indicates that the changes of the rainfall regime between the reference and projection periods are not statistically significant at a 95% confidence level. Therefore, we assume that only small changes are to be expected in the hydrological regimes by 2050.

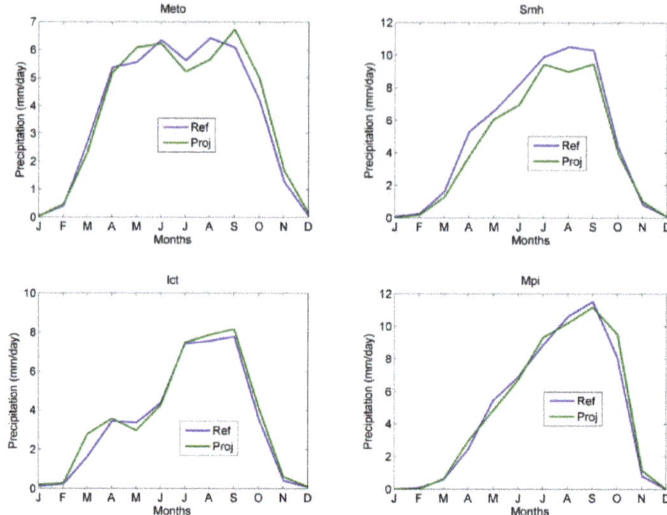

Figure 7. Regional climate models (RCMs) projection of monthly average precipitation (mm/day, 2028–2050) and comparison to the reference period (1988–2010).

Figure 8 reveals that the four RCMs projected an increase of the air temperature by 2050. This could amplify the evaporative loss in the future and potentially induce a decrease in surface runoff. This warming is relatively uniform throughout the year and ranges between +1 and +1.5 °C (a little beyond this range for METO) in accordance with some previous studies [55,59].

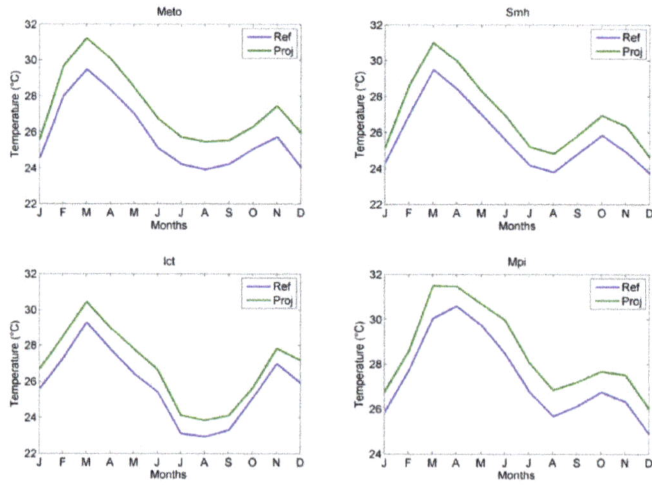

Figure 8. Projection of the monthly near-surface mean air temperature (2028–2050) and comparison to the reference period (1988–2010) for the four RCMs.

Overall, there is a good fit between the seasonal cycles of precipitation and air temperature simulated by the four RCMs, but some of these models still feature biases. The precipitation biases are more dependent on the RCM than the driving GCM. Sylla et al. made the same conclusion [54,59] over West Africa. Indeed, the strongest overestimations of rainfall during the monsoon seasons were obtained with the MPI and SMH experiments. These RCMs were not driven by the same GCM

(ECHAM5 for MPI and HadCM3 for SMH). In addition, both ICT and MPI are forced by ECHAM5 but the ICT simulates a more realistic annual precipitation cycle than MPI. Thus, the projection biases originate from both the RCMs and their driven GCMs.

4.4. Frequency Analysis of Rainfall Distribution in the Mono River Basin

The cumulative frequency of daily rainfall intensity according to the four RCMs is illustrated in Figure 9. This figure compares the distributions of rainfall intensity between observations and the RCM simulations in the reference period, and between the RCM simulations in the reference and projection periods. For most RCMs, the sill of the cumulative daily distribution is obtained after 200 days, except METO, for which the sill is obtained after 250 days. For SMH and ICT, the distribution of daily rainfall frequencies agreed well with the observations, with many days recording less than 20 mm, and a number of wet days (about 200) similar to the observations. METO shows many wet days with a very high number of medium intensity days (5–10 mm). By contrast, the MPI model underestimates precipitation occurrence, especially for intensity values around 5 to 15 mm/day. The study of Tossou et al. [64] also provides a characterization and analysis of rainfall variability in the Mono river watershed complex.

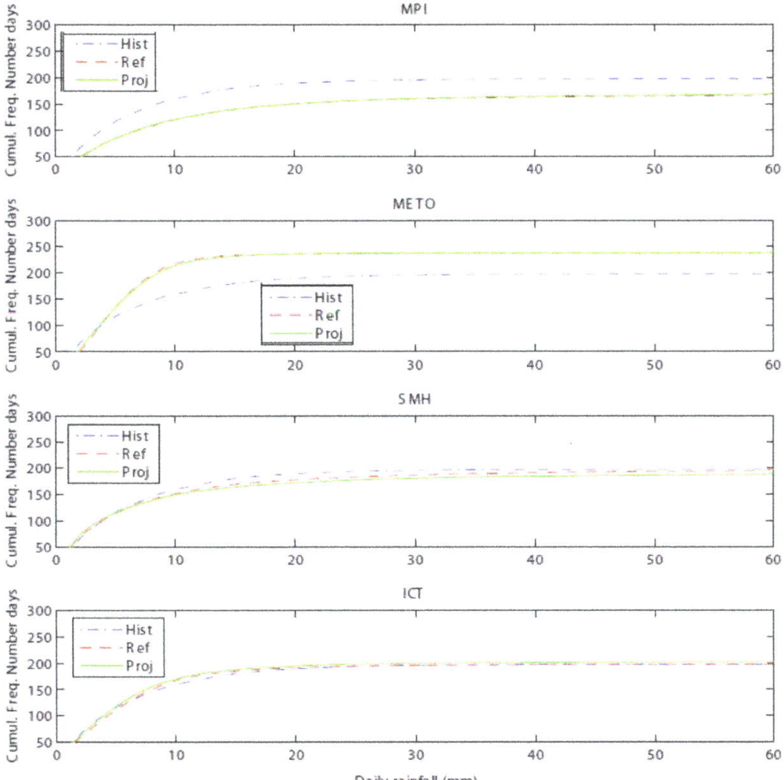

Figure 9. Cumulative distributions of daily precipitation from the four RCMs and comparison to observations (interpolated rainfall field). Cumul. Freq. Number Days = cumulative frequency of the number of days. NB: The empirical cumulative frequency of the number of rainfall days consists of adding up the sum of the number of days with more than 1 mm of rain.

In the projection period, the four RCMs projected a similar distribution of precipitation in the study area as in the reference period, except the SMH model, for which a small change in the distribution of precipitation is displayed, i.e., a slightly reduced number of rain days beyond 10 mm. This close relationship between the distributions from simulations and observations is consistent with previous findings [3,21,27,46,65,66] and confirms that the RCMs satisfactorily reproduced rainfall at a local scale as stated by Amoussou [2] and other works [4,6,67].

A detailed analysis was performed to overcome the bias challenges regarding the number of rainy days and to better describe the future changes in rainfall distribution. We computed the deciles of the observed daily rainfall and examined for each RCM the contribution of these 10 rainfall classes to the total rainfall as shown by López-Moreno and Beniston [27] in their study on the evolution of rainfall intensity in the Pyrenees. The computation was made for both the reference period (1988–2010) and the projection period (2028–2050). This approach helps to quantify the evolution of the influence of precipitation belonging to different classes of the frequency distribution [27,68–71].

Figure 10 displays for the Mono river basin at the outlet of Nangbéto the rainfall distribution from the four RCMs according to the thresholds of the 10 quantiles as obtained from the observations. The threshold values are provided at the bottom of Figure 10.

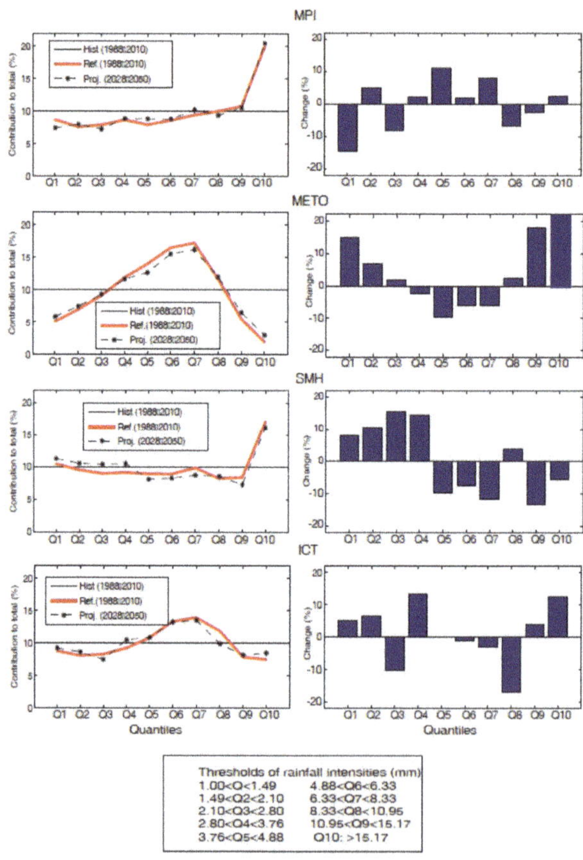

Figure 10. Comparison of the contribution of the 10 classes of quantile of daily precipitation defined from observations (interpolated rainfall field) to total precipitation under present-day and future climates (SRES-A1B scenario) (left), and changes between the reference and projection periods (right).

Figure 10 indicates that apart from SMH, all RCMs underestimate the distribution of rainfall quantiles, especially the contribution of low rainfall (first four quantiles). This reveals an underestimation of the frequency of low rainfall. For the fifth decile to the eighth decile, the MPI and SMH simulations are close to the observed values. Meanwhile, both ICT and METO overestimate the rainfall in the basin, but this overestimation is stronger with METO. ICT and METO also underestimate the heavy rains (Q9 and Q10), unlike SMH and MPI, which simulate an increase of the daily rainfall (Q10) in the Mono river basin. Between the projection period (2028–2050) and the reference period (1988–2010), the analysis shows a relatively stable distribution of rainfall for most rainfall quantiles from MPI and ICT. For SMH, a strong decrease is found in the contribution of rainfall from the fifth quantile. Meanwhile, an increase of the contribution of the high rainfall intensities to the total rainfall is expected with METO.

4.5. Analysis of Extreme Precipitation under Present Day and Future Climates

The hydrological and societal importance of high-intensity rainfall coupled with the disparities between the RCMs regarding the reproduction of the 10th decile of daily rainfall (Q10 > 15 mm) led us to investigate the 99th percentile of daily rainfall in order to better understand the dynamics of heavy rainfall in the Mono river basin. We also analyzed the 5-day cumulative rainfall that likely contributes to flood generation downstream of the basin. Recurrent flood events were experienced in recent years in the Mono river basin [40] because of the high risk of soil saturation that induces a fast runoff. Table 3 displays the 99th percentile of precipitation for the reference and projection periods (1988–2010 and 2028–2050).

Table 3. 99th percentile values of daily precipitation in the Mono river basin.

Precipitation (mm)	Hist	MPI	METO	SMH	ICT
			Ref_1988–2010		
99th percentile	29.92	86.66	17.98	75.22	31.32
			Proj_2028–2050		
99th percentile	-	85.70	19.36	71.39	38.11

Ref = reference period for observations and RCM simulations; Proj = Projection period for the RCMs.

The values in Table 3 show that in the reference period, about 30 mm/day of rainfall is equivalent to the 99th percentile in this basin, i.e., the highest 1% of the precipitation days according to the historical data. This recorded heavy rainfall intensity followed or preceded by rainy days could trigger catastrophic floods in the Mono river basin.

ICT simulation of heavy rainfall is the closest to observations. METO underestimated the heavy rainfall whereas SMH and MPI largely overestimated these rainfall extremes. A 99th percentile greater than 75 mm/day is too high for a catchment of 15,680 km^2. However, this is not excluded because the rainfall of a year can precipitate in almost a day, as was the case in Ouagadougou (Burkina-Faso) in 2009 with nearly 290 mm.

In simulation as in projection, the ICT model simulates the heavy rains, but MPI and SMH overestimate them while the low rains are recorded by METO. With the reference data used, the MPI model especially simulates the very heavy rainfall the best, as well as projection periods. In both the reference and projection periods, the evolution is almost identical whatever the climate model [72,73].

In order to comprehend the seasonal distribution of these heavy rains, we also computed per month the 99th percentile of daily rainfall (Figure 11).

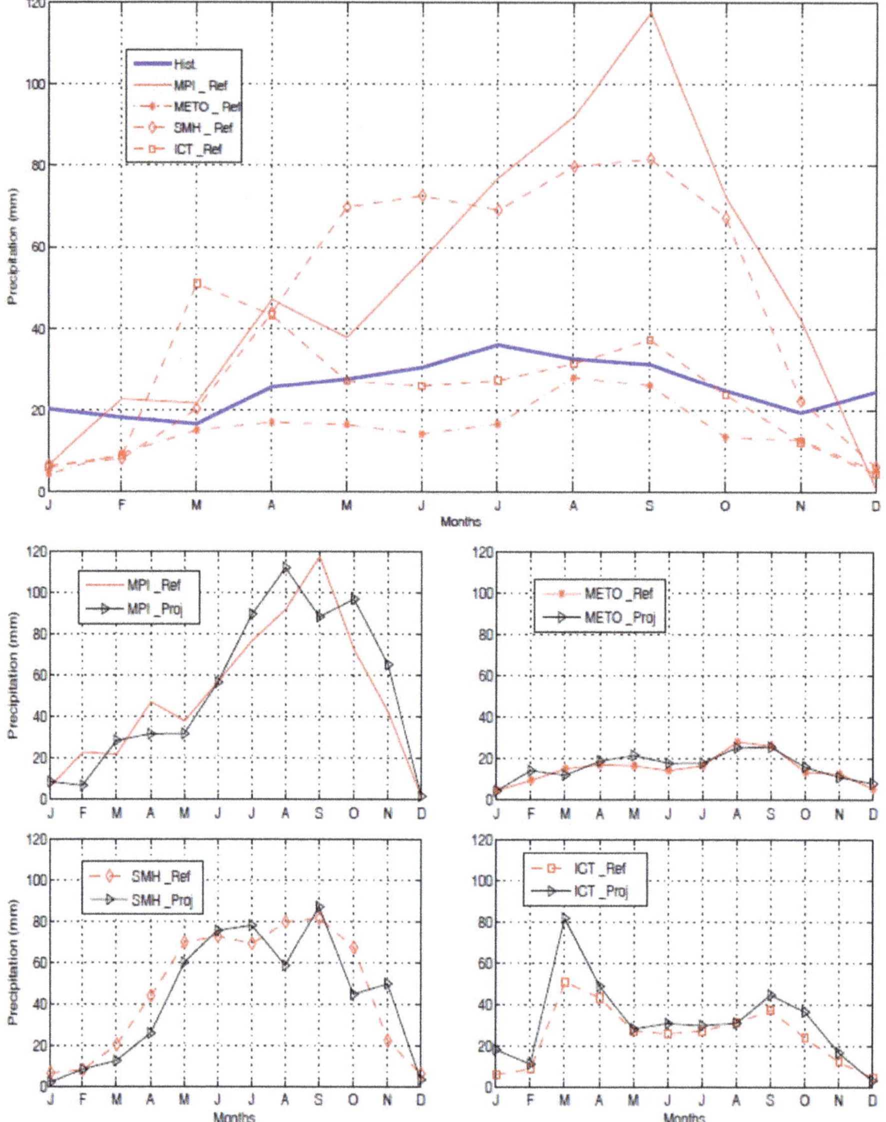

Figure 11. Monthly variation of the 99th percentiles of daily precipitation in the Mono river basin. Top panel: comparison between observation and simulation during the reference period. Bottom panels: comparison between the reference and the projection periods.

The distribution of the 99th percentiles of daily rainfall shows that all RCMs underestimated the heavy rainfall from November to February (dry season). However, MPI overestimated the heavy rainfall in February and November while SMH overestimated only that of November in the dry season. The high rainfall observed in December and January is only sporadic because the occurrence of rainfall events during the dry season is extremely rare. SMH and MPI also overestimated the rainfall intensity in the core rainy season months (a maximum of 65 mm in September for MPI). This hinders a realistic estimation of the hydrological risks.

In the projection period, we will focus on ICT simulations and to a lesser extent on METO simulations as these two climate models stand out as being the most realistic for simulating heavy rainfall events in the basin. For these two climate models, the heavy rainfall intensities obtained in the projection period are close to those of the reference period even if an increase is noticeable in many months. ICT shows a sharp one-off increase of rainfall intensity in March. However, SMH and MPI predicted a slight decrease in the intensity of the heavy rainfall events from February to August and from August to October, respectively, but these two climate models are the least realistic for the simulation of the heavy rainfall intensities.

The monthly variation of the 99th percentile rainfall maxima shows that the most intense heavy rainfall events occur at the core of the rainy season. However, the so-called dry months sometimes record strong rainfall. Therefore, an analysis of the return periods of the heavy rainfall events can help to better understand the distribution and occurrence of the extreme events in the Mono river basin, and the occurrence of high streamflow in the basin upstream of the dam, and the resulting floods. This result agrees with the study of Kodja [74] in tthe Ouémé river basin, which is located in the same climatic zone as our study basin.

4.6. Analysis of the Return Periods of Rainfall Extreme Using a GEV Model

The annual maxima of daily rainfall in the Mono river basin upstream of the outlet of Nangbéto (observed and simulated values) were adjusted to a generalized extreme value (GEV) distribution. The values of the three parameters, namely k, sigma (ς), and mu (μ)) of the GEV distribution (Figure 12), show a similarity between the observed and simulated data for the sigma and mu parameters but strong differences for the k parameter, where K refers to the shape parameter, sigma the scale parameter, and mu stands for the location parameter [75]. Nevertheless, these three parameters were always positive (for observations and simulations in the reference and projection periods) (Figure 12), thus pointing to a Frechet distribution (extreme value distribution type II).

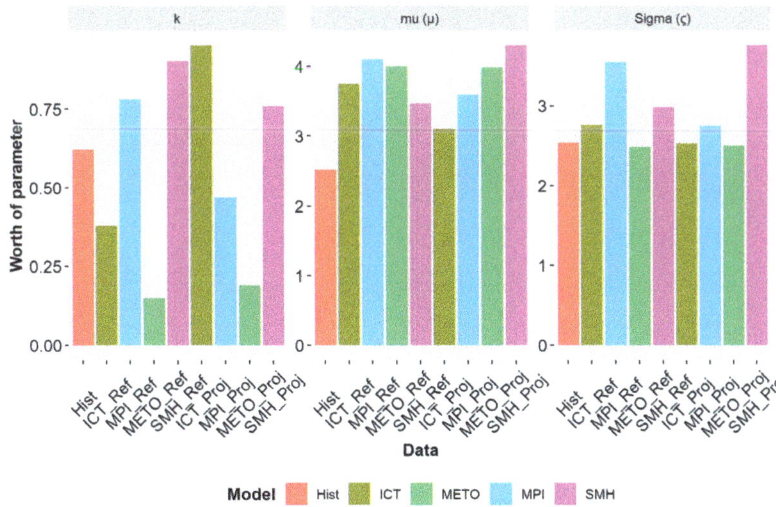

Figure 12. Temporal evolution of the three parameters (k, sigma, mu) of the Generalized Extreme Value (GEV) distribution for different sets of data.

Furthermore, when k > 0, the GEV distribution is the type II, or Frechet, extreme value distribution. The variance of GEV is not finite when k ≥ 1/2. The GEV distribution has positive density only for values of X, such that k × (X-mu)/sigma > −1 [75].

Figure 13 shows the evolution of the annual maxima of observed daily rainfall. The analysis reveals a strong increase in the annual maximum of daily precipitation that is statistically significant at the 95% confidence level. The frequency of high streamflow is not increasing, and this could be explained by the increase in temperature, which induces more evapotranspiration and thus compensates for the impact on high streamflow. This has also been proved by Ntajal et al. [49] downstream of the Mono river basin. However, an in-depth analysis is still needed to prove this assumption because, in principle, very high streamflow is sometimes generated from rapid runoff in response to water excess in soil with a low retention capacity [41,71,76]. This occurs in a very short time period, during which the increase of the evaporative demand is not really a constraint.

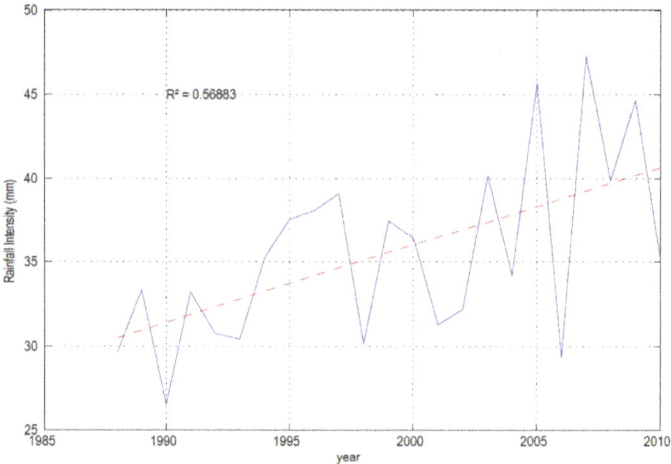

Figure 13. Interannual variation of the daily rainfall maxima from 1988 to 2010 in the Mono river basin at the outlet of Nangbéto.

This result is consistent with the evolution of discharge maxima flowing into the Nangbéto dam from 1988 to 2010, which pointed to an increase of the annual maxima, which is significant at $p = 0.0185$ according to a Mann–Kendall test [41,77,78].

Figure 14 compares the distributions of the quantiles of daily rainfall maxima observed over 23 years (1988–2010) to those adjusted by the GEV.

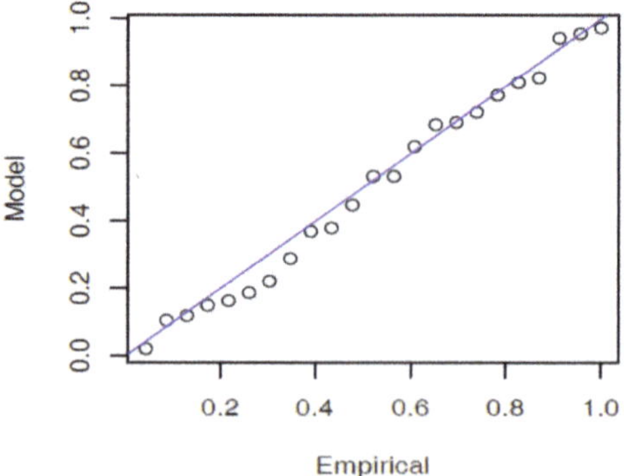

Figure 14. Comparison of the annual maxima of the quantiles of observed and simulated rainfall with a GEV model.

The annual maxima of daily rainfall are well adjusted to the GEV distribution. The analysis of the return periods of the extreme rainfall events, as displayed in Figure 15, shows that the rainfall events of an intensity higher than 40 mm are likely to occur at a recurrence interval of about 5 years.

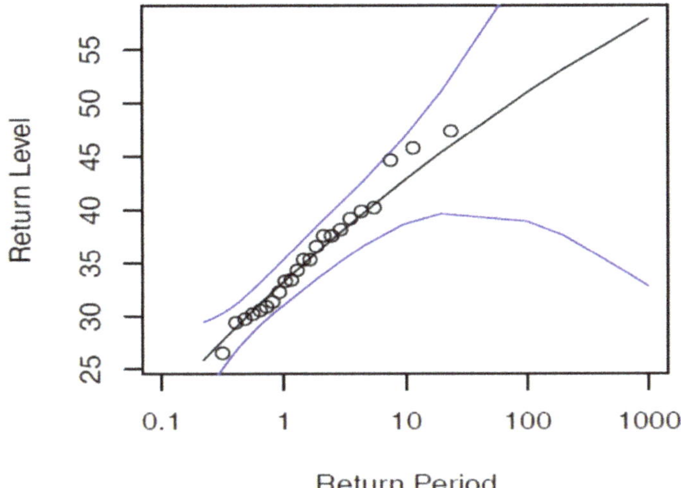

Figure 15. Estimation of the return periods of the annual maxima of rainfall events (mm). The circle represents the observed values over 23 years (1988–2010) and the blue curves refer to the confidence level.

The uncertainty of the rainfall events occurring in a 50-year recurrence interval is higher because of the short length of the available time series and the trend shown in Figure 13.

Table 4 presents the return periods for the annual maxima of the daily rainfall simulated with ICT, which stands out as the best of the four models for the reproduction of intense rainfall. The rarest and thus most intense rainfall events remain poorly simulated, as illustrated by the divergence of the 10-day recurrence values (45 mm in the observations, 77 mm in the model). Nevertheless, the projections show a strong increase in rainfall intensities for all return periods described in Table 4. The daily precipitation increases with the return period in the basin, which is in accordance with Ntajal et al.'s [49] and Kodja's [74] studies in the Mono and Ouémé basins, respectively. Ntajal et al. [49] compared the return period at the annual time scale and showed that GEV gives overestimation than the Goodrich exponential distribution model. This could also justify the overestimation of the model.

Table 4. Annual daily precipitation (mm) for different return periods in the Mono river basin.

Return Periods (Years)	Hist	ICT_Ref	ICT_Proj
1	32.5	40	90
2	35	48	100
5	40	52.5	120
10	45	77	200
20	47.5	89	289

This could induce high magnitude floods, which may pose a critical risk of flood inundation and thus cause damages with severe economic and societal consequences in the lower valley of the Mono river basin. Considering these return periods, the amounts of heavy rains projected are alarming. This confirms the increase in intense rainfall events predicted by the regional climate models.

5. Conclusions

The comparative analysis of different regional climate models simulating air temperature and precipitation in the Mono river basin and the West African sub-region leads to three major conclusions regarding their ability to reproduce the present-day climate.

First, apart from the few exceptions noted below, the realism of the simulated climate is more controlled by the regional climate model (RCM) than the driven global climate model (GCM). For the same forcing GCM (ECHAM5 or HadCM3 in this study), the precipitation fields are substantially different from one RCM to the others used for the dynamical downscaling. This strong model sensitivity may arise from the RCM structure (model approximations) or from the choice of parameterization techniques made to run each model (e.g., convective scheme).

Second, though the seasonal cycles of air temperature and precipitation are well reproduced by the RCMs, most of them still feature systematic errors related to the magnitude of these variations. In general, the selected RCMs underestimate air temperature in the range 1–3 °C for the Mono river basin. Two RCMs largely overestimate rainfall (by a factor of 1.5 to 2) during the rainy season. This overestimation is due to very high rainfall intensities that could not be compensated by the underestimation of the number of rainy days. The most realistic models regarding the reproduction of the actual seasonal cycle are those that simulate both the frequency of rainfall events and the average intensity of daily rainfall well (ICT and to a lesser extent METO). Finally, almost all RCMs introduced bias with respect to the spatial variation of the climatic variables being analyzed. The north–south gradients of precipitation between the Sahel and the Guinean regions are well reproduced by all RCMs, except METO, although the simulated values for precipitation are not always correct. There are also errors in the southern part of the Guinean region regarding the reproduction of the drought band diagonally stretching from southern Ghana to southern Benin (Dahomey Gap). It was not reproduced by the RCMs forced with ECHAM5 GCM at their boundaries (MPI and ICT).

Third, and this is implicitly derived from the conclusion above, the hierarchy of RCMs in terms of the quality of their simulations differs according to the climate variable considered. For example, MPI tends to offer the most accurate reproduction of air temperature (especially their spatial distribution) and ICT is the most realistic for the precipitation patterns. This conclusion still needs to be taken with caution because though ICT satisfactorily reproduces the rainfall regimes and the frequency distribution of the rainfall events, the spatial distribution of the average annual precipitation is better simulated by the SMH model. Indeed, ICT fails to reproduce the relative dryness of the Ghanaian-Togolese coast in the south of the Guinean region. As well, ICT still features bias regarding the reproduction of the return periods of the annual rainfall maxima, for which it overestimated the frequency despite its good performance for the reproduction of the distribution of daily rainfall. Thus, we conclude that the choice of an appropriate RCM should be based on the research objectives, especially for impact studies.

Though these conclusions call for caution in using climate model simulations to assess future climate change in the sub-region, some lessons can nevertheless be drawn. The four RCMs unanimously project a +1 to +1.5 °C temperature increase in the sub-region by 2028–2050 as compared to the 1988–2010 reference period. The projected trends for cumulated precipitation are null or very moderate and diverge from one RCM to the others. Only SMH projected a significant decrease of precipitation in the Mono river basin. The simulations of the intense rainfall events (99th percentiles) also display contrasted results, but the two regional climate models that better reproduce these extreme rainfall events in the reference period have also unanimously projected a significant increase in their intensity by 2028–2050.

For the intense rainfall events, the Mono river basin at the outlet of Nangbéto recorded between 1988 and 2010 a significant increase in the intensity of the maximum daily rainfall averaged over the watershed, and this increase may be a risk factor. The heaviest rainfall events (99th percentile) occur mainly during the months of July to September, with a peak in September. This fits well with the discharge peaks in the basin. Thus, the intense rainfall events recorded in the tropical domain of the basin could lead to probable risks of flood inundations in the lower valley of the Mono river basin due to an overflow of the river.

The ICT simulations were the closest to the observations in terms of the reproduction of the cumulated precipitation and frequency distribution of daily rainfall. ICT is thus more robust for the analysis of the intense rainfall events and thereby flood forecasting in the Mono river basin. The estimation of extreme rainfall events using ICT showed stability except for the rainfall events occurring in October–November, which slightly increase. The adjusted GEV model satisfactorily predicted the heavy rainfall events, but there are uncertainties beyond the 90th percentile. The adjustment of a Weibull-type GEV distribution is well suited here for the prediction of extreme rainfall events (shape parameter of the GEV distribution = 0), although it is heavy-tailed at the 99th percentile.

The annual maxima of daily precipitation are increasing. Therefore, the impact of heavy rainfall events on discharge will increase under climate change conditions. This knowledge is important to develop an early warning system for the Mono river basin for the sustainability of the ecosystems. This calls for further research to better understand the physical mechanisms underlying the changes in extreme precipitation events and thus improve climate model simulations for short- and medium-term use. Moreover, for future study, a bias correction could be done for those datasets before quantifying the magnitudes of changes due to their coarser resolution.

Author Contributions: E.A. conducted this research. Conceptualization: E.A., A.D., H.A., S.O., P.C. and H.S.T.V.; Methodology and modelling and data processing: E.A., H.A., H.S.T.V., A.D., K.K.; Writing—Original Draft Preparation, E.A., H.A., H.S.T.V., S.O., A.D., M.B., G.M., C.H.; Writing—Review and Editing, P.C., G.M., C.H. All authors have read and agreed to the published version of the manuscript.

Funding: The APC was funded by les universités des auteurs ayant contribué à la réduction des coûts de publications.

Acknowledgments: The authors thank the reviewers for their comments which improve the quality of the manuscript. We thank the Climatology Research Center (CRC) of Dijon (France) for providing us AMMA-Ensemble data. We thank LAPA-MF -African Centre of Excellence on Climate Change, Biodiversity and Sustainable Development, the Institute of Research for Development (IRD, France) and Institute of Geosciences for Environment (IGE, University Grenoble Alpes) for providing the facility (the Regional Climate Modelling Platform) to perform this study at the University Félix Houphouët Boigny (Abidjan, Côte d'Ivoire).

Conflicts of Interest: The authors declare no conflict of interest.

References

1. Brown, O.L.I.; Hammill, A.; Mcleman, R. Climate change as the 'new' security threat: Implications for Africa. *Int. Aff.* **2007**, *83*, 1141–1154. [CrossRef]
2. Amoussou, E. Variabilité Pluviométrique et Dynamique Hydro-Sédimentaire du Bassin Versant du Complexe Fluvio-Lagunaire Mono-Ahémé-Couffo (Afrique de l'ouest). Ph.D. Thesis, Université de Bourgogne, Dijon, France, January 2010.
3. Mladjic, B.; Sushama, L.; Khaliq, M.N.; Laprise, R.; Caya, D.; Roy, R. Canadian RCM projected changes to extreme precipitation characteristics over Canada. *J. Clim.* **2011**, *24*, 2565–2584. [CrossRef]
4. Tramblay, Y.; Amoussou, E.; Dorigo, W.; Mahé, G. Flood risk under future climate in data sparse regions: Linking extreme value models and flood generating processes. *J. Hydrol.* **2014**, *519*, 549–558. [CrossRef]
5. Dilley, M.; Chen, R.S.; Deichmann, U.; Lerner-Lam, A.L.; Arnold, M. *Natural Disaster Hotspots: A Global Risk Analysis*; World Bank: Washington, DC, USA, 2005; Volume 7.
6. Crétat, J.; Vizy, E.K.; Cook, K.H. How well are daily intense rainfall events captured by current climate models over Africa? *Clim. Dyn.* **2014**, *42*, 2691–2711. [CrossRef]
7. Panthou, G.; Lebel, T.; Vischel, T.; Quantin, G.; Sane, Y.; Ba, A.; Ndiaye, O.; Diongue-Niang, A.; Diopkane, M. Rainfall intensification in tropical semi-arid regions: The Sahelian case Rainfall intensification in tropical semi-arid regions: The Sahelian case. *Environ. Res. Letts.* **2018**, *13*, 064013. [CrossRef]

8. Niasse, M.; Coalition, I.L. Climate-Induced Water Conflict Risks in West Africa: Recognizing and Coping with Increasing Climate Impacts on Shared Watercourses. In *Human Security and Climate Change*; PRIO, CICERO and GECHS: Asker, Norway, 2005; pp. 21–23. Available online: https://www.researchgate.net/profile/Madiodio_Niasse/publication/237699436_Climate-Induced_Water_Conflict_Risks_in_West_Africa_Recognizing_and_Coping_with_Increasing_Climate_Impacts_on_Shared_Watercourses/links/5440fe550cf2a76a3cc60e7c.pdf (accessed on 1 January 2019).

9. Loughlin, J.O.; Witmer, F.D.W.; Linke, A.M.; Laing, A.; Gettelman, A.; Dudhia, J. Climate variability and con fl ict risk in East. *Proc. Natl. Acad. Sci. USA* **2012**, *109*, 18344–18349.

10. Brown, O.; Crawford, A. Climate change: A new threat to stability in West Africa? Evidence from Ghana and Burkina Faso Climate change: A new threat to stability in West Africa? Evidence from Ghana and Burkina Faso 1. *Afr. Secur. Stud.* **2010**, *17*, 39–57. [CrossRef]

11. Collier, P.; Conway, G.; Venables, T. Climate change and Africa. *Oxf. Rev. Econ. Policy* **2008**, *24*, 337–353. [CrossRef]

12. Oyebande, L. Climate Change Impact on Water Resources at the Transboundary Level in West Africa: The Cases of the Senegal, Niger and Volta Basins. *Open Hydrol. J.* **2013**, *4*, 163–172. [CrossRef]

13. Barbe, L. Rainfall Variability in West Africa during the Years 1950–1990. *J. Clim.* **2002**, *15*, 187–202. [CrossRef]

14. Dezetter, A.; Servat, E.; Paturel, J.E.; Mahé, G.; Dieulin, C. Using general circulation model outputs to assess impacts of climate change on runoff for large hydrological catchments in West Africa Using general circulation model outputs to assess impacts of climate change on runoff for large hydrological catchments. *Hydrol. Sci. J.* **2009**, *6667*, 1.

15. Houghton, J.T.; Ding, Y.D.; Griggs, D.J.; Noguer, M.; van der Linden, P.J.; Dai, X.; Maskell, K.; Johnson, C.A. *Climate Change 2001: The Scientific Basis*; The Press Syndicate of the University of Cambridge: Cambridge, UK, 2001.

16. IPCC (Intergovernmental Panel on Climate Change). Summary for Policymakers. In *Climate Change 2007: The Physical Science Basis*; Contribution of Working Group I to the Fourth Assessment Report of the Intergovernmental Panel on Climate Change; Solomon, S., Qin, M.M., Chen, M.M.Z., Averyt, K.B., Tignor, M., Miller, H.L., Eds.; Cambridge University Press: Cambridge, UK; New York, NY, USA, 2007; pp. 1–18.

17. Paeth, H.; Hall, N.M.; Gaertner, M.A.; Alonso, M.D.; Moumouni, S.; Polcher, J.; Ruti, P.M.; Fink, A.H.; Gosset, M.; Lebel, T.; et al. Progress in regional downscaling of West African precipitation Progress in regional downscaling of west African precipitation. *Atmos. Sci. Lett.* **2016**, *12*, 75–82. [CrossRef]

18. Louis, J.-F. A Parametric Model of Vertical Fluxes in the atmosphere. *Bound. Layer Meteorol.* **1979**, *17*, 187–202. [CrossRef]

19. Dayan, U.; Nissen, K.; Ulbrich, U. Review Article: Atmospheric conditions inducing extreme precipitation over the eastern and western Mediterranean. *Nat. Hazards Earth Syst. Sci.* **2015**, *15*, 2525–2544. [CrossRef]

20. Katz, R.W.; Parlange, M.B.; Naveau, P. Statistics of extremes in hydrology. *Adv. Water Ressour.* **2002**, *25*, 1287–1304. [CrossRef]

21. Fowler, H.J.; Blenkinsop, S.; Tebaldi, C. Linking climate change modelling to impacts studies: Recent advances in downscaling techniques for hydrological. *Int. J. Climatol.* **2007**, *27*, 1547–1578. [CrossRef]

22. Lebel, T.; le Barbé, L.; Delclaux, F.; Polcher, J. From GCM scales to hydrological scales: Rainfall variability in West Africa. *Stoch. Environ. Res. Risk Assess.* **2000**, *14*, 275–295. [CrossRef]

23. Paeth, H.; Scholten, A.; Friederichs, P.; Hense, A. Uncertainties in climate change prediction: El Niño-Southern Oscillation and monsoons. *Glob. Planet. Chang.* **2008**, *60*, 265–288. [CrossRef]

24. Camberlin, P.; Fauchereau, N.; Gabriel, S. Empirical predictability study of October–December East African rainfall. *Q. J. R. Meteorol. Soc.* **2002**, *128*, 2239–2256.

25. Girmes, H.; Robin, P. Improving Seasonal Forecasting in the Low Latitudes. *Mon. Weather Rev.* **2006**, *134*, 1859–1879.

26. Paeth, H.; Diederich, M. Postprocessing of simulated precipitation for impact research in West Africa. Part II: A weather generator for daily data. *Clim. Dyn.* **2011**, *36*, 1337–1348. [CrossRef]

27. López-Moreno, J.I.; Beniston, M. Daily precipitation intensity projected for the 21st century: Seasonal changes over the Pyrenees. *Theor. Appl. Climatol.* **2009**, *95*, 375–384. [CrossRef]

28. Olume, V. Hydrological Processes in Regional Climate Model Simulations of the Central United States Flood of June–July 1993. *J. Hydrometeorol.* **2003**, *4*, 584–598.

29. Fowler, H.J.; Ekström, M.; Kilsby, C.G.; Jones, P.D. New estimates of future changes in extreme rainfall across the UK using regional climate model integrations. 1. Assessment of control climate. *J. Hydrol.* **2005**, *300*, 212–233. [CrossRef]

30. Dee, D.P.; Uppala, S.M.; Simmons, A.J.; Berrisford, P.; Poli, P.; Kobayashi, S.; Andrae, U.; Balmaseda, M.A.; Balsamo, G.; Bauer, D.P.; et al. The ERA-Interim reanalysis: Configuration and performance of the data assimilation system. *Q. J. R. Meteorol. Soc.* **2011**, *137*, 553–597. [CrossRef]

31. Vose, R.S.; Schmoyer, R.L.; Steurer, P.M.; Peterson, T.C.; Heim, R.; Karl, T.R.; Eischeid, J.K. *The Global Historical Climatology Network: Long-Term Monthly Temperature, Precipitation, Sea Level Pressure, and Station Pressure Data*; Oak Ridge National Lab., TN (United States). Carbon Dioxide Information Analysis Center: Washington, DC, USA, 1992.

32. Pendergrass Angeline, The Climate Data Guide: GPCP (Monthly): Global Precipitation Climatology Project, National Center for Atmospheric Research Staff (Eds.), Last Modified 02 July 2016. 2016. Available online: https://climatedataguide.ucar.edu/climate-data/gpcp-monthly-global-precipitation-climatology-project. (accessed on 2 January 2018).

33. New, M.; Lister, D.; Hulme, M.; Makin, I. A high-resolution data set of surface climate over global land areas. *Clim. Res.* **2002**, *21*, 1–25. [CrossRef]

34. Simmons, A.J.; Jones, P.D.; Bechtold, V.C.; Beljaars, A.C.M.; Ka, P.W. Comparison of trends and low-frequency variability in CRU, ERA-40, and NCEP//NCAR analyses of surface air temperature. *J. Geophys. Res.* **2004**, *109*, 1–18. [CrossRef]

35. Hewitt, C.D.; Griggs, D.J. Ensembles-based predictions of climate changes and their impacts. *Eos* **2004**, *85*, 566. [CrossRef]

36. Shahid, S. Trends in extreme rainfall events of Bangladesh. *Theor. Appl. Climatol.* **2011**, *104*, 489–499. [CrossRef]

37. Sciences, E.; Carolina, N.; Columbia, B.; Sciences, E.; Carolina, N. Avoiding Inhomogeneity in Percentile-Based Indices of Temperature Extremes. *J. Clim.* **2005**, *18*, 1641–1651.

38. Villarini, G.; Scoccimarro, E.; Gualdi, S. Projections of heavy rainfall over the central United States based on CMIP5 models. *Atmos. Sci. Lett.* **2013**, *14*, 200–205. [CrossRef]

39. Zolina, O.; Simmer, C.; Belyaev, K.; Kapala, A.; Gulev, S. Improving Estimates of Heavy and Extreme Precipitation Using Daily Records from European Rain Gauges. *J. Hydrometeorol.* **2009**, *10*, 701–716. [CrossRef]

40. Amoussou, E.; Camberlin, P.; Totin, V.S.H.; Pérard, J. Événements hydroclimatiques et risque d'inondation au sud-ouest du Bénin. In Proceedings of the 24ème colloque de l'Association Internationale de Climatologie, Rovereto, Italy, 6–10 September 2011; p. 2010.

41. Amoussou, E.; Tramblay, Y.; Totin, H.S.V.; Mahé, G. Dynamique et modélisation des crues dans le bassin du Mono à Nangbéto (Togo/Bénin). *Hydrol. Sci. J.* **2014**, *59*, 2060–2071. [CrossRef]

42. Fisher, R.A.; Tippett, L.H.C. Limiting forms of the frequency distribution in the largest particle size and smallest member of a sample. *Proc. Camb. Phil. Soc.* **1928**, *24*, 180–190. [CrossRef]

43. Nelder, J.A.; Mead, R. Nelder1965.pdf. *Comput. J.* **1965**, *7*, 308–313. [CrossRef]

44. Coles, S. *An Introduction to Statistical Modeling of Extreme Values*; Springer: London, UK, 2011; Volume 208.

45. Chu, P.S.; Zhao, X.; Ruan, Y.; Grubbs, M. Extreme rainfall events in the Hawaiian Islands. *J. Appl. Meteorol. Climatol.* **2009**, *48*, 502–516. [CrossRef]

46. Cunnane, C. *Statistical Distributions for Flood Frequency Analysis*; WMO Operational Hydrolofy report no. 33; the Secretariat of the Word Meteorological Organization: Geneva, Switzerland, 1989; pp. 581–582.

47. Jenkinson, A.F. The frequency distribution of the annual maximum (or minimum) values of meteorological elements. *Q. J. R. Meteorol. Soc.* **1955**, *81*, 158–171. [CrossRef]

48. El Adlouni, S.; Quarda, T.B.M.J. Comparaison des méthodes d'estimation des paramètres du modèle GEV non stationnaire. *Rev. des Sci. l'Eau* **2008**, *21*, 35–50. [CrossRef]

49. Ntajal, J.; Lamptey, B.L.; Mianikpo, J.; Kpotivi, W.K. R Ainfall T Rends and F Lood F Requency a Nalyses in the L Ower M Ono R Iver B Asin in T Ogo, W Est a Frica. *Int. J. Adv. Res.* **2016**, *4*, 1–11.

50. Davison, A. *Bootstrap Methods and their Application, no. 1.*; Cambridge University Press: Cambridge, UK, 1997.

51. Stedinger, J.R.; Vogel, R.M.; Foufoula-Georgiou, E. Frequency Anlysis of Extreme Events. In *Handbook of Hydrology*; Maidment, D.D.M.-H., Ed.; McGraw-Hill: New York, NY, USA, 1993; p. 68.

52. Lamptey, B.L. Comparison of gridded multisatellite rainfall estimates with gridded gauge rainfall over West Africa. *J. Appl. Meteorol. Climatol.* **2008**, *47*, 185–205. [CrossRef]

53. Gbobaniyi, E.; Sarr, A.; Sylla, M.B.; Diallo, I.; Lennard, C.; Dosio, A.; Dhiédiou, A.; Kamga, A.; Klutse, N.A.; Hewitson, B.; et al. Climatology, annual cycle and interannual variability of precipitation and temperature in CORDEX simulations over West Africa. *Int. J. Climatol.* **2014**, *34*, 2241–2257. [CrossRef]

54. Diallo, I.; Sylla, M.B.; Giorgi, F.; Gaye, A.T.; Camara, M. Multimodel GCM-RCM ensemble-based projections of temperature and precipitation over West Africa for the Early 21st Century. *Int. J. Geophys.* **2012**, *2012*. [CrossRef]

55. Riede, J.O.; Posada, R.; Fink, A.H.; Kaspar, F. What's on the 5th IPCC Report for West Africa? In *Adaptation to Climate Change and Variability in Rural West Africa*; Springer: Cham, Switzerland, 2016; pp. 7–23.

56. Ruti, P.M.; Williams, J.E.; Hourdin, F.; Guichard, F.; Boone, A.; Van Velthoven, P.; Favot, F.; Musat, I.; Rummukainen, M.; Domínguez, M.; et al. The West African climate system: A review of the AMMA model inter-comparison initiatives. *Atmos. Sci. Lett.* **2011**, *12*, 116–122. [CrossRef]

57. Lafore, J.P.; Flamant, C.; Giraud, V.; Guichard, F.; Knippertz, P.; Mahfouf, J.F.; Mascart, P.; Williams, E.R. Introduction to the AMMA Special Issue on 'Advances in understanding atmospheric processes over West Africa through the AMMA field campaign'. *Q. J. R. Meteorol. Soc.* **2010**, *136*, 2–7. [CrossRef]

58. Angelina, A.; Djibo, A.G.; Seidou, O.; Sanda, I.S.; Sittichok, K. Modifications du régime d'écoulement du fleuve Niger à Koulikoro sous changement climatique. *Hydrol. Sci. J.* **2015**, *60*, 1709–1723. [CrossRef]

59. Sylla, M.B.; Nikiema, P.M.; Gibba, P.; Kebe, I.; Klutse, N.A.B. Climate Change over West Africa: Recent Trends and Future Projections. In *Adaptation to Climate Change and Variability in Rural West Africa*; Yaro, J.A., Hesselberg, J., Eds.; Springer International Publishing: Berlin, Germany, 2016; pp. 1–244.

60. Mahe, G. Mémoire Habilitation à Diriger des Recherches Variabilité pluie-débit en Afrique de l'Ouest et Centrale au 20ème siècle: Changements Hydro-Climatiques, Occupation du sol et Modélisation Hydrologique Université des Sciences et Techniques Gil Mahé. HDR Dissertation, Université des Sciences et Techniques Montpellier, Montpellier, France, 2015.

61. Yabi, I.; Afouda, F. Extreme rainfall years in Benin (West Africa). *Quat. Int.* **2012**, *262*, 39–43. [CrossRef]

62. Kasei, R.; Diekkrüger, B.; Leemhuis, C. Drought frequency in the Volta basin of West Africa. *Sustain. Sci.* **2010**, *5*, 89–97. [CrossRef]

63. Hourdin, F.; Musat, I.; Guichard, F.S.; Ruti, P.M.; Favot, F.; Filiberti, M.A.; Pham, M.; Grandpeix, J.Y.; Polcher, J.; Marquet, P.; et al. Amma-Model intercomparison project. *Bull. Am. Meteorol. Soc.* **2010**, *91*, 95–104. [CrossRef]

64. Tossou, E.M.; Ndiaye, M.L.; Traore, V.B.; Sambou, H.; Kelome, N.C.; SY, B.A.; Diaw, A.T. Characterisation and Analysis of Rainfall Variability in the Mono-Couffo River Watershed Complex, Benin (West Africa). *Resour. Environ.* **2017**, *7*, 13–29.

65. Fowler, H.J.; Wilby, R.L. Detecting changes in seasonal precipitation extremes using regional climate model projections: Implications for managing fluvial flood risk. *Water Resour. Res.* **2010**, *46*, 1–17. [CrossRef]

66. Frei, C.; Scho, R.; Fukutome, S.; Vidale, P.L. Future change of precipitation extremes in Europe: Intercomparison of scenarios from regional climate models. *J. Geophys. Res.* **2006**, *111*. [CrossRef]

67. Beniston, M.; Stephenson, D.B.; Christensen, O.B.; Ferro, C.A.; Frei, C.; Goyette, S.; Halsnaes, K.; Holt, T.; Jylhä, K.; Koffi, B.; et al. Future extreme events in European climate: An exploration of regional climate model projections. *Clim. Chang.* **2007**, *81*, 71–95. [CrossRef]

68. Burn, D.H. The use of resampling for estimating confidence intervals for single site and pooled frequency analysis/Utilisation d' un rééchantillonnage pour l' estimation des intervalles de confiance lors d' analyses fréquentielles mono et multi-site intervals f. *Hydrol. Sci. Sci. Hydrol.* **2003**, *48*, 25–38. [CrossRef]

69. Caya, D.; Laprise, R. A semi-implicit semi-Lagrangian regional climate model: The Canadian RCM. *Mon. Weather Rev.* **1999**, *127*, 341–362. [CrossRef]

70. Hall, M.J.; Van den Boogaard, H.F.; Fernando, R.C.; Mynett, A.E. The construction of confidence intervals for frequency analysis using resampling techniques to cite this version: HAL Id: Hal-00304907 The construction of confidence intervals for frequency analysis using resampling techniques. *Hydrol. Earth Syst. Sci.* **2004**, *8*, 235–246. [CrossRef]

71. Sylla, M.B.; Gaye, A.T.; Jenkins, G.S. On the fine-scale topography regulating changes in atmospheric hydrological cycle and extreme rainfall over West Africa in a regional limate model projections. *Int. J. Geophys.* **2012**. [CrossRef]

72. Ekström, M.; Fowler, H.J.; Kilsby, C.G.; Jones, P.D. New estimates of future changes in extreme rainfall across the UK using regional climate model integrations. 2. Future estimates and use in impact studies. *J. Hydrol.* **2005**, *300*, 234–251. [CrossRef]

73. Emori, S.; Hasegawa, A.; Suzuki, T.; Dairaku, K. Validation, parameterization dependence, and future projection of daily precipitation simulated with a high-resolution atmospheric GCM. *Geophys. Res. Lett.* **2005**, *32*, 1–4. [CrossRef]

74. Kodja, D.J. Indicateurs des Évènements Hydroclimatiques Extrêmes dans le Bassin Versant de l'Ouémé à l'exutoire de Bonou en Afrique de l'Ouest. Ph.D. Dissertation, Université of Montpellier, Montpellier, France, 2018.

75. Embrechts, P.; Klüppelberg, C.; Mikosch, T.; Emberchts, P.; Klüppelberg, C.; Mikosch, T. Modelling Extremal Events. In *Statistical Methods for Extremal Events*; Springer: Berlin/Heidelberg, Germany; NewYork, NY, USA; London, UK; Paris, France; Tokyo, Janpan; Hong Kong, China; Barcelona, Spain; Budapest, Hungary, 1997; pp. 283–370.

76. Zwiers, F.W.; Kharin, V.V. Changes in the Extremes of the Climate Simulated by CCC GCM2 under CO_2 Doubling. *J. Clim.* **1998**, *2*, 2200–2222. [CrossRef]

77. Amoussou, E.; Camberlin, P.; Mahe, G. Impact de la variabilité climatique et du barrage Nangbéto sur l' hydrologie du système Mono-Couffo (Afrique de l' Ouest). *Hydrol. Sci. J.* **2012**, *57*, 805–817.

78. Kendall, M.; Stuart, A. *Analysis Multivariate*; Griffin, C., Ed.; Charles Griffim & Company Ltd.: London, UK, 1975.

Article

Are the Fouta Djallon Highlands Still the Water Tower of West Africa?

Luc Descroix [1,2,*], Bakary Faty [3], Sylvie Paméla Manga [2,4,5], Ange Bouramanding Diedhiou [6], Laurent A. Lambert [7], Safietou Soumaré [2,8,9], Julien Andrieu [1,9], Andrew Ogilvie [10], Ababacar Fall [8], Gil Mahé [11], Fatoumata Binta Sombily Diallo [12], Amirou Diallo [12], Kadiatou Diallo [13], Jean Albergel [14], Bachir Alkali Tanimoun [15], Ilia Amadou [15], Jean-Claude Bader [16], Aliou Barry [17], Ansoumana Bodian [18], Yves Boulvert [19], Nadine Braquet [20], Jean-Louis Couture [21], Honoré Dacosta [22], Gwenaelle Dejacquelot [23], Mahamadou Diakité [24], Kourahoye Diallo [25], Eugenia Gallese [23], Luc Ferry [20], Lamine Konaté [26], Bernadette Nka Nnomo [27], Jean-Claude Olivry [19], Didier Orange [28], Yaya Sakho [29], Saly Sambou [22] and Jean-Pierre Vandervaere [30]

1. Museum National d'Histoire Naturelle, UMR PALOC IRD/MNHN/Sorbonne Université, 75231 Paris, France; Julien.ANDRIEU@univ-cotedazur.fr
2. LMI PATEO, UGB, St Louis 46024, Senegal; s.manga4555@zig.univ.sn (S.P.M.); s.soumare913@zig.univ.sn (S.S.)
3. Direction de la Gestion et de la Planification des Ressources en Eau (DGPRE), Dakar 12500, Senegal; bakaryfaty@gmail.com
4. Département de Géographie, Université Assane Seck de Ziguinchor, Ziguinchor 27000, Senegal
5. UFR des Sciences Humaines et Sociales, Université de Lorraine, 54015 Nancy, France
6. Master SPIBES/WABES Project (Centre d'Excellence sur les CC) Bingerville, Université Félix Houphouët Boigny, 582 Abidjan 22, Côte d'Ivoire; bouramanding@gmail.com
7. Doha Institute for Graduate Studies, BP 200592 Doha, Qatar; llambert@qu.edu.qa
8. Ecole Polytechnique de Thiès, Laboratoire des Sciences et Techniques de l'Eau et de l'Environnement, LaSTEE, Thiès BP A10, Senegal; afall@ept.sn
9. ESPACE, Université Côte d'Azur, 06204 Nice, France
10. UMR G-EAU, IRD (AgroParisTech, Cirad, INRAE, MontpellierSupAgro, University of Montpellier), Dakar 12500, Senegal; andrew.ogilvie@ird.fr
11. IRD HSM, Case MSE, Université Montpellier, 300 avenue Emile Jeanbrau, 34090 Montpellier, France; gilmahe@hotmail.com
12. CERE Centre d'Etudes et de Recherches sur l'Environnement, UGANC Conakry, Conakry 001, Guinea; binetasombily@yahoo.fr (F.B.S.D.); damirou2013@gmail.com (A.D.)
13. Institut Universitaire des Hautes Etudes IUHEG, Conakry et GEOPROSPECT, Conakry 001, Guinea; Kadiatoudiallo5459@gmail.com
14. Montpellier SupAgro, UMR LISAH, IRD, INRA, University of Montpellier, 34090 Montpellier, France; jean.albergel@ird.fr
15. NBA (ABN) Niger Basin Authority, Niamey 8000, Niger; balkaly83@gmail.com (B.A.T.); i_amadou@yahoo.fr (I.A.)
16. UMR G-EAU, IRD (AgroParisTech, Cirad, INRAE, MontpellierSupAgro, Univ Montpellier), Univ Montpellier, 34196 Montpellier, France; jean-claude.bader@ird.fr
17. Direction Nationale de l'Hydraulique, Conakry 001, Guinea; barryaliou55@yahoo.fr
18. Laboratoire Leïdi "Dynamique des Territoires et Développement", Université Gaston Berger (UGB), BP 234-Saint Louis 46024, Senegal; ansoumana.bodian@ugb.edu.sn
19. Senior Investigator of IRD; yj.boulvert@gmail.com (Y.B.); jean-claude.olivry@wanadoo.fr (J.-C.O.)
20. UMR G-EAU, IRD (AgroParisTech, Cirad, INRAE, MontpellierSupAgro, Univ Montpellier), Conakry 001, Guinea; nadine.braquet@ird.fr (N.B.); luc.ferry3435@gmail.com (L.F.)
21. Independent Consultant, 34980 Combaillaux, France; jean-louis.couture@wanadoo.fr
22. Département de Géographie, Faculté des Lettres et Sciences Humaines, Université Cheikh Anta Diop, Dakar 12500, Senegal; honore.dacosta@ucad.edu.sn (H.D.); sambousaly@gmail.com (S.S.)

23 Grdr Migrations, Citoyenneté, Développement, 93100 Montreuil, France; gwenaelle.dejacquelot@grdr.org (G.D.); eugenia.gallese@grdr.org (E.G.)

24 OMVS Organisation Pour la Mise en Valeur du Bassin du Fleuve Sénégal, Dakar 12500, Senegal; diakitemm2@gmail.com

25 FPFD, Fédération des Paysans du Fouta Djallon BP 52 PITA, Timbi Madina 7311, Guinea; kourahoye@yahoo.fr

26 OMVG Organisation pour la Mise en Valeur du Bassin du Fleuve Gambie, Dakar 12500, Senegal; lkonate@omvg.sn

27 Institut de Recherches Géologiques et Minières, Centre de Recherches Hydrologiques, Yaoundé BP 4110, Cameroon; bnnomo@gmail.com

28 IRD, Cirad, INRAE, Montpellier SupAgro, UMR Eco&Sols, University of Montpellier, 34196 Montpellier, France; didier.orange@ird.fr

29 SENASOL, Service National des Sols, Conakry 001, Guinea; mouctaryaya3@gmail.com

30 UMR IGE Univ Grenoble/Alpes, IRD, CNRS, Grenoble INP, Université Grenoble Alpes, 38058 Grenoble, France; jean-pierre.vandervaere@univ-grenoble-alpes.fr

* Correspondence: luc.descroix@ird.fr; Tel.: +33-678-92-06-84

Received: 31 July 2020; Accepted: 10 September 2020; Published: 22 October 2020

Abstract: A large share of surface water resources in Sahelian countries originates from Guinea's Fouta Djallon highlands, earning the area the name of "the water tower of West Africa". This paper aims to investigate the recent dynamics of the Fouta Djallon's hydrological functioning. The evolution of the runoff and depletion coefficients are analyzed as well as their correlations with the rainfall and vegetation cover. The latter is described at three different space scales and with different methods. Twenty-five years after the end of the 1968–1993 major drought, annual discharges continue to slowly increase, nearly reaching a long-term average, as natural reservoirs which emptied to sustain streamflows during the drought have been replenishing since the 1990s, explaining the slow increase in discharges. However, another important trend has been detected since the beginning of the drought, i.e., the increase in the depletion coefficient of most of the Fouta Djallon upper basins, as a consequence of the reduction in the soil water-holding capacity. After confirming the pertinence and significance of this increase and subsequent decrease in the depletion coefficient, this paper identifies the factors possibly linked with the basins' storage capacity trends. The densely populated areas of the summit plateau are also shown to be the ones where vegetation cover is not threatened and where ecological intensification of rural activities is ancient.

Keywords: Fouta Djallon; water tower; depletion (or recession) coefficient; runoff coefficient; soil water holding capacity; basement; sandstone; Fula society

---

1. Introduction: Statements and State of the Art

The Fouta Djallon receives some of the greatest rainfall in West Africa. Although rainfall decreases inland from the coast, where Conakry receives more than 4000 mm yearly, to the mountains, the highlands, which separate the coastal basins from the great "sudano-sahelian" basins of Senegal and Niger Rivers, have annual precipitation ranging from 1800 to 2300 mm. On the northern slope, annual rainfall ranges from 1300 to 2000 mm. This is sufficient to ensure significant streamflows on both the coastal and continental hillslopes of the Fouta Djallon and the "Guinean Dorsale", its natural extension southeastward [1,2]. At the continental scale, the Fouta Djallon is one of the two great "water towers" of Northern sub-Saharan Africa (Figure 1), the other one being in Ethiopia [3,4].

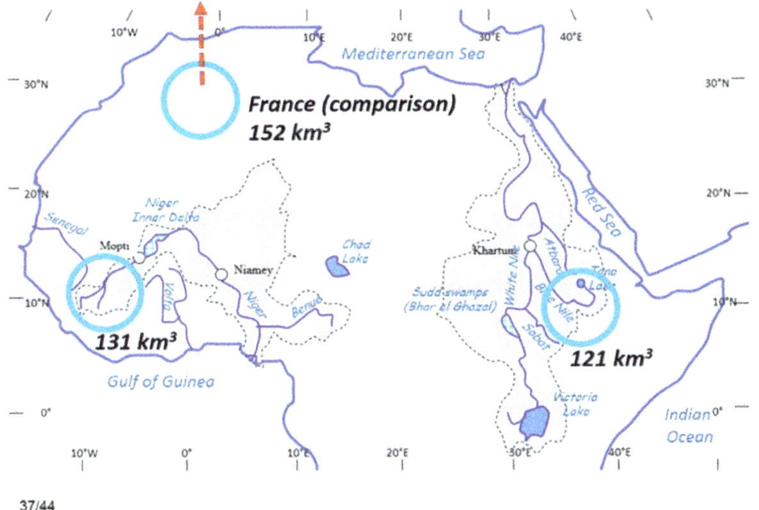

37/44

(**a**)

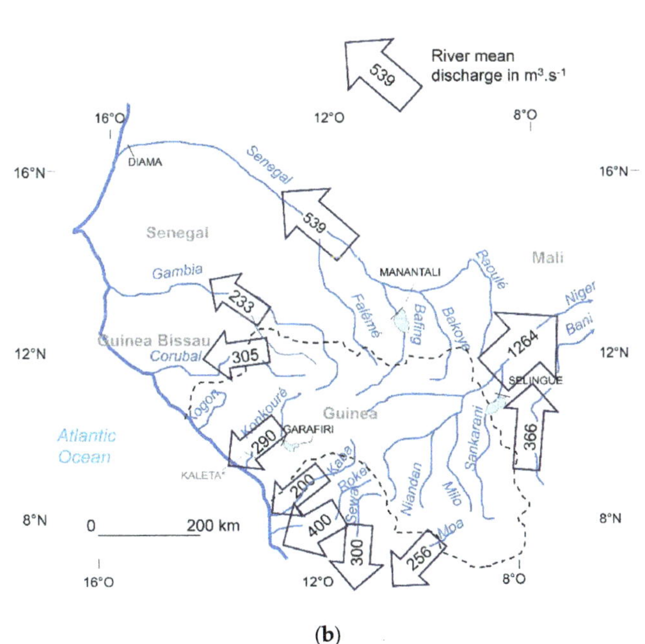

(**b**)

Figure 1. *Cont.*

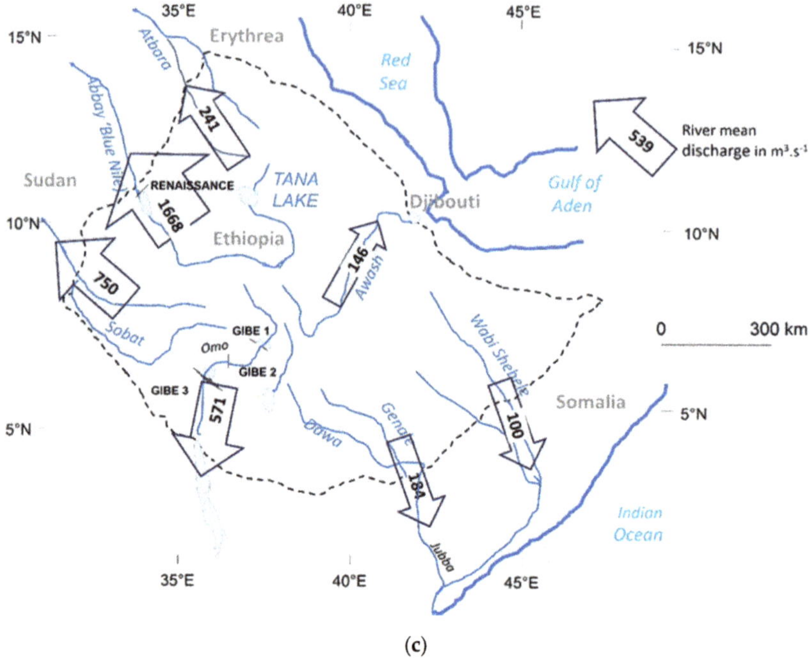

(c)

Figure 1. The two main natural water towers of northern sub-Saharan Africa and their 1951–2000 mean annual discharges; (**a**) comparison of the water towers (total discharges of French rivers is given for comparison purposes); (**b**) the Guinean water tower; (**c**) the Ethiopian water tower.

The Fouta Djallon highlands are unanimously recognized as the 'water tower' of West Africa as they constitute the source of many transboundary rivers and their tributaries in the region. Table 1 gives the dependence rate of some countries on the FD rivers. In this paper, the acronym 'FD' refers to the Fouta Djallon and the broader Guinean Dorsale (Figure 2) that runs southeastwards from the Fouta Djallon mountains, in Guinea, to the Nimba mountains, on the border with Côte d'Ivoire. By contrast, the acronym 'FDss' refers to the Fouta Djallon stricto sensu, which only constitutes the northwestern side of the Guinean Dorsale.

Table 1. Dependence rate (%) from water coming of the Fouta Djallon highlands of the concerned West African countries [5].

Country	% of Surface Water Resources Originating in the Fouta Djallon	Concerned River Basins
Mauritania	96	Senegal
Gambia	63	Gambia
Senegal	60	Senegal, Gambia, Anyamba/Geba
Guinea-Bissau	48	Corubal/Koliba, Anyamba/Geba
Mali	55	Senegal, Niger
Niger	70	Niger

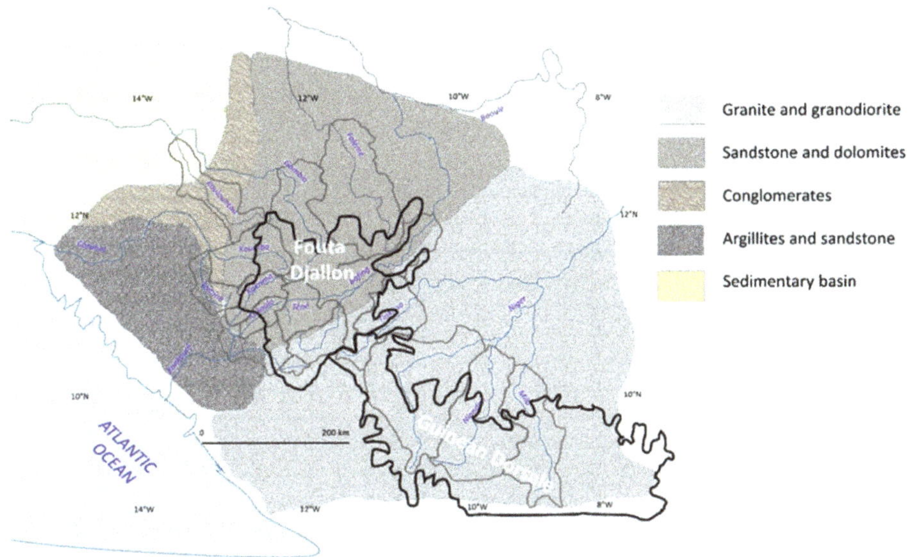

Figure 2. The main geological parts of the Fouta Djallon and Guinean Dorsale ensemble.

The range gathers the sources of the three main West Sudano-Sahelian rivers: the Gambia River, the sources of the Bafing-Falémé complex which form the Senegal River after the confluence of the Bafing and the Bakoye in Mali, and those of the Niger River and its Guinean tributary on the left bank, the Tinkisso [2]. The range also feeds the two main rivers of the Atlantic side of Guinea, the Kouba/Koumba and the Tominé, which become the Koliba and then the Rio Corubal in Guinea Bissau, as well as the Konkouré complex, and the main rivers crossing Sierra Leone and parts of Liberia.

1.1. What Does Not Change

The geological context of the Fouta Djallon differs from a natural water tower. The entire range can be divided into a Precambrian and mostly granitic basement in the south east and a sedimentary ensemble in the north west. The latter is mostly composed of sandstone with doleritic sills and intercalations of argillites (Figure 2). The separation follows a virtual Conakry–Bamako line [6]. In the FDss, the granite is widely covered by hundreds of meters of sandstone, dolomites, and argillites, with frequent sills of dolerites, none of which constitutes a good natural reservoir. Mamedov et al. (2010) [7] estimated the sandstone-quartzite-argillite thickness as ranging from 1500 to 4000 m. The FDss is itself divided into three geological contexts: a small granitic area southward, the western sandstone area (soft sandstone and shale with some injections of dolerites), and the northwestern doleritic FD, which is also mostly composed of sandstone and shale, but which features a large number of outcropping sills and dolerite veins. Boulvert [6] noticed that these thick sandstone plateaus are separated by the incisions (of more than 700 m) of the so-called 'Atlantic southern rivers' (i.e., the Kokoulo, the Kakrima, and the Koumba), highlighted by spectacular sandstone cornices, gorges, and waterfalls. The Guinean Dorsale is mostly an Archean granitic basement with few sills of dolerites and few outcrops of gneiss and quartzite [2]. All the Upper Niger basin and a part of the Konkouré River basin constitute a granitic basement, locally covered by sandstone and the typical 'lateritic pan'. The common characteristics of the whole mountain area of Guinea is the great extension of the 'bowe', i.e., lateritic plateaus composed of ferruginous and overall aluminous crusts [8].

However, no clear difference can be noticed in their hydrological behavior, most probably due to their similarly high levels of imperviousness. Infiltration exists at the FD scale (faults, cracks, and fractures, very few permeable rocks, local shale, and green rocks), but not at the sample scale. Most of the rocks are either completely impervious (granite, dolerites, quartzite) or mainly impervious (sandstone), except their alterites. In the entire FD, the rocks have a very low water storage capacity [2]. However, Boulvert [8] highlighted the extension of "paleo-crypto karsts" in the Bowe area of Guinea, which covers most of the Fouta Djallon. Therefore, the role of possible occulted streamflows remains to be analyzed.

The main characteristics of the FDss, which are common to the entire FD, can be summarized by the following [9]:

- the total absence of a generalized water table and of continuous groundwater reservoir [9]. There are only small and localized aquifers, well known by the local people [10].
- the very low mineral concentration of water (lower than 50 mg L^{-1}). The average of dissolved load in total sampling observed by Orange (1990a) [9] is only 35 mg L^{-1}, due to the sandstone and/or granite predominance and the importance of stable lateritic crusts [9]. This relatively old finding remains overall coherent and in line with a series of small sampling tests made in the FDss by the authors in March and May 2019. In 28 samples taken at an elevation of less than 800 m, the mineral concentration of water was on average 55 mg·L^{-1} (standard deviation = 74, with values ranging from 6.5 to 414), and in 14 samples taken at an elevation higher than 800 m, the mineral concentration was on average 24.5 mg·L^{-1} (standard deviation = 14.8, values from 6.5 to 53).
- the stationary geochemical regime: the alteration profile deepening (4.6 mm·yr^{-4}) is slightly higher than soil erosion (4.2 mm·yr^{-4}) [9,10].

The FD's overall impermeability strongly impacts surface water behavior. All of the geological context display the same functioning: very weak discharges during the dry season (from December to June), because they are not supported by groundwater. In contrast, at the occurrence of the first monsoon rainfall events (in June or July), discharges increase suddenly to peak in September, with the rainfall peak [2].

The impervious geology of most of the FD mountains [9,11] leads to high runoff and recession coefficients in the main rivers, due to a very low soil water-holding capacity on the one hand and due to the above-mentioned absence of a generalized water tables on the other.

The monsoon lasts five to six months a year in the core part of the highlands, and annual precipitation is very low during the dry season. The low yet permanent streamflows of the main rivers are then fed by the emptying of the eight thousand small wetlands (commonly settled on alterites), of the thin aquifers, and of the free water of the lowlands alongside rivers and streams of the whole Fouta Djallon and Guinean Dorsale.

1.2. What Could Change

D'Aubreville (1947) [12] and then Maignien (1966) [13] were impressed by the great extension of "ferrallitic pans in Guinea". The former thought that the formation of these "bowe" (sing. "bowal"), or "bowalisation" (as the process), is still ongoing, yet most of them have proven to be fossils [8]. These are not evidence of any land degradation processes and results.

During the 1940s and 50s, geographers [14–16] such as Richard-Molard (1943), Pouquet (1956, 1960), agronomists [17] like Sudres (1947), or foresters [12] (e.g., D'Aubreville, 1947) were studying the impact of deforestation and erosion before the 70s [8].

The most densely populated area is the summit of the FDss (around Labé and Mali town) and its historical core. This location is very rainy, but it allows very poor water storage. People encounter severe problems in drinking water supply and even greater challenges to obtain water to irrigate a second or third yearly crop and produce enough food for the entire population.

As observed in the Sahel, the decrease in discharge during the prolonged rainfall deficit (1968–1993) of the Sahel's so-called "Great Drought" is approximately twice that of the rainfall [18]. During the drought, the authors highlighted that the severe rainfall deficit provoked for the first time in the Senegal River an exponential increase of the recession (or depletion) coefficient, which was stable for 70 years (a = 0.0186 d^{-1}) and which strongly increased to reach 0.038 in 1984 [19]. This long "Great Drought" (Figure 3) led to a lasting degradation of the hydrological system. In Guinean areas, the decrease in discharges lasted 10 to 15 years. The increase in recession coefficients also lasted for years beyond the end of the drought. In 1997, Bricquet et al. [20] showed that despite the observed rainfall recovery, discharges did not begin to increase again at that time. This was due to the necessary refilling of the various natural storages such as surface and sub-surface reservoirs, and streamflows recovered over more than a decade [21]. Figure 4 [1] provides the spatial distribution of rainfall.

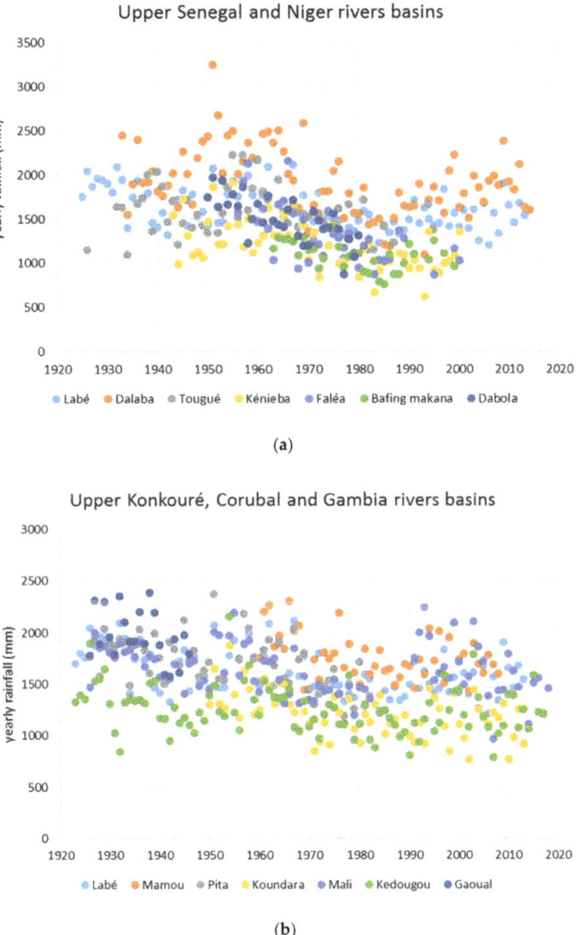

Figure 3. Rainfall evolution in the Fouta Djallon mountains since 1923; (**a**) Senegal and Niger upper basins. (**b**) Konkouré, Corubal, and Gambia rivers basins. Dots represent the annual total rainfall (ATR), in mm.

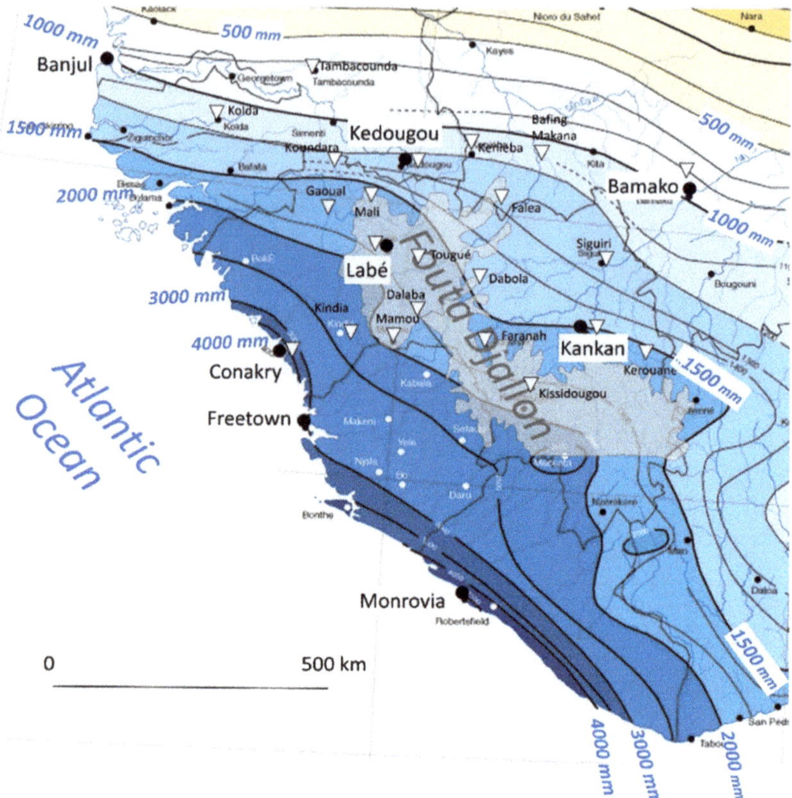

Figure 4. Rainfall spatial distribution (1950–2000) [1] and location of rain gauges (white triangles) used to unskew CHIRPS data.

However, Gomis observed in 2000 [22] that the rainfall recovery led to a decrease in the recession coefficient of the Gambia River. In the Bafing basin (Upper Senegal basin), this slow and progressive discharge recovery was also characterized by a decrease in the recession coefficient, from 0.025 d^{-1} in 1972–1994 to 0.022 d^{-1} during the 1995–2005 period at the Bafing Makana stream gauge station [23].

In some basins, dams have been built or are being planned, and they will strongly modify the annual regime and allow hydroelectric power generation, such as in the Konkouré River basin [24].

This paper aims to study the recent evolution of the FD as a natural water tower through the analysis of runoff and depletion coefficients, along other factors such as rainfall and land cover dynamics.

2. Material and Methods

This study of the permanency of the Guinean water tower is based on the analysis of the hydrological behavior evolution of 16 basins and sub-basins across these highlands. Basins and their downstream gauging stations are indicated in Figure 5.

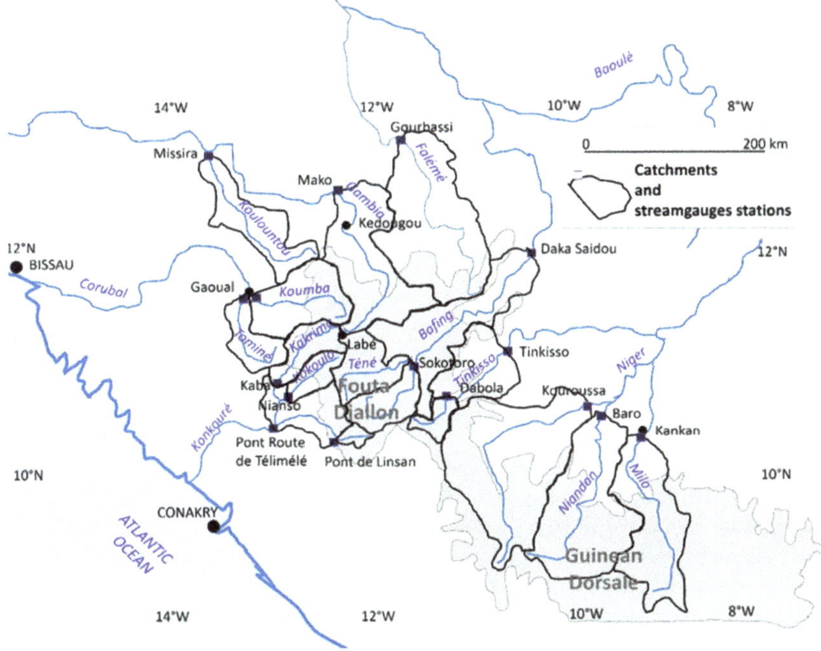

Figure 5. Location of the 16 basins and sub-basins considered in this study.

Excel file in supplementary materials gives the list of the 16 considered basins and the availability of their hydrological data, where vertical black lines separate the three climatic periods: 1947–1969, 1970–1993 and 1994–2019.

Depletion coefficient (or recession coefficient). The recession coefficient is used after the Maillet law (1905) and as cited by Tallaksen (1995) [25] as follows:

$$Q_t = Q_0\, e^{-\alpha.t} \tag{1}$$

Q_t = discharge at t time; Q_0 = initial discharge at the beginning of the receding waters; α = recession index (= DP). The latter is the inverse of the elapsed time in d^{-1} and is expressed as

$$\alpha = -\ln(Q_t/Q_0)/t \tag{2}$$

The depletion coefficient (DC) used here are the average value of the DC for three durations: one month, two months, and four months (three on the smaller basins). The determined value of the DC is the average value of the DC for each duration around the maximal value of the DC for this duration. The considered period is the one when, on average for the period, the DC is the highest; thus, it is the same for all the years of the period, in the ensuing "Results" part. In a "Supplementary Material" part, we considered, for each station, and for each year, and for three durations (1, 2, and 4 months), the period when the DC reaches its highest value.

3. Discharges and Runoff Coefficients

Daily discharge values are provided by the basin agencies (ABN/NBA Niger Basin Authority, for the Niger basin; OMVS—Organisation pour la Mise en Valeur du bassin du fleuve Sénégal- for the Senegal basin; OMVG—Organisation pour la Mise en Valeur du bassin du fleuve Gambie- for the Gambia and Corubal basins) and the National Hydraulic Service of Guinea for the Konkouré basin.

Rainfall data of 23 stations are provided by SIEREM (Système d'Informations Environnementales sur les Ressources en Eau et leur Modélisation) [26,27], AMMA Catch (AMMA = African Monsoon Muldisciplinary Analysis) (http://www.amma-catch.org/), ANACIM (Agence Nationale de l'Aviation CIvile et de la Météorologie) (Senegalese data), and from the CHIRPS (Climate Hazards Group InfraRed Precipitation with Station data) data system [28] for recent years (since 1980, 1994, or 2000 according to the basins). The CHIRPS products used here have a spatial resolution of 0.25° and a daily temporal resolution. The covered area is from 50° N to 50° S. They combine satellite data at 0.25° to soil rainfall data [28]. In order to improve the data accuracy, the CHIRPS data were unskewed again using observed rainfall data for a few stations (see Figure 4): Labé, Koundara, Mamou, Dalaba, Faranah, Tougué, Siguiri, Mali, Gaoual, Conakry, Kindia, Dabola, Kissidougou, Kankan, and Kerouané in Guinea; Kedougou, Tambacounda, and Kolda in Senegal; Kayes, Kenieba, Bafing Makana, Bamako, and Falea in Mali.

4. Land Use/Land Cover Changes (LULCC) are Determined at Three Scales

4.1. At the Regional Scale

At the regional scale, the tree cover variations between 2000 and 2019 over the 16 basins were extracted from the Global Forest Change (GFC2019) datasets produced by the University of Maryland, based on Landsat 4, 5, 7, and 8 imagery. These provide 30 m spatial resolution data of tree gains and tree losses compared to land use cover of the year 2000. Cloud-free observations for each pixel, based on quality assessment layers, are used to create composite imagery in bands 3, 4, 5, and 7 during the growing season. Image interpretation relies on a decision tree supervised learning approach and training data, including very high resolution Quickbird imagery as described in Hansen et al. (2013) [29]. Google Earth Engine© (Google, Silicon Valley, CA, USA) geoprocessing capabilities [30] were used here to reduce the time to extract and combine pixel data for the 16 basins.

4.2. At the Basin Scale

At the basin scale, a special analysis was conducted in the upper Bafing basin. At the Upper Bafing basin scale, changes in vegetation were searched using a time series analysis (TSA) of the normalized difference vegetation index (NDVI) based on a MODIS (Moderate-Resolution Imaging Spectroradiometer; EOS Earth Observing System, NASA, Washington, D.C., USA) (MODQ13 dataset (250 m resolution; 16 days temporal resolution; from February 2000 to February 2020). Mann-Kendall's tau test is an efficient trend detection method [31] and has been used for this data type for other regions of western Africa [32–34]. Mann-Kendall's tau test can be submitted to null hypothesis (P value) testing as the data come from a population with independent realizations and are identically distributed. A significance at 0.001 has been retained for this paper meaning that a sample as extreme as that observed would occur 1 time in 1000 if the null hypothesis is true.

4.3. At the Local Scale

At the local scale, some small wetlands of river sources are compared in the central area of the FD. Here, we compare the current status (via Google Earth© images; Google, Silicon Valley, CA, USA) with aerial imagery acquired by the IGN (Institut National de l'Information Géographique et Forestière, Saint Mandé, France) 1953 mission over West Africa.

5. Results

5.1. Changes of Depletion Coefficients

Figure 6a–e and Table 2A–D gather the results about changes in the DC (depletion coefficient). In these figures, the DC is calculated for a period of four months (and three months for the two Sokotoro and Pont de Linsan smaller basins) for each year on the same date, i.e., the average date of the maximum depletion of the discharge (in the Supplementary Materials, Figures S1a,b–S16a,b illustrate the evolution of the DC at the highest yearly depletion rate of each year).

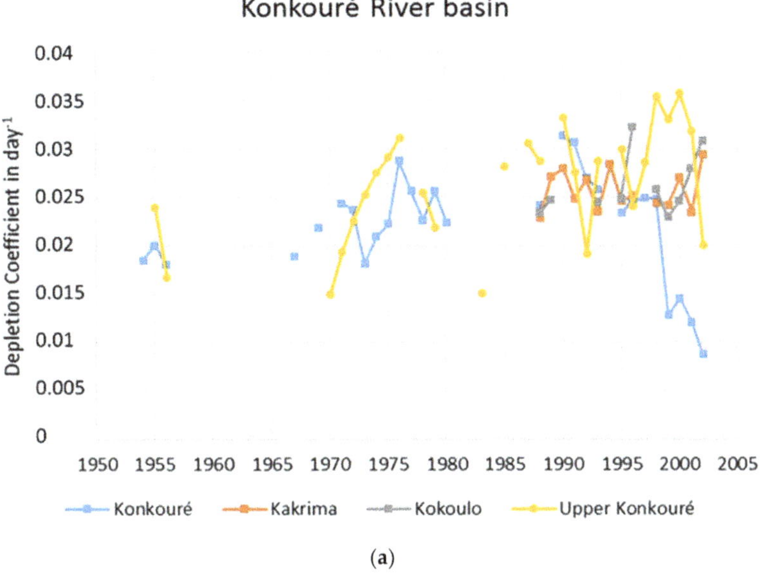

(**a**)

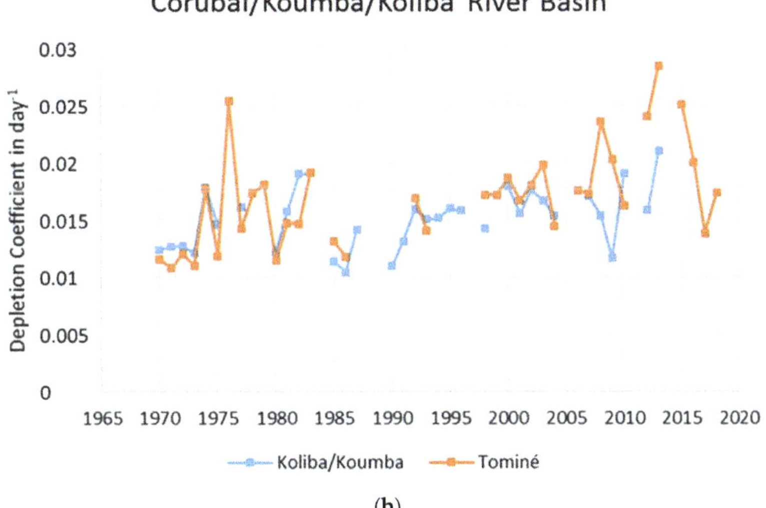

(**b**)

Figure 6. *Cont.*

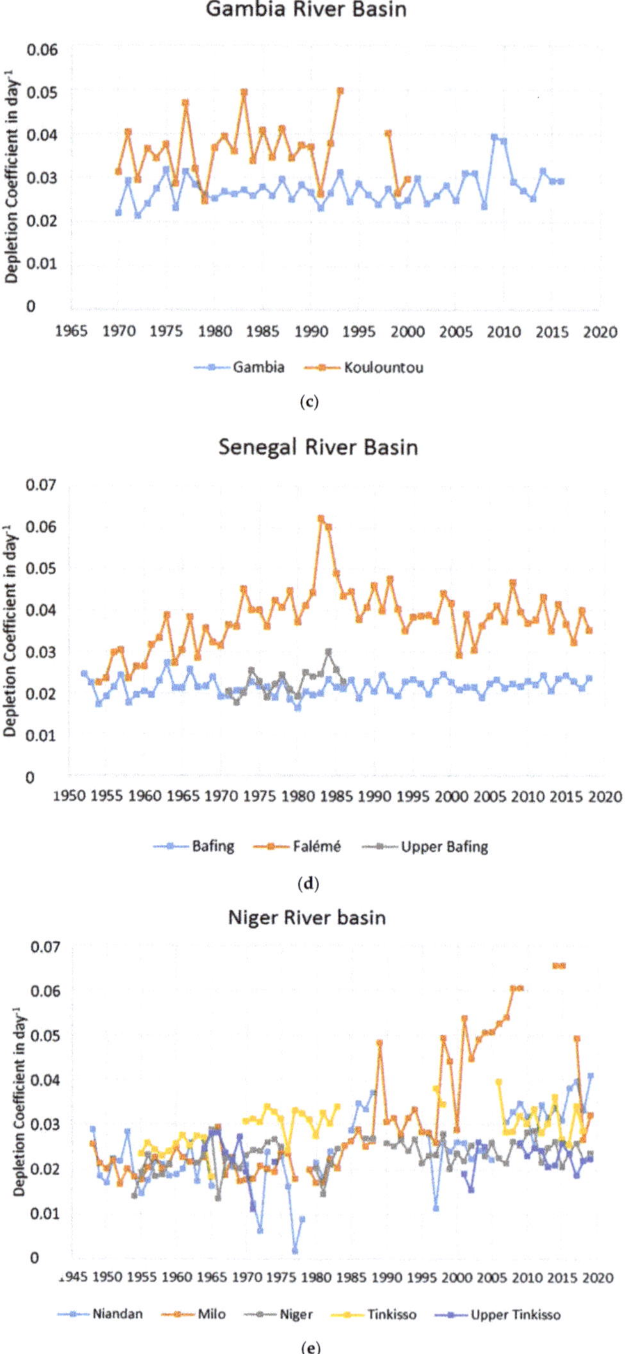

Figure 6. Dynamic of depletion coefficients for the period 1947–2018, by basin; (**a**) Konkouré River basin; (**b**) Koliba/Tominé/Corubal River basin; (**c**) Gambia River basin; (**d**) Senegal River basin; (**e**) Niger River basin.

Table 2. Depletion coefficient by basins, by periods, and for different durations.

A	Decreasing DC		4 Months Depletion Coef			2 Months Depletion Coef			1 Month Depletion Coef			Yearly
			Mean 1947–70	Mean 71–93	Mean 94–2019	Mean 1947–70	Mean 71–93	Mean 94–2019	Mean 1947–70	Mean 71–93	Mean 94–2019	Deforestation %
Senegal	Gourbassi	Faleme coef	0.0339	0.0460	0.0412	0.0433	0.0533	0.0465	0.0550	0.0622	0.0554	0.099
		day	28-November	25- November	1-December	27-October	3-November	3-Nov	18-October	23-October	18-October	
Senegal	Daka Saidou	Bafing coef	0.0238	0.0243	0.0237	0.0298	0.0321	0.0317	0.0352	0.0395	0.0392	0.197
		day	29-November	8-November	3-December	7-November	5-November	17-November	5-November	18-October	11-November	
Niger	Dabola	Tinkisso coef	0.0264	0.0228	0.0252	0.0381	0.0366	0.0351	0.0485	0.0456	0.0424	0.873
		day	3-December	20-November	9-December	4-November	25-October	15-November	27-October	17-October	2-November	

B	Stabilized DC		4 Months Depletion Coef			2 Months Depletion Coef			1 Month Depletion Coef			Yearly
			Mean 1947–70	Mean 71–93	Mean 94–2019	Mean 1947–70	Mean 71–93	Mean 94–2019	Mean 1947–70	Mean 71–93	Mean 94–2019	Deforestation %
Gambia	Missirah	Koulountou coef		0.0407	0.0394		0.0434	0.0443		0.0525	0.0625	0.339
		day		21-November	3-December		6-November	15-November		24-October	8-November	
Gambia	Mako	Gambia coef		0.0294	0.0308		0.0381	0.0401		0.0485	0.0500	0.427
		day		29-November	4-November		27-Oct	5-November		22-October	29-October	
Corubal	Gaoual K	Koumba (Koliba) coef		0.0201	0.0197		0.0284	0.0267		0.0388	0.0371	0.836
		day		30-November	9-November		30-October	9-November		31-October	14-November	
Senegal	Sokotoro	Bafing coef		0.0269			0.0314			0.0403		2.229
		day		27-November			16-November			30-Oct		
Niger	Kouroussa	Niger coef	0.0249	0.0269	0.0271	0.0300	0.0321	0.0312	0.0355	0.0383	0.0366	0.936
		day		27-December	18-December	28-December	25-December	14-December	1-December	26-November	10-December	21-November

C	Increasing DC < 20%		4 Months Depletion Coef			2 Months Depletion Coef			1 Month Depletion Coef			Yearly
Ddef			Mean 1947–70	Mean 71–93	Mean 94–2019	Mean 1947–70	Mean 71–93	Mean 94–2019	Mean 1947–70	Mean 71–93	Mean 94–2019	Deforestation %
Konkouré	Nianso	Kokoulo coef		0.0297	0.0305		0.0281	0.0334		0.0368	0.0418	1.858
		day		30-November	12-December		14-November	24-November		1-November	19-November	
Konkouré	Kaba	Kakrima coef		0.0278	0.0295		0.0320	0.0344		0.0399	0.0400	1.724
		day		9-December	2-December		29-November	17-November		15-November	8-November	
Corubal	Gaoual T	Tominé coef		0.0201	0.0241		0.0279	0.0318		0.0393	0.0428	2.273
		day		30-November	4-December					30-October	6-November	

Table 2. *Cont.*

D	Increasing	DC	> 20%	4 Months Depletion Coef	2 Months Depletion Coef	1 Month Depletion Coef	Yearly						Deforestation %
				mean 1947–70	mean 71–93	mean 94–2019	mean 1947–70	mean 71–93	mean 94–2019	mean 1947–70	mean 71–93	mean 94–2019	
Niger	Tinkisso	Tinkisso	coef	0.0269	0.0333	0.0365	0.0342	0.0374	0.0390	0.0459	0.0465	0.0505	0.235
			day	18-December	21-December	1-Jan	24-November	19-November	2-December	14-November	2-November	19-November	
Niger	Baro	Niandan	coef	0.0230	0.0279	0.0314	0.0279	0.0325	0.0383	0.0356	0.0393	0.0443	0.626
			day	14-December	21-December	22-December	10-December	5-December	27-December	30-November	25-November	24-December	
Niger	Kankan	Milo	coef	0.0227	0.0256	0.0321	0.0285	0.0310	0.0419	0.0347	0.0397	0.0517	0.634
			day	8-December	6-December	15-November	16-November	11-November	14-November	2-November	9-November	11-November	
Konkouré	Pt Rte Télimélé	Konkouré	coef	0.0132	0.0212	0.0245	0.0153	0.0252	0.0268	0.0191	0.0281	0.0359	0.881
			day	7-January	26-December	13-December	8-December	3-December	29-November	26-November	19-November	11-November	
Konkouré	Pt de Linsan	Upper K.	coef	0.0234	0.0303	0.0334	0.0294	0.0343	0.0376	0.0365	0.0473	0.0460	3.731
			day	7-November	14-November	22-November	22-October	6-November	7-November	23-October	31-October	11-November	

Basins are classified into four classes, according to the temporal evolution of the depletion coefficient (DC): decreasing, stabilized, increasing until 20% from the first to third period, and increasing above 20%. Deforestation is here one of the tree cover variations between 2000 and 2019 over the 16 basins, using data of Maryland University (the GFC2019), according to the method proposed by [29] Hansen et al. (2013).

Data covering the 1947–2019 timeline was divided into three periods, based on previous segmentation by several authors ([35–38]):

-   the first one is the 'hyper humid' 1947–1969 period, when rainfall was significantly higher than the long-term average across the whole of West Africa.
-   the second period is the well-known "Great Drought" of West Africa (1970–1993). The rainfall decreased significantly in 1968, but we arbitrarily made the second period begin in 1970 because a significant number of stream gauge stations have data starting from 1970. The drought is considered to last until 1993.
-   the last period spans the years 1994–2019. The year 1994 is when annual rainfall again reached again its long-term mean value. During this period, rainfall is approximately of the long-term trend.

Rainfall predictions in West Africa indicate a regional increase in rainfall during the ongoing 21st century due to the activation of monsoons as a result of the increase in the difference between the continental temperature and SST (sea surface T°) [39]. However, locally, in the western part of West Africa, rainfall is expected to decrease significantly after 2030 or 2035 (Senegambia, Mauritania, and western Mali), while further south, an increase in ATR (annual total rainfall) across most of the Guinean territory is expected [40,41].

The dynamics of the depletion coefficients (see Figure 6a–e and Table 2A–D; see also the Supplementary Materials) shows that:

-   the evolution through the three periods is not easy to observe for each basin due to the lack of data (some of the data begins in 1970). The only basins where an analysis over the entire period of 1950–2019 is possible are the largest ones (Senegal and Niger Rivers).

From period 1 to period 2:

-   an increase in the DC between period 1 and period 2 is clearly observed in the Upper Niger (at Kouroussa, where it begins in 1965, and at Tinkisso, where an increase in the DC appeared after 1970) and in the Senegal basin. This is clear in the Faleme basin (and probably in the upper Bafing at Sokotoro, though data are lacking). This increase is also observed in the Corubal basin (but particularly over the first drought period, 1970–1975). Despite the lack of data, such a trend also appears in the Gambia River basins (Mako and Missira stream gauges, the second one showing much higher variability). However, it is very difficult to observe this increasing trend in the Konkouré basin, and clearly, the Bafing basin at Daka Saidou does not show any such trends across all available observations.
-   only the Niandan at Baro and, to a lesser extent, the upper Tinkisso at Dabola (Figure 5e, but with very limited data) showed a decrease in the DC.

From period 2 to period 3:

-   except the Bafing at Daka Saidou, the three main 'Sudano-Sahelian' rivers (the Gambia at Mako, the Niger at Kouroussa, as well as the Faleme at Gourbassi) show a decrease in the DC between the dry 1970–1993 period and the following 1994–2019 period. The DC of Koliba Koumba River also decreased slowly after the drought.
-   for 8 out of the 16 basins, a decrease (3) or a stabilization (5) in the DC is observed after 1994 and the rainfall recovery. For the other eight stations, an increase in the DC is ongoing. The increase over the whole period is lower than 20% in three basins (Table 2C) and higher than 20% in five others (Table 2D).
-   some basins (Niandan, Milo, Gambia, Tominé) show a marked increase in the DC at the end of the last period: in all of these basins, depletion was higher during the 2005–2015 period than during the Great Drought period.

- Extreme values can be observed in the last years (e.g., in the Milo basin), and in 1983–1984 in the Faleme. As described by Olivry (1986), this happens when the higher value of depletion coefficient occurs when the river dries up completely. Consequently, the DC becomes in some cases the drying up coefficient.

General observations

- in some basins, the depletion coefficient can be different depending on the method selected or on the context.
- except the increase in the DC during the drought and decrease after the drought in the large basins, there are no general trends to detect, reflecting great variability across the basins.
- overall, contrary to previous observations by Descroix [5] and Descroix et al. [42], which were based on the behavior of the Upper Niger basin tributaries, there is not a general trend of increasing DC, which could demonstrate a general reduction in the soil water-holding capacity. This trend only appears in the Konkouré basin and in the Milo and Niandan basins.

In Table 2, the last column provides the evolution of forest cover by basin based on the Global Forest Change (GFC2019) methodology.

The day when the maximal depletion occurs seems to follow a general behavior, occurring earlier when the depletion coefficient increases (mostly during the drought), later during the first period (yet with numerous gaps in the data), and overall during the last period. This follows an intuitive pattern, with a depletion stage of the river starting later when discharges are abundant and earlier during the dry years. The exception is the case of the Konkouré basin, where the upper basin (Pont de Linsan station) is the one wherein the discharge has decreased the most. The same evolution is shown at the downstream station "Pont de la route de Télimélé", but this station has been influenced since 1999 by the filling and management of the Garafiri dam. This is why its DC decreased strongly after 1999.

5.2. Evolution of Runoff Coefficients

Table 3 gathers the basins' hydrological characteristics and their changes over the three determined periods: discharge (Q), specific discharge (SD), runoff coefficient (RC), as well as the maximal discharge of the year and its day of occurrence.

Table 3. Hydrological characteristics of the 16 classified basins.

B	Class B			Discharge $m^3 \cdot s^{-1}$			Specific Discharge $l.s^{-1} \cdot km^{-2}$			Runoff Coefficient			Max Daily Discharge/+ Day		
	Station	Basin	Area km^2	47–70	71–93	94–2019	47–70	71–93	94–2019	47–70	71–93	94–19	47–70	71–93	94–2019
Senegal	Daka Saidou	Bafing	15700	306.6	182.4	225.9	19.5	11.6	14.4	0.390	0.264	0.310	1750/16 September	1074/29 August	1363/4 September
Senegal	Sokotoro	Up Bafing	1750		29.6			16.9			0.354			136/5 September	
Niger	Tinkisso	Tinkisso	6370	89.0	55.0	87.1	14.0	8.6	13.7	0.296	0.197	0.308	337/31 August	244/19 September	332/20 September
Niger	Dabola	Up Tinkisso	1260	14.9	14.1	14.5	11.8	11.2	11.5	0.244	0.226	0.242	67.1/4 September	65.1/5 September	64.2/21 September
Niger	Kouroussa	Niger	16560	256.0	165.0	181	15.5	10.0	10.9	0.286	0.201	0.212	1188/29 September	728/15 September	840/25 September
Niger	Baro	Niandan	12770	274.2	205.2	205.4	21.5	16.1	16.1	0.348	0.281	0.284	1192/21 September	865/13 September	922/7 September
Niger	Kankan	Milo	9620	206.4	149.4	144.9	21.0	15.5	15.1	0.349	0.284	0.262	771/18 September	665/10 September	712/19 September

A	Class A			Discharge $m^3 \cdot s^{-1}$			Specific Discharge $l.s^{-1} \cdot km^{-2}$			Runoff Coefficient			Max Daily Discharge/+ Day		
	Station	Basin	Area km^2	47–70	71–93	94–2019	47–70	71–93	94–2019	47–70	71–93	94–2019	47–70	71–93	94–2019
Senegal	Gourbassi	Faleme	17100	166.5	64.8	100.4	9.7	3.8	5.9	0.225	0.102	0.149	1322/1 January	746/6 September	977/10 September
Gambia	Mako	Gambia	10540		79.3	116.6		7.6	11.2		0.184	0.247		699/6 September	922/10 September
Gambia	Missirah	Koulountou	6200		27.3	32.9		4.4	5.3		0.118	0.131		170/17 September	209/11 September

C	Class C			Discharge $m^3 \cdot s^{-1}$			Specific Discharge $l \cdot s^{-1} \cdot km^{-2}$			Runoff Coefficient			Max Daily Discharge/+ Day		
	Station	Basin	Area km^2	47–70	71–93	94–2019	47–70	71–93	94–2019	47–70	71–93	94–2019	47–70	71–93	94–2019
Konkouré	Pt de Linsan	Konkouré	402	17.5	13.1	12.2	43.5	32.5	30.5	0.655	0.528	0.476	113/20 August	88.9/26 August	93.3/6 September
Konkouré	Nianso	Kokoulo	2260		60.8	62.1		26.9	27.5		0.486	0.439		533/28 August	441/29 August
Konkouré	Kaba	Kakrima	3190		72.4	77.6		22.7	24.3		0.407	0.392		468/27 August	496/26 August
Konkouré	Pt Télimélé	Konkouré	10210	427.4	255.0	262.0	41.9	25.0	25.7	0.604	0.410	0.401	1971/19 August	1409/23 August	1422/5 September
Corubal	Gaoual T	Tominé	3300		98.1	124.2		29.7	37.6		0.471	0.529		565/30 August	642/13 September
Corubal	Gaoual K	Koumba	6200		174.1	201.5		28.5	33.0		0.500	0.533		879/6 September	972/8 September

Basins are classified according to the specific discharges (SD) and runoff coefficient (RC). Class A: SD below 12, RC below 0.2. Class B: SD ranging from 12 to 22, RC ranging from 0.2 to 0.4. Class C: SD above 22, RC above 0.4.

The general dynamic follows the rainfall trends with a decrease in streamflow between the first period (1947–1969) and the second (1970–1993), followed by an increase in discharge between the second period and the third (1994–2019). Yet two important exceptions have to be highlighted:

- the Konkouré basin, where runoff did not rise again after the dry period, but in this basin, data is not available after 2002, preventing analysis of the long term and obscuring a possible recovery of streamflows. However, since rainfall increased significantly between period 2 and 3, a decrease in the runoff coefficient is noticed.
- the Upper Niger basin where runoff increased only in the Tinkisso basin (except for its upper part) over the last period.

This could be explained by a longer time being necessary to refill the natural water reservoir.

The maximum daily discharge observed each year shows approximately the same trends (Table 3, last column): the day when this maximum is observed is in most cases later when the discharge increases and earlier when streamflows are low.

5.3. Any Explanation from the LULCC?

The analysis of the relationship between, on the one hand, runoff and depletion coefficients and, on the other hand, land use/land cover changes (LULCC) during the period of 2000–2019 is made with the data provided by the 30 m Global Forest Change datasets [29] (Figures 7 and 8). Considering no local field validation has been done and the inconsistencies in some results, the forest evolution shown here must be considered as relative and not as absolute.

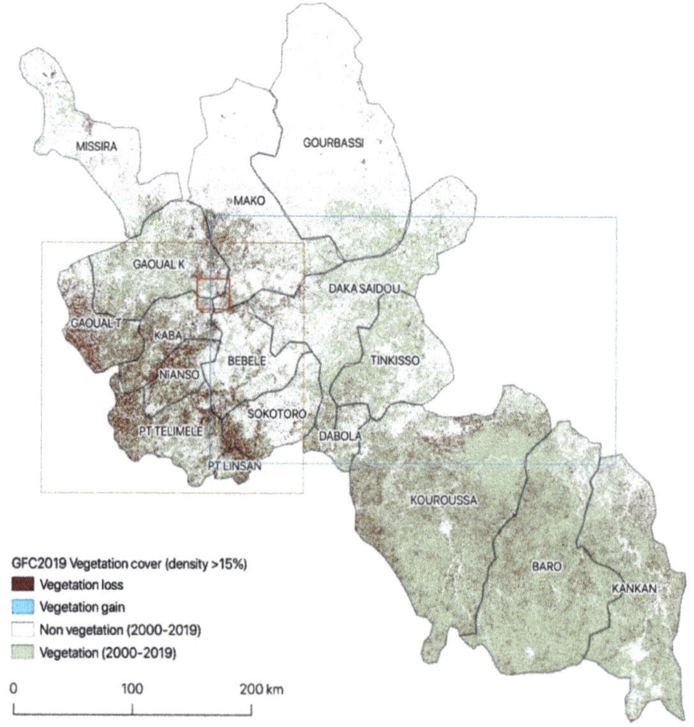

Figure 7. Vegetation (forest density higher than 15%) in 2000 and 2019 and vegetation loss and gain (GFC (Global Forest Change) product [29]), for the 16 basins. The empty brown rectangle is the area covered by the map in Figure 8.

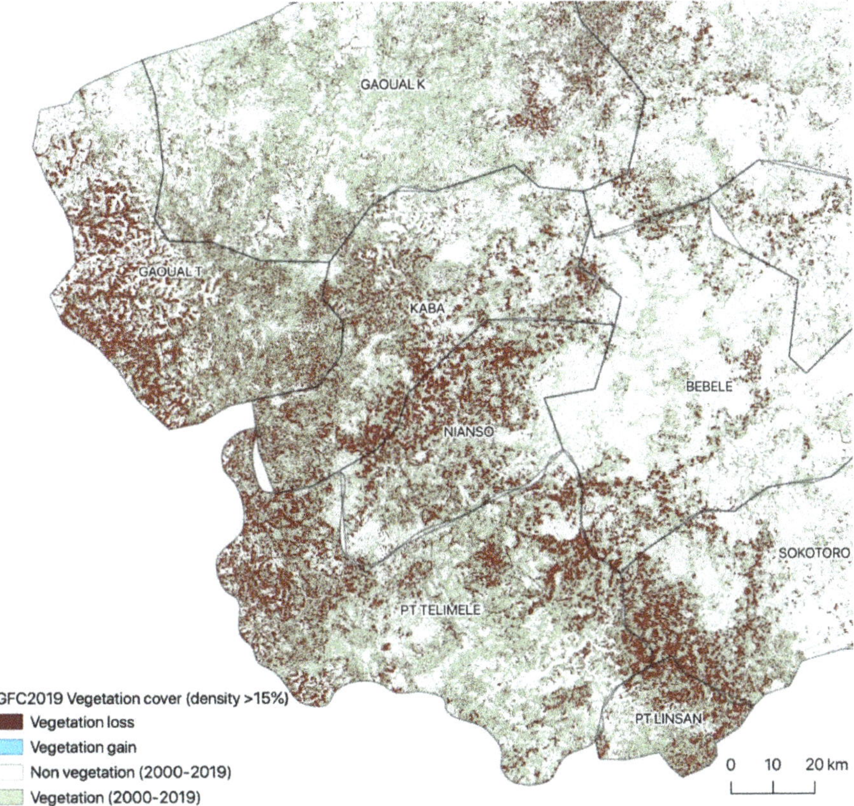

Figure 8. Vegetation (density higher than 15%) in 2000 and 2019 and vegetation losses and gains (2001-2019) (GFC product, [29]) for the western basins of the FD, where vegetation losses are the highest. The area represented is the one of the empty brown square in Figure 7.

One of the most important trends is the higher deforestation rate measured in the Upper Niger basin (Kouroussa stream gauge) and the Konkouré basin (mostly in the basins of Pont de Linsan and Pont de la Route de Télimélé stations, secondarily Kaba and Nianso stations). These basins are the ones where:

- RC paradoxically decreases or remains low during the last period, which is characterized by rainfall recovery, possibly because natural reservoirs continue to be filled. Can this hypothesis however still be valid after nearly 25 years of rainfall recovery?
- DC increased strongly (downstream Konkouré basin), moderately (Kaba and Nianso stations), or inversely, and remained unchanged (Upper Niger at Kouroussa station).
- in Table 3, the basins where DC increased more than 20% are clearly not the ones where the deforestation rate was the highest (except the small Pt de Linsan–Upper Konkouré River basin).

6. Analysis of LULCC

6.1. LULCC at the Regional Scale

GFC2019 rasters reveal that the whole south area of the FD has a high value of vegetation cover; only the Upper Bafing basin is extensively deforested, corresponding to the most densely populated area: this is the extension of the 18th and 19th century Peul/Fulani confederation of small theocratic states which dominated West Africa. Deforestation in recent decades on the western side of the FD is not necessarily due to the extension of this high density of the upper plateaus area. It is more likely due to wood-charcoal extraction/forest exploitation. A very low number of blue pixels appear in Figures 7 and 8, reflecting the low vegetation gain detected by GFC2019, which is partly due to methodological reasons of vegetation density and vegetation height. The Bebele basin (Tene River at Bebele stream gauge station), which is presented in Figures 7 and 8, is not considered in this analysis, due to its very limited hydrological data set.

The northern areas on Figure 7 correspond to the driest Sudanian zone, covered with bushes, savannah, and low forest cover. Forest vegetation density falls below 15% here, and this limited vegetation is excluded by the vegetation cover threshold employed here.

6.2. LULCC at the Upper Bafing Basin Scale

Figure 9 is a map of the Mann-Kendall tau after applying, for each pixel, the P value test at 0.001 and classifying in three categories: significant positive trend, stable or not significant trend, and significant negative trend.

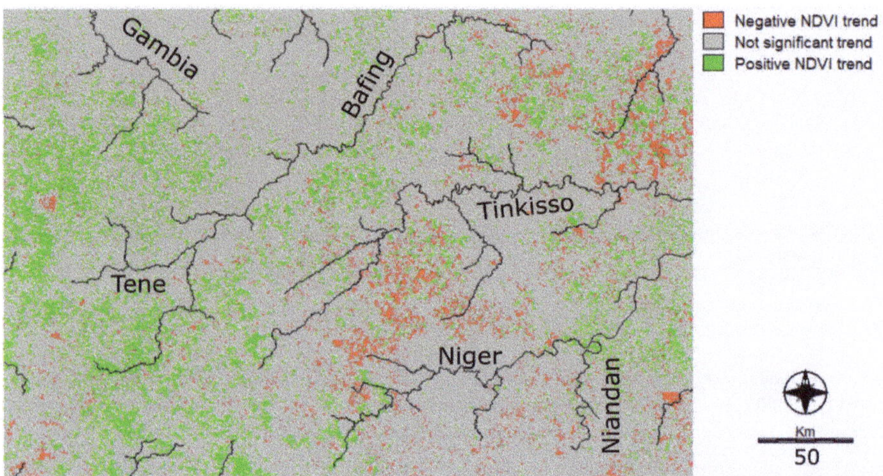

Figure 9. Evolution of vegetation cover (NDVI—normalized difference vegetation index) in the Upper Bafing Basin from February 2000 to February 2020. The area represented is the one of the empty blue rectangle in Figure 7.

Eighty-six percent of the zone covered by the map (Figure 9) has shown stable values of NDVI from 2000 to 2020 or non-significant trends according to the P value test at 0.001, and 11% of the zone has shown a significant positive trend (greening; biomass increase) for the same period. Only 3% has shown a significant negative trend (browning; biomass decrease).

Unlike the whole Fouta Djalon region, the Upper Bafing area is not concerned by important changes in its vegetation over the last 20 years, and the main change is a positive one for vegetation cover, even if it is hard to say without field validation if it corresponds to higher vegetation cover, higher biomass (either for spontaneous vegetation or for crops), or simple betterment of vegetative activities (i.e., phenology with longer green season).

The basins are not very different regarding this analysis. Stability ranges from 84.1% in the Bafing basin to 90.5% in the Niger basin (Figure 10). Negative trends range from 0.4% in the Gambia basin to 4.4% in the Tinkisso basin. Positive trends range from 6.3% to 14.6% in the Niger and Bafing basins, respectively.

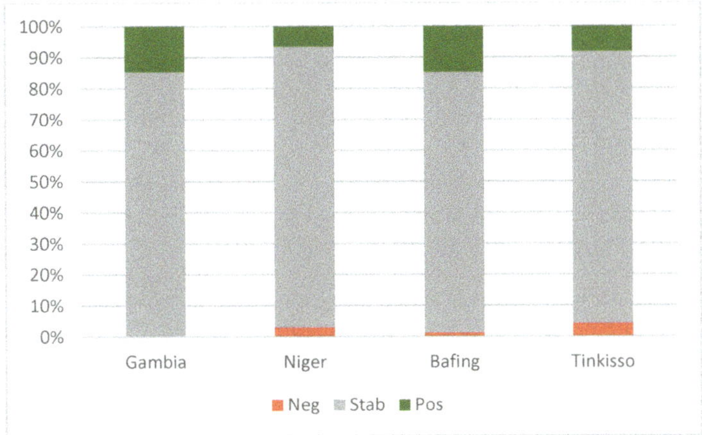

Figure 10. Part of the vegetation trends on the four main basins in the Bafing River basin.

6.3. LULCC at the Sources Scale

The comparison between the 1953 aerial picture (IGN—see Section 4.3 aerial mission in Guinea) and the Google Earth satellite scene (2019) does not demonstrate any apparent deforestations (Figure 11). The quality of both products being very different, it is difficult to do a classical numerical comparison, but the current situation (2019, on the right) seems a priori to be at least as forested as the one in 1953 (Figure 11). This area is named here the "three sources area": effectively, the two creeks (brooks) flowing through the north are the main sources of the Kouba (Koumba) River and the main upper reach of the Koliba River, which becomes the Rio Corubal in Guinea Bissau. The one flowing eastward is the Dima, the upper Gambia River. 'Ore Dima' means "the sources of the Dima River" in Pulaar language (Ore is the head). The other two rivers flowing southwards and the one flowing through the west are the main sources of the Kakrima River, the main tributary of the Konkouré River. This area is mostly composed by a doleritic bedrock.

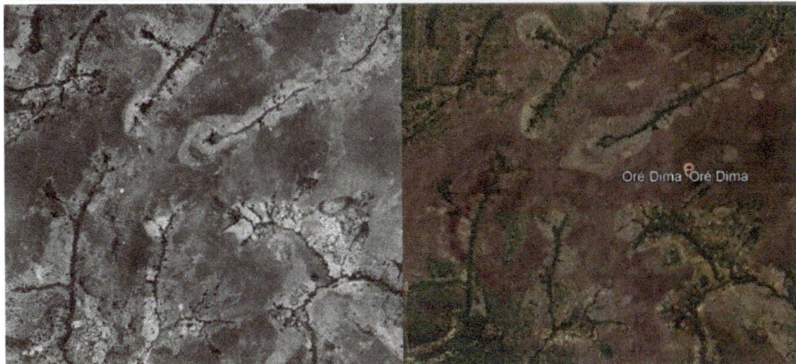

Figure 11. The so-called "three sources" area (15 km northward from Labé, the main city of the Fouta Djallon) in 1953 (IGN, **left**) and in 2019 (Google Earth©, **right**). No apparent deforestation. The extension of the picture is 5 km from the west to the east by 4.5 km from the north to the south. The area represented is the one of the red square in Figure 7.

Changes in vegetation cover evolution over time appear different at these three spatial scales but are difficult to compare considering the different analytical methodologies which are here applied across different areas, periods, and resolutions. However, clearly, some areas are being deforested, though results do not conclude that there is an overall deforestation, contrary to what is commonly described in the literature. Our three studies notably disagree in identifying negative trends in vegetation cover.

6.4. Is There Any Synthesis Provided by Data Analysis?

A PCA (Principal Component Analysis) is realized in order to provide clearer information about the variables and observations (basins) sampled in the FD. Table 4 details the variables considered in this analysis. The observations (sampling) are the 16 basins studied.

Table 4. List of the variables considered in the PCA analysis.

Q	Mean Yearly Discharge in m^3/s
QS	specific discharge ($l \cdot s^{-1} \cdot km^{-2}$)
QDX	maximum daily discharge
DC4	depletion coef in 4 months
ED41	DC4 evolution period 1 to per 2
ED42	DC4 evolution period 2 to per 3
DC1	depletion coef in 1 month
ED11	DC1 evolution period 1 to per 2
ED12	DC1 evolution period 2 to per 3
RAD	rainfall depth
RUD	runoff depth
RC	runoff coefficient
ER1	RC evolution period 1 to per 2
ER2	RC evolution period 2 to per 3
A	area
DEFO	% deforestation 2000–2020

No geological data are accounted for in this study, because it appears that there are no significant differences in the hydrological functioning of the different outcrops in the FD.

In both Figures 12 and 13:

- axis 1 (component 1) explains 46.4% of the total variance; it is mostly provided by QS, RAD, RUD, and RC (The definition of all abbreviations is given in Table 4).
- axis 2 (component 2) explains 23.8% of the variance, provided by Q, QDX, and A.
- axis 3 (component 3) explains 11.5% of the variance, using DC4, ED41, ED11, and ER1.
- axis 4 (component 4) explains 7.6% of the variance, provided by ED42 and ED12.

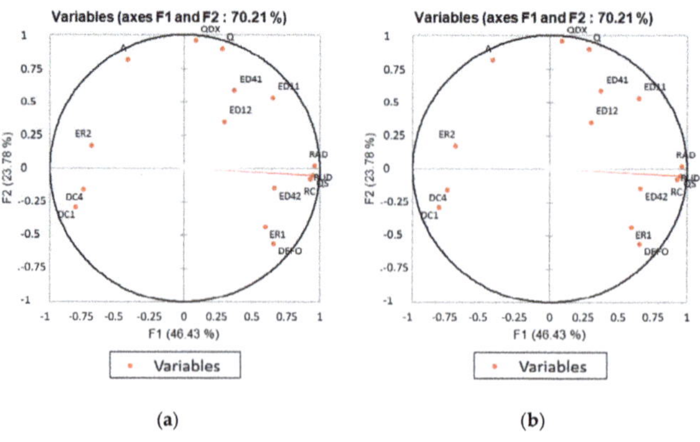

Figure 12. Variables spaces of the PCA (Principal Component Analysis) analysis; (**a**) components 1 and 2; (**b**) components 3 and 4.

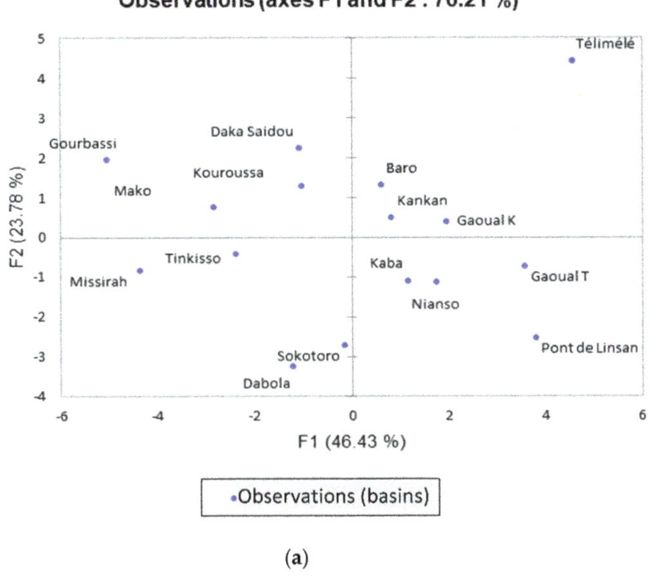

(**a**)

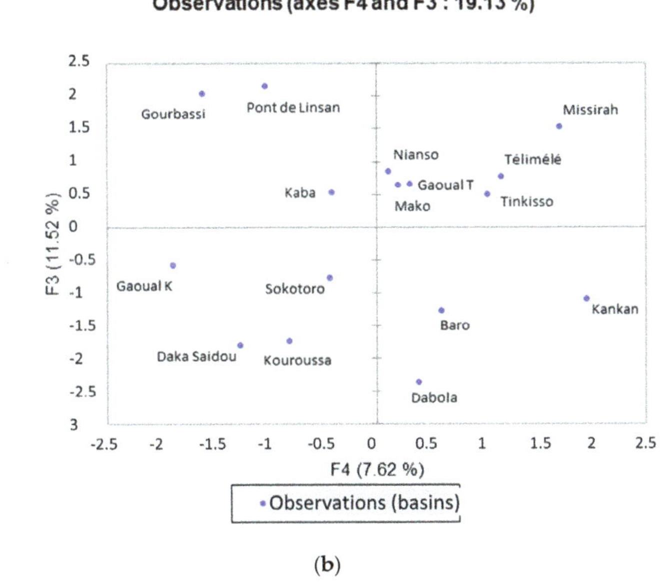

(**b**)

Figure 13. Observations spaces of the PCA analysis; (**a**): components 1 and 2; (**b**) components 3 and 4.

Figure 12a shows some redundancy of, on the one hand, hydrological variables (RAD, RUD, QS, RC) and, on the other, the depletion coefficient, separated from their evolution in two different clusters. They are, by contrast, gathered according to axis 3 and axis 4 (Figure 12b). The LULCC is considered here through the regional mapping. Although its results are different (and locally contradictory) with the ones provided by the other LULCC analyses, we considered these to be suitable as relative values rather than as absolute values. The LULCC is also then determined with the same data and methods for the 16 basins. Deforestation as shown in Figure 7 is correlated with the increase in runoff between period 1 and 2, but negatively correlated with the increase in runoff between period 2 and 3. Relative deforestation is clearly correlated with the increase in the DC (depletion coefficient) in four months during the second period, the one when rainfall recovery should have allowed a significant regreening. However, it is not so clearly correlated with the DC in one month (Figure 12a); yet following components 3 and 4 (Figure 12b), they seem more clearly correlated.

Figure 13 shows firstly a great dispersion of samples (basins) in all the statistical spaces, as evidence of their great geographical and hydrological diversity. Figure 13a seems logically dominated by runoff, and then, the basins of the western slope of the FDss and the ones coming off the Guinean Dorsale are clearly opposed to the ones of the northern part, where streamflow is weaker. No general trend appears in Figure 13b.

7. Discussion

The following points must be highlighted:

7.1. The DC

A strong interannual increase in the DC (see also [5,42]) is observed for the two main tributaries of the Upper Niger River from 1995. However, this is not observed in the Upper Niger (Kouroussa) nor in the Tinkisso River (Figure 6e). Otherwise, the DC, according to the three durations tested here, can be similar (e.g., in the Niandan River) or very different (e.g., in the Milo River)

A great deal was devoted to the DC during the 1990s and the 2000s by Mahé et al. [43,44] (1997 and 2001) and Olivry [18,19,21] (1987, 1993, 2002), as well as Bricquet et al. [20] (1997), Bamba et al. [45] (1996), and Sangaré et al. [46] (Figure 14). The latter showed that the DC increased from the beginning of the 1970s in all the basins, even if more slowly for the Sankarani River (one of the other main tributaries of the Upper Niger River, not investigated here). The exception was then the Milo River, the one whose DC has further increased in recent years. The DC of the Milo River only started its increase at the beginning of the 1980s [39]. A possible role of the basins' water-holding capacity remains to be investigated because the geological context is probably more complex than the schematic regionalization in Figure 2. The Upper Niger River basin is mostly a granitic basement one, however, Brunet Moret et al. (1986) [47] showed that there are sometimes sandstone outcrops in the lower basins of the Milo River. This could have served as a better water reservoir than granite and thus reduced the impact of the first drought spell in the 1970s. Another explanation can be provided by the rainfall's variation coefficient, which is lower in the Upper Milo basin than in most places in West Africa, as was noticed by Mahé et al. [44], most probably due to the specific relief orientation in the area. The Upper Milo basin runs through elevated areas both in the South and in the East, which seem to play a role in the reduction of the interannual variability, as is also observed in Central Ghana [44]. As can be seen (Figure 14) the DC of the Sankarani river also showed a reduced variability compared to that of other Upper Niger River tributaries, during the Great Drought. In the Senegal River basin, Bodian et al. (2020) [48] observed that the runoff recovery after the Great Drought lasted for five or ten years and is more significant for yearly and maximum discharge and less for low waters, which remain very weak. The same authors concluded that the high value of the DC for the Faleme River at Gourbassi (0.62) at the beginning of the 1980s (Figure 6d) can be attributed to a quasi-total drying up of the streamflow. A similar observation can be made for the Milo River during the 2010s (Figure 6e), but the reason remains to be explained since neither rainfall evolution nor the LULCC are satisfactory explaining factors for this behavior.

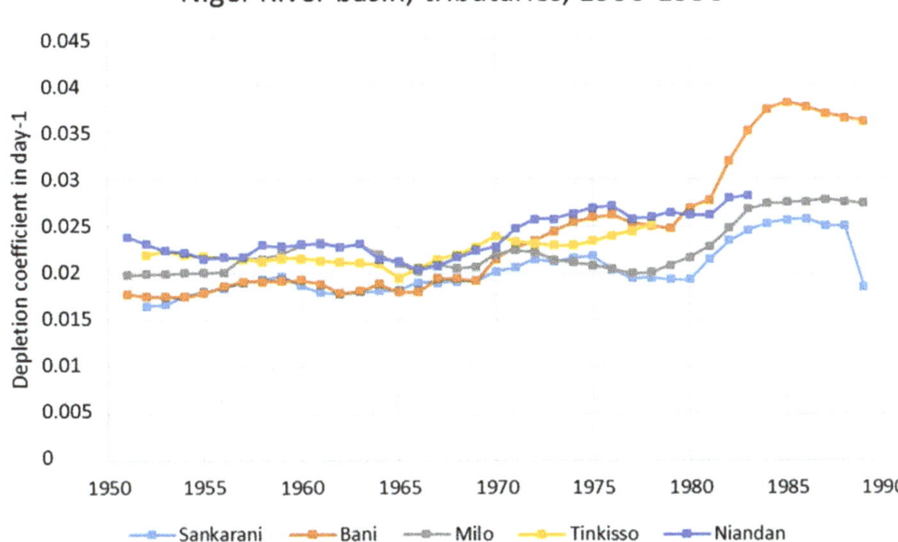

Figure 14. Evolution of DC in some Sudano-Guinean tributaries of the Niger River before and during the drought [41].

7.2. LULCC (Land Use/Land Cover Change)

As it has been shown above, vegetation cover in West Africa does not decline systematically though the population keeps on increasing at a rapid pace [5]. The Fouta Djallon is one of the best examples of this trend. A negative and Afro-pessimistic belief often links African deforestation with demographic growth. This stereotype anticipated catastrophic scenarios based on extensive peasant practices, which, against the background of strong demographic growth, should threaten the integrity of the Fouta Djallon ecosystems and therefore endanger the whole 'West African water tower'. In fact, local research has opened the debate on the reality and validity of such claims [49]. Brandt et al. (2017) [50] showed localized reductions in woodlands across West Africa but that farmland in Sahelian areas also promotes woody cover, contradicting simplistic ideas of a high negative correlation between population density and woody cover.

This strengthens the Boserupian point of view [51] (Boserup, 1965), increasingly accepted in the African scientific literature, as in Kenya [52], Niger [53], Northern Côte d'Ivoire [54], Northeastern Nigeria [55] or the whole Sudano-Sahelian strip [5,42]. This perspective is opposed to the catastrophist, Neo-Malthusian, and Afro-pessimistic general consensus of the second half of the 20th century.

Finally, it is worth noting that the 'tapades' traditional intensive gardening and tree cropping practices of the Fouta Djallon, which have played the role of man-made sponges though they have mostly been managed by women, are largely being abandoned. These practices should be rehabilitated in order to promote a sustainable agricultural intensification as the most useful way to infiltrate and store water in a catchment [5].

7.3. Some Hydrological Consequences

Except during the drought, the increase in the DC has only been observed in a few basins in recent years (the Konkouré and southern tributaries of the Upper Niger). The partial recovery in the RC (runoff coefficient), as well as the observed increase—at the regional scale—of the occurrence of extreme rainfall events [36,42,56,57], has consequences for flooding frequency and intensity.

This increases streamflow irregularities, accentuating low waters as well as floods. This could have impacts on hydraulic and civil engineering equipment and infrastructures. As illustrated in Figures 15 and 16, the bridges are threatened by some of these extreme flooding events.

Figure 15. Bridge of the Koundara-Youkounkoun road, northern Guinea, on the Koulountou River. Built in the 1950s, it was flooded by the streamflow in August 2018, depositing whole trees (bottom) and wood debris (on the bridge deck).

Figure 16. Pieces of branches blocked under the bridge deck of the new bridge of Komba on the Komba River (Upper Koliba/Corubal basin) after the flood of 2018. On the small picture: the new (2012) and, in the back, the ancient bridge (1950s).

However, as this paper attempted to demonstrate, it is not necessarily the sign of a decrease in soil water-holding capacity attributable to LULCC. Depletion coefficients of the main rivers providing freshwater resources to the Sahel are decreasing due to the rainfall recovery. West African agro-ecosystems seem to be developing resilience, which could invert the observed trend and constitute rural resiliency models and sustainable development practices.

8. Conclusions

The Fouta Djallon seems to have suffered from a reduction in water-holding capacity during the drought (1970s and 1980s). However, after this dry period, in some basins and mostly in those flowing towards Sahelian areas, the expected decrease in the DC due to rainfall recovery is observed (the Niger at Kouroussa, the Falémé at Gourbassi, probably the Gambia and Corubal basin rivers). The Bafing at Daka Saidou did not show any increases in DC during the drought.

In the Upper Niger basin (the Milo and Niandan rivers), the Konkouré, and partially the Koliba/Tominé basins, a significant increase in the DC was noticed during the end of the recent 1994-2019 period. However, in the Upper Niger basin, the DC increased more in the basins where relative deforestation was lower (Milo and Niandan) and decreased slightly in the basin where relative deforestation was stronger (Upper Niger at Kouroussa). Following the PCA analysis, only the increase in the DC in four months between period 2 and period 3 is clearly correlated with relative deforestation. Besides the well-known and well-studied increase in rainfall intensity (increase in extreme events occurrence and increase in rainfall hourly mean intensity), the LULCC could be a discriminant factor of any increase in runoff and depletion coefficients, particularly in the Konkouré basin where deforestation remains an important dynamic. The PCA analysis only discriminates the southwestern basins (the Konkouré and Koliba/Tominé basins).

Contrary to popular belief, there is evidence of successful rural intensification practices in West Africa, which are generally characterized by an increasing vegetation cover and, locally, increasing tree cover. This kind of landscape should logically improve the soil and basins' water-holding capacity and eventually lead to a reduction in the DC (depletion coefficient). In the core of the FD (i.e., the FDss), the 'tapades' intensive system is disappearing, yet no deforestation is observed due to several combined patterns.

Since the sole increase in extreme rainfall events is unlikely to be responsible for the increase in the DC, LULCC seems to be a factor in the reduction in the basin water-holding capacity. However, it is statistically apparent only in the Konkouré and in the Koliba/Tominé basins, where annual rainfall is higher and where the runoff coefficients are significantly higher than in all the other basins, making their susceptibility to changes in runoff conditions more influential than in other basins.

The Fouta Djallon, as the West African water tower, does not appear as threatened. However, there is a stark difference between, on the one hand, the southwestern basins (the Konkouré and Upper Corubal), where LULCC seems to impact the basin water-holding capacity, and, on the other hand, basins such as the Gambia, Senegal, and Niger basins, which have been impacted more by the Great Drought but which are at least partially recovering their natural water storage capacity. This is an optimistic observation for the future of the West African Sahel, whose basins depend the most on the Fouta Djallon for its supply of fresh water resources.

Supplementary Materials: The following are available online at http://www.mdpi.com/2073-4441/12/11/2968/s1. Excel 1: 16 considered basins and the availability of their hydrological data. Figure S1a: Evolution of depletion coefficient, Konkouré River basin. Figure S1b: Evolution of rainfall and runoff coefficient, Konkouré River basin. Figure S2a: Evolution of depletion coefficient, Upper Konkouré River basin. Figure S2b: Evolution of rainfall and runoff coefficient, Konkouré River basin. Figure S3a: Evolution of depletion coefficient, Kokoulo River basin. Figure S3b: Evolution of rainfall and runoff coefficient, Kokoulo River basin. Figure S4a: Evolution of depletion coefficient, Kakrima River basin. Figure S4b: Evolution of rainfall and runoff coefficient, Kakrima River basin. Figure S5a: Evolution of depletion coefficient, Tominé River basin. Figure S5b: Evolution of rainfall and runoff coefficient, Tominé River basin. Figure S6a: Evolution of depletion coefficient, Koliba/Koumba River basin. Figure S6b: Evolution of rainfall and runoff coefficient, Koliba/KoumbaRiver basin. Figure S7a: Evolution of depletion coefficient, Koulountou River basin. Figure S7b: Evolution of rainfall and runoff coefficient, Koulountou

River basin. Figure S8a: Evolution of depletion coefficient, Gambia River basin. Figure S8b: Evolution of rainfall and runoff coefficient, Gambia River basin. Figure S9a: Evolution of depletion coefficient, Faleme River basin. Figure S9b: Evolution of rainfall and runoff coefficient, Faleme River basin. Figure S10a: Evolution of depletion coefficient, Upper Bafing River basin. Figure S10b: Evolution of rainfall and runoff coefficient, Upper Bafing River basin. Figure S11a: Evolution of depletion coefficient, Bafing River basin Figure S11b: Evolution of rainfall and runoff coefficient, Bafing River basin. Figure S12a: Evolution of depletion coefficient, Upper Tinkisso River basin. Figure S12b: Evolution of rainfall and runoff coefficient, Upper Tinkisso River basin. Figure S13a: Evolution of depletion coefficient, Tinkisso River basin. Figure S13b: Evolution of rainfall and runoff coefficient, Tinkisso River basin. Figure S14a: Evolution of depletion coefficient, Upper Niger River basin. Figure S14b: Evolution of rainfall and runoff coefficient, Upper Niger River basin. Figure S15a: Evolution of depletion coefficient, Niandan River basin. Figure S15b: Evolution of rainfall and runoff coefficient, Niandan River basin. Figure S16a: Evolution of depletion coefficient, Milo River basin. Figure S16b: Evolution of rainfall and runoff coefficient, Milo River basin.

Author Contributions: Conceptualization, L.D., B.F., S.P.M. and A.B.D.; data curation, B.F., S.P.M., A.B.D., S.S. (Safietou Soumaré), F.B.S.D., A.D., K.D. (Kadiatou Diallo), J.A. (Julien Andrieu), I.A., J.-C.B., A.B. (Aliou Barry), A.B. (Ansoumana Bodian), Y.B., N.B., J.-L.C., H.D., G.D., M.D., K.D. (Kourahoye Diallo), A.F., E.G., L.F., L.K., B.N.N., Y.S., S.S. (Saly Sambou) and B.A.T.; formal analysis, B.F., S.P.M., A.B.D., J.A. (Julien Andrieu), A.O., J.A. (Jean Albergel), and J.-C.B.; funding acquisition, J.A. (Julien Andrieu), I.A., J.A. (Jean Albergel), J.-C.B., A.B. (Aliou Barry), A.B. (Ansoumana Bodian), and Y.B.; investigation, S.P.M., A.B.D., S.S. (Safietou Soumaré), B.F., F.B.S.D., A.D., K.D. (Kadiatou Diallo), and J.-P.V.; methodology, A.B.D., S.S. (Safietou Soumaré), J.A., G.M., A.O., D.O., and B.F.; project administration, L.D., J.A. (Julien Andrieu), G.M., J.A. (Jean Albergel), A.O., and D.O.; resources, F.B.S.D., A.D., and K.D. (Kadiatou Diallo); software, B.F., F.B.S.D., and K.D. (Kourahoye Diallo); supervision, J.-P.V., L.A.L., J.-C.O. and L.D.; validation, G.M., J.A. (Julien Andrieu), A.O., D.O., L.A.L., and J.-P.V.; visualization, F.B.S.D., A.D., and K.D. (Kadiatou Diallo); writing—original draft, L.D., B.F., S.P.M., and A.B.D.; writing—review and editing, L.D., B.F., S.P.M., and A.B.D.; G.M., J.A. (Julien Andrieu), A.O., J.-C.B., D.O., L.A.L., and J.-P.V. All authors have read and agreed to the published version of the manuscript.

Funding: This research was funded by the European Union, via AICS (Agenzia Italiana de Cooperazione per il Sviluppo, grant number ENV/2017/383-744/AP n) 01/2019/WEFE SENEGAL for the OMVS, Senegal River Basin Agency.

Acknowledgments: The authors warmly acknowledge the Rivers Basins Agency (NBA, OMVS and OMVG) as well as the Direction of Hydraulics of Guinea for providing the daily discharge data, and the SIEREM and AMMA Catch databases as well as the ANACIM (Senegalese Met Office) for the daily rainfall data. We kindly acknowledge the reviewers who allowed us to propose a deeply improved version of the manuscript.

Conflicts of Interest: The authors declare no conflict of interest.

References

1. L'Hôte, J.-P.; Mahé, G. *Afrique Centrale et de l'Ouest: Précipitations Moyennes Annuelles (Période 1951–1989)*; IRD (Orstom): Bondy, France, 1996; 1 map.
2. Aurouet, A.; Ferry, L.; Cougny, G.; ECOWAS. *Atlas de l'eau du Massif du Fouta Djalon-Le Château d'eau de l'Afrique de l'Ouest/Fouta Djallon Highland-Water Atlas*; CEDEAO (ECOWAS), IWA, Anteagroup, The World Bank: Washington, DC, USA, 2017; p. 114. Available online: http://documents.banquemondiale.org/curated/fr/844511520436091243/pdf/123986-WP-P150210-FRENCH-PUBLIC.pdf (accessed on 11 April 2020).
3. Gebrehiwot, S.G.; Gardenas, A.I.; Bewket, W.; Seibert, J.; Ilstedt, U.; Bishop, K. The long-term hydrology of East Africa's water tower: Statistical change detection in the watersheds of the Abbay Basin. *Reg. Environ. Chang.* **2013**. [CrossRef]
4. Blanc, P. De l'Egypte à l'Ethiopie, quand la puissance se déplace en Afrique Nilotique. In *Confluences Méditerranée*; L'Harmattan: Paris, France, 2014; ISSN 1148-2664; ISBN 9782343045627. Available online: https://www.cairn.info/revue-confluences-mediterranee-2014-3-page-123.htm (accessed on 21 June 2020).
5. Descroix, L. *Processus et Enjeux d'eau en Afrique de l'Ouest Sahélo-Soudanienne*; Editions Des Archives Contemporaines: Paris, France, 2018; 320p, ISBN 9782813003140. [CrossRef]
6. Boulvert, Y. Approche Synthétique des Aplanissements Cuirassés de Centrafrique et de Guinée (Conakry). 2005. Available online: http://horizon.documentation.ird.fr/exl-doc/pleins_textes/divers19-10/010038259.pdf (accessed on 18 April 2020).
7. Mamedov, V.I.; Boufeev, Y.V.; Nikitine, Y.A. *Geology of the Republic of Guinea*; Geoprospects and State University Lomonossov of Moscow: Moscow, Russia, 2010; 326p.

8.   Boulvert, Y. Repères Historiques sur la Caractérisation du Cuirassement et de la Géomorphologie Guinéenne. Conférence à l'Université de Conakry, Conakry (GIN), 1998/04. Available online: https://numerisud.ird.fr/documents-et-films/publications/Reperes-historiques-sur-la-caracterisation-du-cuirassement-et-de-la-geomorphologie-guineenne (accessed on 18 April 2020).

9.   Orange, D. Hydroclimatologie du Fouta Djallon et Dynamique Actuelle d'un Vieux Paysage Latéritique. Ph.D. Thesis, Université Louis Pasteur de Strasbourg, Strasbourg, France, 1990; 232p. Available online: https://www.persee.fr/doc/sgeol_0302-2684_1992_mon_93_ (accessed on 30 October 2018).

10. Orange, D.; Gac, J.-Y. Reconnaissance géochimique des eaux du Fouta Djalon (Guinée), flux de matières dissoutes et en suspension en Haute-Gambie. *Géodynamique* **1990**, *5*, 35–49.

11. Michel, P. *Les Bassins des Fleuves Sénégal et Gambie, Etude Géomorphologique*; ORSTOM: Bondy, France, 1973; 810p.

12. D'Aubreville, A. Erosion et "bovalisation" en Afrique noire française. *L'Agron. Trop.* **1947**, *2*, 339–357.

13. Maignien, R. Compte-rendu de recherches sur *les* latérites. In *Colloquium on Laterites, Madagascar 1964*; Reports on Natural Ressources, n°4; UNESCO: Paris, France, 1966; 155p.

14. Richard-Molard, J. Les traits d'ensemble du Fouta-Djalon. *Rev. Géograph. Alp.* **1943**, *XXXI*, 199–213. [CrossRef]

15. Pouquet, J. Le plateau de Labé (Guinée française, A.O.F). Remarques sur le caractère dramatique des phénomènes d'érosion des sols et sur les remèdes proposés. *Bull. l'IFAN (Inst. Fr. d'Afr. Noire)* **1956**, *18*, 1–16.

16. Pouquet, J. Quelques types d'évolution du relief en Guinée française: Processus hydrographiques et phénomènes de cuirassement sur les hautes surfaces du Fouta-Dialon (AO.F.). In Proceedings of the Eighteenth Internacional Congress of Geography, Rio de Janeiro, Brazil, 9–18 August 1956; pp. 350–361.

17. Sudres, A. La dégradation des sols au Foutah Djalon. *L'Agron. Trop.* **1947**, *I*, 227–246.

18. Olivry, J.-C. *Synthèse des Connaissances Hydrologiques et Potentiel en Ressources en Eau du Fleuve Niger*; Provisional Report; World Bank, Niger Basin Authority: Niamey, Niger, 2002; 160p.

19. Olivry, J.-C. Les conséquences durables de la sécheresse actuelle sur l'écoulement du fleuve Sénégal et l'hypersalinisation de la Basse-Casamance. In *The Influence of Climate Change and Climatic Variability on the Hydrologic Regime and Water Resources; Proceedings of the Vancouver Symposium*; IAHS Publ. No. 168: Wallingford, UK, 1987; 12p.

20. Bricquet, J.-P.; Bamba, F.; Mahé, G.; Touré, M.; Olivry, J.-C. Evolution récente des ressources en eau de l'Afrique atlantique. *Rev. Sci. l'Eau* **1997**, *3*, 321–337. [CrossRef]

21. Olivry, J.-C.; Bricquet, J.-P.; Mahé, G. Vers un appauvrissement durable des ressources en eau de l'Afrique humide? In *Hydrology of Warm Humid Regions. Proceedings of the 4th Assembly IAHS, Yokohama, Japan, 13–15 July 1993*; Gladwell, J.S., Ed.; Publication IAHS 216: Wallingford, UK, 1993; pp. 67–78.

22. Gomis, D.E.R. *Synthèse Hydrologique Du Fleuve GAMBIE en Amont de Gouloumbou*; Master Memory; UCAD: Dakar, Senegal, 2000; 166p.

23. Bodian, A.; Dacosta, H.; Dezetter, A. Analyse des débits de crues et d'étiages dans le bassin versant du fleuve sénégal en amont du barrage de Manantali. *Clim. Dev.* **2013**, *15*, 46–56.

24. Samoura, K. Contributions Méthodologiques à L'évaluation Environnementale Stratégique De L'exploitation du Potentiel Hydroélectrique des Bassins Côtiers en Milieu Tropical: Cas du Konkouré, en Guinée. Ph.D. Thesis, UQAM, Montreal, QC, Canada, 2011; 304p.

25. Tallaksen, L.M. A review of baseflow recession analysis. *J. Hydrol.* **1995**, *165*, 349–370. [CrossRef]

26. Boyer, J.F.; Dieulin, C.; Rouché, N.; Crès, A.; Servat, E.; Paturel, J.E.; Mahé, G. SIEREM: An environmental information system for water resources. In *Water Resource Variability: Hydrological Impacts. Proceedings of the 5th FRIEND World Conference, Havana, Cuba, November 2016*; IAHS Publ.: La Havana, Cuba, 2006; Volume 308, pp. 19–25.

27. Dieulin, C.; Mahé, G.; Paturel, J.-E.; Ejjiyar, S.; Tramblay, Y.; Rouché, N.; El Mansouri, B. A new 60-year 1940–1999 monthly gridded rainfall data set for Africa. *Water* **2019**, *11*, 387. [CrossRef]

28. Funk, C.; Peterson, P.; Landsfeld, M.; Pedreros, D.; Verdin, J.; Shukla, S.; Husak, G.; Rowland, J.; Harrison, L.; Hoell, A.; et al. The climate hazards infrared precipitation with stations—A new environmental record for monitoring extremes. *Sci. Data* **2015**, *2*, 150066. [CrossRef] [PubMed]

29. Hansen, M.C.; Potapov, P.V.; Moore, R.; Hancher, M.; Turubanova, S.A.; Tyukavina, A.; Thau, D.; Stehman, S.V.; Goetz, S.J.; Loveland, T.R.; et al. High-Resolution Global Maps of 21st-Century Forest Cover Change. *Science* **2013**, *342*, 850–853. [CrossRef] [PubMed]

30. Gorelick, N.; Hancher, M.; Dixon, M.; Ilyushchenko, S.; Thau, D.; Moore, R. Google Earth Engine: Planetary-scale geospatial analysis for everyone. *Remote Sens. Environ.* **2017**, *202*, 18–27. [CrossRef]

31. Gilbert, R.O. *Statistical Methods for Environmental Pollution Monitoring*; Van Nostrand Reinhold: New York, NY, USA, 1987; ISBN 0-442-23050-8.

32. San Emeterio, J.-L.; Alexandre, F.; Andrieu, J.; Génin, A.; Mering, C. Changements socio-environnementaux et dynamiques des paysages ruraux le long du gradient bioclimatique nord-sud dans le sud-ouest du Niger (régions de Tillabery et de Dosso. *VertigO Rev. Electron. Sci. L'Environ.* **2013**, *13*. Available online: http://journals.openedition.org/vertigo/14456 (accessed on 12 October 2015). [CrossRef]

33. Andrieu, J. Phenological analysis of the savanna-forest transition from 1981 to 2006 form Côte d'Ivoire to Benin with NDVI NOAA time series. *Eur. J. Remote Sens.* **2017**. [CrossRef]

34. Andrieu, J. Vegetation change analysis in Côte d'Ivoire during conflicts using a phenological metric and Kendall correlation of two NDVI time series (Text in French). *Tropicultura* **2018**, *36*, 258–270. Available online: http://www.tropicultura.org/text/v36n2/258.pdf (accessed on 9 November 2019).

35. Nicholson, S.E. On the question of the "recovery" of the rains in the West African Sahel. *J. Arid Environ.* **2005**, *63*, 615–641. [CrossRef]

36. Ali, A.; Lebel, T. The Sahelian standardized rainfall index revisited. *Int. J. Climatol.* **2009**, *29*, 1705–1714. [CrossRef]

37. Nicholson, S.E. The West African Sahel: A review of recent studies on the rainfall regime and its interannual variability. *Int. Sch. Res. Not.* **2013**, *2013*, 453251. [CrossRef]

38. Descroix, L.; Diongue Niang, A.; Panthou, G.; Bodian, A.; Sané, T.; Dacosta, H.; Malam Abdou, M.; Vandervaere, J.-P.; Quantin, G. Evolution Récente de la Mousson en Afrique de l'Ouest à Travers Deux Fenêtres (Sénégambie et Bassin du Niger Moyen). *Climatologie* **2015**, *12*, 25–43.

39. Ardoin-Bardin, S.; Dezetter, A.; Servat, E.; Paturel, J.-E.; Mahé, G.; Niel, H.; Dieulin, C. Using general circulation model outputs to assess impacts of climate change on runoff for large hydrological catchments in West Africa. *Hydrol. Sci. J.* **2009**, *54*, 77–89. [CrossRef]

40. Biasutti, M. Forced Sahel rainfall trends in the CMIP5 archive. *J. Geophys. Res. Atmos.* **2013**, *118*, 1613–1623. [CrossRef]

41. IPCC. *Climate Change Synthesis Report*; IPCC, WMO: Geneva, Switzerland, 2014; 167p.

42. Descroix, L.; Guichard, F.; Grippa, M.; Lambert, L.A.; Panthou, G.; Gal, L.; Dardel, C.; Quantin, G.; Kergoat, L.; Bouaïta, Y.; et al. Evolution of surface hydrology in the Sahelo-Sudanian stripe: An updated synthesis. *Water* **2018**, *10*, 748. [CrossRef]

43. Mahé, G.; Bamba, F.; Diabaté, M.; Diarra, A.; Diarra, M. The reduction of the water resources on upper basins of the Niger river: Hydrological balances and analysis of the depletion curves (1951–1989). Poster proceedings, Sustainability of water ressources under increasing uncertainty. In Proceedings of the 5th IAHS Assembly, Rabat, Maroc, 23 April–3 May 1997; pp. 9–12.

44. Mahé, G.; L'Hôte, Y.; Olivry, J.C.; Wotling, G. Trends and discontinuities in regional rainfall of west and central Africa–1951–1989. *Hydrol. Sci. J.* **2001**, *46*, 211–226. [CrossRef]

45. Bamba, F.; Mahé, G.; Bricquet, J.P.; Olivry, J.C. Changements climatiques et variabilité des ressources en eau des bassins du Haut Niger et de la Cuvette Lacustre. In *Réseaux Hydrométriques, Réseaux Télématiques, Réseaux Scientifiques: Nouveaux Visages de l'Hydrologie Régionale en Afrique*; Fritsch, J.M., Paturel, J.E., Servat, E., Eds.; XIIèmes Journées Hydrologiques de l'ORSTOM: Montpellier, France, 1996; 26p.

46. Sangaré, S.; Mahé, G.; Paturel, J.-E.; Bangoura, Y. Bilan hydrologique du fleuve Niger en Guinée de 1950 à 2000. *Sud Sciences et Technologies* **2002**, *9*, 21–33. Available online: https://horizon.documentation.ird.fr/exl-doc/pleins_textes/divers16-04/010034319.pdf (accessed on 18 June 2017).

47. Brunet Moret, Y.; Chaperon, P.; Lamagat, J.-P.; Molinier, M. *Monographie Hydrologique du Fleuve Niger*; Orstom: Paris, France, 1986; 800p.

48. Bodian, A.; Diop, L.; Panthou, G.; Dacosta, H.; Deme, A.; Dezetter, A.; Ndiaye, P.M.; Diouf, I.; Vischel, T. Recent Trend in Hydroclimatic Conditions in the Senegal River Basin. *Water* **2020**, *12*, 436. [CrossRef]

49. André, V.; Pestaña, G. Les visages du Fouta-Djalon. *Les Cah. d'Outre-Mer* **2002**, *217*. Available online: http://com.revues.org/index1038.html (accessed on 11 September 2014). [CrossRef]

50. Brandt, M.; Rasmussen, K.; Peñuelas, J.; Tian, F.; Schurgers, G.; Verger, A.; Mertz, O.; Palmer, J.R.B.; Fensholt, R. Human population growth offsets climate-driven increase in woody vegetation in sub-Saharan Africa. *Nat. Ecol. Evol.* **2017**, *1*, 0081. [CrossRef]

51. Boserup, E. *The Conditions of Agricultural Growth: The Economics of Agrarian Change under Population Pressure*; (Republished 1993: Earthscan Publications: London, UK); Allen and Unwin: London, UK, 1965.

52. Tiffen, M.; Mortimore, M.; Gichuki, F. *More People, Less Erosion: Environmental Recovery in Kenya*; John Wiley & Sons: London, UK, 1994; 311p.

53. Luxereau, A.; Roussel, B. *Changements Ecologiques et Sociaux au Niger*; Etudes Africaines; L'Harmattan Ed: Paris, France, 1997; 239p.

54. Demont, M.; Jouve, P. *Evolution d'Agro-Systèmes Villageois Dans la Region de Korhogo (ord Côte d'Ivoire): Boserup vs. Malthus, Opposition ou Complémentarité? Dynamiques Agraires et Construction Sociale Du Territoire*; Séminaire CNEARC-UTM: Montpellier, France, 2000; pp. 93–108.

55. Mortimore, M.J.; Adams, W.M. Farmer adaptation, change and &crisis' in the Sahel. *Glob. Environ. Chang.* **2001**, *11*, 49–57.

56. Panthou, G.; Vischel, T.; Lebel, T. Recent trends in the regime of extreme rainfall in the Central Sahel. *Int. J. Climatol.* **2014**, *34*, 3998–4006. [CrossRef]

57. Panthou, G.; Lebel, T.; Vischel, T.; Quantin, G.; Sané, Y.; Ba, A.; Ndiaye, O.; Diongue-Niang, A.; Diop Kane, M. Rainfall intensification in tropical semi-arid regions: The Sahelian case. *Environ. Res. Lett.* **2018**, *13*, 064013. [CrossRef]

Article

A Comparative Study of Statistical Methods for Daily Streamflow Estimation at Ungauged Basins in Turkey

Mustafa Utku Yilmaz [1,*] and Bihrat Onoz [2]

[1] Department of Civil Engineering, Kirklareli University, 39100 Kirklareli, Turkey
[2] Department of Civil Engineering, Istanbul Technical University, 34469 Istanbul, Turkey; onoz@itu.edu.tr
*   Correspondence: utkuyilmaz@klu.edu.tr; Tel.: +90-288-214-0514

Received: 20 November 2019; Accepted: 7 February 2020; Published: 9 February 2020

Abstract: In this study, a comparative evaluation of the statistical methods for daily streamflow estimation at ungauged basins is presented. The single donor station drainage area ratio (DAR) method, the multiple-donor stations drainage area ratio (MDAR) method, the inverse similarity weighted (ISW) method, and its variations with three different power parameters (1, 2, and 3) are applied to the two main subbasins of the Euphrates Basin in Turkey to estimate daily streamflow data. Each station in each basin is considered in turn as the target station where there are no streamflow data. The donor stations are selected based on the physical similarities between the donor and target stations. Then, streamflow data from the most physically similar donor station(s) is transferred to the target station using the statistical methods. In addition, the effect of data preprocessing on the estimation performance of the statistical methods is investigated. The preprocessing discussed in this study is streamflow data smoothing using the two-sided moving average (MA). Three statistical methods using the smoothed data by the MA, named as DAR-MA, MDAR-MA, and ISW-MA, are proposed. The estimation performance of the statistical methods is compared by using daily streamflow data with preprocessing and without preprocessing. The Nash–Sutcliffe efficiency (NSE), the ratio of the root mean square error (RMSE) to the standard deviation of the observed data (RSR), the percent bias (PBIAS), and the coefficient of determination (R^2) are used to evaluate the performance of the statistical methods. The results show that MDAR and ISW give improved performances compared to DAR to estimate daily streamflow for 7 out of 8 target stations in the Middle Euphrates Basin and for 4 out of 7 target stations in the Upper Euphrates Basin. Higher NSE values for both MDAR and ISW are mostly obtained with the three most physically similar donor stations in the Middle Euphrates Basin and with the two most physically similar donor stations in the Upper Euphrates Basin. The best statistical method for each target station exhibits slightly greater NSE when the smoothed data by the MA is used for all target stations in the Middle Euphrates Basin and for 6 out of 7 target stations in the Upper Euphrates Basin.

Keywords: data preprocessing; donor selection; drainage area ratio; Euphrates basin; moving average; physical similarity; streamflow estimation; ungauged basins

---

1. Introduction

In recent years, several factors, such as climate change, global warming, drought, population growth, and industrialization, have led to a rapid increase in demand for water. Hence, issues related to the planning and implementation of the water budget become important. Measurements and estimates of streamflow play an important role in the stage of the planning and implementation of the water budget. Since drainage basins in many parts of the world are ungauged or poorly gauged, the International Association of Hydrological Sciences (IAHS) launched a scientific decade from 2003 to 2012 on Predictions in Ungauged Basins (PUB) [1]. It was an effort to improve streamflow estimations for ungauged basins. Streamflow estimation at ungauged and poorly gauged basins is an important issue in growing economies countries such as Turkey because there are a limited number of stations in the streamflow gauging network of Turkey and streamflow estimates are often needed at ungauged basins where water resources projects are planned. Some stations in the river basins of Turkey contain large amounts of missing data during the observation period [2,3]. This lack of adequate data creates significant problems in the water resources projects for Turkey. For these reasons, accurate measurement and analysis of streamflow data and reliable streamflow estimates are needed.

Many methods have been used to improve the reliability and the accuracy in estimations for the development of streamflow estimation methods, and the research in this area still continues. In order to estimate streamflow, several researchers have suggested the artificial intelligence methods such as artificial neural networks [4–6], fuzzy logic [7–9], genetic programming [10–12], and machine learning [13–15]. In addition, artificial intelligence methods have been coupled with the data preprocessing methods to improve streamflow estimation accuracy and reliability in recent studies in the literature [16–19]. For example, Wu and Chau [20] used data preprocessing methods such as moving average (MA) and singular spectrum analysis (SSA) in order to improve the performance of artificial neural networks (ANN). Results showed that the MA was more effective than the SSA when they were coupled with the ANN. Moreover, ANN methods coupled with the MA performed the best among all methods. On the other hand, statistical methods such as interpolation by inverse distance weighted (IDW) [21,22] and kriging [23,24], regression analysis [25,26], flow duration curves [27,28], and information transfer methods [29,30] are widely used in the estimation of streamflow. This study focuses on statistical methods for improving estimation accuracy and reliability for ungauged basins.

Regionalization is a statistical process, which aims to estimate streamflow at ungauged basins. Various regional methods have been used for regional estimation of streamflow for the different time scales (i.e., daily, monthly, or annually) at ungauged basins in the literature [31–35]. Streamflow estimation at ungauged basins where streamflow data are not available requires the transfer of hydrologic information available at a donor station to the target station where only morphological and meteorological characteristics are available [36]. Drainage area ratio (DAR) method is one of the oldest information transfer methods for obtaining streamflow values at the target station from the donor station. This method is straightforward to apply and is in widespread use by hydrologists because it requires no additional information other than the streamflow values at the target station and the drainage areas of the donor and target stations. The DAR method has gained acceptance in Turkey as well, and it is widely used to estimate streamflow for ungauged basins in Turkey [2,3]. In the traditional application of this method, area-normalized streamflow values are transferred from only a single donor station to the target station. In addition to the drainage area, there are some other factors that have a significant influence on the unique streamflow characteristics of a station. Because the DAR method is used with only a single donor station, systematic errors can be encountered in the estimation of a target station [32]. When more streamflow gauging stations are used to estimate streamflow for the target station, this method is referred to as the multiple-donor stations drainage area ratio (MDAR) method [2,32]. The MDAR method assumes that the streamflow estimates at the target station can be computed as the weighted average of the estimates (produced by the DAR method) of the multiple donor stations selected.

The inverse distance weighted (IDW) method is one of the most widely used interpolation methods based on the geographical distance between the donor and target stations [21,22]. This method can be considered as a variant of the DAR method. The IDW method estimates the streamflow value for the target station by taking the geographical distance between the donor station and the target station as the weight. The closer the geographical distance between the donor station and the target station is, the larger the influence on the target station will be. That is, when the distance decreases, the weight coefficient increases. The IDW method, also called an inverse distance to power, is a weighted average interpolator, and the main factor affecting the accuracy of the IDW method is the value of the power parameter. As the power parameter increases, more influence is given to the donor stations close to the target station. In the literature, the value of the power parameter is commonly chosen as 2, which is known as the inverse distance squared weighted [36]. Alternatively, the inverse similarity weighted (ISW) method [37,38], which is similar to the IDW method, can be applied on the basis of multiple donor stations. Unlike the IDW method, the ISW method uses physical similarity instead of the geographical distance between the target and the donor station. The ISW method with three different power parameters (1, 2, and 3) was used for daily streamflow estimation in this study, and area normalized streamflow values are directly transferred to a target station from multiple donor stations.

The streamflow characteristics at the ungauged basin are directly affected by the donor stations. Therefore, the selection of hydrologically similar donor stations is important for estimating streamflow values at the ungauged basin. In practical applications, the donor station is usually selected as the geographically nearest station to the ungauged basin [36,39,40]. However, the geographical distance may not always be correct for the selection of the donor stations [41,42]. In this study, the physical similarities between the donor and the target station were taken into account when selecting the appropriate donor station for the target station. Physical similarity defines which stations are most similar in terms of some physical characteristics such as drainage area, elevation, precipitation, temperature, latitude, and longitude. According to the physical similarity, donor stations were defined for each target station. This procedure is described in detail in the section "Selection of Donor Stations".

In this study, continuous daily streamflow data were used between 1986–2009 for selected streamflow gauging stations in two subbasins of the Euphrates basin. In order to estimate daily streamflow at the ungauged basin, the single-donor station drainage area ratio (DAR) method, the multiple-donor stations drainage area ratio (MDAR) method, and the inverse similarity weighted (ISW) methods were applied. Three different power parameters (1, 2, and 3) of the ISW method were compared to determine their accuracy and suitability for estimating daily streamflow values. In addition, the daily streamflow data were smoothed with symmetric two-sided moving average (MA) filtering in order to reduce noise. The observed (original) data (without data preprocessing) or the smoothed data (with data preprocessing by the MA) were used as inputs of the statistical methods for estimating daily streamflow values at the target station. In the former case, the estimated daily streamflow values at the target station were compared to the observed (original) daily streamflow values at the target station, while in the latter case, the estimated daily streamflow values at the target station were compared to the observed-MA (smoothed) daily streamflow values at the target station. These two approaches were presented to estimate the daily streamflow values with and without MA. It is believed that the results will help decision makers choose the best one for their objectives. In summary, the major objectives of this study can be listed as follows: 1) to test applicability of the statistical methods to two subbasins of the Euphrates basin in Turkey, 2) to evaluate the success of physical similarity approaches in selecting donor stations in this basin, and 3) to investigate the effect of the statistical methods coupled with the data preprocessing method of moving average (MA) on the accuracy of streamflow estimation.

2. Study Area and Data

2.1. Study Area

Turkey is divided into 25 hydrological river basins, where the Euphrates-Tigris (indicated with basin number 21) is regarded as one single basin (Figure 1). Euphrates-Tigris Basin is located in the eastern part of Turkey with a drainage area of 185,000 km^2, which is the largest basin of Turkey. It has also nearly 28.5% of the water potential of Turkey. As the biggest water source of the Euphrates-Tigris Basin, the Euphrates River is the longest and one of the most historically significant rivers of the Middle East. The total length of Euphrates is nearly 2800 km, and 40% of its length is in Turkey, 25% is in Syria, and 35% is in Iraq. The Euphrates River consists of two major tributaries, the Karasu River and the Murat River, which both originate in the Eastern Anatolia mountains of Turkey. These two rivers merge near the Keban Dam, which is one of the largest dams of Turkey. The Euphrates River

Basin is subdivided into the Upper Euphrates, the Middle Euphrates, and the Lower Euphrates basins, which have some distinctive physical features. The water regime of the Euphrates River Basin depends heavily on winter rainfalls and spring snowmelt.

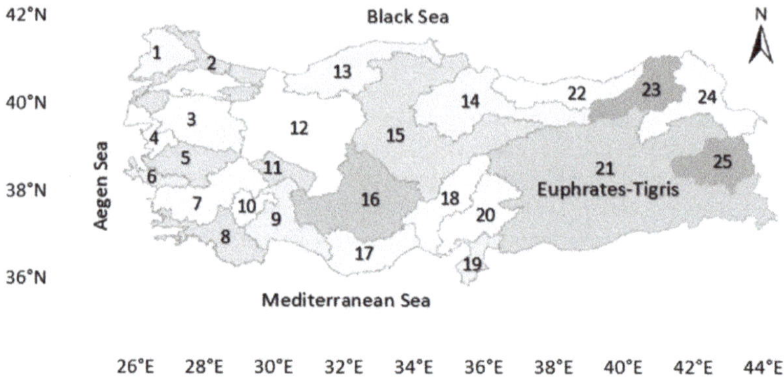

Figure 1. Hydrological river basins in Turkey.

In order to determine the water potential of the Euphrates-Tigris Basin, a large number of streamflow gauging stations were established on the Euphrates River and its tributaries. However, there was a large amount of missing data in the daily streamflow measurements of some streamflow gauging stations. These missing data lead to significant problems in hydrological modeling studies. Statistical estimation methods require the use of daily streamflow time series obtained from a large number of streamflow gauging stations within the study area. Also, the observation period should be the same for all these streamflow gauging stations. In the Euphrates Basin as a case study, a total of 15 streamflow gauging stations, which have 24 years (1986–2009) of common daily streamflow data, was selected. Eight of these stations are located in the Middle Euphrates Basin, and the other seven stations are located in the Upper Euphrates Basin (Figure 2). Moreover, they are not located downstream of a dam.

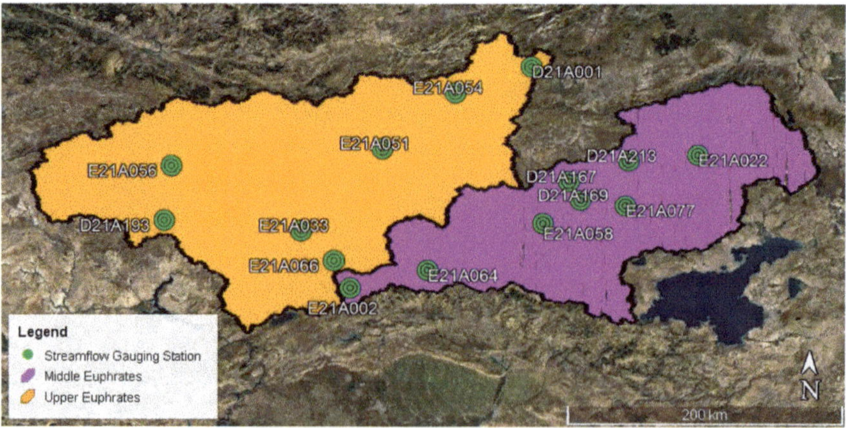

Figure 2. Location of the two selected basins with streamflow gauging stations.

2.2. Hydrological and Meteorological Data

Eight streamflow gauging stations from the Middle Euphrates Basin and seven streamflow gauging stations from the Upper Euphrates Basin were selected for this case study. The stream networks of these two basins and the locations of the selected streamflow gauging stations in each basin are shown in Figure 3. Continuous daily streamflow data of the streamflow gauging stations operated by the General Directorate of State Hydraulic Works (DSI) were used. Each streamflow gauging station contains a 24-year period spanning from 1986 to 2009, and there is no missing data within the streamflow time series. The main characteristics of these streamflow gauging stations are listed in Tables 1 and 2. As shown by Table 1, drainage areas of the stations in the Middle Euphrates Basin vary between 65.3 and 25,515.6 km^2 whereas their elevations range between 852 and 1810 m above sea level. As shown by Table 2, drainage areas of the stations in the Upper Euphrates Basin vary between 233.2 and 15,562 km^2 whereas their elevations range between 840 and 1830 m above sea level.

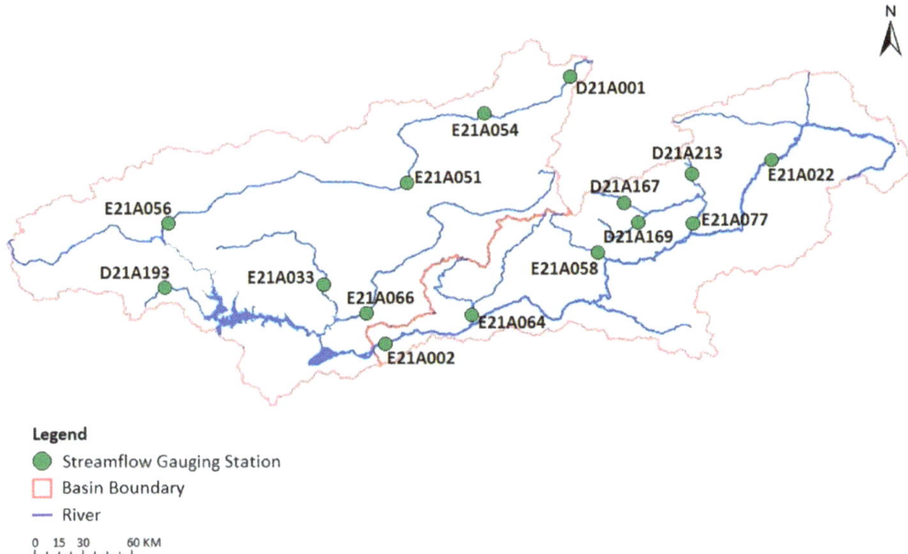

Figure 3. Streamflow gauging stations in the study area.

Table 1. Characteristics of streamflow gauging stations in the Middle Euphrates Basin.

Station Number	Drainage Area (km^2)	Elevation (m)	Long-Term Mean (m^3/s)	Record Period (Years)
D21A167	250	1650	3.55	1986–2009
D21A169	276.1	1600	3.35	1986–2009
D21A213	65.3	1810	0.74	1986–2009
E21A002	25,515.6	852	239.82	1986–2009
E21A022	5882.4	1552	48.20	1986–2009
E21A058	1577.6	1310	18.91	1986–2009
E21A064	2232	990	32.97	1986–2009
E21A077	2995.3	1452	29.94	1986–2009

Table 2. Characteristics of streamflow gauging stations in the Upper Euphrates Basin.

Station Number	Drainage Area (km^2)	Elevation (m)	Long-term Mean (m^3/s)	Record Period (Years)
D21A001	233.2	1830	2.75	1986–2009
D21A193	518.1	1000	6.31	1986–2009
E21A033	3284.8	875	89.38	1986–2009
E21A051	8185.6	1355	60.23	1986–2009
E21A054	2886	1675	19.68	1986–2009
E21A056	15,562	865	153.57	1986–2009
E21A066	5430	840	78.26	1986–2009

Basin characteristics such as geographical, topographical, and climate variables were considered for determining the physical similarity between the donor and target stations. Annual mean total precipitation and annual mean temperature were selected as climatic variables. Concurrent precipitation and temperature data of the meteorological stations operated by the Turkish State Meteorological Service (DMI) were used. The annual mean total precipitation and annual mean temperature values for each streamflow gauging station were calculated by the Thiessen polygon method (Figure 4). Thus, annual mean total precipitation and annual mean temperature values of the drainage area represented by each streamflow gauging station were obtained using the precipitation and temperature data of the meteorological stations. Drainage area, elevation, basin slope, and channel length were selected as topographical variables. Basin slope and channel length for the drainage basin of each streamflow gauging station were extracted using geographic information system (GIS) software. The latitude and longitude were selected as geographical variables because geographically nearby streamflow gauging stations could have similarities in hydrological behavior. They were converted to decimal degrees and then used to calculate the similarity coefficient. Since the latitude and longitude define the geographical location of the streamflow gauging stations, these selected basin characteristics combine the physical similarity approach with the geographical proximity approach [43]. Descriptive statistics of the selected physical characteristics are presented in Table 3 for the Middle Euphrates Basin and in Table 4 for the Upper Euphrates Basin.

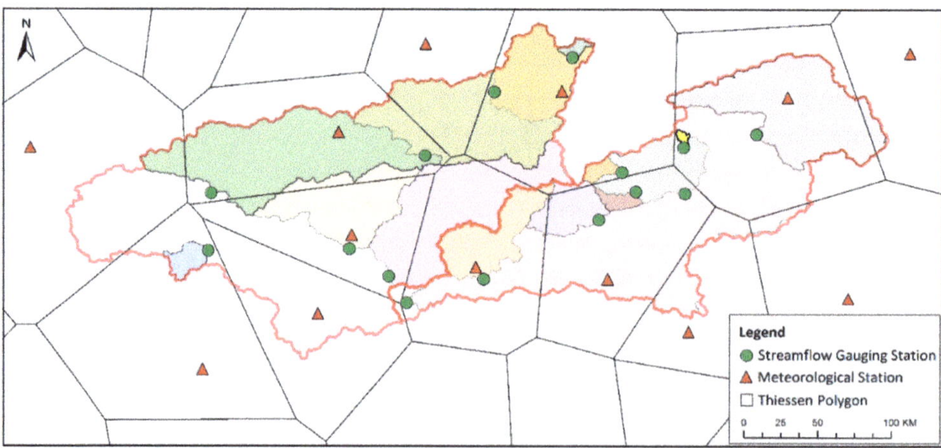

Figure 4. Meteorological stations and Thiessen polygons of the study area.

Table 3. Statistics of physical characteristics used in the Middle Euphrates Basin.

Physical Characteristics	Maximum	Minimum	Mean
Drainage Area (km^2)	25,515.6	65.3	4849.3
Elevation (m)	1810	852	1402
Annual Mean Total Precipitation (mm)	939.50	431.20	679.68
Annual Mean Temperature (°F)	53.60	42.26	47.29
Basin Slope (%)	2.69	0.19	1.21
Channel Length (km)	565.11	14.75	142.27
Latitude (°)	39.54	38.69	39.19
Longitude (°)	42.78	39.93	41.58

Table 4. Statistics of physical characteristics used in the Upper Euphrates Basin.

Physical Characteristics	Maximum	Minimum	Mean
Drainage Area (km^2)	15,562	233.2	5157.1
Elevation (m)	1830	840	1205.7
Annual Mean Total Precipitation (mm)	840.17	374.90	524.15
Annual Mean Temperature (°F)	56.48	42.26	48.52
Basin Slope (%)	2.82	0.16	0.96
Channel Length (km)	381.60	25.10	161.19
Latitude (°)	40.11	38.86	39.44
Longitude (°)	41.39	38.41	39.79

3. Methods

Estimation of daily streamflow time series at the target station consists of the following steps: (1) the selection of hydrologically similar donor stations to the target station and (2) the transfer of the daily streamflow time series from the donor station to target station by using statistical streamflow transfer methods. Proposed flowcharts for streamflow estimation at the target station are illustrated in Figure 5.

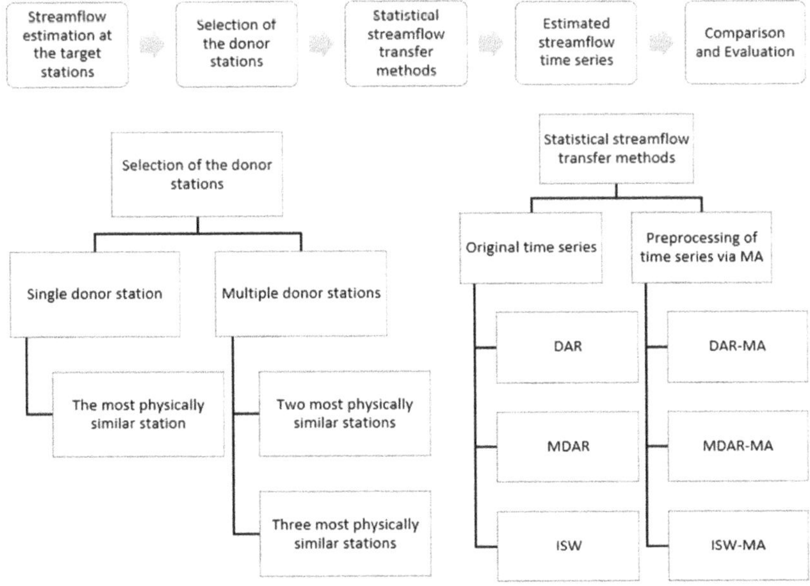

Figure 5. Proposed flowcharts for streamflow estimation at the target station.

3.1. Statistical Information Transfer Methods

For daily streamflow estimation at the target stations, three statistical streamflow transfer methods are considered in this study. These methods include the single-donor station drainage area ratio (DAR) method, the multiple-donor stations drainage area ratio (MDAR) method, and the inverse similarity weighted (ISW) method. Moreover, the variations of the ISW method, in which constant power parameters are modified, are utilized as well. The DAR and the MDAR methods were based on the drainage area of the stations, whereas the ISW method was based on the physical similarity between the donor and target stations. Each statistical method is briefly described below.

3.1.1. Drainage Area Ratio (DAR) Method

The DAR method [39,40] assumes that the streamflow per unit drainage area for the target station equals that at the streamflow gauging station used as a donor station for a given day, as described in Equation (1).

$$Q_{target} = \frac{A_{target}}{A_{donor}} Q_{donor} \tag{1}$$

where Q_{target} is the daily streamflow for the target station, Q_{donor} is the streamflow at a donor station, and A_{target} and A_{donor} are the drainage areas for the target station and the donor station, respectively.

3.1.2. Multiple-Donor Stations Drainage Area Ratio (MDAR) Method

The MDAR method [2,32] generates the streamflow estimation at the target station as the weighted average of the DAR method estimations from the donor stations. The streamflow at the target station from the n donor stations can be calculated for a given day using Equation (2).

$$Q_{target} = \frac{\sum_{i=1}^{n} w_i \hat{Q}_{donor_i}}{\sum_{i=1}^{n} w_i} \tag{2}$$

where w_i is the weight of the donor station i on the target station, $\hat{Q}_{donor_i}$ is the daily streamflow estimations from each donor station, and n is the total number of the donor stations. The values of the weights in Equation (2) which show the similarity between the target station and donor station can be calculated as Equation (3).

$$w_i = \frac{\frac{1}{d_i}}{\sum_{i=1}^{n} \frac{1}{d_i}} \tag{3}$$

where d_i is the similarity distance between the target station and donor station i. The drainage area is frequently considered as the most important variable in many hydrological regionalization studies [39,40,44]. Moreover, the drainage area is also the only scaling factor used in the DAR method for streamflow estimation at the target stations. Therefore, it was used as the similarity distance in this study. It can be calculated using Equation (4).

$$d_i = \left| A_{target} - A_{donor_i} \right| \tag{4}$$

where A_{donor_i} is the drainage area of the donor station i and A_{target} is the drainage area of the target station.

3.1.3. Inverse Similarity Weighted (ISW) Method

The ISW method [37,38] estimates streamflow values at the target station as the weighted average of the streamflow values at n donor stations. The weights are inversely proportional to the power of physical similarity from the target station. The mathematical expression of the ISW method is given by Equation (5).

$$q_{target} = \sum_{i=1}^{n} w_i q_{donor_i} \text{ and } Q_{target} = q_{target} A_{target} \tag{5}$$

where q_{target} is the area normalized streamflow (m^3/s/km^2) at the target station and q_{donor_i} is the area normalized streamflow (m^3/s/km^2) at the donor station i. The weights w_i based on physical similarity can be calculated for all donor stations using Equation (6). The sum of the weights assigned to each donor station is equal to 1.

$$w_i = \frac{\frac{1}{s_i^p}}{\sum_{i=1}^{n} \frac{1}{s_i^p}} \text{ and } \sum_{i=1}^{n} w_i = 1 \tag{6}$$

where s_i is the similarity coefficient between the target station and donor station i and where the exponent p is called a power parameter ($p > 0$). In this study, the estimation performance of the ISW method was evaluated using different power parameters from 1 to 3. For power parameters of 1, 2, and 3, the ISW method was referred to as ISW1, ISW2, and ISW3, respectively.

The similarity coefficient, s, is used to define the physical similarity between the target station and the donor station, which is calculated using Equation (7) [43,45]. Drainage area, elevation, annual mean total precipitation, annual mean temperature, basin slope, channel length, latitude, and longitude were considered as the basin characteristics in order to measure the physical similarity between the donor station and the target station. The station with the lowest similarity coefficient was selected as the donor station. The similarity coefficient was used both to select the donor stations and to transfer streamflow from several donor stations as the weight.

$$s = \sum_{i=1}^{k} \frac{\left| X_i^{donor} - X_i^{target} \right|}{max(X_i) - min(X_i)} \tag{7}$$

where i indicates one of a total of k selected basin characteristics; X_i^{donor} and X_i^{target} are the values of basin characteristic i for the donor station and the target station, respectively; and $max(X_i)$ and $min(X_i)$ are the maximum and the minimum values of basin characteristics over the set of stations considered, respectively.

3.2. Selection of Donor Stations

For transferring the streamflow to the target (ungauged) station, the streamflow values of the donor stations are used. Therefore, the selection of the donor stations is an important step in estimating streamflow at the target station. In this study, the physical similarity approach was considered to identify donor stations. In the physical similarity approach, the station that minimizes the similarity coefficient defined in Equation (7) was used as the donor station. That is, the best donor station was given to the station having the smallest s value. Although all stations are gauged, initially, each station in each basin was considered in turn as a target station and daily streamflow time series were estimated for all stations assumed as a target station within each basin. Subsequently, their actual streamflow time series were used in order to evaluate the performance of the streamflow estimation. When using one donor station, the most physically similar station was identified for each target station. On the other hand, when using more than one donor station, the two or three most physically similar stations were identified for each target station (Figure 5). Therefore, the DAR method was applied to each station using the most physically similar station as the donor station, while the MDAR and ISW methods were applied to each station for two different cases: 1) using the two most physically similar stations as the donor stations and 2) using the three most physically similar stations as the donor stations. The results for these two different cases were compared with the original observations to evaluate which one provides better estimation performance.

3.3. Data Preprocessing

The streamflow data may contain possible errors, and these errors are collectively referred to as noise. As the noise in the data increases, reliable results will be difficult to achieve. In this study, data preprocessing was conducted to remove noise and to improve the reliability of daily streamflow estimates. The data preprocessing discussed here was daily streamflow data smoothing using the moving average (MA). Each daily streamflow time series of all stations was smoothed by a centered (or symmetric two-sided) moving average of length $m = 2k + 1$, i.e., MA(m), and then, the smoothed streamflow time series were used into the statistical methods. Hereafter, the statistical methods, DAR, MDAR, and ISW are referred to as DAR-MA, MDAR-MA, and ISW-MA, respectively. A centered moving average smooths data by replacing each observed daily streamflow value with the average of the current day, previous day, and subsequent days and is defined as Equation (8). For example, a centered moving average of length $m = 3$ (hence $k = 1$), i.e., MA(3) with equal weights, replaces the observed daily streamflow value x_t at time t with the averages of x_{t-1}, x_t, and x_{t+1}.

$$x_t^* = \frac{1}{2k + 1} \sum_{i=-k}^{k} x_{t+i} \tag{8}$$

where x_t^* is the smoothed streamflow value at time t and $m = 2k + 1$ is the number of observed values that are averaged.

In order to smooth daily streamflow data, MA was applied with lengths of 3, 5, 7, 9, and 11 days in this study, and then, it was seen that the larger the length $m = 2k + 1$, the more the streamflow peaks (maximum values) and streamflow valleys (minimum values) were smoothed out. The peaks and valleys of streamflow are not well represented by the relatively high length of MA(5), MA(7), MA(9), and MA(11). For a better representation of streamflow peaks and valleys, MA(5), MA(7), MA(9), and MA(11) were excluded from the rest of the study.

3.4. Evaluation Criteria

A jackknife (leave one out) procedure was used for evaluating the performance of each method. In this procedure, each station in each basin was considered in turn as a target station and was removed from the database. This procedure was repeated for all stations considered for this study.

The performance of each statistical method was evaluated in terms of the Nash–Sutcliffe efficiency (NSE) [46], the ratio of the root mean square error (RMSE) to the standard deviation of the observed data (RSR) [47], the percent bias (PBIAS), and the coefficient of determination (R^2) between the estimated and observed streamflow values. NSE, RSR, PBIAS, and R^2 were calculated as follows:

$$NSE = 1 - \frac{\sum_{i=1}^{n}\left(X_i^{obs} - X_i^{est}\right)^2}{\sum_{i=1}^{n}\left(X_i^{obs} - \overline{X}^{obs}\right)^2} \tag{9}$$

$$RSR = \frac{RMSE}{\sqrt{\frac{\sum_{i=1}^{n}\left(X_i^{obs} - \overline{X}^{obs}\right)^2}{n}}} = \frac{\sqrt{\sum_{i=1}^{n}\left(X_i^{obs} - X_i^{est}\right)^2}}{\sqrt{\sum_{i=1}^{n}\left(X_i^{obs} - \overline{X}^{obs}\right)^2}} = \sqrt{1 - NSE} \tag{10}$$

$$PBIAS = \frac{\sum_{i=1}^{n}\left(X_i^{obs} - X_i^{est}\right)}{\sum_{i=1}^{n}\left(X_i^{obs}\right)} x100 \tag{11}$$

$$R^2 = \left(\frac{\sum_{i=1}^{n} \left(X_i^{obs} - \overline{X}^{obs}\right)\left(X_i^{est} - \overline{X}^{est}\right)}{\sqrt{\sum_{i=1}^{n} \left(X_i^{obs} - \overline{X}^{obs}\right)^2} \sqrt{\sum_{i=1}^{n} \left(X_i^{est} - \overline{X}^{est}\right)^2}} \right)^2 \tag{12}$$

where X_i^{obs} is the ith observed daily streamflow value; X_i^{est} is the ith estimated daily streamflow value; $\overline{X}^{obs}$ and $\overline{X}^{est}$ are the mean of observed and estimated daily streamflow values, respectively; and n is the total number of observed daily streamflow values.

The NSE values range between $-\infty$ and $+1$, where a value of 1 indicates a perfect agreement between estimated and observed streamflow values. The values closer to 1 indicate an increasingly better agreement, whereas the values far from 1 indicate an increasingly poor agreement. The RSR standardizes RMSE using the standard deviation of the observed data. The RSR varies from the optimal value of 0 to a large positive value. The lower the RSR, the better the performance of the method [47]. PBIAS measures the average tendency of the estimated values to be larger or smaller than corresponding observed values. The optimal value of PBIAS is 0, and the closer it is to 0, the more accurate the estimated values are to the observed values. Negative PBIAS values indicate overestimation, while positive PBIAS values indicate underestimation [47]. The coefficient of determination (R^2) is the square of the correlation coefficient according to Pearson. The R^2 values range from 0 to 1, with higher values indicating better agreement between estimated and observed values. Generally, R^2 values greater than 0.5 are considered acceptable [47].

Moriasi et al. [47] suggested performance ratings of recommended statistics such as NSE, RSR, and PBIAS for monthly streamflow. According to Moriasi et al. [47], the performance of the method is considered satisfactory when the NSE is greater than 0.5, the RSR is less than 0.7, and the PBIAS ranges are less than ±25%. However, NSE values lower than 0.5 for daily streamflow can still be considered satisfactory [48]. Therefore, some of the constraints for the recommended statistics can be relaxed for daily streamflow. In this study, the adjusted performance ratings of the NSE and PBIAS statistics for the daily time scale proposed by Kalin et al. [49] were used to evaluate the performance of the statistical methods (Table 5).

Table 5. Performance ratings of the Nash–Sutcliffe efficiency (NSE) and percent bias (PBIAS) statistics for daily streamflow.

Performance Rating	NSE	abs(PBIAS) %
Very good	NSE ≥ 0.7	abs(PBIAS) ≤ 25
Good	0.5 ≤ NSE < 0.7	25 < abs(PBIAS) ≤ 50
Satisfactory	0.3 ≤ NSE < 0.5	50 < abs(PBIAS) ≤ 70
Unsatisfactory	NSE < 0.3	abs(PBIAS) > 70

4. Results and Discussion

4.1. Middle Euphrates Basin

The statistical methods described above were applied on each of eight streamflow gauging stations in the Middle Euphrates Basin, which were considered in turn as the target station. The donor station or stations based on physical similarity was selected to transfer daily streamflow data to the target station. Table 6 shows the sequence of the donor stations for each target station in the Middle Euphrates Basin which was determined according to the similarity coefficient.

In order to estimate daily streamflow at the target stations, the DAR method was applied by using the most physically similar station to each target station. In order to test the applicability of the donor station selection criteria for the study area, the NSE values were determined for the donor stations identified by the physical similarity and compared with the NSE values obtained from the donor stations traditionally selected as the geographically nearest stations. On the other hand, the

MDAR and ISW methods were applied by using the two and the three most physically similar stations to each target station. In order to test the effect of different power parameter selection in the use of the ISW method on the accuracy of daily streamflow estimation, the ISW method was applied with power parameters of 1, 2, and 3. In addition, the comparisons between the statistical methods with and without the MA were carried out to indicate the effectiveness of the MA-based preprocessing on the accuracy of daily streamflow estimation. Daily streamflow values estimated using both observed and smoothed data from the donor stations were compared with observed data at the target station. According to the selection criteria of the donor stations, the NSE values obtained for the target stations in the Middle Euphrates Basin are given in Table 7 for DAR and MDAR and in Table 8 for ISW with three different power parameters (1, 2, and 3). Performance evaluations of the best statistical method without and with the MA for each target station in the Middle Euphrates Basin are presented in Table 9 and in Table 10, respectively.

Table 6. Physically similar donor stations for each target station in the Middle Euphrates Basin.

Target Station	Donor Station						
	1st	2nd	3rd	4th	5th	6th	7th
D21A167	D21A213	D21A169	E21A022	E21A058	E21A077	E21A064	E21A002
D21A169	E21A058	E21A077	D21A167	D21A213	E21A022	E21A064	E21A002
D21A213	D21A167	E21A022	D21A169	E21A058	E21A077	E21A064	E21A002
E21A002	E21A064	E21A058	E21A077	E21A022	D21A169	D21A167	D21A213
E21A022	D21A213	E21A077	D21A167	E21A058	D21A169	E21A064	E21A002
E21A058	E21A077	D21A169	E21A064	E21A022	D21A167	D21A213	E21A002
E21A064	E21A058	E21A077	D21A169	E21A002	E21A022	D21A167	D21A213
E21A077	E21A058	D21A169	E21A022	E21A064	D21A213	D21A167	E21A002

Table 7. NSE values for the drainage area ratio (DAR) and multiple-donor stations drainage area ratio (MDAR) methods in the Middle Euphrates Basin.

Target Station	The Geographically Nearest Donor Station	The Most Physically Similar Donor Station	Two Most Physically Similar Donor Stations	Three Most Physically Similar Donor Stations
	DAR	DAR	MDAR	MDAR
D21A167	0.313	−0.187	0.312	0.316
D21A169	0.581	0.850	0.852	0.595
D21A213	0.718	0.354	0.369	0.608
E21A002	−0.171	−0.171	0.433	0.729
E21A022	−0.205	−0.205	0.622	0.718
E21A058	0.826	0.814	0.894	0.839
E21A064	0.649	0.724	0.694	0.706
E21A077	0.300	0.647	0.651	0.781

Table 8. NSE values for the inverse similarity weighted (ISW) with powers of 1, 2, and 3 in the Middle Euphrates Basin.

Target Station	Two Most Physically Similar Donor Stations			Three Most Physically Similar Donor Stations		
	ISW1	ISW2	ISW3	ISW1	ISW2	ISW3
D21A167	0.136	0.060	−0.009	0.317	0.184	0.065
D21A169	0.844	0.849	0.852	0.816	0.825	0.833
D21A213	0.501	0.463	0.429	0.608	0.556	0.502
E21A002	0.418	0.396	0.374	0.696	0.664	0.630
E21A022	0.432	0.375	0.316	0.636	0.591	0.535
E21A058	0.893	0.892	0.892	0.904	0.907	0.906
E21A064	0.696	0.701	0.706	0.712	0.713	0.714
E21A077	0.641	0.651	0.656	0.774	0.743	0.716

Table 9. Performance evaluation of the best statistical method without the moving average (MA) for each target station in the Middle Euphrates Basin.

Target Station	Method	NSE	RSR	PBIAS
D21A167	ISW1	0.317 [3]	0.827	23.265 [1]
D21A169	ISW3	0.852 [1]	0.385	6.459 [1]
D21A213	ISW1	0.608 [2]	0.626	−6.122 [1]
E21A002	MDAR	0.729 [1]	0.520	−30.114 [2]
E21A022	MDAR	0.718 [1]	0.531	−39.089 [2]
E21A058	ISW2	0.907 [1]	0.304	3.123 [1]
E21A064	DAR	0.724 [1]	0.525	18.865 [1]
E21A077	MDAR	0.781 [1]	0.468	−11.015 [1]

[1] Very good, [2] good, [3] satisfactory, and [4] unsatisfactory.

Table 10. Performance evaluation of the best statistical method with the MA for each target station in the Middle Euphrates Basin.

Target Station	Method	NSE	RSR	PBIAS
D21A167	ISW1-MA	0.326 [3]	0.821	23.265 [1]
D21A169	ISW3-MA	0.876 [1]	0.352	6.459 [1]
D21A213	ISW1-MA	0.618 [2]	0.618	−6.122 [1]
E21A002	MDAR-MA	0.767 [1]	0.483	−30.114 [2]
E21A022	MDAR-MA	0.725 [1]	0.524	−39.089 [2]
E21A058	ISW2-MA	0.921 [1]	0.282	3.123 [1]
E21A064	DAR-MA	0.744 [1]	0.506	18.865 [1]
E21A077	MDAR-MA	0.817 [1]	0.428	−11.015 [1]

[1] Very good, [2] good, [3] satisfactory, and [4] unsatisfactory.

For 2 out of 8 target stations (i.e., E21A002 and E21A022), the geographically nearest and the most physically similar station were the same. For 3 out of the remaining 6 target stations (i.e., D21A169, E21A064, and E21A077), higher NSE values were obtained using DAR with the most physically similar station as the donor station. For 3 out of the target stations (i.e., D21A167, D21A213, and E21A058), higher NSE values were obtained using DAR with the geographically nearest station as the donor station. According to these results, for half of the target stations in the study area, the geographical distance seems to be a good selection criterion as the donor station; however, for the remaining half of the target stations, geographical distance cannot identify the best donor station. Therefore, donor station selection criteria can provide different estimated results that vary from basin to basin.

As can be seen in Table 7, for all target stations other than E21A064, higher NSE values were obtained with MDAR compared with DAR. Especially for D21A167, E21A002, and E21A022, negative NSE values obtained with DAR using the most physically similar donor station improved considerably when the three most physically similar donor stations were used with MDAR. The performance of the DAR method was unsatisfactory for D21A167, E21A002, and E21A022. This was mostly due to the significant increase in the drainage area ratio between the donor and target stations. D21A213 has the smallest drainage area (65.3 km^2) in the Middle Euphrates Basin and was determined as the most physically similar donor station for both D21A167 (250 km^2) and E21A022 (5882.4 km^2). Moreover, E21A002 has the largest drainage area (25,515.6 km^2) in the Middle Euphrates Basin. Its drainage area is more than four times the next largest station. For all target stations other than D21A169 and E21A058, MDAR using of the three most physically donor stations produced better NSE values than that using the two most physically similar donor stations. For D21A169, the NSE value decreased from 0.852 to 0.595 when the three most physically similar donor stations were used instead of the two most physically similar donor stations (see Table 7). In case of the use of the three most physically similar donor stations, the third most physically similar donor station for D21A169 was determined as D21A167. The NSE value obtained for D21A169 using the DAR method and utilizing D21A167 was lower than the NSE values obtained from the other two donor stations (i.e., E21A058 and E21A077). The drainage area of the donor station D21A167 is very close to the target station D21A169. On the other hand, the drainage areas of the other two donor stations are much larger than D21A169. Hence, the weight of donor station D21A167 for streamflow estimation of D21A169 is significantly larger compared to the other two. Consequently, the NSE value obtained for D21A169 using MDAR with the three most physically similar donor stations is predominantly influenced by donor station D21A167. Similarly, for E21A058, the decrease in the NSE (i.e., from 0.894 to 0.839) was due to the same reason as for D21A169.

As can be seen in Table 8, in case of the use of the two most physically similar donor stations, the best performance results were obtained with ISW1 for 5 out of 8 target stations (i.e., D21A167, D21A213, E21A002, E21A022, and E21A058). On the other hand, in case of the use of the three most physically similar donor stations, the best performance results were obtained with ISW1 for 5 out of 8 target stations (i.e., D21A167, D21A213, E21A002, E21A022, and E21A077). In both cases, the most reasonable estimation results were mostly obtained when ISW1 was applied instead of ISW2 and ISW3. Moreover, the NSE values mostly improved when the three most physically similar donor stations were used instead of the two most physically similar donor stations.

As can be seen in Table 9, for all target stations other than for E21A064, the MDAR and the ISW methods resulted in higher NSEs compared to the DAR method. For 6 out of 8 target stations, the results can be rated as "very good" for NSE according to the performance ratings in Table 5. For 7 out of 8 target stations, the RSR values were considered satisfactory (i.e., less than 0.7) according to the performance ratings recommended by Moriasi et al. [47]. The negative PBIAS values for D21A213, E21A002, E21A022, and E21A077 demonstrate that the method overestimated daily streamflow, while positive PBIAS values for D21A167, D21A169, E21A058, and E21A064 demonstrate underestimation. For all target stations, the statistical methods with the MA tend to achieve slightly higher NSE values. However, the PBIAS values of the target stations did not change when the statistical methods with the MA were used.

For the target station E21A058 as the example, the estimated daily streamflow values from the statistical methods without MA were compared to the observed (original) daily streamflow values in the hydrograph and scatter plots in Figure 6. The remarkably better agreement between observed and estimated daily streamflow values by three statistical methods (i.e, DAR, MDAR, and ISW) was obtained for E21A058 compared to the other target stations in the Middle Euphrates Basin. ISW2 gave a coefficient of determination (R^2) of 0.91, which was higher than the R^2 values of 0.87 and 0.90 obtained by using the DAR and MDAR, respectively. The NSE values for these methods ranged from 0.814 to 0.907, and the best NSE value was achieved by ISW2. The best NSE performance for E21A058 was obtained using ISW2 with the three most physically similar donor stations.

On the other hand, for the target station E21A058 as the example, the estimated daily streamflow values from the statistical methods with MA were compared with observed-MA (smoothed) daily streamflow values in the hydrograph and scatter plots in Figure 7. The statistical methods with MA performed slightly better for E21A058.

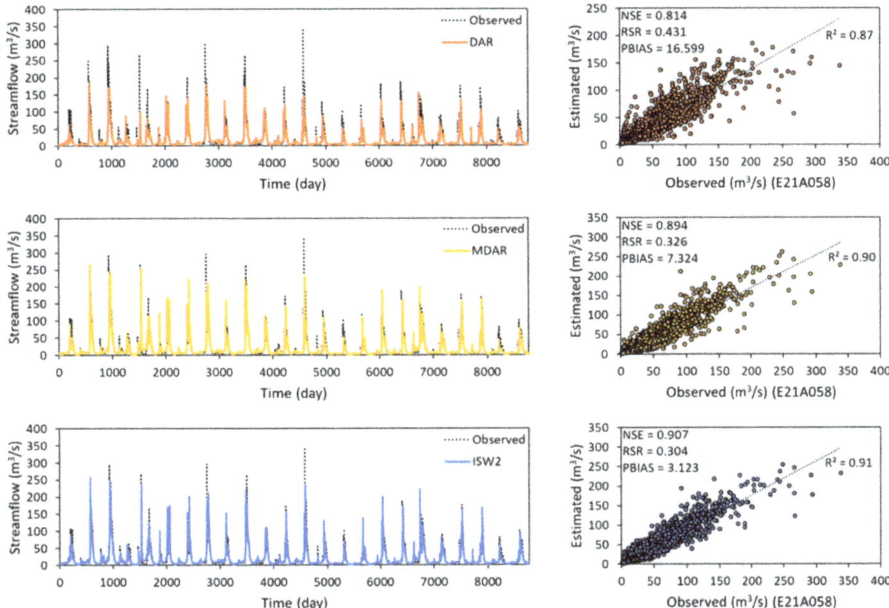

Figure 6. Comparison of daily streamflow estimated using DAR (top row), MDAR (mid row), and ISW2 (bottom row) with observed (original) daily streamflow at station E21A058 in the Middle Euphrates Basin.

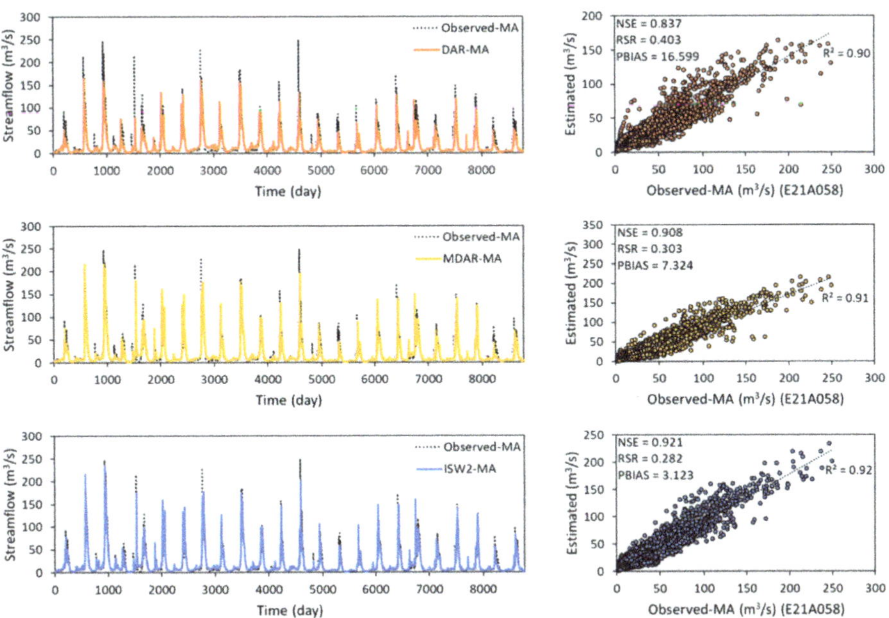

Figure 7. Comparison of daily streamflow estimated using DAR-MA (top row), MDAR-MA (mid row), and ISW2-MA (bottom row) with observed-MA (smoothed) daily streamflow at station E21A058 in the Middle Euphrates Basin.

4.2. Upper Euphrates Basin

Using the same procedure applied for the Middle Euphrates Basin, the statistical methods were applied on each of seven streamflow gauging stations in the Upper Euphrates Basin for the purpose of estimating daily streamflow. Table 11 shows the sequence of the donor stations for each target station in the Upper Euphrates Basin which was determined according to the similarity coefficient.

Table 11. Physically similar donor stations for each target station in the Upper Euphrates Basin.

Target Station	Donor Station					
	1st	**2nd**	**3rd**	**4th**	**5th**	**6th**
D21A001	E21A054	E21A051	D21A193	E21A033	E21A033	E21A066
D21A193	E21A033	E21A066	E21A056	E21A051	E21A051	E21A054
E21A033	E21A066	D21A193	E21A051	E21A056	E21A056	D21A001
E21A051	E21A054	E21A056	E21A066	E21A033	E21A033	D21A193
E21A054	E21A051	D21A001	E21A033	E21A056	E21A056	D21A193
E21A056	E21A051	E21A066	E21A033	D21A193	D21A193	D21A001
E21A066	E21A033	E21A051	E21A056	D21A193	D21A193	D21A001

According to the selection criteria of the donor stations, the NSE values obtained for the target stations in the Upper Euphrates Basin were given in Table 12 for DAR and MDAR and in Table 13 for ISW with three different power parameters (1, 2, and 3). Performance evaluations of the best statistical method without and with the MA for each target station in the Upper Euphrates Basin are presented in Table 14 and in Table 15, respectively.

For 2 out of the 3 target stations (i.e., E21A054 and E21A056) for which the geographically nearest and the most physically similar stations were not the same, higher NSE values were obtained using DAR with the most physically similar station (see Table 12). For E21A054 and E21A056, the NSE values improved considerably when the most physically similar station was used. These results indicate that the physical similarity may be a better selection criterion for the donor station in the study area compared to the geographical distance between the stations.

Table 12. NSE values for the DAR and MDAR methods in the Upper Euphrates Basin.

Target Station	The Geographically Nearest Donor Station	The Most Physically Similar Donor Station	Two Most Physically Similar Donor Stations	Three Most Physically Similar Donor Stations
	DAR	**DAR**	**MDAR**	**MDAR**
D21A001	0.619	0.619	0.630	−0.328
D21A193	0.299	−0.646	−0.053	0.089
E21A033	0.446	0.446	0.377	0.319
E21A051	0.883	0.883	0.932	0.362
E21A054	0.159	0.885	0.585	−5.128
E21A056	−3.199	0.769	0.415	−0.586
E21A066	−0.067	−0.067	0.698	0.732

Table 13. NSE values for ISW with powers of 1, 2, and 3 in the Upper Euphrates Basin.

Target Station	Two Most Physically Similar Donor Stations			Three Most Physically Similar Donor Stations		
	ISW1	**ISW2**	**ISW3**	**ISW1**	**ISW2**	**ISW3**
D21A001	0.633	0.629	0.626	0.597	0.626	0.631
D21A193	0.015	−0.085	−0.185	0.307	0.217	0.108
E21A033	0.416	0.439	0.444	0.353	0.412	0.435
E21A051	0.931	0.928	0.921	0.811	0.877	0.905
E21A054	0.753	0.775	0.794	0.244	0.580	0.719
E21A056	0.404	0.541	0.640	−0.788	−0.306	0.083
E21A066	0.479	0.191	0.033	0.645	0.331	0.097

Table 14. Performance evaluation of the best statistical method without the MA for each target station in the Upper Euphrates Basin.

Target Station	Method	NSE	RSR	PBIAS
D21A001	ISW1	0.633 [2]	0.606	40.537 [2]
D21A193	ISW1	0.307 [3]	0.832	−51.370 [3]
E21A033	DAR	0.446 [3]	0.744	47.035 [2]
E21A051	MDAR	0.932 [1]	0.261	−10.006 [1]
E21A054	DAR	0.885 [1]	0.339	−7.892 [1]
E21A056	DAR	0.769 [1]	0.480	25.436 [2]
E21A066	MDAR	0.732 [1]	0.518	−22.121 [1]

[1] Very good, [2] good, [3] satisfactory, and [4] unsatisfactory.

Table 15. Performance evaluation of the best statistical method with the MA for each target station in the Upper Euphrates Basin.

Target Station	Method	NSE	RSR	PBIAS
D21A001	ISW1-MA	0.646 [2]	0.595	40.537 [2]
D21A193	ISW1-MA	0.305 [3]	0.834	−51.371 [3]
E21A033	DAR-MA	0.453 [3]	0.740	47.035 [2]
E21A051	MDAR-MA	0.939 [1]	0.248	−10.006 [1]
E21A054	DAR-MA	0.897 [1]	0.322	−7.892 [1]
E21A056	DAR-MA	0.780 [1]	0.469	25.436 [2]
E21A066	MDAR-MA	0.749 [1]	0.501	−22.120 [1]

[1] Very good, [2] good, [3] satisfactory, and [4] unsatisfactory.

As can be seen in Table 12, for all target stations other than E21A033, E21A054, and E21A056, higher NSE values were obtained with MDAR compared with DAR. Especially for E21A066, the NSE value improved considerably with MDAR as compared with DAR. For D21A001, E21A051, E21A054, and E21A056, the NSE values decreased when MDAR was applied using the three most physically similar donor stations instead of the two most physically similar donor stations (see Table 12). In case of use of the three most physically similar donor stations, the third most physically similar donor stations for the target stations D21A001, E21A051, and E21A054 were determined as D21A193, E21A066, and E21A033, respectively. All NSE values obtained for these target stations using the DAR method utilizing their third most physically similar donor stations were negative. The drainage areas of these target stations and their third most physically similar donor stations are very close to each other. Therefore, the weight of their third most physically similar donor stations for streamflow estimation of these target stations is significantly larger. As a result, the NSE values obtained for the target stations D21A001, E21A051, and E21A054 using MDAR with three most physically similar donor stations are predominantly influenced by their third most physically similar donor stations. On the other hand, for E21A056, the reason is slightly different from the others. The NSE value obtained for E21A056 using the DAR method utilizing its third most physically similar donor station E21A033 was too low (i.e., −10.029). Although the contribution of the donor station E21A033 is not much more than the other two, this leads to poor estimation performance for E21A056 when MDAR was applied using the three most physically similar donor stations.

As can be seen in Table 13, in case of the use of the two most physically similar donor stations, the best performance results were obtained with ISW1 for 4 out of 7 target stations (i.e., D21A001, D21A193, E21A051, and E21A066). On the other hand, in case of the use of the three most physically similar donor stations, the best performance results were obtained with ISW3 for all target stations other than D21A193 and E21A066. As the power parameter increased from 1 to 3, the NSE values mostly improved when the three most physically similar donor stations were used, whereas the NSE values mostly decreased when the two most physically similar donor stations were used. Moreover, the NSE values mostly improved when the two most physically similar donor stations were used instead of the three most physically similar donor stations.

As can be seen in Table 14, for all target stations other than for E21A033, E21A054, and E21A056, the MDAR and the ISW methods resulted in higher NSE compared to the DAR method. For 4 out of 7 target stations, the results can be rated as "very good" for the NSE according to the performance ratings in Table 5. For 5 out of 7 target stations, the RSR values were considered satisfactory (i.e., less than 0.7) according to the performance ratings recommended by Moriasi et al. [47]. The negative PBIAS values for D21A193, E21A051, E21A054, and E21A066 demonstrate that the method overestimated daily streamflow, while positive PBIAS values for D21A001, D21A033, and E21A056 demonstrate underestimation. For all target stations other than D21A193, the statistical methods with MA tend to achieve slightly higher NSE values. However, the PBIAS values of the target stations did not change when the statistical methods with MA were used.

For the target station E21A051 as the example, the estimated streamflow values from the statistical methods without the MA were compared to observed (original) streamflow values in the hydrograph and scatter plots in Figure 8. The remarkably better agreement between observed and estimated streamflow values by three statistical methods was obtained for E21A051 compared to the other stations in the Upper Euphrates Basin. Both MDAR and ISW1 gave a coefficient of determination (R^2) of 0.94, which was higher than the R^2 values of 0.89 by using DAR. The NSE values for these methods ranged from 0.883 to 0.932, and the best NSE value was achieved by MDAR. The best NSE performance for E21A051 was obtained using MDAR with the two most physically similar donor stations.

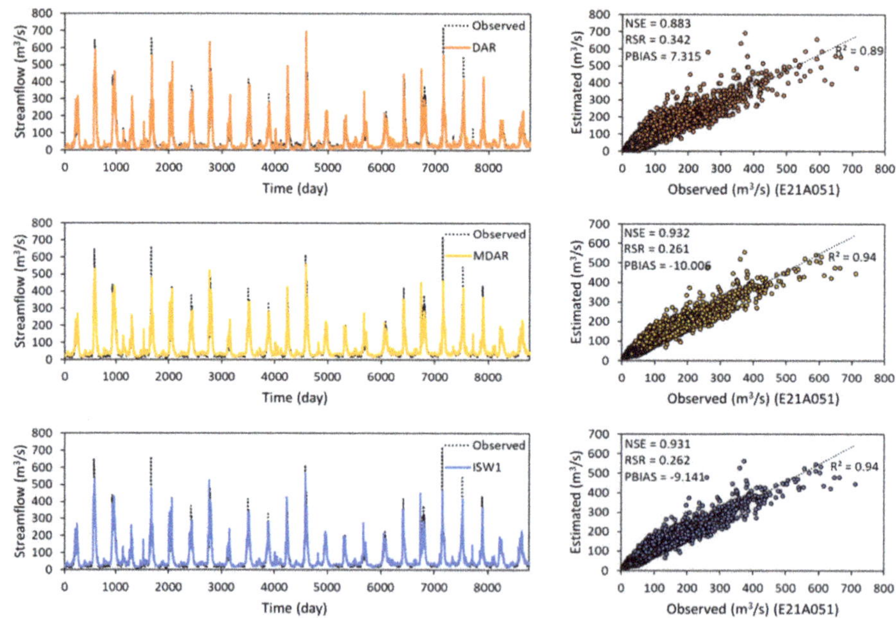

Figure 8. Comparison of daily streamflow estimated using DAR (top row), MDAR (mid row), and ISW1 (bottom row) with observed (original) daily streamflow at station E21A051 in the Upper Euphrates Basin.

On the other hand, for the target station E21A051 as the example, the estimated streamflow values from the statistical methods with the MA were compared to observed-MA (smoothed) streamflow values in the hydrograph and scatter plots in Figure 9. The statistical methods with the MA performed slightly better for E21A051.

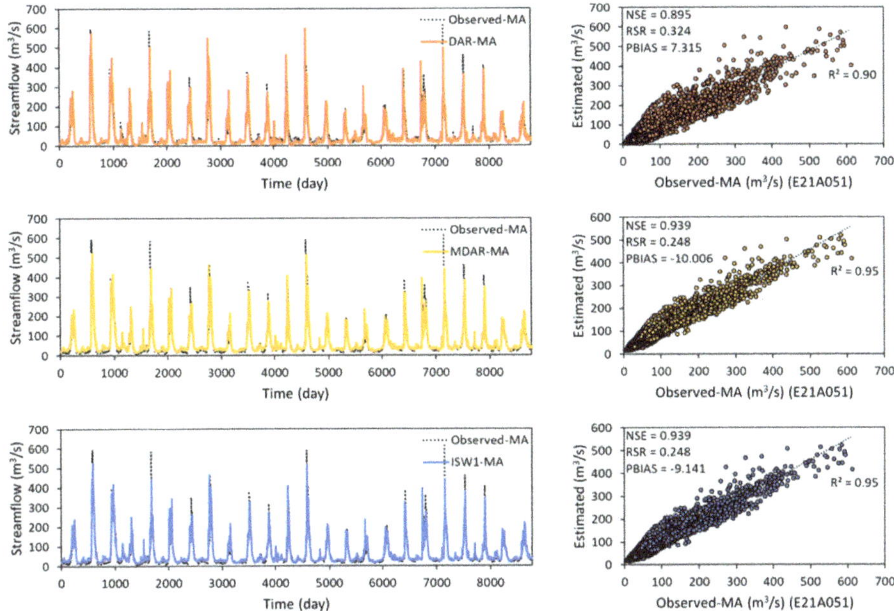

Figure 9. Comparison of daily streamflow estimated using DAR-MA (top row), MDAR-MA (mid row), and ISW1-MA (bottom row) with observed-MA (smoothed) daily streamflow at station E21A051 in the Upper Euphrates Basin.

5. Conclusions

This study provided a comparative evaluation of three statistical methods, DAR, MDAR, and ISW, which estimate daily streamflow at ungauged basins. These statistical methods were applied to two study basins: The Middle and Upper Euphrates basins in Turkey. DAR was implemented with the most physically similar donor station determined using the similarity coefficient. On the other hand, the two and the three most physically similar donor stations were used with both MDAR and ISW. By using three different power parameters (1, 2, and 3) in ISW, the effect of the selection of different power parameters on the accuracy of the daily streamflow estimation was tested. In addition, this study investigated the effects of the statistical methods using the smoothed data by the MA on the accuracy and reliability of daily streamflow estimation. Three statistical methods using the smoothed data by the MA, named DAR-MA, MDAR-MA, and ISW-MA, were proposed. The performance of each statistical method was evaluated in terms of the NSE, RSR, PBIAS, and R^2 between the observed and estimated daily streamflow. When the estimated daily streamflow values at the target station were obtained from the statistical methods using the observed (original) daily streamflow values at the donor station(s), they were compared to the observed (original) daily streamflow values at the target station. On the other hand, when the estimated daily streamflow values at the target station were obtained from the statistical methods using the observed-MA (smoothed) daily streamflow values at the donor station(s), they were compared to the observed-MA (smoothed) daily streamflow values at the target station. These two approaches were presented to estimate the daily streamflow values with and without MA. It is believed that the results will help decision makers choose the best one for their objectives.

In the Middle Euphrates Basin, the DAR method resulted in negative NSE values. indicating unsatisfactory performance for 3 out of 8 target stations when the most physically similar donor station was used. These negative NSE values obtained with DAR improved considerably when the three most physically similar donor stations were used with MDAR. Higher NSE values were mostly obtained from both MDAR and ISW used with the three most physically similar donor stations instead of the two most physically similar donor stations. ISW with a power parameter of 1 (i.e., ISW1) mostly outperformed compared to ISW2 and ISW3, when both the two and the three most physically similar donor stations were used. The results obtained for 8 target stations in the Middle Euphrates Basin indicated that ISW for 4 stations, MDAR for 3 stations, and DAR for 1 station performed best in estimating daily streamflow. For all but one target station, the NSE values obtained were greater than 0.6, indicating good or very good performance. For all target stations, the performance of the best statistical method for each target station slightly improved when the smoothed data by the MA was used.

In the Upper Euphrates Basin, for one target station, the NSE value improved from a negative to over 0.7 when MDAR was applied instead of DAR. Higher NSE values were mostly obtained from both MDAR and ISW used with the two most physically similar donor stations instead of the three most physically similar donor stations. ISW1 used with the two most physically similar donor stations and ISW3 used with the three most physically similar donor stations gave better performance than the others. The results obtained for 7 target stations in the Upper Euphrates Basin indicated that DAR for 3 stations, MDAR for 2 stations, and ISW for 2 stations performed best in estimating daily streamflow. For all but two target stations, the NSE values obtained were greater than 0.6, indicating good or very good performance. For 6 out of 7 target stations, the performance of the best statistical method for each target station slightly improved when the smoothed data by the MA was used.

The overall results suggest that, besides the statistical method selection, the selection of appropriate donor stations is an important step to achieve better streamflow estimates at target stations. Also important is that increasing the number of donor stations can also improve or decrease estimation performance. Besides, the estimation performance of the statistical methods can vary from basin to basin. Moreover, data preprocessing can have a positive effect on the estimation performance of statistical methods.

Finally, since obtaining reliable and accurate streamflow estimations is very important in water resource studies, the statistical methods used in the study can be easily applied for decision making and design in many water resources projects that have difficulty in obtaining data.

Author Contributions: Conceptualization, M.U.Y. and B.O.; methodology, M.U.Y. and B.O.; software, M.U.Y.; formal analysis, M.U.Y. and B.O.; investigation, M.U.Y. and B.O.; writing—original draft preparation, M.U.Y.; writing—review and editing, M.U.Y. and B.O.; visualization, M.U.Y. and B.O.; supervision, B.O. All authors have read and agreed to the published version of the manuscript.

Funding: This study was supported by Research Fund of Istanbul Technical University. Project Number: 40717.

Acknowledgments: This study was supported by the Research Fund of Istanbul Technical University under the project "Improvement of Streamflow Estimation in Ungauged Basins" (Project number: 40717). The authors thank the General Directorate of State Hydraulic Works (DSI) and the Turkish State Meteorological Service (DMI) for providing data.

Conflicts of Interest: The authors declare no conflict of interest.

References

1. Sivapalan, M.; Takeuchi, K.; Franks, S.W.; Gupta, V.K.; Karambiri, H.; Lakshmi, V.; Liang, X.; McDonnell, J.J.; Mendiondo, E.M.; O'Connell, P.E.; et al. IAHS decade on predictions in ungauged basins (PUB), 2003–2012: Shaping an exciting future for the hydrological sciences. *Hydrol. Sci. J.* **2003**, *48*, 857–880. [CrossRef]

2. Ergen, K.; Kentel, E. An integrated map correlation method and multiple-source sites drainage-area ratio method for estimating streamflows at ungauged catchments: A case study of the Western Black Sea Region, Turkey. *J. Environ. Manage.* **2016**, *166*, 309–320. [CrossRef] [PubMed]

3. Yilmaz, M.U.; Onoz, B. Evaluation of statistical methods for estimating missing daily streamflow data. *Tek. Derg.* **2019**, *30*, 9597–9620. [CrossRef]

4. Besaw, L.E.; Rizzo, D.M.; Bierman, P.R.; Hackett, W.R. Advances in ungauged streamflow prediction using artificial neural networks. *J. Hydrol.* **2010**, *386*, 27–37. [CrossRef]

5. Huo, Z.; Feng, S.; Kang, S.; Huang, G.; Wang, F.; Guo, P. Integrated neural networks for monthly river flow estimation in arid inland basin of Northwest China. *J. Hydrol.* **2012**, *420*, 159–170. [CrossRef]

6. Noori, N.; Kalin, L. Coupling SWAT and ANN models for enhanced daily streamflow prediction. *J. Hydrol.* **2016**, *533*, 141–151. [CrossRef]

7. Chang, F.J.; Chen, Y.C. A counterpropagation fuzzy-neural network modeling approach to real time stream-flow prediction. *J. Hydrol.* **2001**, *245*, 153–164. [CrossRef]

8. Ozger, M. Comparison of fuzzy inference systems for streamflow prediction. *Hydrol. Sci. J.* **2009**, *54*, 261–273. [CrossRef]

9. Toprak, Z.F.; Eris, E.; Agiralioglu, N.; Cigizoglu, H.K.; Yilmaz, L.; Aksoy, H.; Coskun, H.G.; Andic, G.; Alganci, U. Modeling monthly mean flow in a poorly gauged basin by fuzzy logic. *CLEAN-Soil Air Water* **2009**, *37*, 555–564. [CrossRef]

10. Khu, S.T.; Liong, S.Y.; Babovic, V.; Madsen, H.; Muttil, N. Genetic programming and its application in real-time runoff forecasting. *J. Am. Water Resour. Assoc.* **2001**, *37*, 439–451. [CrossRef]

11. Maity, R.; Kashid, S.S. Hydroclimatological approach for monthly streamflow prediction using genetic programming ISH. *J. Hydraul. Eng.* **2009**, *15*, 89–107. [CrossRef]

12. Mehr, A.D.; Kahya, E.; Olyaie, E. Streamflow prediction using linear genetic programming in comparison with a neuro-wavelet technique. *J. Hydrol.* **2013**, *505*, 240–249. [CrossRef]

13. Lin, J.Y.; Cheng, C.T.; Chau, K.W. Using support vector machines for long-term discharge prediction. *Hydrol. Sci. J.* **2006**, *51*, 599–612. [CrossRef]

14. Solomatine, D.P.; Maskey, M.; Shrestha, D.L. Instance-based learning compared to other data-driven methods in hydrological forecasting. *Hydrol. Process.* **2008**, *22*, 275–287. [CrossRef]

15. Yilmaz, A.; Muttil, N. Runoff estimation by machine learning methods and application to Euphrates Basin in Turkey. *J. Hydrol. Eng.* **2014**, *19*, 1015–1025. [CrossRef]

16. Wu, C.L.; Chau, K.W.; Li, Y.S. Methods to improve neural network performance in daily flows prediction. *J. Hydrol.* **2009**, *372*, 80–93. [CrossRef]

17. Di, C.; Yang, X.; Wang, X. A Four-stage hybrid model for hydrological time series forecasting. *PLoS ONE* **2014**, *9*, e104663. [CrossRef]

18. Mehr, A.D.; Kahya, E. A Pareto-optimal moving average multigene genetic programming model for daily streamflow prediction. *J. Hydrol.* **2017**, *549*, 603–615. [CrossRef]

19. Zhou, J.; Peng, T.; Zhang, C.; Sun, N. Data pre-analysis and ensemble of various artificial neural networks for monthly streamflow forecasting. *Water* **2018**, *10*, 628. [CrossRef]

20. Wu, C.L.; Chau, K.W. Prediction of rainfall time series using modular soft computing methods. *Eng. Appl. Artif. Intell.* **2013**, *26*, 997–1007. [CrossRef]

21. Waseem, M.; Ajmal, M.; Kim, U.; Kim, T.W. Development and evaluation of an extended inverse distance weighting method for streamflow estimation at an ungauged site. *Hydrol. Res.* **2016**, *47*, 333–343. [CrossRef]

22. Razavi, T.; Coulibaly, P. Improving streamflow estimation in ungauged basins using a multi-modelling approach. *Hydrol. Sci. J.* **2016**, *61*, 2668–2679. [CrossRef]

23. Huang, W.C.; Yang, F.T. Streamflow estimation using kriging. *Water Resour. Res.* **1998**, *34*, 1599–1608. [CrossRef]

24. Farmer, W.H. Ordinary kriging as a tool to estimate historical daily streamflow records. *Hydrol. Earth Syst. Sci.* **2016**, *20*, 2721–2735. [CrossRef]

25. Tencaliec, P.; Favre, A.C.; Prieur, C.; Mathevet, T. Reconstruction of missing daily streamflow data using dynamic regression models. *Water Resour. Res.* **2015**, *51*, 9447–9463. [CrossRef]

26. Masselot, P.; Dabo-Niang, S.; Chebana, F.; Ouarda, T.B. Streamflow forecasting using functional regression. *J. Hydrol.* **2016**, *538*, 754–766. [CrossRef]

27. Swain, J.B.; Patra, K.C. Streamflow estimation ungauged catchments using regional flow duration curve: comparative study. *J. Hydrol. Eng.* **2017**, *22*, 04017010. [CrossRef]

28. Burgan, H.I.; Aksoy, H. Annual flow duration curve model for ungauged basins. *Hydrol. Res.* **2018**, *49*, 1684–1695. [CrossRef]

29. Farmer, W.H.; Vogel, R. Performance-weighted methods for estimating monthly streamflow at ungauged sites. *J. Hydrol.* **2013**, *477*, 240–250. [CrossRef]

30. Yilmaz, M.U. Performance-Weighted Methods for Estimating Monthly Streamflow: An Application for Middle Part of Euphrates Basin. Master's Thesis, Istanbul Technical University, Institute of Science and Technology, Istanbul, Turkey, 2014.

31. Merz, R.; Blöschl, G. Regionalisation of catchment model parameters. *J. Hydrol.* **2004**, *287*, 95–123. [CrossRef]

32. Shu, C.; Ouarda, T.B.M.J. Improved methods for daily streamflow estimates at ungauged sites. *Water Resour. Res.* **2012**, *48*. [CrossRef]

33. Hrachowitz, M.; Savenije, H.H.G.; Blöschl, G.; McDonnell, J.J.; Sivapalan, M.; Pomeroy, J.W.; Arheimer, B.; Blume, T.; Clark, M.P.; Ehret, U.; et al. A decade of Predictions in Ungauged Basins (PUB)—A review. *Hydrol. Sci. J.* **2013**, *58*, 1198–1255. [CrossRef]

34. Alipour, M.H.; Kibler, K. A framework for streamflow prediction in the world's most severely data-limited regions: test of applicability and performance in a poorly-gauged region of China. *J. Hydrol.* **2018**, *557*, 41–54. [CrossRef]

35. Alipour, M.H.; Kibler, K.M. Streamflow prediction under extreme data scarcity: a step toward hydrologic process understanding within severely data-limited regions. *Hydrol. Sci. J.* **2019**, *64*, 1038–1055. [CrossRef]

36. Patil, S.; Stieglitz, M. Controls on hydrologic similarity: role of nearby gauged catchments for prediction at an ungauged catchment. *Hydrol. Earth Syst. Sci.* **2012**, *16*, 551–562. [CrossRef]

37. Heng, S.; Suetsugi, T. Comparison of regionalization approaches in parameterizing sediment rating curve in ungauged catchments for subsequent instantaneous sediment yield prediction. *J. Hydrol.* **2014**, *512*, 240–253. [CrossRef]

38. Yang, X.; Magnusson, J.; Rizzi, J.; Xu, C.Y. Runoff prediction in ungauged catchments in Norway: comparison of regionalization approaches. *Hydrol. Res.* **2017**, *49*, 487–505. [CrossRef]

39. Emerson, D.G.; Vecchia, A.V.; Dahl, A.L. *Evaluation of Drainage-Area Ratio Method Used to Estimate Streamflow for the Red River of the North Basin, North Dakota and Minnesota*; U.S. Geological Survey Scientific Investigations Report 2005–5017; U.S. Geological Survey: Reston, VA, USA, 2005.

40. Asquith, W.H.; Roussel, M.C.; Vrabel, J. *Statewide Analysis of the Drainage-Area Ratio Method for 34 Streamflow Percentile Ranges in Texas*; U.S. Geological Survey Scientific Investigations Report 2006–5286; U.S. Geological Survey: Reston, VA, USA, 2006.

41. Oudin, L.; Andréassian, V.; Perrin, C.; Michel, C.; Le Moine, N. Spatial proximity, physical similarity, regression and ungaged catchments: A comparison of regionalization approaches based on 913 French catchments. *Water Resour. Res.* **2008**, *44*. [CrossRef]

42. Zelelew, M.B.; Alfredsen, K. Transferability of hydrological model parameter spaces in the estimation of runoff in ungauged catchments. *Hydrol. Sci. J.* **2014**, *59*, 1470–1490. [CrossRef]

43. Arsenault, R.; Brissette, F.P. Continuous streamflow prediction in ungauged basins: the effects of equifinality and parameter set selection on uncertainty in regionalization approaches. *Water Resour. Res.* **2014**, *50*, 6135–6153. [CrossRef]

44. He, Y.; Bárdossy, A.; Zehe, E. A review of regionalisation for continuous streamflow simulation. *Hydrol. Earth Syst. Sci.* **2011**, *15*, 3539–3553. [CrossRef]

45. Burn, D.H.; Boorman, D.B. Estimation of hydrological parameters at ungauged catchments. *J. Hydrol.* **1993**, *143*, 429–454. [CrossRef]

46. Nash, J.E.; Sutcliffe, J.V. River flow forecasting through conceptual models part I–A discussion of principles. *J. Hydrol.* **1970**, *10*, 282–290. [CrossRef]

47. Moriasi, D.N.; Arnold, J.G.; Van Liew, M.W.; Bingner, R.L.; Harmel, R.D.; Veith, T.L. Model evaluation guidelines for systematic quantification of accuracy in watershed simulations. *Trans. ASABE* **2007**, *50*, 885–900. [CrossRef]

48. Christiansen, D.E.; Haj, A.E.; Risley, J.C. *Simulation of Daily Streamflow for 12 River Basins in Western Iowa Using the Precipitation-Runoff Modeling System*; U.S. Geological Survey Scientific Investigations Report 2017-5091; U.S. Geological Survey: Reston, VA, USA, 2017.

49. Kalin, L.; Isik, S.; Schoonover, J.E.; Lockaby, B.G. Predicting water quality in unmonitored watersheds using artificial neural networks. *JEQ* **2010**, *39*, 1429–1440. [CrossRef]

Article

UAV and LiDAR Data in the Service of Bank Gully Erosion Measurement in Rambla de Algeciras Lakeshore

Radouane Hout [1,*], Véronique Maleval [1], Gil Mahe [2], Eric Rouvellac [1], Rémi Crouzevialle [1] and Fabien Cerbelaud [1]

1   Department of Geography, Faculté des Lettres et Sciences Humaines, Limoges University, 39E rue Camille Guérin, 87036 Limoges CEDEX, France; veronique.maleval@unilim.fr (V.M.); eric.rouvellac@unilim.fr (E.R.); remi.crouzevialle@unilim.fr (R.C.); fabien.cerbelaud@unilim.fr (F.C.)
2   IRD, UMR HSM IRD, University Montpellier, 34090 Montpellier, France; gil.mahe@ird.fr
*   Correspondence: radouane.hout@unilim.fr; Tel.: +33-7-69-21-47-85

Received: 30 July 2020; Accepted: 24 September 2020; Published: 1 October 2020

Abstract: The Rambla de Algeciras lake in Murcia is a reservoir for drinking water and contributes to the reduction of flooding. With a semi-arid climate and a very friable nature of the geological formations at the lakeshore level, the emergence and development of bank gullies is favored and poses a problem of silting of the dam. A study was conducted on these lakeshores to estimate the sediment input from the bank gullies. In 2018, three gullies of different types were the subject of three UAV photography missions to model in high resolution their low topographic change, using the SfM-MVS photogrammetry method. The combination of two configurations of nadir and oblique photography allowed us to obtain a complete high-resolution modeling of complex bank gullies with overhangs, as it was the case in site 3. To study annual lakeshore variability and sediment dynamics we used LiDAR data from the PNOA project taken in 2009 and 2016. For a better error analysis of UAV photogrammetry data we compared spatially variable and uniform uncertainty models, while taking into account the different sources of error. For LiDAR data, on the other hand, we used a spatially uniform error model. Depending on the geomorphology of the gullies and the configuration of the data capture, we chose the most appropriate method to detect geomorphological changes on the surfaces of the bank gullies. At site 3 the gully topography is complex, so we performed a 3D distance calculation between point clouds using the M3C2 algorithm to estimate the sediment budget. On sites 1 and 2 we used the DoD technique to estimate the sediment budget as it was the case for the LiDAR data. The results of the LiDAR and UAV data reveal significant lakeshore erosion activity by bank gullies since the annual inflow from the banks is estimated at 39 T/ha/year.

Keywords: Rambla de Algeciras; semi-arid; lake; lakeshores; silting; bank gullies; UAV; LiDAR; DoD; M3C2

---

1. Introduction

Gully erosion is a very common form of water erosion, this geomorphological process is triggered after the concentration of surface or groundwater in narrow areas. This drags soil particles and rock fragments from these narrow flow paths to considerable depths. The steep slopes, the low density of plant cover, the overexploitation of arable lands, overgrazing as well as the torrential flow regime due to irregular rainfall, are all factors favorable to the initiation and even the acceleration of the gully erosion phenomenon [1–3].

This process of global erosion occurs in urban, forest and even rural areas [4]. That is why we find the term "gully" depicted in several languages, it is called, "Ravine" in French, "wadi" in Arabic, "'cárcava" in Spain, "lavaka" in Madagascar, "donga" in South Africa, "voçoroca" in Brazil and "barranco" in Argentina [4].

Globally, gullying accounts for 10% up 94% of the total sediment yield produced by water erosion [5], with an average contribution of 50% to 80% in arid areas [6]. In Western European countries, the same process generates between 30 and 80% [5] of the sediment.

Bank gullies are themselves a very particular form of gullies which are formed after a height drop from terraces or river banks, the lake banks, and which develop by headward retreat in erodible hillslopes. Bank gullies are themselves a very particular form of gullies which are formed after a height drop from terraces or river banks, the lake banks, and which develop by headward retreat in erodible hillslopes.

Furthermore, Vanmaercke et al (2016) [7] have demonstrated that gullying and bank gullies are the main sources of sediment loss. It is also important to note that they also have off-site effects [8] such as degradation of water quality, siltation of dams, pollution, degradation of agricultural land, reduction of soil moisture, hydrological dysfunction, and even fertility transfers [9]. Vanmaercke et al (2016) [7] confirmed that the recession rate of bank gullies heads varies between 0.01 m/year and 135.2 m/year with an average of 5 m/year, and a median of 0.89 m/year in Spain. In most landscapes that are subject to different climatic conditions and different anthropogenic activities, we can observe the presence of various forms of gully erosion, namely ephemeral gullies, permanent or classic gullies and bank gullies [5].

In the literature, researchers address the question of gullying according to different approaches:

- Approach based on the quantitative estimation of the sediment balance, using direct measurements in the field, namely, measuring tape, total station (TS), DGPS (differential GPS), drone (UAV), LiDAR, etc. [10];
- Indirect approach based on the modeling of the gully's sub-processes, by numerical or statistical gully forecasting models such as the SCS (Soil Conservation Service) model [11], the Thompson models [12,13], AnnAGNPS [14], EGEM (Ephemeral Gully Erosion Model) [15], GULTEM (model to Predict Gully Thermoerosion and Erosion) [16], or CHILD (Channel-Hillslope Integrated Landscape Development) [17].

The work of Evans [18] has shown that the activity of bank gullies lead to the evolution of hillslope-channel coupling related to flow erosion. In Morocco, the work of Maleval [19] demonstrated the importance of sedimentary inputs from the bank gullies in the silting of the SMBA dam. Monitoring and mapping of gullies for lakeshores remain one of the very important objectives in the management of drinking water reservoirs. Very high resolution topographic acquisitions have considerably improved our ability to better understand the functioning and evolution of geomorphological changes in the shoreline. The emergence of new drone photography techniques in recent years has greatly facilitated digital terrain modeling and geomorphological analysis. This ease is mainly due to the appearance of civil drones (Unmanned Airborne Vehicles, UAVs) and to SfM-MVS 3D photogrammetry algorithms, but also to the development of the computational capacity of computers. This method is increasingly being used for modeling and quantifying gullying in several regions of the world and in various climatic contexts, with work carried out in Morocco [20], Spain [21], China [22], Italy [23], Ethiopia [24], the United States [25], Australia [26], etc.

In this paper, the gullies for lakeshores of the Rambla de Algaciras dam have been mapped using UAV and LiDAR. Our objective was to understand the influence of gullies for lakeshores in the silting of artificial dams using 4D models and photo-interpretation of aerial photos. A detailed methodology for the collection, processing and validation of very high resolution 4D models has been developed to quantify sediment budgets and geomorphological changes of lake bank gullies in the short and medium term. The topographic complexity of certain bank gullies was taken into account in this work through a combination of oblique and vertical acquisition of aerial photos.

In this article we also measured the accuracy of digital terrain models taken by UAVs with the k-fold cross-validation method, to estimate the sediment budget on bank gullies using the DoD method and the M3C2-PM algorithm. The conclusions of this study will guide dam managers to implement several strategies to take into account the contributions of bank gullies which are generally underestimated in the hydro-sediment balances.

2. Methodology

2.1. Study Site

In the south of Spain, in Alhama de Murcia, the Rambla de Algeciras reservoir was built in 1995 on the Rambla de Algeciras river. It has a torrential character, it rises at the gully of Valdelaparra and flows into the Guadalentín river. The dam is 80 m high and the area of the lake is 235.50 ha and 42.13 hm^3 (Figure 1). The dam was built under the project "Plan General de Defensa contra Avenidas de la Cuenca del Segura", intended to develop and regularize water resources in the Segura watershed in south-eastern Spain. The Rambla de Algeciras dam is located in the catchment area by the same name, southwest of the city of Murcia at the northern limit of the depression of the Guadalentín river (left bank). Surrounded by las sierras del Cura, de Espuña and La Muela. This watershed covers an area of 44.91 km^2, and 15 km, with an altitude that varies between 200 and 1320 m above sea level. It undergoes a high rate of erosion reaching 100 t/ha/year (Global Atlas of the Murcia region 2002), linked to the torrential nature of precipitation and the detrital nature of the soil composed of marls and polygenic conglomerates.

Locally, Lake Rambla de Algeciras has a semi-arid climate, with short episodes of intense precipitation. Annual rainfall 393 mm, 60% of which is concentrated over 3 to 4 days/year.

The lakeshore and substrate of the Rambla de Algeciras dam are made up of different friable formations of the Tortonian period, on the banks of the lake we find blue loams interspersed with whitish calcareous loams and gypsiferous loams in discordance with calcarenite and molasses formations, on the right bank of the dam we note the presence of alternating soft and consolidated layers of polygenic conglomerates in a marly sand matrix.

The choice of gullies aimed to represent as faithfully as possible the erosion activity on the lakeshore, in a semi-arid climatic context. This choice takes into account several parameters such as the shape, the exposure, the head of the gullies, the density of the plant cover, and the lithology. Based on the diachronic photo-interpretation of aerial photos (PNOA), three gullies models were selected (Figure 1).

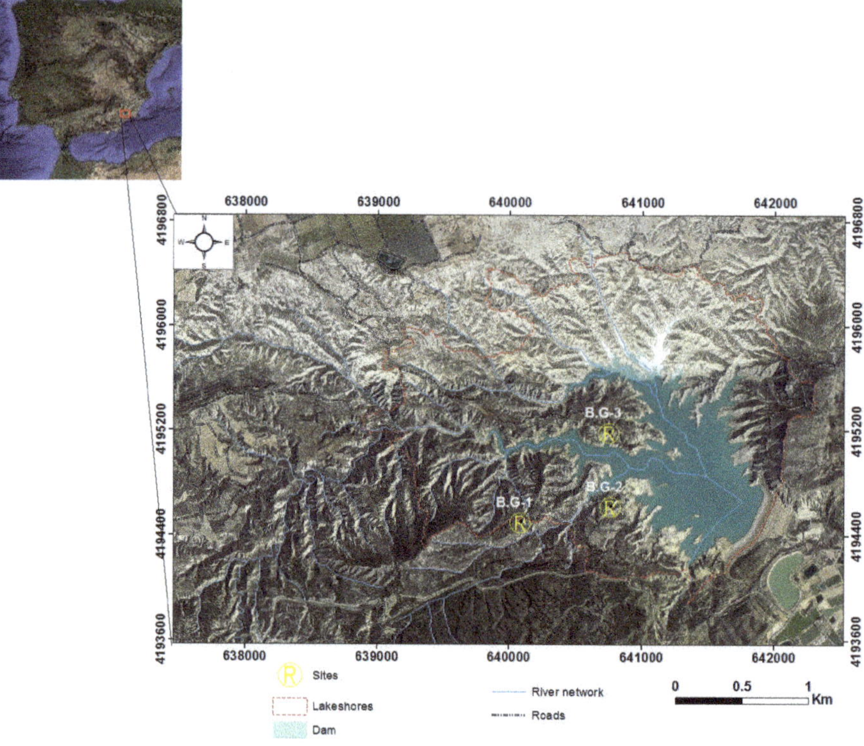

Figure 1. Location of the study area.

2.1.1. Site 1

The bank gully (B.G-1) (Figure 2C), extends over approximately 130 m. It is V-shaped in the upper and middle part, then takes a U-shape downstream from the gully, where the slopes are steep and exposed to the East (low grass cover) or to the West (bare soil). Upstream of the gully (head) is covered by tree layers, planted as part of the restoration project (Proyecto de correccion y repoblacion forestal de las cuencas de las Ramblas de Belen, Librillas y Algeciras. Provincia de Murcia). The North-south oriented gullies have been reforested and are difficult to access. Thus we studied the B.G-1 which is located 250 m from the reservoir. Access to the bank gully of the lakeshore protected by the forest cover and is very difficult. It is for this reason that we chose site B.G-1 which is located at 250 m from the lake.

2.1.2. Site 2

The second bank gully (B.G-2)(B.G-1) (Figure 2D) is U-shaped. It is oriented East-west with very steep slopes facing North (herbaceous vegetation) and South (bare soil). It is 110 m long, and receives tributary gullies.

2.1.3. Site 3

The third bank gully (B.G-3)(B.G-1) (Figure 2B) has a complex shape and is the result of a combination of tunnel gullying (pipig). The slopes of the gully (B.G-3) covered by bare soil. According to photo-interpretation, the B.G-3 gully began to develop after the lake level receded. It's likely to turn into badlands.

Figure 2. Photos of the lakeshore of Rambla de Algeciras. (**A**) Lakeshore. (**B**) Third bank gully. (**C**) Bank gully (B.G.-1). (**D**) Second bank gully.

2.2. Precipitation Analysis

2.2.1. Rainfall Extremes

Rare and very rare extreme events were identified using the prec90p index, of the STARDEX project (Statistical and Regional Dynamical downscaling of Extremes for European regions). The prec90p index is used to determine heavy rain events and intense rain when the cumulative daily precipitation values are greater than or equal to 90 and 95 percentiles, respectively. Between 2009 and 2016 there were two heavy showers (Figure 3).

- 7 June 2011, with cumulative precipitation of 81 mm in 2 h and a high intensity of precipitation which exceeds 40 mm/h.
- 28 September 2012, with cumulative precipitation was 84 mm produced in 4 h and an intensity of 21 mm/h.

These two rain events are considered as extreme events and automatically play an important role in erosion. The modification of the precipitation regime with the decrease in the amount of precipitation and rainy days as well as the greater frequency of rare and very rare events during the eight years of topographic surveys can be explained by the effect of climate change in the Mediterranean area [27].

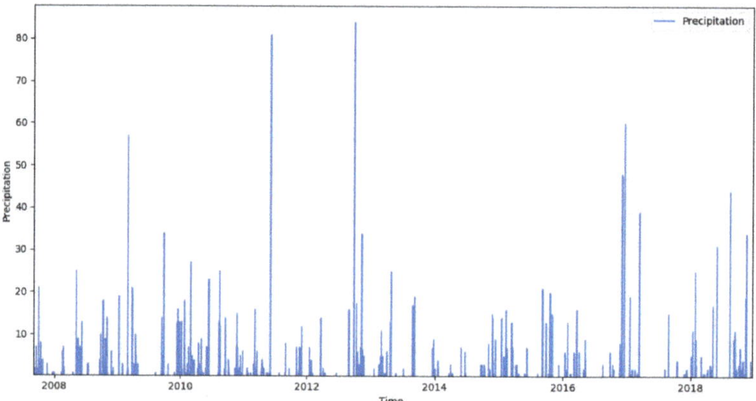

Figure 3. Rain event between 2009 and 2019 in the station of Rambla de Algeciras.

2.2.2. Rainfall Erosivity Factor "R-Factor"

The R-factor was estimated using equation (Equation (1)) (zone-2) proposed by ICONA for the region of Murcia. The monthly analysis of the factor R traces the evolution of the aggressiveness of the rains on the gullies between the study period of 2009 and 2016 as well as for the year 2018. Between 2009 and 2016 the devastating activity of the rains is concentrated annually in March and September. This activity is linked to the increase in intense rain events during this period. Throughout the period 2009 to 2016, the R-factor peaks are recorded in June 2011 and September 2012, which coincides with an anomaly in the kinetic energy explained by the recording of two extreme phenomena. The year 2018 is relatively wet therefore implying a natural increase in the factor R. From the graph (Figure 4) we can see that the average value of R, calculated for the wet season, is quite low compared to that of the dry season. During the wet season the values of (R) are moderate with an average of 8.8 MJ cm ha^{-1} h^{-1} year^{-1}, through contribution of the season which recorded an average of 17 MJ cm ha^{-1} h^{-1} year^{-1}. The lowest values are recorded between February and April. Conversely, during the dry season, the highest cumulative value of R is 102 MJ cm ha^{-1} h^{-1} year^{-1}, and it is associated with very intense precipitation in August and November. R: Climate aggressiveness (MJ cm ha^{-1} h^{-1} year^{-1}) PMEX: The maximum monthly value (mm) MR: Cumulative precipitation between October–May (mm) MV: Cumulative precipitation between June–September (mm):

$$R = e^{-1.235} \times \text{PMEX}^{1.297} \times \text{MR}^{-0.511} \times \text{MV}^{0.366} \times F_{24}^{0.414};$$

(1)

where

R: Rainfall erosivity factor (MJ cm ha^{-1} h^{-1} year^{-1})
PMEX: Maximum monthly pre-cipitation (mm)
MR: Total rainfall from October to May (mm)
MV: Total rainfall from June–September (mm)
F_{24}: Ratio of the square of the maximum annual rainfall in 24 hours (mm) to the sum of the maximum monthly rainfall in 24 hours (mm) $F_{24} = \dfrac{\left(P_{24h,annuel}\right)^2}{\sum_{i=1}^{12} P_{24h,i}}$.

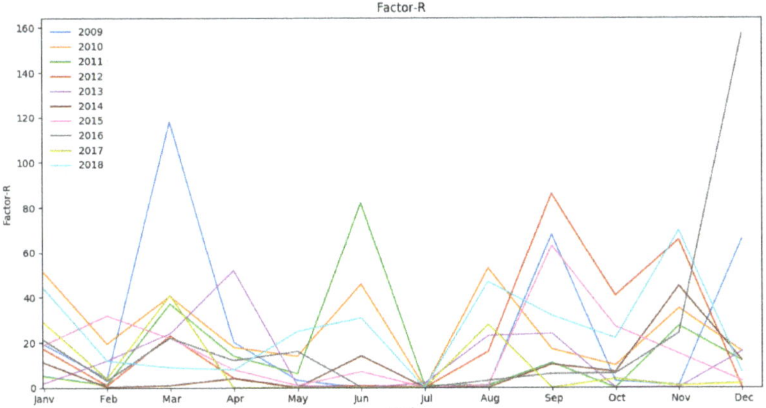

Figure 4. R-factor.

2.3. UAV Data Processing

2.3.1. UAV Data Collection

The aerial photography missions were carried out easily and at low cost by the quadriotor drone DJI Inspire 1 v2 (Figure 5A), equipped with the metric and multispectral camera ZENMUSE X3 without distortion or aspect and with a focal distance of 20 mm (Figure 5D). The DJI Inspire1 V2 is sold "Ready-To-Fly", and is designed to have a maximum speed of 22 m/s (ATTI mode, no wind), and a maximum flight time of approximately 18 min. It can reach 2500 m altitude in flight. In addition, it is equipped with a satellite positioning system which uses GLONASS and GPS system.

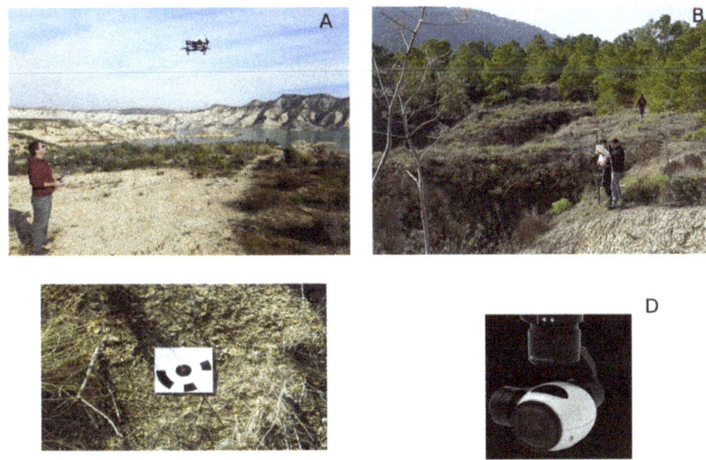

Figure 5. (**A**) Drone DJI Inspire 1 v2. (**B**) GPS Trimble Geo 7x. (**C**) Ground control points. (**D**) Camera ZENMUSE X3.

Flight mission planning was done in navigation mode via the Pix4Dcapture mobile application. After the definition of some flight parameters (flight plan, flight altitude, recovery rate, shooting angle, starting and arrival point of the UAV) the application automatically pilots the UAV. The acquisitions of aerial photos by drone were carried out according to two configurations, for the two gullies (B.G 1 and 2) the photos were taken nadir acquisitions. However, by geomorphological constraints of having gully (B.G-3) we combined the two modes of acquisition nadir and oblique at (30°) in order to model in 3D the overhanging areas of the gully (B.G-3). To ensure a successful photographic reconstruction we have chosen a longitudinal overlap of 80% and a lateral overlap of 60% to avoid gaps between the different photographs. A minimum of 10 ground control points (GCPs) (Figure 5C) (Table 1) is required on each site in the middle and at the stable edge, and the same targets were used in each drone mission (except for those which were covered by the sediments), in order to obtain optimal results. The surveys were carried out by the GPS Trimble Geo 7x (Figure 5B), which includes the following parameters: the ellipsoid GRS 1980, the datum: European Terrestrial Reference System 1989, coordinate system: ETRS89 / UTM zone 30N, with a horizontal accuracy of 1 cm + 1 ppm, and vertical of 1.5 cm + 2 ppm. We used the Stop&Go kinematic approach for the collection of GCPs points. We then made subsequent corrections using RINEX data from a reference station located 30 km from our study area.

Table 1. Characteristics of UAV photo missions.

Bank Gullies	Missions	Local Time	Flight Height (m)	Cloud Cover (Octas)	Number of Images	Density (pts/m^2)	Coverage (Ha)	Number of GCPs	Angle of Acquisition (°)
	22/01/18	11 h	27.6	02/08	436	517	2.95	7	90
B.G-1	04/06/18	15 h	36.4	01/08	304	408	2.34	7	90
	18/12/18	16 h	27.5	03/08	422	524	2.7	7	90
	22/01/18	15 h	31.3	02/08	646	425	4.3	13	90
B.G-2	04/06/18	11 h	29.4	01/08	465	455	1.79	13	90
	18/12/18	15 h	30	03/08	384	397	2.02	10	90
	22/01/18	17 h	37.4	02/08	741	394	2.28	16	90 + 30
B.G-3	04/06/18	10 h	28.9	01/08	520	433	1.28	14	90 + 30
	18/12/18	10 h	33.3	03/08	602	328	2.5	11	90 + 30

2.3.2. Processing

The aerial photo processing is based on the SfM-MVS photogrammetry technique. We have chosen the Agisoft PhotoScan commercial software to process all the photos taken by the drone, eventually generating dense point clouds and orthomosaics. This software generates exportable datasets and thus has the most comprehensive and easy to use workflow compared to other commercial and open source photogrammetry software's. The following steps on PhotoScan consisted of establishing dense point clouds which we will then export in ASCII or PTS "XYZ RGB" format (Figure 6).

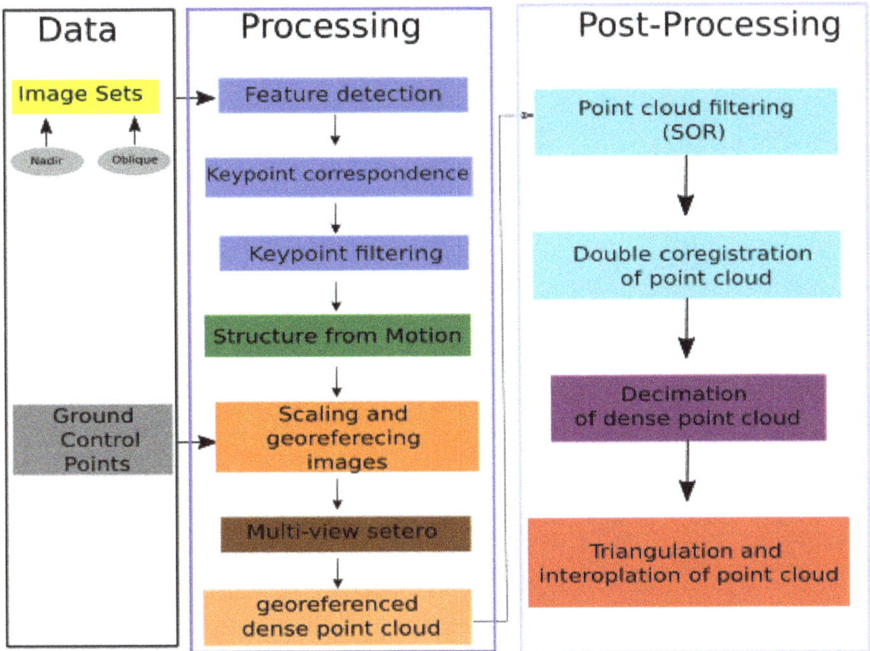

Figure 6. A schematic of a workflow that was applied to UAV data.

2.3.3. Post-Processing

The SfM-MVS technique is recognized as the fastest and inexpensive technique for 3D modeling. Post-processing was applied to the raw point clouds obtained by the SfM-MVS technique before generating the digital terrain model (Figure 6). The post-processing step was carried out by three open source external software: CloudCompare software (http://www.cloudcompare.org/), ToPCAT (https://code.google.com/archive/p/point-cloud-tools/downloads), and SAGA GIS software

Filtering and Coregistration of Data

CloudCompare open source software is designed for processing and comparing 3D point clouds.The dense 3D point clouds obtained were imported in ASCII XYZRGB formats into the CloudCompare software. We then filtered the point clouds using the SOR filter. Once the point clouds were filtered, we applied the ICP coregistration algorithm to match the homologous points of the multi-temporal point clouds, so that each dense point cloud would refer to the same area [28–32].

The Decimation and Interpolation of Point Clouds

Generating a digital terrain model from dense point clouds directly derived from MVS algorithms often requires large computing resources. The interpolation of the points for this kind of processing drains the machines in terms of memory, hence the interest for the decimation of the clouds of dense points which has the potential to reduce the number of points while preserving the topographic complexity of the soil surface. Persistent in the point cloud, we used the ToPCAT geospatial toolbox (Topographic Point Cloud Analysis Toolkit). ToPCAT is an intelligent analysis and decimation algorithm that breaks down the very dense point cloud into a set of non-overlapping meshes, suitable for detecting morphological changes. It then calculates the minimum elevation (Z_{min}) of each mesh. (Z_{min}) is considered to be the elevation of bare ground points [33–36]. The (Z_{min}) elevation obtained was then converted to TIN (Triangular Irregular Networks) in order to generate the digital terrain model under the SAGA GIS software.

2.4. LiDAR-PNOA Data Processing

In the Murcia region, two ALS airborne LiDAR missions were conducted, on 24 November 2009 and between 20 and 25 August 2016 (Figure 7). During the first mission, data were collected using a Leica ALS50-II sensor, and for the second mission, data were collected using a Leica ALS60-II sensor.

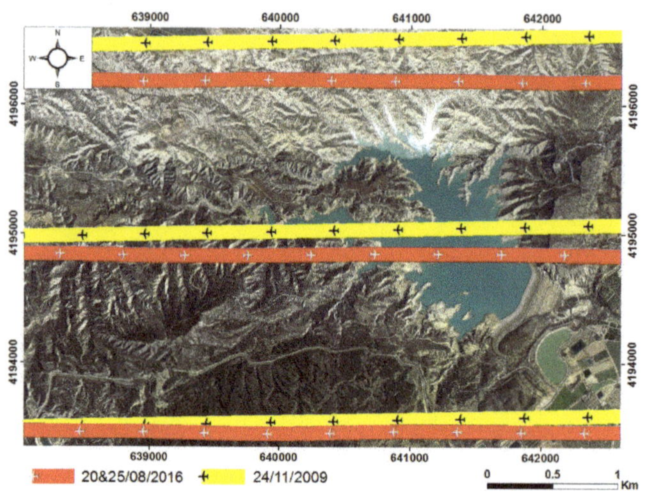

Figure 7. LiDAR flight turnings.

The LAZ files (binary file compression format LAS version 1.1) that we used come from the ETRS89 UTM Zone 30 geodetic reference system for X, Y and Z coordinates in the EGM08-REDNAP system. Before the classification of the LiDAR point cloud, we prepared the LiDAR raw data in the open source software LASTools. We first converted the LAZ files into LAS files using the las2las algorithm, then we applied a vertical and a horizontal alignment between the flight lines using the lasoverlap algorithm, and finally we transformed the point cloud into EGM08-REDNAP orthometric coordinates using the ascii file <EGM08-REDNAP.asc > and the lasheight algorithm.

Subsequently, we removed outliers and noise using the SOR filter in the CloudCompare software. After pre-processing the point cloud, we chose the MCC-LiDAR v2.15 (Multiscale Curvature Classification) application, an open source command line tool for processing LiDAR data. It classifies data points as ground or non-ground, using the MCC algorithm, to obtain a digital model of bare terrain. Due to the low density of PNOA LiDAR point clouds ranging from 0.5 to 1 point/m^2 , we did not apply a mesh to the cloud.

Finally, the ground points were triangulated and then interpolated with the linear interpolation algorithm under the SAGA GIS software, in order to obtain two ground DTMs of 1 m resolution for 2009 and 2016.

2.4.1. Error Modeling

Cross Validation: K-fold Cross-Validation

It is likely that the roughness-based estimates are incorrect, For quantify spatially variable SfM-DEM uncertainty we used cross-validation, then the results of this technique will be integrated from a FIS error model more representative of the different sources of uncertainty. This statistical technique of resampling without replacement consists in dividing the decimated point cloud into validation (20%) and training data (80%) (Figure 8). The clouds are divided repeatedly with the k-folds (k-fold cross validation) algorithm with a k = 5 (5 subsets of validation point cloud). The training points are interpolated to build a DEM, which compared the validation points in order to quantify the uncertainty of DEM-SfM. We chose the Scikit-Learn (v-0.22.0) free software machine learning library to apply k-fold cross validation on the point cloud decimated by the ToPCAT algorithm. 80% of the (Z_{min}) points were randomly selected to generate a DEM with a resolution of 5 cm. Then we used the remaining 20% of (Z_{min}) to compare them with the DEM. The average absolute difference is taken as an indicator of the altimetric uncertainty of the data. The cross-validation test was repeated five times to check the consistency of the results.

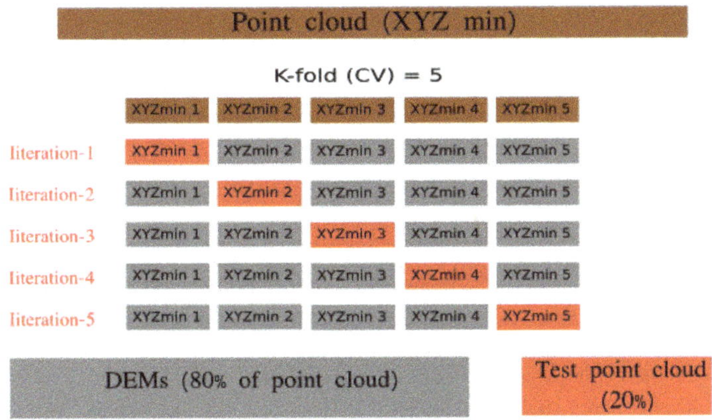

Figure 8. K-fold cross-validation model.

Fuzzy Inference System (FIS)

Within the SfM-MVS data, the sources of error vary according to the type of site. In the case of bank gullies, the main sources of error are roughness, slope of the terrain, camera geometry, interpolation error and point cloud density. In this case we were interested in a more comprehensive error modeling that takes into account the different sources of uncertainty. To meet the modelling requirements we adopted a spatially variable FIS (Fuzzy Inference System) error model, which is more reliable model errors in the SfM-MVS-4D data [35,37].

We defined four variables as the error sources of the SfM-MVS data to model with the FIS, namely slope, point density, interpolation error, and 3D point cloud error.Slope is the parameter most representative of the topographic complexity of the terrain, and it is the easiest parameter to derive from the DTM or point clouds. The slope (degrees) was generated directly from the decimated 'Z_{min}' point clouds with a resolution of 5 cm.

The modeling of the bare ground roughness from the value of the standard deviation of the decimated dense point clouds σ_{Zmin} can be considered to take into account the main source of error of the SfM-MVS-4D data. Terrain roughness affects the ability of the SfM-MVS algorithm to reconstruct the dense point cloud, since the error due to roughness depends on the position of the scan, which is never the same for multi-date missions, as well as the nature and occupancy of the terrain. We recall that vegetation is absent on our study gullies. This facilitates the calculation of roughness.

Then we also generated the interpolation error from a simple absolute difference between the altitude of the TIN node and the DTM-SfM $Z_{DTM} - Z_{NodesTIN}$. Finally, for the error variable of the 3D point clouds we used the results of cross-validation.

The matrix data of the different variables were reclassified into four classes ('low', 'medium', 'high' and 'extreme') (Table 2).

Table 2. A four input fuzzy inference system for elevation uncertainty.

Rule	Input				Output
	Slope (%)	Point Density (Pts/m^2)	3-D Point Quality	Interpolation Error (m)	Elevation Uncertainty (m)
1	Low	Low	NA	NA	High
2	Low	Medium	NA	NA	Medium
3	Low	High	NA	NA	Low
4	Medium	Low	NA	NA	High
5	Medium	Medium	NA	NA	Medium
6	Medium	High	NA	NA	Low
7	High	Low	NA	NA	High
8	High	Medium	NA	NA	High
9	High	High	NA	NA	Medium
10	Extreme	NA	NA	NA	Extreme
11	NA	NA	High	High	Extreme

2.5. Geomorphic Change Detection

An error model makes it possible to associate the control measurements made in the field with an accuracy estimate (example standard deviation in X, Y and Z). When applying the SfM-MVS method to create a digital terrain model, the errors generated are related to several factors such as the position and precision of the control points, camera distortions, resolution, number of photos, processing software, topographic complexity, interpolation methods, surface humidity, vegetation, etc., [38–40] . Several sources of error also affect the quality of DTMs for LiDAR data. These sources of error include terrain complexity, vegetation, weather, variation in point density, data filtering process and interpolation technique [41].

2.5.1. DEM of Difference (DoD)

This approach measures the vertical distance between two DEMs, through a simple subtraction between the pixels of these two DEMs according to the Equation (2). But before calculating the difference between the DEMs, a threshold value must be set for the minLOD (Equation (2)) [42,43]. This threshold means that substraction values lower than minLoD are revealed as being uncertainties of the two numerical models. We can thus assign a value for the minLoD with confidence intervals that we can set. The user must define this threshold (statistic) according to the probability that he wishes to have so that the detection of changes can be real (Equation (2)) [44]. This fast method allows to determine the uncertainties related to the roughness of the terrain, the quality of the point clouds and the acquisition of the point clouds. However, great uncertainty in the quantification of sediment budgets occurs when the DEMs are not precise (since the minLoD threshold will be higher).

$$minLoD = \sqrt{\varepsilon_{2i}^2 + \varepsilon_{1i}^2} \ \ or \ \ minLoD = t\sqrt{\varepsilon_{2i}^2 + \varepsilon_{1i}^2}; \tag{2}$$

where

- $t = \frac{Z_{2i} - Z_{1i}}{\sqrt{\varepsilon_{2i}^2 + \varepsilon_{1i}^2}}$
- ε_1 = The individual error of pixel i of first DTM
- ε_2 = The individual error of pixel i of second DTM
- t = is the critical value for a given confidence interval (1.28 for 80% CI, 1.96 for 95% CI); and z_2 and z_1 are the elevations in a given cell of DEMs.

2.5.2. Multi-scale Model to Model Cloud Comparison-Precision Maps (M3C2-PM)

M3C2-PM is an integrated approach in the CloudCompare software. It is used to directly compare dense point clouds in 3D instead of comparing DTMs or Mesh. The technique calculates the distance in the best perpendicular direction to the surface, with a confidence interval normally set at 95% (to assess the statistical significance of small topographic changes) and a measurement accuracy estimated from the point cloud roughness, allowing us to detect real topographic changes in elevation and planimetry between two degraded surfaces by gullying (erosion or sedimentation). The combination of three factors makes this the preferred method for determining the uncertainty of our digital terrain models: It can work perfectly even with changes in point density and even in the case of missing data. It calculates the distance between point clouds along the normal direction of the surface (the calculation of uncertainty on quasi-vertical areas). It estimates a spatially variable uncertainty for each point cloud and in accordance with terrain roughness and coregistration error [31,45]. To apply the M3C2 method, two parameters must be defined:

- 'D': Scale and orientation of the normal are used to calculate the normal area of each point which depends on the roughness and geometry of the point clouds. In the case of bank gullies we first applied a local model of <Quadric >shape along the preferential axis to orient the reference point cloud towards the comparison point cloud. Then we chose the value of D/2 = 10 cm to look for neighbouring points in the reference point cloud while taking into account the terrain roughness.
- 'd': Projection scale, or the diameter of the cylinder in which the average distance of each surface of the point clouds is calculated. In this study we used a value of d = 20cm.

3. Results

3.1. Data Processing

3.1.1. Data Precision

In this section we will focus mainly on the application and comparison of different error quantification techniques related to the collection and processing of SfM-MVS-4D data. Working on gullies of complex topography, the quantification of errors related to SfM-MVS data along three types of bank gullies allows on the one hand to analyze the reliability of the methodology adopted for data collection, and on the other hand to quantify with great precision the sediment budget over a relatively wet year.

Georeferencing Errors

The mean uncertainty values of the GCPs in XYZ were 1.5 cm (X), 2.4 cm (Y) and 5 cm (Z) (Table 3). These DGPS data processing results highlighted a flaw in the methodology for collecting GCP point's network. The points installed at the bottom of the gullies have a low accuracy compared to the others. They don't exceed 4 cm at best. This can be explained by the poor reception of the GNSS signal at the bottom of the gullies. The low precision of certain points reduced the general network precision of the GCPs, but this accuracy of the GCPs does not call into question the precision of the results since the three study sites were covered by a network of photos of high geometry that testify, by the total absence of systematic errors, to the shape of the surfaces such as the appearance of domes. In our case study, to georeference a network of photos with a high overlap, we used a network of GCPs well distributed spatially but of low precision. This greatly improved the georeferencing of the point clouds, where the value of the mean vertical RMES XY error is less than 2 cm for the 1st and 2nd sites and does not exceed 1.3 cm for the third (Table 3). The recessing of the geometry of the aerial photo network, either using a high rate of overlap, or with the combination of two modes of oblique and nadir acquisition, allowed us to obtain very precise internal measurements but weakly georeferenced in the external coordinate system.

Since the GCPs placed at the bottom of the gullies caused a degradation in the quality of their point network, it is necessary to ensure that the error in the SfM-MVS-4D results is acceptable.

Table 3. Errors in X, Y and Z of the three sites.

Sites	B.G-1			B.G-2			B.G-3		
Missions	M-I	M-II	M-III	M-I	M- II	M-III	M-I	M-II	M-III
X (cm)	1.93	2.96	2.17	3.26	4.94	4.47	1.46	1.01	1.18
Y (cm)	2.19	3.13	2.37	3.36	4.53	4.46	1.27	1.60	1.82
Z (cm)	5.32	5.13	4.87	5.3	5.79	5.67	1.36	1.44	1.98

Accuracy analysis of the control points showed a strong degradation in the accuracy of the control points depending on the depths of the gullies. This is due to the poor reception of the GNSS signal at the bottom of the gullies, which led us not to use GCPs for point cloud validation. In this study we chose to evaluate the quality of the data by three methods.

The Uncertainty of Coregistration

We presented in the methodology the procedure used for the coregistration of dense point clouds using the ICP algorithm which allows to considerably increase the accuracy of georeferencing of SfM-MVS-4D data in an external georeferencing system. One of the main sources of error in detecting changes between two point clouds is uncertainty in coregistration. Even with low coregistration uncertainties between each pair of dense point clouds, the risk of the uncertainty amplifying increases with the temporal resolution of the point cloud collection. To reduce the risk, we proceeded by coregistration in pairs of point clouds. Here we seek to assess the quality of dense point clouds from a dual coregistration (GCP and ICP). Data uncertainty is modeled using three spatially variable models: FIS and cross-validation.

Cross Validation

The training and validation data are constructed by resampling without replacement, which allows us to calculate the uncertainty on the different cloud points. This method is more robust, more stable and more thorough than Bootstrapping (resampling with discount). To our knowledge, this method has never been used to quantify the uncertainty of point clouds. The quantification of the uncertainty of the point clouds with the cross validation tools was successfully carried out for the three bank gullies. Analysis of the uncertainty results obtained by the cross-validation method shows that the absolute difference can go up to 0.5 m, but the frequency distribution shows an dissymmetry with an absolute average for the three gullies not exceeding 0.08 m and a standard deviation of 0.09. The repetitive analysis of uncertainty on the three gullies has brought to light aspects of the spatial structure of uncertainty. The visual interpretation of the results (Figure 9) aimed that the high uncertainty values tend to be concentrated on the talwegs (main and secondary), and on some parts of the slopes. On the other hand, the low values are generally found in the flat areas and at the tops of the gullies. The uncertainty values are plotted against the slope and the roughness. On the Figure 9, it appears that all the high values of uncertainties are strongly correlated with the steep slopes and the zones of strong roughness what met in relation to the spatial distribution of the uncertainties with the topographic complexity and the roughness of ground. It must be concluded that roughness and slope are the main sources of errors. In several studies carried out, in order to determine the accuracy of the SfM-MVS data, the DGPS surveys of the GCPs are compared with the final product (DEM-SFM-MVS). This confusion is deliberately maintained, as if the uncertainty of the SfM-MVS data was only related to the quality of the DGPS surveys, and that the geometry of the photos taken could not improve the accuracy of the data. In our case of river gullies, the possibility of using a high number of photos, of high overlap, of including oblique photos as well as the double coregistration of point clouds with the ICP algorithm allowed us to compensate and to improve the georeferencing weaknesses by the GCPs.

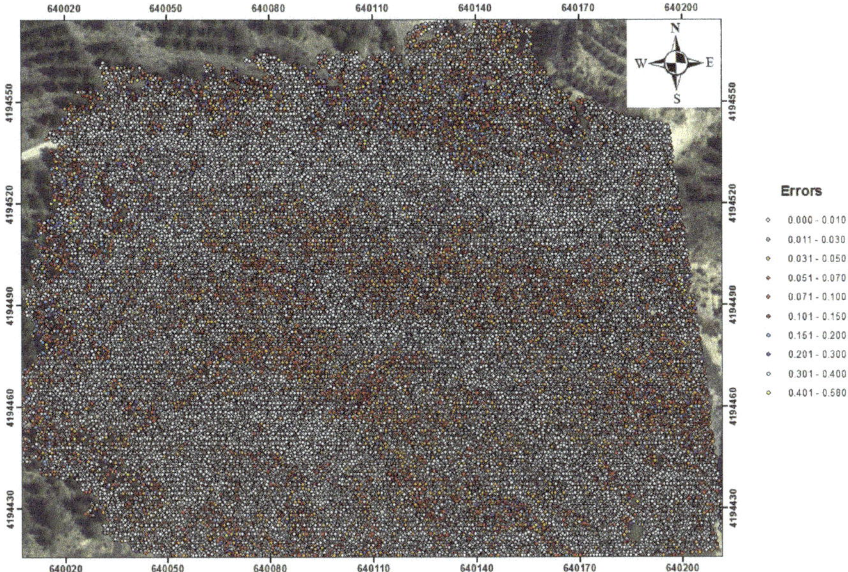

Figure 9. Example of point cloud uncertainty estimated by cross-validation on the gully B.G-1.

FIS

The results of the detection of geomorphological changes by the DoD technique are influenced by the type of error model used (spatially uniform or variable). Spatially uniform models can lead to overestimation or underestimates of topographic changes.

The FIS model, on the other hand, allows the changes to be reflected thanks to its better resolution, which leads to a better interpretation of the results. This is due to its ability to exclude noise, which is more important than for other models, also it leaves a morphological signature. Finally the results are transcribed in the form of a histogram, often bimodal. This tool facilitates the quantification of real changes in topographic models.

For the current study, on the three bank gullies, we detected geomorphological changes in the gullies with spatially variable FIS models. As the resolution of the data influences the detection of changes, we chose to use this model with a resolution of 5 cm. This technique allowed us to detect all possible changes, even the smallest ones, for our digital terrain models.

3.2. Detection of Geomorphological Changes from UAV Data

All change detection measurements over a period of time represent the result of the aggregation of all geomorphological processes. Positive values are assigned to all sedimentation processes, while negative values are appropriate for all erosion processes. According to the change map established by the DoD technique and the FIS propagated error, on both sides of the B.G-1 gully, erosion occurs with a significant difference between the two. This difference is mainly due to the exposure of the two slopes. The results of the two change detection techniques are similar but differ slightly in the quantification of sediment budgets. This small difference corresponds to the error model used by each technique. The two change detections are implemented by spatially variable error models. M3C2 uses roughness and coregistration error as error model variables. The FIS combines all the potential sources of errors (slope, point density, interpolation error, DTM-SfM-MVS error) to obtain a single error model. However, it is possible that the M3C2 technique underestimates the SfM-MVS error model.

3.2.1. DOD in Site 1

In general, monitoring of the sediment balance on the bank gully (B.G-1) allows us to notice significant changes during the year 2018. The bank gully (B.G-1) is located 250 m from the reservoir with a V shape.

Upstream the B.G-1 gully is covered by tree strata forming a protective layer. In this area of the gully, the changes are not significant, but photo-interpretation of the PNOA orthophotographs shows a strong degradation of the vegetation cover due to the digging of the head of the gully. The orthophotographs generated by the SfM-MVS data showed significant immobilization of the wood planks on the gully bed (Figure 10). On the south-western south facing slope, the annual rate of erosion between the 1st and 3rd missions varied between 30 and 60 cm/year with vertical erosion. Conversely, a low erosion rate of no more than 10 cm/year was recorded on the north-northeast facing slope. However, the rate of erosion is not constant on the same slope. That said, the rate of erosion or sedimentation varies according to the slope, roughness, and exposure of each slope (Figure 10).

It should be noted that the sediment balance on the gully (B.G-1) during the wet season shows a dominance of erosion activity over the entire gully. The south-western south slope shows the highest erosion rates of the gully. On the other hand, there are spatial variations in the erosion rate within the same slope, highlighting the contrast in the existence of land use variation upstream of the gully.

On the gully bed this dynamic is reversed. The transfer of sediment to the bed has generated a fattening of the bed. Between the first and the third mission, the calculated sedimentation rate is of the order of 353.85 T/ha/year (442.31 m^3/ha/year) (Figure 10) which does not reach the reservoir.

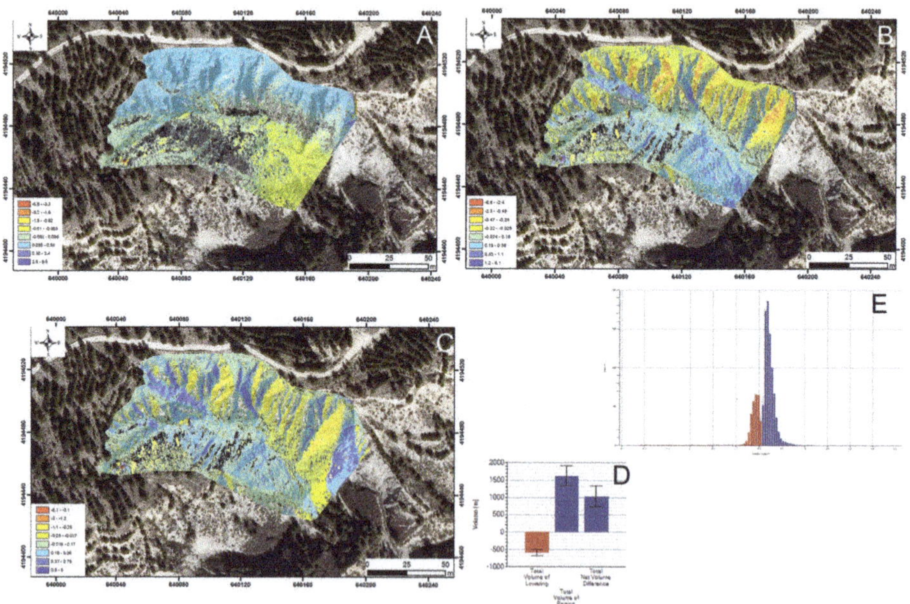

Figure 10. Bank gully (B.G-1)(**A**) DoD Mission I and II. (**B**) DoD Mission II and III. (**C**) DoD Mission I and III. (**D**) Graphe of DoD volume. (**E**) Bar chart of DoD volume change distribution (erosion volume in red, deposition volume in blue.

3.2.2. DOD in Site 2

The U-shaped B.G-2 gully, located near the reservoir, is highly exposed to erosion. The results measured by the DoD method have further highlighted the variation in the erosion rate between the two sides of the gully. The results show a strong activity of the annual sedimentation rate on the north-facing slope as opposed to the south-facing slope where the annual erosion rate exceeds 90 cm of digging, Similarly, on the map, we observe significant changes at the head of the gully. These changes are reflected by an erosion rate of up to 2 m/year (Figure 11).

On the gully bed we observe the same sedimentation phenomenon as in Gully B.G-1 since this sector represents a sedimentary basin of the elements eroded by the banks and the head of the gully. The results of the DoD method show that the annual sedimentation rate at the bottom of the gully is of the order of 4264.1 T/ha/year (5330.1 m^3/ha/year) (Figure 11D). At the time of maximum filling of the lake the bottom of the gully corresponds to the bottom of the lake.

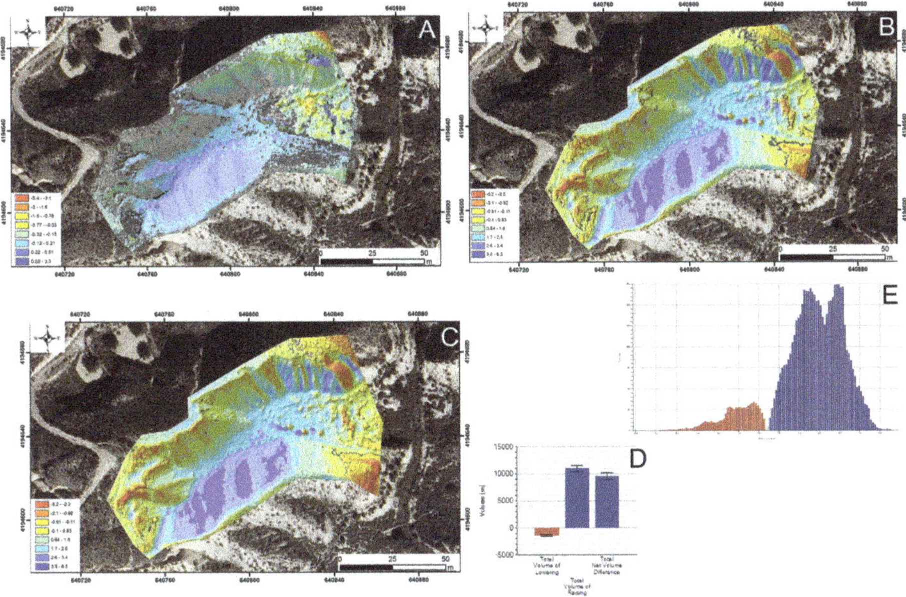

Figure 11. Bank gully (B.G-2)(**A**) DoD Mission I and II. (**B**) DoD Mission II and III. (**C**) DoD Mission I and III. (**D**)Graphe of DoD volume. (**E**) Bar chart of DoD volume change distribution (erosion volume in red, deposition volume in blue.

3.2.3. DOD in Site 3

For the gully (B.G-3), the result of the sediment balance reveals that the rate of erosion is slightly lower than the rate of accumulation during the wet season. This is because the water level decreased by 1 m between the first and second field missions, which limits the sediment transport capacity by littoral drift and waves, but also that, in its bed, the gully experienced a strong sedimentation activity after the mixing of the lake water (regression deposits). Thus, during this season, the lakeshores were less watered. The sedimentation rate is the highest on all the slopes. The volume of sediment decreases from upstream to downstream of the gully bed. During the dry season, the banks were watered more, resulting in a high erosive potential (Figure 4) and a higher rate of slope erosion, estimated at 1772 m^3. Thus, the dry season of 2018 contributed to the triggering of the gullying process after episodes of heavy rainfall. The particles eroded on the slopes are transported by the water to the gully bed and the lake. The rate of sediment accumulated in the gully bed is 1993 m^3. The sedimentation rate of the gully (B.G-3) is estimated at 138.3 T/ha/year (172.7 m^3/ha/year) (Figure 12).

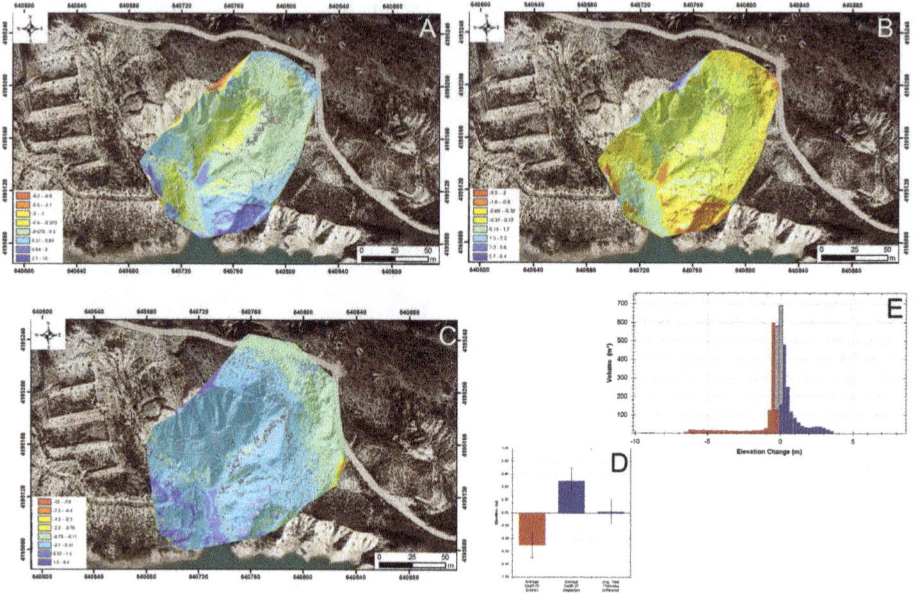

Figure 12. Bank gully (B.G-3)(**A**) DoD Mission I and II. (**B**) DoD Mission II and III. (**C**) DoD Mission I and III. (**D**) Graphe of DoD volume. (**E**) Bar chart of DoD volume change distribution (erosion volume in red, deposition volume in blue.

3.2.4. M3C2 in Site 3

The Gully (B.G-3) is formed by overhanging slopes. To quantify the sedimentation rate of this particular morphology we used a 3D comparison of the point clouds using the M3C2 algorithm, in order to take into account the share of sediments coming from the overhanging areas. This method seems to us more complete than the DoD method, which allowed us to highlight the difference between 2.5 D digital terrain models. This gully is typical of heterogeneous soil formations rich in clay. It is developed on the concave bank of the Rambla de Algeciras where the reservoir constitutes a sedimentary basin of products from the eroded slopes of the gully. The gully (B.G-3) is the result of a combined erosion of the slopes and tunnels. This complex shape is representative of the fragile geomorphological bank gullies. The tunnels formed on Gully B.G-3 are related to the high clay contents in the upper soil horizon (river deposits). In this part of the gully the water carries away the clays and forms tunnels that develop until the ground collapses. From this we can deduce several results about the erosion rate illustrated in particular by distance measurements. Concerning the tunnel zones, the changes are almost zero. It appears that they remain stable, but these areas are very dangerous because their evolution is unpredictable. At the bottom of the gully, the sedimentation rate is very high, exceeding 250 m^3 over an area of 2.28 ha. We notice that the erosion rate on both sides of the gully is similar, which seems logical for an area with steep slopes and no vegetation cover. It is clear from the distance measurements that the sedimentation rate at the bed of gully B.G-3 is higher than for the other gullies.

3.3. Detection of Geomorphological Changes from LiDAR-PNOA Data

The Figure 13 shows DoD values for all bank gullies along the lakeshore between 2009 and 2016. The value (minLOD = 0.28 m) was determined according to the equation (Equation 2). According to the gross DoD, about 45% of the lakeshore area has been gully eroded and the rest of the area has been sedimented.The histogram (Figure 14B) of the distribution of elevation changes in the figure reflects the positive bias of the eroded areas.

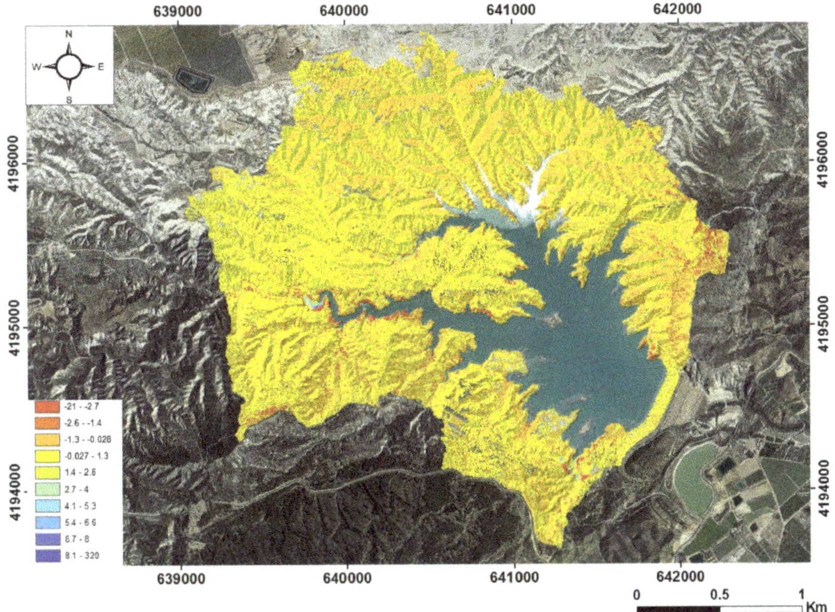

Figure 13. DoD in the lakeshore area for the survey PNOA 2009 and PNOA 2016.

Interpretations of the signature of topographic changes can be justified, however, taking into account the volume and area of the areas that have changed. The histogram (Figure 14A) shows that the total surface of the gullies undergoes changes in altitude due to erosion. On the histogram, we can visualize an imbalance between the rate of sedimentation and erosion on the lakeshore. This unbalanced sediment balance is represented by the asymmetric distribution of the elevation values around the value 0 with an erosion bias. Visualization of the histogram of the distribution of eroded or sedimented volumes on the lakeshore provides a more detailed view of the quantity of sediment deposited in the lake: between 2009 and 2016, 177,641 m^3 (39 T/ha/year) of sediment was deposited (Figure 14A).

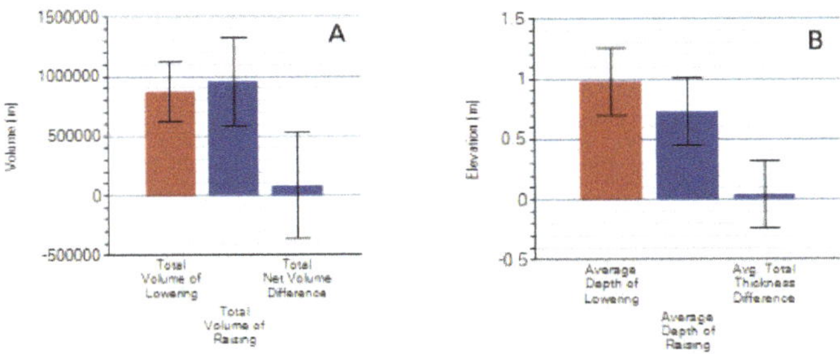

Figure 14. (**A**) Bar chart of DoD volume change distribution. (**B**) Bar chart of DoD elevation change distribution.

From the histogram (Figure 15), it can be seen that the highest morphological changes are in the eroded part with moderate concentrations but with a large amplitude reaching -0.98 m over a surface area of 900,000 m^2. The peak of the changes is not quite in the same area. This peak corresponds to a dominance of low amplitude accumulation processes of 0.7 m over an area of 1,300,000 m^2.

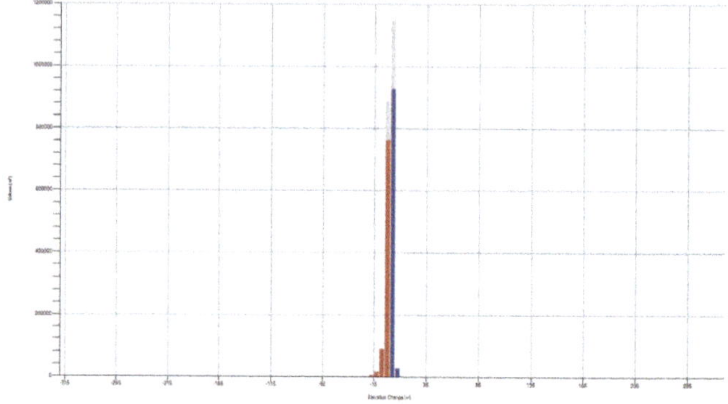

Figure 15. Histogram of elevation distribution.

From the multimodal distribution of the histogram (Figure 15), it is easy to conclude the complexity of the morphodynamic events of the embankment gullies, especially since the significant temporal variability of the R-factor agrees on this point.

With the LiDAR PNOA data two sediment budgets were established. The first calculated balance is really the result of the mean RMSEZ error of 0.2 m. The second is inherent to the actual cubic content of the balance with statistical thresholds that vary little. The results of these two balances vary slightly at the gully scale, but on all lakeshores the difference between the results of these two balances is significant.

4. Discussion

4.1. Data Acquisition

The acquisitions of SfM-MVS-4D data by UAV met our needs in terms of accuracy, and frequency of monitoring the gullies along the lakeshore. Throughout the shooting, the flight altitude remained at a quasi-constant level at each site during the three missions, allowing us to keep the same pixel resolution for each gully. The UAV flew over the gullies at an altitude varying between 28 and 37 m, which allowed us to guarantee adequate resolution for monitoring topographic changes in the gullies. We also noted that the choice of these altitudes allowed us to ensure a small B/h ratio of the order of 0.6, in order to minimize as much as possible the surface of the hidden areas.

The results of the dense point clouds obtained by a vertical shot provide significant information on the topography of the gullies (B.G-1 & 2) compared to the gully (B.G-3). However, it should be noted that the photos taken on the gully (B.G-3) with a vertical camera do not cover the entire studied area. The overhang, having a vertical geometry, could not be taken into consideration. Moreover, the photos taken with an oblique camera were very useful to link the base and the top of the gully overhang slopes. Several previous studies have shown that the systematic errors of the dome effect significantly decrease or even eliminate after the use of an oblique and nadir image acquisition configuration, regardless of the image overlap rate, the camera angle and the configuration of the oblique and nadir images [46,47]. On the gullies (B.G-1 and B.G-2) the aerial photos are taken at nadir. The application of the integrated approach of the two flight configurations, i.e. nadir (90°) and oblique (30°) acquisition, reinforces the geometry of the photo network on the gully (B.G-3). This technique of oblique photo acquisition increases the overlap rate between the photos, making it possible on the one hand to densify the density of the point clouds and on the other hand to fill the void created by the topographical complexity of the gullies (overhang). This therefore reduces georeferencing errors linked mainly to the low precision of the GCPs. We noted that the integrated approach has considerably reduced the RMSE in XYZ of the GCPs on the gully (B.G-3) with an average that does not exceed 1.4 cm in X, 1.2 cm in Y and 1.3 cm in Z during the three missions, bearing in mind that the GCPs went from 16 points to 12 points during the last mission. For the gullies (B.G-1; 2) it is the opposite, since the RMSE errors are high and vary between 2 and 5 cm.

4.2. Data Processing

Vertical and oblique photos are somewhat contrasting since the angle between the two shooting modes fluctuates very strongly along the same swath. Alignment in the PhotoScan software of the photos from two angles failed. The alignment results of the two shooting modes contain only the sparse point cloud from the nadir photos. To solve this problem we realigned all the photos using the common GCPs on both photos and the geolocation data of the photos before applying the bundle adjustment model.

Then we included the following camera parameters: Fit F, Cx, Cy, K1, K2, K3, P1, P2 (Table 4) which improved the camera model and reduced the RMSE errors of the photos as well as the deviations at the GCP point networks. The parameters were selected based on correlation analysis where high correlation parameters were eliminated to avoid over-configuration of the camera model. In this case study Mean precision across the camera is 0.10, this precision value will allow us to evaluate the correlation between the camera parameters. The low correlation between the parameters indicates a good image network except for the radial parameters, however when the correlation between F and Cy is very high this is explained by the absence of large camera rotation and a slight weakness of the image network. In our case study, the value is very close to the accuracy which illustrates the good image network taken in the field.

Table 4. Correlation between the camera parameters.

	F	Cx	Cy	K1	K2	K3	P1	P2
F	1	**−0.61**	**−0.13**	−0.07	0.06	−0.08	**0.51**	**0.22**
Cx		1	−0.04	−0.05	0.8	−0.06	**0.33**	−0.03
Cy			1	0.01	0.02	−0.02	**0.17**	**0.76**
K1				1	**−0.80**	**0.90**	−0.03	0.02
K2					1	**−0.89**	0.02	0.00
K3						1	0.00	0.01
P1							1	**0.17**
P2								1

4.3. Error Modeling

We took advantage of the Lidar PNOA data collected between 2009 and 2016 to continue the analysis of the morpho-sedimentary dynamics of the three gullies. Similarly, the work was enriched by taking into account the spatio-temporal variability of the dynamics of all the bank gullies as a function of topographical influence and climate action. We present and analyze here the results of the raw DoD method obtained from the difference between the two DEMs Lidar 2009 and 2016 of the PNOA project. however, the error of the two Lidar missions PNOA 2009 and 2016 is of the order of 0.28 m this value is considered spatially homogeneous and uniform for two main reasons:

(1) The low density of the Lidar point clouds (between 0.5 and $1/m^2$): the denser the point cloud is, the more robust it is to accurately determine terrain parameters such as slope or roughness.

(2) The absence of GCPs: the georeferencing errors of the GCPs are essential to estimate an error model of the Lidar PNOA data.

To quantify the uncertainty of the SfM-MVS data accurately, it is necessary to validate the SfM-MVS data by a comparison with previous SfM-MVS data or with a very high precision field truth (DGPS, TS). The method of validation is often dictated by the extent, scale, field survey method, data format, and morphological characteristics of the study site.

In this paper we first chose to quantify spatially variable SfM DEM uncertainty, using the cross-validation method instead of Bootstrapping which has already been used by Wheatonand [35] and Rossi [48]. Then the uncertainty values will be used to build an error model that takes into account the different sources of uncertainty using the FIS model.

Several works have shown that the vertical accuracy on the GCPs control points is about two to three times higher than the accuracy obtained by the products derived from the SfM-MVS data (DTM and orthophotography) [49,50]. The table (Table 5) shows that the FIS error propagation model estimates higher vertical errors on GCPs points than those obtained after georeferencing (Table 3). The accuracy of the SfM-MVS data obtained during this study is comparable to that of the GNSS/TS data [35,51].

Table 5. FIS errors in X, Y and Z of the GCPs.

Sites	B.G-1			B.G-2			B.G-3		
Missions	M-I	M-II	M-III	M-I	M- II	M-III	M-I	M-II	M-III
X (cm)	3.62	3.63	4.02	6.21	6.33	6.52	2.63	2.21	2.54
Y (cm)	4.45	6.12	3.24	4.06	6.87	6.77	2.31	3.12	3.05
Z (cm)	10.53	12.69	9.35	11.55	12.81	11.31	2.80	3.32	3.54

4.4. Detection of Geomorphological Changes

From the Figure 16 we observe a ramified gully system on the entire lakeshore, with a gully density varying between $18Km/Km^2$ and $59\ Km/Km^2$ with an annual erosion rate of 39 T/ha/year these values are extreme are of the same order as some areas that reach values varying between 15% and 60% [52–55].

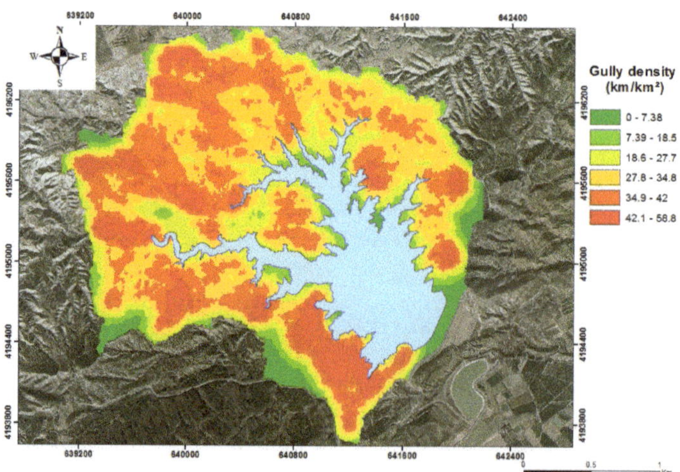

Figure 16. Bank gullies density.

Globally, the volume eroded by gullies varies between 0.002 and 47,430 m^3/year with an average of 358.6 m^3/year. In the Murcia region, research on the gully erosion is limited. As a result, there is very little data on the gully erosion budget. Indeed, in the literature, we find two references on the sediment balance of gullies and bank gullies in the Guadalentin watershed downstream of the

Rambla de Algeciras watershed, namely the work of L. Vandekerckhove [56] and Marzolff [57]. Before comparing the results of this study with those of Vandekerckhove and Marzolff, it should be pointed out that the very complex morphology, the very friable pedology and the strong anthropic activity in the Rambla de Algeciras catchment area allowed us to consider that the Guadalentin basin is less vulnerable to gully erosion than the Rambla de Algeciras basin. Despite this observation, the volumes calculated in this work are close to the results estimated in this study. The annual volume calculated by Vandekerckhove over some time varies between 40 and 43 years based on aerial photos and field measurements vary between 13 and 3467 m^3/year. Similarly, Marzolff's work in 2011 using aerial photos taken by the UAV showed that the rate varies between 0.5 and 100 m^3/year. On the gully (B.G-3) we have estimated the volume eroded from the calculation of the distance between the 3D SfM-MVS point clouds according to the direction of the surface normals, rather than the difference between two 2.5 D rasters. This technique takes into account the volume of sediments eroded on the gully overhang area. The M3C2 algorithm calculates the distance between the multi-temporal point clouds, which allows to estimate the eroded volume with high accuracy. The results of the M3C2 technique thus testify on the good choice of an oblique and nadir acquisition configuration on the gully (B.G-3) to generate accurate and complete SfM-MVS data.

The analysis of the geomorphological changes estimated by the UAV and LiDAR are very different, with the results of the LiDAR data the percentage of the eroded area over the whole period 2009–2016 is not more than (45%) however the deposit areas are more important with a percentage of (55%). On the other hand, the estimates of the sedimentary balance during the year 2018 with the UAV data allow us to observe an erosion activity that affected on average more than 60% of the surface of the studied gullies. The results of the LiDAR data appeared aberrant but the monitoring of the fluctuation of the lake level explains the strong sedimentation activity on the lake banks between 2009 and 2016. The rise of lake level is related to the high trend in rainfall inputs between 2010 and 2015 (Figure 17). It should also be noted that the lake level decreased from 260 m in 2014 to 240 m in 2018. This is related to the decrease in rainfall inputs between 2015 and 2017. This water level fluctuation has caused the settling of regression deposits. After the period of maximum lake filling at 260.12 m (4 May 2014), the lakeshore has experienced silt sedimentation in flat areas and gully beds. These sediments have four origins:

- Gullying (Figure 18)
- Bluff slumping (Figure 19)
- Biogenic (biological activity)
- Detritic (torrential inputs of Rambla)

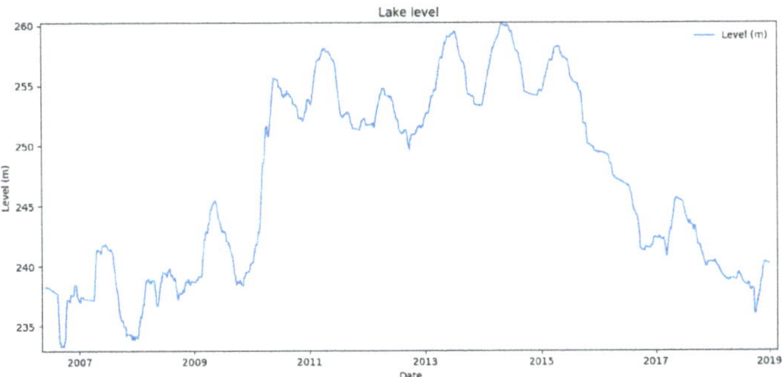

Figure 17. Lake level fluctuation between 2006 and 2019.

The photo-interpretation of PNOA orthophotography illustrates an imbalance in the morphodynamics of the lakeshore.

When the lake level reached the rills and small gullies, the vertical movement of the lake water overwhelmed them and caused them to lose their erosion ability (Figure 18). This has led to the temporal disappearance of these small forms of erosion. They were then replaced by the sedimentary activity of waves loaded with suspended matter (a process linked to longshore drift).

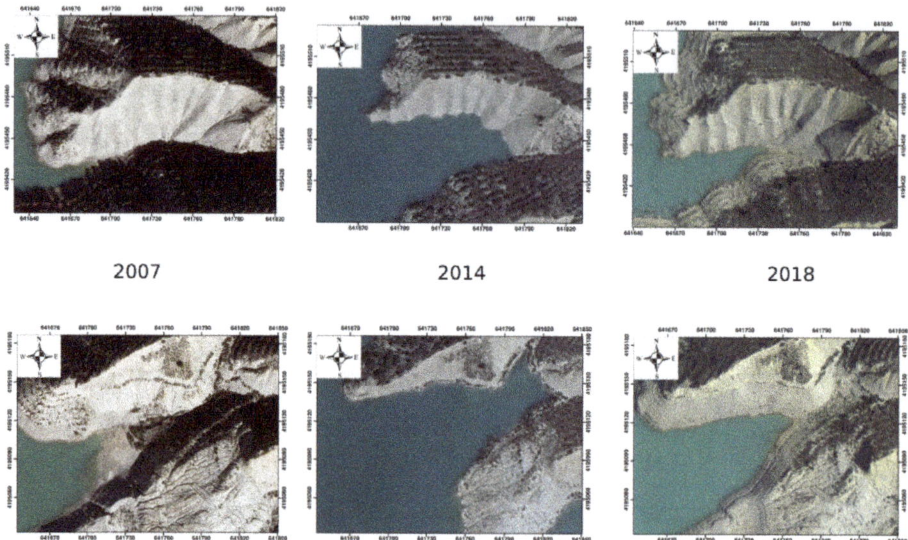

2007   2014   2018

Figure 18. Temporal disappearance of the small bank gullies.

The missing data on the solid transport at gauging stations tributaries of the Rambla, prevent from quantifying the sediment inputs into the Rambla de Algeciras, as it has been done in Morocco by Ezzaouini et al [58], and unfortunately also, we have not been able to install sedimentometres [59] in the lake to collect and analyze the nature and percentage of detrital and biogenic sediments, mainly due to time and money issues.

The bluff slumping of the lakeshore, observed through photo-interpretation of the PNOA aerial orthophotographs between 2009 and 2016, is a phenomenon triggered during the maximum filling of the lake. The former overhanging areas are destabilized by moisture and weight. This phenomenon is localized in particular on the meanders and concave banks of the Rambla (Figure 19).

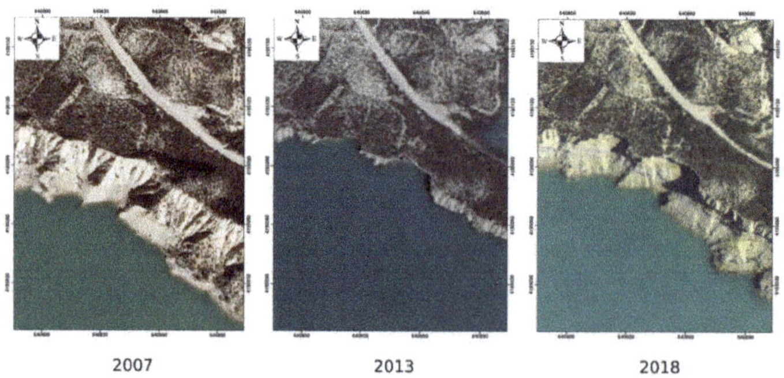

2007   2013   2018

Figure 19. Bluff slumping.

5. Conclusions

The friable nature of the soil, the fluctuation of the lake level, anthropic activity and climate changes marked by increasingly rare but very violent rainfall episodes have made it possible to carve out several forms of gullies on the lakeshores of Rambla de Algeciras. The main goal of this paper is to demonstrate whether the SfM-MVS-4D model generated from photos taken by a UAV and the LiDAR PNOA data, are able to accurately quantify the sediment budget on the banks gullies of the lakeshores. The answer is yes. Indeed, for bank gully monitoring on lakeshores, we detected geomorphological changes on three types of banks gullies. The models for the detection of geomorphological changes were constructed using multi-temporal numerical terrain models coupled with spatially variable uncertainty models. The rate of geomorphological changes detected on the three gullies in 2018 is mainly related to orientation, land use and the shape of the gullies.

The aerial photos taken by the UAV are capable of generating SfM-MVS-4D models with very high spatial and temporal resolution, which allows to highlight the movement of sediments along the slopes of the gullies. The use of the nadir and oblique configuration for taking pictures has improved the accuracy of the SfM-MVS models taken on very steep slopes or with overhangs. This accuracy was measured using several techniques, namely georeferencing error estimation and the statistical cross-validation approach and finally the FIS technique. This was useful for detecting geomorphological changes on banks gullies. The accuracy of the SfM-MVS data varies between 1 and 5 cm.

The data from the PNOA project (orthophotographs and LiDAR) provided reliable and accurate data to study the effects of lake level fluctuation, climate change and land use change on the initiation of lake bank gullying. The DOD technique applied on the LiDAR data from the PNOA allowed the estimation of an annual sedimentation rate of 39 T/ha/year (49 m^3/ha/year) from the gullies of the Rambla de Algeciras lakeshores.

We also note that the approach proposed in this paper will be deployed in Morocco on the Sidi Mohamed Ben Abdellah dam, which shares the same bank gully on lakeshores issues and geo-environmental characteristics as our study site.

Author Contributions: Conceptualization, V.M. and R.H.; Data curation, R.C.; Funding acquisition, R.C. and F.C.; Methodology, V.M., R.H. and F.C.; Project administration, G.M. and E.R.; Resources, V.M, E.R. and R.C.; Software, F.C.; Validation, R.H.; Writing—original draft, R.H.; Writing—review and editing, V.M, G.M. and R.H. All authors have read and agreed to the published version of the manuscript.

Funding: This research received no external funding.

Acknowledgments: This work has been financed with the following funds: GEOLAB laboratory and CASBA project and is gratefully acknowledged. The authors also want to thank Confederación Hidrográfica del Segura in Murcia, Mr Pablo Albaladejo and all colleagues, engineer and technicians participating in the field campaigns supporting successful data acquisition. Constructive and valuable feedback given from the anonymous reviewers helped to increase the quality of the paper.

Conflicts of Interest: The authors declare no conflict of interest.

References

1. Hentati, A.; Kawamura, A.; Amaguchi, H.; Nakagawa, N. Erosion assessment at small hillside river basins in semiarid region of Tunisia. *Annu. J. Hydraul. Eng. Jsce* **2010**, accepted under publication.
2. Antipolis, S. Les menaces sur les sols dans les pays méditerranéens (Etude bibliographique). *Les Cahiers du 690 Plan Bleu* **2003**, *2*, 44–48.
3. Inoubli, N. Ruissellement et Éronsion Hydrique en Milieu Méditerranéen Vertique: Approche Expérimentale et Modélisation. Ph.D. Thesis, Montpellier, SupAgro, Montpellier, France, 2016.
4. Castillo, C.; Gómez, J. A century of gully erosion research: Urgency, complexity and study approaches. *Earth-Sci. Rev.* **2016**, *160*, 300–319. [CrossRef]
5. Poesen, J.; Nachtergaele, J.; Verstraeten, G.; Valentin, C. Gully erosion and environmental change: importance and research needs. *Catena* **2003**, *50*, 91–133. [CrossRef]

6.   Bull, L.J.; Kirkby, M.J. *Dryland Rivers: Hydrology and Geomorphology of Semi-Arid Channels.*JohnWiley & Sons: Chichester, UK, 2002.

7.   Vanmaercke, M.; Poesen, J.; Van Mele, B.; Demuzere, M.; Bruynseels, A.; Golosov, V.; Bezerra, J.F.R.; Bolysov, S.; Dvinskih, A.; Frankl, A.; et al. How fast do gully headcuts retreat? *Earth-Sci. Rev.* **2016**, *154*, 336–355. [CrossRef]

8.   Nazari Samani, A.; Tavakoli Rad, F.; Azarakhshi, M.; Reza Rahdari, M.; Rodrigo-Comino, J. Assessment of the sustainability of the territories affected by gully head advancements through aerial photography and modeling estimations: A case study on Samal Watershed, Iran. *Sustainability* **2018**, *10*, 2909. [CrossRef]

9.   Valentin, C. *Les Sols au Coeur de la Zone Critique 5: DÉgradation et Réhabilitation*; ISTE Group: Santa Clara, CA, USA, 2018; Volume 5.

10.   Wu, Y.; Cheng, H. Monitoring of gully erosion on the Loess Plateau of China using a global positioning system. *Catena* **2005**, *63*, 154–166. [CrossRef]

11.   Mockus, V. *National Engineering Handbook*; U.S. Soil Conservation Service: Washington, DC, USA, 1964; Volume 4.

12.   Thompson, J.R. Quantitative effect of watershed variables on rate of gully-head advancement. *Trans. Asae* **1964**, *7*, 54-0055. [CrossRef]

13.   Seginer, I. Gully development and sediment yield. *J. Hydrol.* **1966**, *4*, 236–253. [CrossRef]

14.   Yuan, Y.; Bingner, R.; Rebich, R. Evaluation of AnnAGNPS on Mississippi delta MSEA watersheds. *Trans. Asae* **2001**, *44*, 1183. [CrossRef]

15.   Nachtergaele, J.; Poesen, J.; Vandekerckhove, L.; Oostwoud Wijdenes, D.; Roxo, M. Testing the ephemeral gully erosion model (EGEM) for two Mediterranean environments. *Earth Surf. Process. Landf. J. Br. Geomorphol. Res. Group* **2001**, *26*, 17–30. [CrossRef]

16.   Sidorchuk, A. Dynamic and static models of gully erosion. *Catena* **1999**, *37*, 401–414. [CrossRef]

17.   Tucker, G.; Lancaster, S.; Gasparini, N.; Bras, R. The channel-hillslope integrated landscape development model (CHILD). In *Landscape Erosion and Evolution Modeling*; Springer: Cham, Switzerland, 2001; pp. 349–388.

18.   Evans, M.; Lindsay, J. High resolution quantification of gully erosion in upland peatlands at the landscape scale. *Earth Surf. Process. Landf.* **2010**, *35*, 876–886. [CrossRef]

19.   Maleval, V. *Les Apports Sédimentaires Directs au Barrage SMBA*; Workshop SIGMED, 2012. Available online: https://hal-bnf.archives-ouvertes.fr/GEOLAB/hal-00924798v1/ (accessed on 29 September 2020).

20.   D'Oleire Oltmanns, S.; Marzolff, I.; Peter, K.D.; Ries, J.B. Unmanned aerial vehicle (UAV) for monitoring soil erosion in Morocco. *Remote Sens.* **2012**, *4*, 3390–3416. [CrossRef]

21.   Castillo, C.; James, M.; Redel-Macías, M.; Pérez, R.; Gómez, J. SF3M software: 3-D photo-reconstruction for non-expert users and its application to a gully network. *Soil* **2015**, *1*, 583. [CrossRef]

22.   Wang, R.; Zhang, S.; Pu, L.; Yang, J.; Yang, C.; Chen, J.; Guan, C.; Wang, Q.; Chen, D.; Fu, B.; et al. Gully erosion mapping and monitoring at multiple scales based on multi-source remote sensing data of the Sancha River Catchment, Northeast China. *Isprs Int. J. Geo-Inf.* **2016**, *5*, 200. [CrossRef]

23.   Di Stefano, C.; Ferro, V.; Palmeri, V.; Pampalone, V.; Agnello, F. Testing the use of an image-based technique to measure gully erosion at Sparacia experimental area. *Hydrol. Process.* **2017**, *31*, 573–585. [CrossRef]

24.   Frankl, A.; Stal, C.; Abraha, A.; Nyssen, J.; Rieke-Zapp, D.; De Wulf, A.; Poesen, J. Detailed recording of gully morphology in 3D through image-based modelling. *Catena* **2015**, *127*, 92–101. [CrossRef]

25.   Christian, P.; Davis, J. Hillslope gully photogeomorphology using structure-from-motion. *Z. Geomorphol. Suppl. Issues* **2016**, *60*, 59–78. [CrossRef]

26.   Koci, J.; Jarihani, B.; Leon, J.X.; Sidle, R.C.; Wilkinson, S.N.; Bartley, R. Assessment of UAV and ground-based structure from motion with multi-view stereo photogrammetry in a gullied savanna catchment. *Isprs Int. J. Geo-Inf.* **2017**, *6*, 328. [CrossRef]

27.   Stocker, T.F.; Qin, D.; Plattner, G.K.; Tignor, M.; Allen, S.K.; Boschung, J.; Nauels, A.; Xia, Y.; Bex, V.; Midgley, P.M.; et al. Climate change 2013: The physical science basis. *Contrib. Work. Group Fifth Assess. Rep. Intergov. Panel Clim. Chang.* **2013**, *1535*.

28.   Zhang, Z. Iterative point matching for registration of free-form curves and surfaces. *Int. J. Comput. Vis.* **1994**, *13*, 119–152. [CrossRef]

29.   Stöcker, C.; Eltner, A.; Karrasch, P. Measuring gullies by synergetic application of UAV and close range photogrammetry—A case study from Andalusia, Spain. *Catena* **2015**, *132*, 1–11. [CrossRef]

30. Micheletti, N.; Chandler, J.H.; Lane, S.N. Investigating the geomorphological potential of freely available and accessible structure-from-motion photogrammetry using a smartphone. *Earth Surf. Process. Landf.* **2015**, *40*, 473–486. [CrossRef]

31. Lague, D.; Brodu, N.; Leroux, J. Accurate 3D comparison of complex topography with terrestrial laser scanner: Application to the Rangitikei canyon (NZ). *Isprs J. Photogramm. Remote. Sens.* **2013**, *82*, 10–26. [CrossRef]

32. Mosbrucker, A.R.; Major, J.J.; Spicer, K.R.; Pitlick, J. Camera system considerations for geomorphic applications of SfM photogrammetry. *Earth Surf. Process. Landf.* **2017**, *42*, 969–986. [CrossRef]

33. Cucchiaro, S.; Maset, E.; Fusiello, A.; Cazorzi, F. 4D-SFM photogrammetry for monitoring sediment dynamics in a debris-flow catchment: Software testing and results comparison. In Proceedings of the 2018 ISPRS TC II Mid-term Symposium "Towards Photogrammetry 2020". International Society for Photogrammetry and Remote Sensing, Riva Del Garda, Italy, 4–7 June 2018; Volume 42, pp. 281–288.

34. Vericat, D.; Smith, M.; Brasington, J. Patterns of topographic change in sub-humid badlands determined by high resolution multi-temporal topographic surveys. *Catena* **2014**, *120*, 164–176. [CrossRef]

35. Wheaton, J.M.; Brasington, J.; Darby, S.E.; Sear, D.A. Accounting for uncertainty in DEMs from repeat topographic surveys: improved sediment budgets. *Earth Surf. Process. Landf. J. Br. Geomorphol. Res. Group* **2010**, *35*, 136–156. [CrossRef]

36. Brasington, J.; Vericat, D.; Rychkov, I. Modeling river bed morphology, roughness, and surface sedimentology using high resolution terrestrial laser scanning. *Water Resour. Res.* **2012**, *48*. [CrossRef]

37. Bangen, S.; Hensleigh, J.; McHugh, P.; Wheaton, J. Error modeling of DEMs from topographic surveys of rivers using fuzzy inference systems. *Water Resour. Res.* **2016**, *52*, 1176–1193. [CrossRef]

38. Lane, S.N.; Chandler, J.H. The generation of high quality topographic data for hydrology and geomorphology: new data sources, new applications and new problems. *Earth Surf. Process. Landf. J. Br. Geomorphol. Res. Group* **2003**, *28*, 229–230. [CrossRef]

39. Milan, D.J.; Heritage, G.L.; Large, A.R.; Fuller, I.C. Filtering spatial error from DEMs: Implications for morphological change estimation. *Geomorphology* **2011**, *125*, 160–171. [CrossRef]

40. Passalacqua, P.; Belmont, P.; Staley, D.M.; Simley, J.D.; Arrowsmith, J.R.; Bode, C.A.; Crosby, C.; DeLong, S.B.; Glenn, N.F.; Kelly, S.A.; et al. Analyzing high resolution topography for advancing the understanding of mass and energy transfer through landscapes: A review. *Earth-Sci. Rev.* **2015**, *148*, 174–193. [CrossRef]

41. Victoriano, A.; Brasington, J.; Guinau, M.; Furdada, G.; Cabré, M.; Moysset, M. Geomorphic impact and assessment of flexible barriers using multi-temporal LiDAR data: The Portainé mountain catchment (Pyrenees). *Eng. Geol.* **2018**, *237*, 168–180. [CrossRef]

42. Brasington, J.; Rumsby, B.; McVey, R. Monitoring and modelling morphological change in a braided gravel-bed river using high resolution GPS-based survey. *Earth Surf. Process. Landf. J. Br. Geomorphol. Res. Group* **2000**, *25*, 973–990. [CrossRef]

43. Brasington, J.; Langham, J.; Rumsby, B. Methodological sensitivity of morphometric estimates of coarse fluvial sediment transport. *Geomorphology* **2003**, *53*, 299–316. [CrossRef]

44. Lane, S.N.; Westaway, R.M.; Murray Hicks, D. Estimation of erosion and deposition volumes in a large, gravel-bed, braided river using synoptic remote sensing. *Earth Surf. Process. Landf. J. Br. Geomorphol. Res. Group* **2003**, *28*, 249–271. [CrossRef]

45. James, M.R.; Robson, S.; Smith, M.W. 3-D uncertainty-based topographic change detection with structure-from-motion photogrammetry: precision maps for ground control and directly georeferenced surveys. *Earth Surf. Process. Landf.* **2017**, *42*, 1769–1788. [CrossRef]

46. Nesbit, P.R.; Hugenholtz, C.H. Enhancing UAV–SFM 3D model accuracy in high-relief landscapes by incorporating oblique images. *Remote. Sens.* **2019**, *11*, 239. [CrossRef]

47. James, M.R.; Robson, S. Mitigating systematic error in topographic models derived from UAV and ground-based image networks. *Earth Surf. Process. Landf.* **2014**, *39*, 1413–1420. [CrossRef]

48. Rossi, R.K. *Evaluation of Structure-from-Motion from a Pole-Mounted Camera for Monitoring Geomorphic Change.* Ph.D. Thesis, Utah State University, Logan, UT, USA, 2018.

49. Fernández, T.; Pérez, J.L.; Cardenal, J.; Gómez, J.M.; Colomo, C.; Delgado, J. Analysis of landslide evolution affecting olive groves using UAV and photogrammetric techniques. *Remote Sens.* **2016**, *8*, 837. [CrossRef]

50. Fernández, T.; Pérez, J.L.; Colomo, C.; Cardenal, J.; Delgado, J.; Palenzuela, J.A.; Irigaray, C.; Chacón, J. Assessment of the evolution of a landslide using digital photogrammetry and LiDAR techniques in the Alpujarras region (Granada, southeastern Spain). *Geosciences* **2017**, *7*, 32. [CrossRef]

51. Rumsby, B.; Brasington, J.; Langham, J.; McLelland, S.; Middleton, R.; Rollinson, G. Monitoring and modelling particle and reach-scale morphological change in gravel-bed rivers: Applications and challenges. *Geomorphology* **2008**, *93*, 40–54. [CrossRef]

52. Dewitte, O.; Daoudi, M.; Bosco, C.; Van Den Eeckhaut, M. Predicting the susceptibility to gully initiation in data-poor regions. *Geomorphology* **2015**, *228*, 101–115. [CrossRef]

53. Martinez-Casasnovas, J. A spatial information technology approach for the mapping and quantification of gully erosion. *Catena* **2003**, *50*, 293–308. [CrossRef]

54. Perroy, R.L.; Bookhagen, B.; Asner, G.P.; Chadwick, O.A. Comparison of gully erosion estimates using airborne and ground-based LiDAR on Santa Cruz Island, California. *Geomorphology* **2010**, *118*, 288–300. [CrossRef]

55. Nadal-Romero, E.; Revuelto, J.; Errea, P.; López-Moreno, J. The application of terrestrial laser scanner and SfM photogrammetry in measuring erosion and deposition processes in two opposite slopes in a humid badlands area (central Spanish Pyrenees). *Soil* **2015**, *1*, 561. [CrossRef]

56. Vandekerckhove, L.; Poesen, J.; Govers, G. Medium-term gully headcut retreat rates in Southeast Spain determined from aerial photographs and ground measurements. *Catena* **2003**, *50*, 329–352. [CrossRef]

57. Marzolff, I.; Ries, J.B.; Poesen, J. Short-term versus medium-term monitoring for detecting gully-erosion variability in a Mediterranean environment. *Earth Surf. Process. Landf.* **2011**, *36*, 1604–1623. [CrossRef]

58. Ezzaouini, M.A.; Mahé, G.; Kacimi, I.; Zerouali, A. Comparison of the MUSLE Model and Two Years of Solid Transport Measurement, in the Bouregreg Basin, and Impact on the Sedimentation in the Sidi Mohamed Ben Abdellah Reservoir, Morocco. *Water* **2020**, *12*, 1882. [CrossRef]

59. Maleval, V.; Pitois, F. Fonctionnement hydro-sédimentaire et bilan sédimentaire du lac de Saint-Germain-de-Confolens (Charente, France): paramètres géomorphologiques à prendre en considération dans la gestion du lac. *Phys.-Géo. Géo. Phys. Environ.* **2017**, 197–227. [CrossRef]

Article

Understory Limits Surface Runoff and Soil Loss in Teak Tree Plantations of Northern Lao PDR

Layheang Song [1,2], Laurie Boithias [1,*], Oloth Sengtaheuanghoung [3], Chantha Oeurng [2], Christian Valentin [4], Bounthan Souksavath [5], Phabvilay Sounyafong [6], Anneke de Rouw [4], Bounsamay Soulileuth [6], Norbert Silvera [4], Bounchanh Lattanavongkot [5], Alain Pierret [6] and Olivier Ribolzi [1]

[1] GET, Université de Toulouse, CNRS, IRD, UPS, 31400 Toulouse, France; layheangsong@itc.edu.kh (L.S.); olivier.ribolzi@ird.fr (O.R.)

[2] Research and Innovation Center (RIC), Faculty of Hydrology and Water Resources Engineering, Institute of Technology of Cambodia, Phnom Penh 12156, Cambodia; chantha@itc.edu.kh

[3] Department of Agricultural Land Management (DALaM), Ministry of Agriculture and Forestry, Vientiane 01000, Laos; oloth_s@hotmail.com

[4] Institute of Ecology and Environmental Sciences of Paris (iEES-Paris), Sorbonne Université, Univ Paris Est Creteil, IRD, CNRS, INRA, 75005 Paris, France; christian.valentin@ird.fr (C.V.); anneke.de_rouw@ird.fr (A.d.R.); norbert.silvera@ird.fr (N.S.)

[5] Luang Prabang Teak Program (LPTP-TFT), Luang Prabang Provincial Forestry Section, Luang Prabang 06000, Laos; b.souksavath@earthworm.org (B.S.); chanh.lptp@gmail.com (B.L.)

[6] IRD, IEES-Paris UMR 242, c/o National Agriculture and Forestry Research Institute, Vientiane 01000, Laos; phabvilay_laopdr@yahoo.com (P.S.); sbounsamay@gmail.com (B.S.); alain.pierret@ird.fr (A.P.)

* Correspondence: laurie.boithias@get.omp.eu; Tel.: +33-561-332-712

Received: 14 July 2020; Accepted: 15 August 2020; Published: 19 August 2020

Abstract: Many mountainous regions of the humid tropics experience serious soil erosion following rapid changes in land use. In northern Lao People's Democratic Republic (PDR), the replacement of traditional crops by tree plantations, such as teak trees, has led to a dramatic increase in floods and soil loss and to the degradation of basic soil ecosystem services such as water filtration by soil, fertility maintenance, etc. In this study, we hypothesized that conserving understory under teak trees would protect soil, limit surface runoff, and help reduce soil erosion. Using 1 m^2 microplots installed in four teak tree plantations in northern Lao PDR over the rainy season of 2017, this study aimed to: (1) assess the effects on surface runoff and soil loss of four understory management practices, namely teak with no understory (TNU; control treatment), teak with low density of understory (TLU), teak with high density of understory (THU), and teak with broom grass, *Thysanolaena latifolia* (TBG); (2) suggest soil erosion mitigation management practices; and (3) identify a field visual indicator allowing a rapid appraisal of soil erosion intensity. We monitored surface runoff and soil loss, and measured teak tree and understory characteristics (height and percentage of cover) and soil surface features. We estimated the relationships among these variables through statistics and regression analyses. THU and TBG had the smallest runoff coefficient (23% for both) and soil loss (465 and 381 g·m^{-2}, respectively). The runoff coefficient and soil loss in TLU were 35% and 1115 g·m^{-2}, respectively. TNU had the highest runoff coefficient and soil loss (60%, 5455 g·m^{-2}) associated to the highest crusting rate (82%). Hence, the soil loss in TBG was 14-times less than in TNU and teak tree plantation owners could divide soil loss by 14 by keeping understory, such as broom grass, within teak tree plantations. Indeed, a high runoff coefficient and soil loss in TNU was explained by the kinetic energy of rain drops falling from the broad leaves of the tall teak trees down to bare soil, devoid of plant residues, thus leading to severe soil surface crusting and soil detachment. The areal percentage of pedestal features was a reliable indicator of soil erosion intensity. Overall, promoting understory, such as broom grass, in teak tree plantations would: (1) limit surface runoff and improve soil infiltrability, thus increase soil water stock available for both root absorption and groundwater recharge; and (2) mitigate soil loss while favoring soil fertility conservation.

Keywords: overland flow; inter-rill erosion; teak tree plantation; understory; broom grass; South–East Asia; land management; soil erosion

---

1. Introduction

Mountain areas of the humid tropics are characterized by steep slopes and heavy rains [1]. Hence these regions are prone to high surface runoff and soil erosion [2]. Conversion of natural forest to e.g., agricultural land exacerbates runoff production and soil erosion, leading to physicochemical and biological changes of the altered ecosystems [3,4].

On-site effects of increased soil erosion include the reduction of soil quality impacting the sustainability of agricultural production [5], and economics, due to the loss of ecosystem services [6]. The tremendously higher rate of soil loss compared to its formation rate threatens food production and environmental quality (water, soil, and air) [7–9].

Off-site effects comprise floods, depletion of groundwater recharge, degradation of stream environments, and downstream sedimentation [5,10]. In addition, the adsorption of organic and inorganic matter on soil particles and suspended sediments plays a leading role in the transport of nutrients [11], radionuclides [12], metals [13], pesticides [14], and bacterial pathogens [15,16]. Higher contamination of rivers by fecal bacteria is often correlated to higher in-stream suspended sediment concentration [17,18]. Sediment loads cause massive accumulation of fine sediments to river beds and cause the siltation of irrigation canals and dam reservoirs [19–21], thus reducing their life spans. All of these off-site effects lead to high economic and ecological costs (i.e., sedimentation, flooding, landslides, water eutrophication, biodiversity loss, land abandonment, destruction of infrastructures) [6].

Natural forests are known for their protective effect against erosion [22]. Forests have favorable hydrodynamic properties for surface infiltration [23], subsurface drainage [24] due to the biological activity, and the development of macropores in the litter and underneath. A fraction of rainfall, known as throughfall, is intercepted by the canopy and the underlying vegetative strata [25]. The amount of water and the kinetic energy of drops corresponding to throughfall is lower than that of rainfall, thus reducing soil erosion. Kinetic energy is the main factor initiating soil erosion; raindrops hit the soil surface and disaggregate the soil structure [26] resulting in redistribution of soil material by splash effect [5]. Kinetic energy is controlled by rainfall (amount, drop size, fall velocity), vegetation characteristics (height, cover, residues) [27], and slope [26].

In the mountainous region of northern Lao People's Democratic Republic (PDR), widespread agricultural practices are known to increase surface runoff, soil erosion, and in-stream suspended sediment concentrations compared to natural forest [3,23]. In particular, commercial perennial monocultures, such as teak tree plantations, lead to a drastic surface runoff increase [28] and exacerbate soil loss along hillslopes [29,30]. In teak tree plantations, such as the old teak trees (high timbers with broad leaves) growing in the Houay Pano catchment, surface runoff and soil erosion are likely related to recurrent understory and leaf litter suppression by burning at the end of the dry season, leaving the soil bare and exposed to the kinetic energy of the raindrops [31].

Land use, understory cover, and soil surface condition, are the main drivers influencing the infiltration of water into the soil [32,33]. Living plant roots modify both mechanical and hydrological characteristics of the soil matrix and negatively influence the soil erodibility [34,35]. Understory and plant residues on soil surface are known to efficiently attenuate the effect of splash and thus soil detachment [36–40]. Soil surface features have a strong impact on surface runoff and soil erosion: soil aggregation limits surface runoff generation and soil loss [41,42], while crusted soils can be self-protecting from erosion because of their high surface shear stress resistance, but they also promote surface runoff, hence downstream erosion, especially on non-crusted areas [26,30]. For example, in northern Lao PDR, splash in teak tree plantations increases soil erosion, and the absence of understory enhances the effect of splash [29]. By adapting the model developed by [43], Patin et al. [31] showed that plant residues and weed cover at soil surface were the main attenuation factors of soil erosion. In northern Thailand, soil erosion in rubber tree plantations decreased when understory was grown [44]. Indeed, in northern Lao PDR headwater catchments, the runoff coefficient is approximately 55% under teak trees at the plot scale [30], whereas it nearly doubled from 16% to 31% within 13 years at the catchment scale [29], resulting from the absence of understory in the teak tree plantation area. In particular, mature teak trees with limited understory were shown to export respectively 5.5- and 31-times more water and soil than broom grass (*Thysanolaena latifolia*) at the plot scale [30].

Hence, in this paper, we hypothesized that conserving understory such as broom grass, which provides income to farmers through broom making and selling [45], protects soil, limits surface runoff, and helps to reduce soil erosion in teak tree plantations. The overarching goal of this study, conducted in the teak tree plantations of Ban Kokngew, northern Lao PDR, was thus to find the best strategy to contribute to sustainable agricultural production. In this context, the three objectives of our 1 m^2 microplot experiment performed during the June to October monsoon period of 2017 were to: (1) assess the effects on surface runoff and soil loss of four understory management practices, namely teak with no understory (TNU; control treatment), teak with low density of understory (TLU), teak with high density of understory (THU), and teak with broom grass (TBG); (2) suggest soil erosion mitigation management practices; and (3) identify a field visual indicator allowing a rapid appraisal of soil erosion intensity.

2. Materials and Methods

2.1. Study Area and Experimental Plots

We conducted the experiment in 2017 in teak tree plantations surrounding Ban Kokngew, a village located in Luang Prabang Province, northern Lao PDR, and predominantly situated over Acrisol soil and Carboniferous and Permian limestones (Figure 1). The climate is sub-tropical humid and is characterized by a monsoon regime with a dry season from November to May, and a rainy from June to October. Mean annual rainfall recorded at Luang Prabang from 1960 to 2006 was 1268 mm, about 76% of which falls during the rainy season. The mean annual temperature is 25.3 °C. Mean annual reference evapotranspiration is 1116 mm. The study area belongs to the mountainous region of northern Laos PDR. More specifically, the area is located within the "Luang Prabang mountain rain forest" ecoregion [46]. The area has been experiencing dramatic land use changes in the last decade with the introduction of the teak tree plantations [29,47].

We selected this area because it presents, over short distances, diversely managed teak tree plantations. This area used to be shifting cultivation land. This last decade, teak has gradually replaced most of the fields and spontaneous forest regrowth because of land degradation, lack of labor and the expectation of profitability. Teak timber is valued for its durability and water resistance; it is used for furniture and construction in the rapidly developing city of Luang Prabang. Farmers spontaneously adopted different management practices within their plantations, and we selected four sites corresponding to actual contrasted situations that we intended to test and compare.

Aside from the most common situation which is teak with no understory (TNU), i.e., teak tree plantations where soil is kept bare, often by burning the leaf litter and understory, and which represented our control situation, we considered the three following alternative treatments: teak trees grown with high density understory (THU), teak trees grown with low density and/or periodically pruned understory (TLU), and teak trees grown with broom grass, *Thysanolaena latifolia* (TBG). In each treatment, we set up 1 m^2 microplots (Figure 2) with six replicates per treatment.

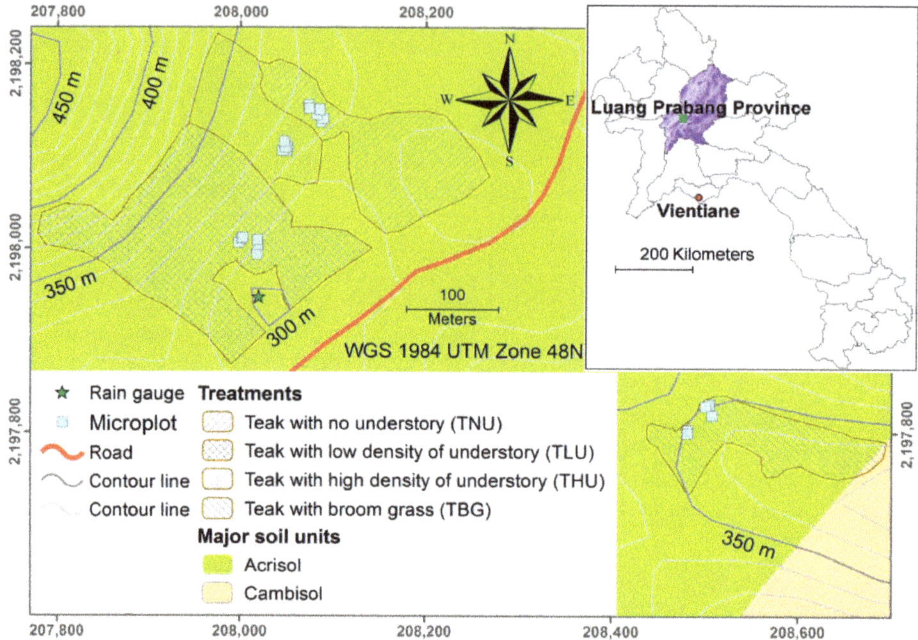

Figure 1. Study site in Ban Kokngew, Luang Prabang Province, Lao People's Democratic Republic, with location of experimental microplots, treatments, and soil types.

The age of the teak trees in the four treatments varied between 12 years in TBG and 18 years in TLU (Table 1). Elevation above sea level ranged between 316 m in TLU and 358 m in TBG, whereas TNU and THU were both at 325 m. The slopes of the microplots ranged between 39% in TNU and 46% in TBG. The slope difference between TNU and TBG was not considered a limitation for the comparison of treatments in this study since the effect of slope is known to be imperceptible for these slope ranges [31]. The size of the teak tree plantations was 1.87 ha, 3.78 ha, 1.21 ha, and 1.32 ha in TNU, TLU, THU, and TBG, respectively. We installed a rain gauge near the treatments TNU, TLU, and THU, and approximately 500 m from TBG (Figure 1).

Table 1. Characteristics of the four experimental sites in 2017, measured on 14 December 2017, in Ban Kokngew, Luang Prabang Province, Lao People's Democratic Republic. TNU: teak with no understory; TLU: teak with low density of understory; THU: teak with high density of understory; TBG: teak with broom grass.

Treatment	Teak Height (m)	Teak Stem Diameter (cm)	Teak Cover (%)	Tree Density (tree·ha^{-1})	Teak Age (Years)	Altitude (m)	Slope (%)	Latitude (°)	Longitude (°)
TNU	22	15.3	35	1200	15	325	39.2	19.8577	102.21269
TLU	22	17.0	30	1000	18	316	40.7	19.85641	102.21202
THU	20	15.4	70	800	15	325	41.5	19.85735	102.21237
TBG	20	16.4	60	1000	12	358	45.5	19.85475	102.21666

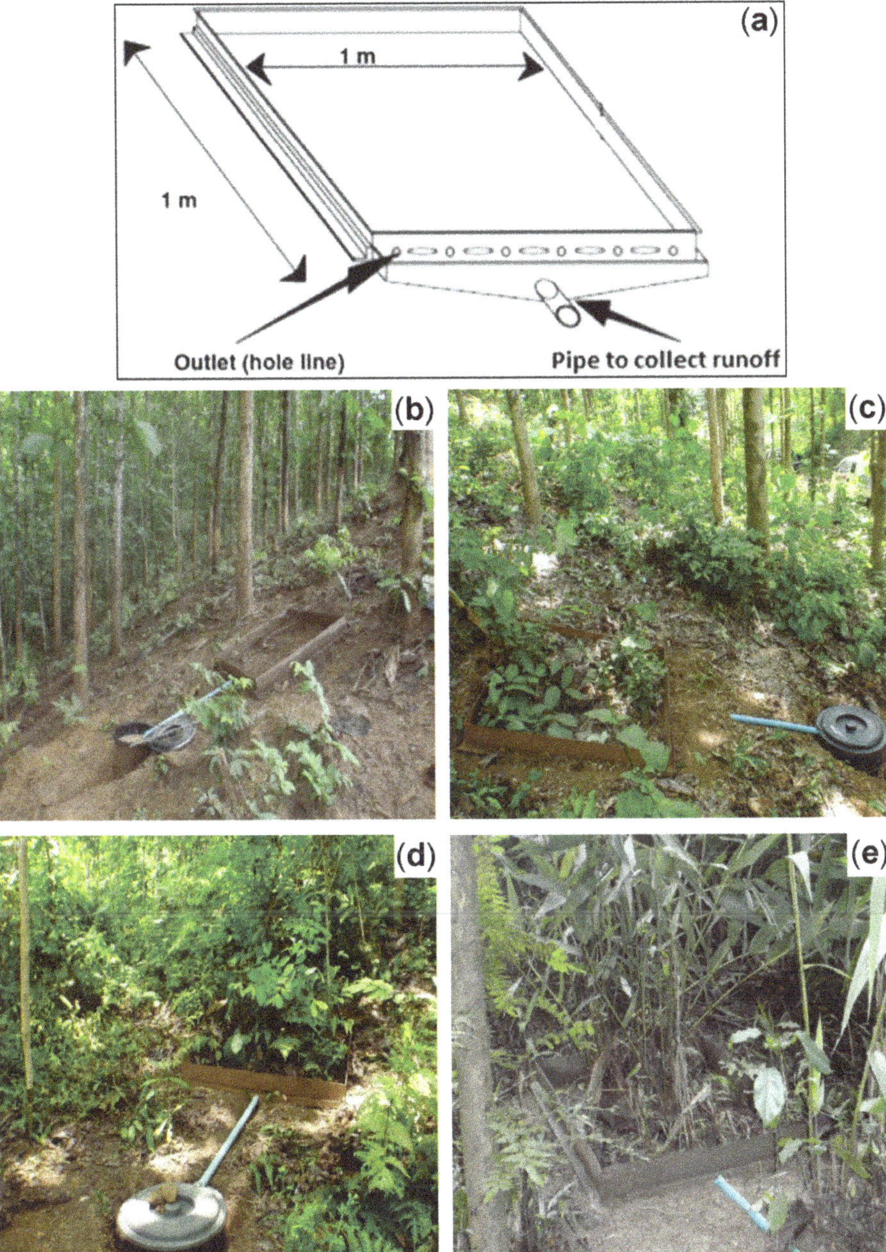

Figure 2. (**a**) Sketch of microplot of 1 × 1 m metal frame connected to a bucket through a pipe for surface runoff and sediment collection (source: [31]). (**b**) Microplot of TNU: teak with no understory. (**c**) Microplot of TLU: teak with low density of understory. (**d**) Microplot of THU: teak with high density of understory. (**e**) Microplot of TBG: teak with broom grass. Microplots were installed on 9th and 10th of May 2017, in Ban Kokngew, Luang Prabang Province, Lao People's Democratic Republic.

2.2. Teak Trees and Understory Structure Assessment

We measured teak tree height, canopy cover, and stem diameter at 1.6 m height in 10×10 m plots enclosing the microplots. At the time of measurement, on 14 December 2017, i.e., during the dry season, some teak trees were already shedding their leaves (30% and 35% cover in TLU and TNU, respectively) but in other plantations, many teak leaves were still attached (70% and 60% cover in THU and TBG, respectively). Average teak heights ranged from 20 m to 22 m. Average diameter at 1.6 m height was 16 cm. The original planting density had been 2000 tree·ha^{-1} in TNU and TBG, from which the actual density was obtained by thinning 40% and 50%, respectively. The original planting density had been 1600 tree·ha^{-1} in THU and only 1200 tree·ha^{-1} in TLU; here the actual densities were obtained by thinning 50% and 16%, respectively (Table 1).

We considered two categories of understory: one consisting of weed and low vegetation, hereafter referred to as "understory" and another consisting of purposely planted broom grass (*Thysanolaena latifolia*) monocrop, hereafter referred to as "broom grass". Both kinds of understories are spontaneous vegetation, but farmers enhance broom grass propagation by cutting. The TBG treatment in this study was made of replanted broom grass. Farmers cut the inflorescences of broom grass to make brooms and regularly prune the grass. We described the structure of the understory in each of the four selected treatments during the 2017 rainy season by combining the use of visual inspections (percentage of cover assessment) and measuring tapes (girth, height). We estimated the mean understory height and cover in a representative area of 18 m^2 encompassing the three microplot replicates on 4 June and 27 October 2017.

2.3. Rainfall Measurements

We measured rainfall by using a rain gauge (Campbell BWS200 equipped with ARG100, 0.2 mm capacity tipping-bucket; Figure 1).

2.4. Surface Runoff, Soil Loss, and Soil Surface Features Assessment

We collected surface runoff through a small channel at the lower side of each microplot, connected to a large plastic bucket by a plastic pipe (Figure 2a). The height of the buckets was 0.45 m while their diameter at the bottom was 0.32 m and the diameter at the top was 0.38 m. We emptied the buckets after every major rainfall event, or after a series of two to ten smaller rainfall events. We calculated the runoff coefficient as the ratio between the total surface runoff depth and the total rainfall depth, expressed in percentage. We also computed a cumulated runoff coefficient over the rainy season.

Soil loss was the total weight of sediment collected each time the buckets were emptied. It was measured after flocculation, filtration, and oven dehydration. We calculated suspended sediment concentration by dividing sediment mass by surface runoff depth and we cumulated soil loss over the rainy season.

Soil surface features were assessed at the beginning (4 June) and at the end (27 October) of the rainy season using the method proposed by [48] and extensively used by [23,26,28–33,44,49–54]. We calculated the average of each soil surface feature for each treatment by first calculating the averages of the two measurements dates per treatment, and then by calculating the averages among the six replicates per treatment. Surface features include understory, residues (leaves, branches, seeds), constructions by soil macro-organisms like earthworms and termites, moss and algae, charcoals, free aggregates, free gravel, and three types of crust: structural, erosion, and gravel crusts. In these soils, structural crusts result from the packing of highly stable micro-aggregates [32,50]. Compacted by raindrops and smoothed by surface runoff, this structural crust gradually transforms into an erosion crust characterized by a thin and very compacted smooth plasmic layer [55]. When they include gravels, structural or erosion crusts become a gravel crust [51]. Additionally, we assessed the percentage areas of soil corresponding to pedestal features [5].

2.5. Statistical Analysis and Modelling

We performed non-parametric Wilcoxon tests (R version 3.5.3, The R Foundation for Statistical Computing) to compare the distribution between paired groups of the four treatments for three variables (surface runoff depth, suspended sediment concentration, and runoff coefficient).

We conducted a correlation analysis between the measured variables, namely Rc: seasonal runoff coefficient; Sl: seasonal soil loss; and soil surface features expressed in areal percentage: Fa: free aggregates; Fg: free gravel; Tc: total crust; Sc: structural crust; Ec: erosion crust; Gc: gravel crust; Cha: charcoals; Res; residues; Wor: worm casts; Alg: algae; Mos: mosses; Ped: pedestals; Und: understory. To meet the distributional and variance assumptions required for linear statistical models [56], we transformed the variables prior to the analysis; variables expressed in percent were normalized using the arcsine of the square root, which is a classical transformation for percentages, while Sl was scaled with logarithm transformation [31]. The main objective of these transformations was to make each distribution symmetrical. After transformation, we calculated Pearson correlation coefficients and significance levels (XLSTAT Premium version 20.1.1., Addinsoft, Paris, France) in order to test the correlation between variables.

We also calculated partial least squares regression (PLSR) analysis [57] on the measured variables (XLSTAT Premium version 20.1.1.) in order to model Rc and Sl depending on soil surface features: Fa: free aggregates; Fg: free gravel; Tc: total crust; Cha: charcoals; Res; residues; Wor: worm casts; Alg: algae; Mos: mosses; Ped: pedestals; Und: understory. PLSR has the advantage to be little sensitive to multi-collinearity and can be used with datasets where the number of observations is close to the number of variables, or even smaller. The importance of each projected variable is estimated by the variable importance in the projection number (VIP). In order to limit the uncertainty related to the variables that bring little information to the model, and consequently to limit the distortion of the results, we discarded the VIP values below 0.8 [58,59].

3. Results

3.1. Rainfall

Accumulated rainy season rainfall was 1133 mm from 4 June to 15 October 2017. A total of 22 major rainfall events occurred during the same period (Figure S1). Minimum rainfall depth was 17 mm whereas maximum was 93 mm. Average rainfall depth was 52 mm. About 36% of the major rainfall events occurred in July, which represents 33% of the accumulated rainy season rainfall depth.

3.2. Height and Cover of Teak Trees and of Understory

The average height of the teak trees ranged from 20 m in THU and TBG to 22 m in TNU and TLU. Teak cover was 30% and 35% in TLU and TNU, respectively, and was 60% and 70% in TBG and THU, respectively. The average height of understory varied between 0.6 m in TNU and 4 m in TLU. Understory cover varied between 30% in TNU and 90% in THU (Figure 3).

3.3. Soil Surface Features and Pedestal Features

The average height of pedestal features ranged from 1.2 cm in TBG to 2.1 cm in TNU (Figure 4a,b). The pedestal features cover was high in TNU (50%) compared to the other treatments (3–6%; Figure 4b).

Figure 4c illustrates the percentage area of total crust (erosion crust, structural crust, and gravel crust), free aggregates, free gravel, charcoals, and residues in each treatment. Soil surface across the treatments excluding TNU shared similar conditions, such as total crust (8.5%), free aggregates (32.5%), and residues (58.5%). On the contrary, TNU exhibited high total crust (82.5%), free gravel (9.3%), and little residues (2.91%). Structural crust accounts for more than 90% of total crust in each treatment and charcoals are negligible (less than 0.03%). Nevertheless, erosion crust was 0.54% in TNU, and negligible in the other three treatments (less than 0.05%).

Figure 5 shows the relationship between the percentage of cover of pedestal features and both surface runoff (logarithm model, $R^2 = 0.82$) and soil loss (linear model, $R^2 = 0.91$). The percentage of cover of pedestal features exhibits a consistent increase with increasing surface runoff and soil loss.

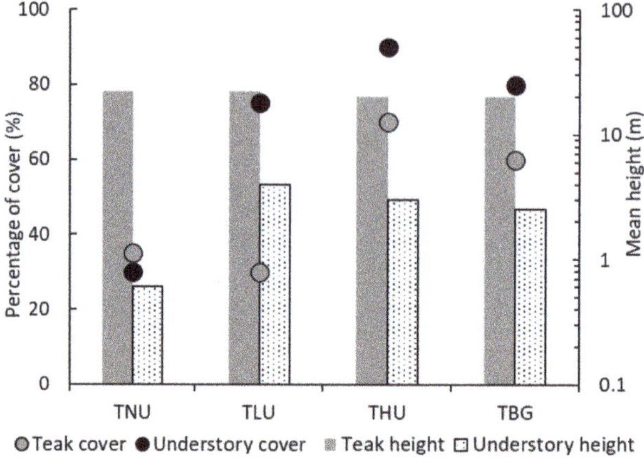

Figure 3. Percentage of cover (%) by teak trees and understory, and mean height (m) of teak trees and understory in each treatment measured on 14 December 2017, in Ban Kokngew, Luang Prabang Province, Lao People's Democratic Republic. TNU: teak with no understory; TLU: teak with low density of understory; THU: teak with high density of understory; TBG: teak with broom grass.

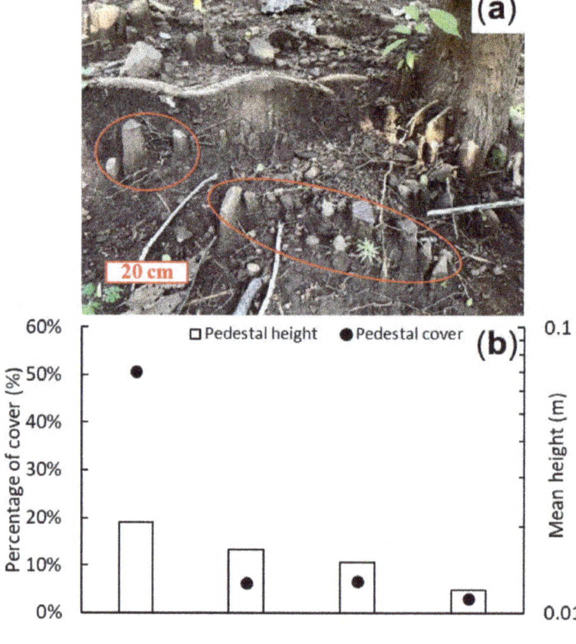

Figure 4. *Cont.*

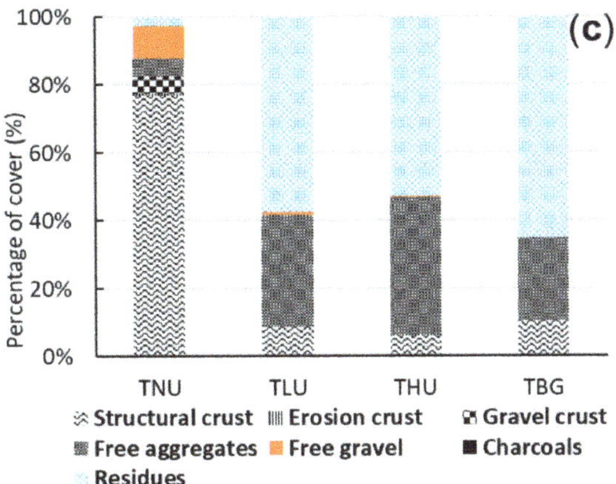

Figure 4. (**a**) Example of pedestal features circled by red lines. (**b**) Percentage of cover of pedestal features (%) and pedestal features' height (m; logarithmic scale). (**c**) Cumulative percentage areas (%) of the soil surface features in 2017 in Ban Kokngew, Luang Prabang Province, Lao People's Democratic Republic. TNU: teak with no understory; TLU: teak with low density of understory; THU: teak with high density of understory; TBG: teak with broom grass.

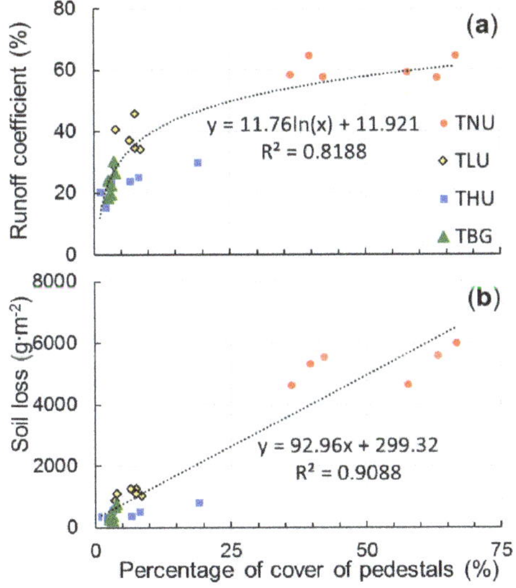

Figure 5. (**a**) Relationship between runoff coefficient (%) and percentage of cover of pedestal features (%). (**b**) Relationship between soil loss (g·m^{-2}) and percentage of cover of pedestal features (%) measured in 2017 in Ban Kokngew, Luang Prabang Province, Lao People's Democratic Republic. TNU: teak with no understory; TLU: teak with low density of understory; THU: teak with high density of understory; TBG: teak with broom grass.

3.4. Relationship between Surface Runoff and Soil Loss across Four Treatments

Figure 6 depicts the linear correlation between accumulated surface runoff and accumulated soil loss on log scale during the rainy season in 2017 for all the replicates. Total surface runoff averaged over the rainy season ranged from approximately 170 mm in THU and TBG to 480 mm in TLU, while it reached up to 680 mm in TNU. The average value in the four treatments was 370 mm. TBG and THU provided the same amount of surface runoff. Soil loss varied between 249 g·m^{-2} in TBG to 6012 g·m^{-2} in TNU, with an average of 1848 g·m^{-2} in all treatments. The highest sediment concentration was 9.60 g·L^{-1} in TNU, while the other three treatments had an average of 2.34 g·L^{-1}.

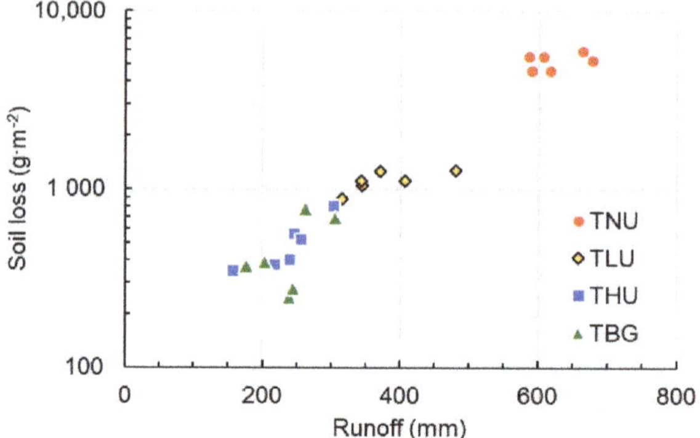

Figure 6. Surface runoff (mm) and soil loss (g.m^{-2}, logarithmic scale) measured from 4 June to 15 October 2017, in the four treatments with six replicates in Ban Kokngew, Luang Prabang Province, Lao People's Democratic Republic. TNU: teak with no understory; TLU: teak with low density of understory; THU: teak with high density of understory; TBG: teak with broom grass.

3.5. Effect of Understory on Soil Loss and Surface Runoff Generation

The Wilcoxon test applied to the cumulative surface runoff, the average suspended sediment concentration, and the runoff coefficient for the four treatments highlighted three significantly (p-value > 0.05) different categories of treatments: from the little erosive treatments (THU and TBG) to highly erosive treatments (TNU; Figure 7).

The surface runoff pattern matched the rainfall pattern, while the soil detachment pattern did not perfectly match rainfall pattern (Figure S1). The highest median values of surface runoff and runoff coefficient (44 mm mostly in TNU and about 110% in TBG and TNU, respectively) were observed from 15 July to 15 August and this pattern was less clear for soil loss. We found the maximum median value of soil loss in TNU (551 g·m^{-2}). For one replicate of TNU, the soil loss reached 1054 g·m^{-2} (8 July).

Figure S2 shows the cumulative surface runoff and cumulative soil loss in relation to cumulative rainfall over the 2017 rainy season for the different treatments. Surface runoff and soil loss in TBG were 242 mm and 381 g·m^{-2}, respectively. Surface runoff and soil loss in THU were 242 mm and 465 g·m^{-2}, respectively. Surface runoff and soil loss in TLU were 358 mm and 1115 g·m^{-2}, respectively. TNU produced more surface runoff (612 mm) and much more soil loss (5455 g·m^{-2}) than the other treatments. Hence, the surface runoff in TNU was approximately 2.5-times higher than in THU and TBG, and 1.7-time higher than in TLU. The soil loss in TNU was 13- and 14-times higher than in THU and TBG, respectively. TNU had the sharpest rise of soil loss among all the treatments. Median runoff coefficients for TBG, THU, TLU, and TNU were 23%, 23%, 25%, and 60%, respectively.

3.6. Runoff Coefficient and Soil Loss in Relation to Soil Surface Features and Understory Cover

Table 2 shows the relationship between the surface runoff coefficient, soil loss, and soil surface features areal percentages. Erosion crust, charcoals, worm casts, algae, and mosses show weak correlations with the surface runoff coefficient while the other soil surface features and understory cover provided Pearson correlation coefficient (r) above 0.7. The features with lower r had lower percentages, which may cause poor relation with the surface runoff coefficient. Free aggregates, residues and understory had a strong and negative correlation with surface runoff with r between −0.72 and −0.91. Free gravel, total crust, structural crust, gravel crust, and pedestal features cover exhibited a strong correlation with surface runoff (r = 0.84, 0.89, 0.89, 0.98, and 0.89, respectively).

We also found a strong correlation between soil loss and surface runoff (r = 0.97). Soil surface features and understory cover exhibited significant inter-correlation except for charcoals, worm casts, algae, and mosses, which were weakly related to other types of soil surface features or understory cover.

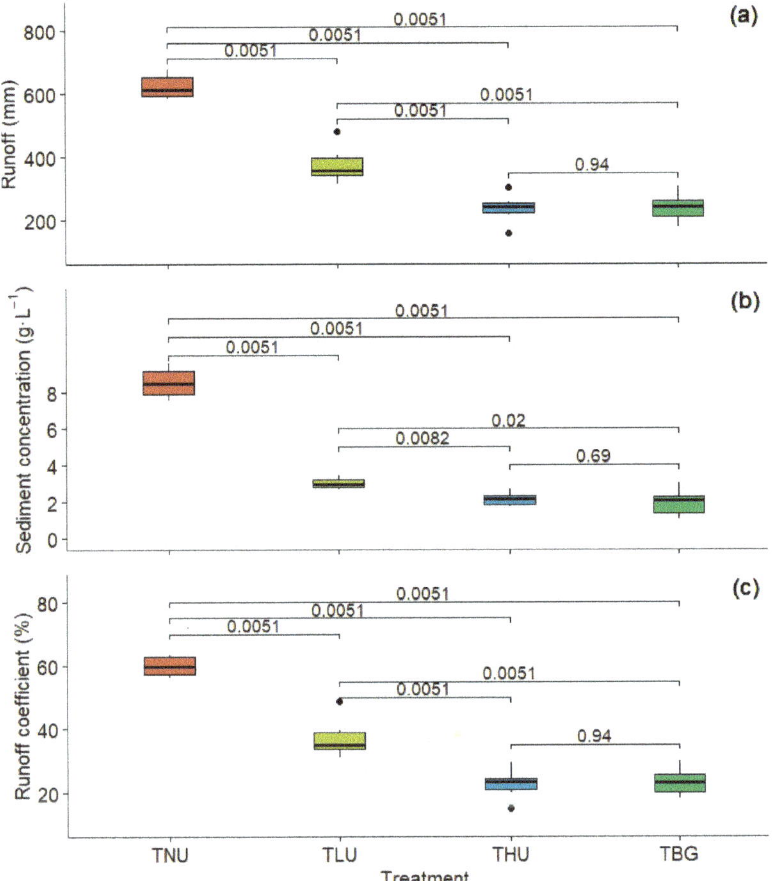

Figure 7. Boxplots of (**a**) cumulative surface runoff (mm), (**b**) average suspended sediment concentration (g·L^{-1}), and (**c**) runoff coefficient (%) in each treatment measured from 4 June to 15 October 2017, in Ban Kokngew, Luang Prabang Province, Lao People's Democratic Republic. Each rainfall bar represents the accumulated rainfall over the period prior to the sampling. Each boxplot contains the extreme of the lower whisker (vertical line), the lower hinge (thin line), the median (bold line), the upper hinge (thin line), the extreme of the upper whisker (vertical line), and the outliers (black dots) with *p*-values from Wilcoxon tests between two groups of treatments. The whiskers extend to the most extreme data point, which is no more than 1.5-times the interquartile range from the box. TNU: teak with no understory; TLU: teak with low density of understory; THU: teak with high density of understory; TBG: teak with broom grass. The runoff coefficient is the ratio in percentage between total surface runoff depth and total rainfall depth.

Table 2. Correlation (Pearson coefficient) between seasonal runoff coefficient (Rc), seasonal soil loss (Sl), and soil surface features: Fa: free aggregates; Fg: free gravel; Tc: total crust; Sc: structural crust; Ec: erosion crust; Gc: gravel crust; Cha: charcoals; Res; residues; Wor: worm casts; Alg: algae; Mos: mosses; Ped: pedestals; Und: understory. All variables were measured in 2017 in Ban Kokngew, Luang Prabang Province, Lao People's Democratic Republic.

Variables	Rc	Sl	Fa	Fg	Tc	Sc	Ec	Gc	Cha	Res	Wor	Alg	Mos	Ped
Sl	0.97 ****													
Fa	−0.72 ****	−0.77 ****												
Fg	0.84 ****	0.86 ****	−0.72 ****											
Tc	0.89 ****	0.90 ****	−0.87 ****	0.84 ****										
Sc	0.89 ****	0.90 ****	−0.87 ****	0.83 ****	0.99 ****									
Ec	0.62 **	0.64 ***	−0.58 **	0.44 *	0.67 ***	0.66 ***								
Gc	0.87 ****	0.90 ****	−0.78 ****	0.90 ****	0.91 ****	0.90 ****	0.70 ***							
Cha	0.48 *	0.44 *	−0.39	0.32	0.52 **	0.52 **	0.49 *	0.43 *						
Res	−0.87 ****	−0.89 ****	0.77 ****	−0.88 ****	−0.97 ****	−0.97 ****	−0.66 ***	−0.94 ****	−0.53 **					
Wor	−0.47 *	−0.42 *	−0.07	−0.46 *	−0.27	−0.27	−0.15	−0.37	−0.13	0.39				
Alg	0.55 **	0.53 **	−0.31	0.49 *	0.45 *	0.45 *	0.16	0.44 *	0.28	−0.47 *	−0.23			
Mos	−0.26	−0.34	0.27	−0.25	−0.39	−0.38	−0.32	−0.37	−0.10	0.42 *	0.26	−0.13		
Ped	0.89 ****	0.91 ****	−0.80 ****	0.82 ****	0.94 ****	0.94 ****	0.73 ****	0.91 ****	0.54 **	−0.92 ****	−0.30	0.38	−0.31	
Und	−0.91 ****	−0.93 ****	0.89 ****	−0.86 ****	−0.98 ****	−0.98 ****	−0.63 **	−0.91 ****	−0.49 *	0.95 ****	0.26	−0.47 *	0.36	−0.92 ****

Significance level: **** $p < 0.0001$; *** $p < 0.001$; ** $p < 0.01$; * $p < 0.05$.

Figure 8a shows PLSR biplot of inter-correlation between runoff coefficient, soil loss, understory, and soil surface features. The runoff coefficient and soil loss were positively correlated with total crust, pedestal features cover, free gravel, charcoals, and algae, but were negatively correlated with understory, residues, and free gravel. Worm casts and mosses had no significant relation with the other variables.

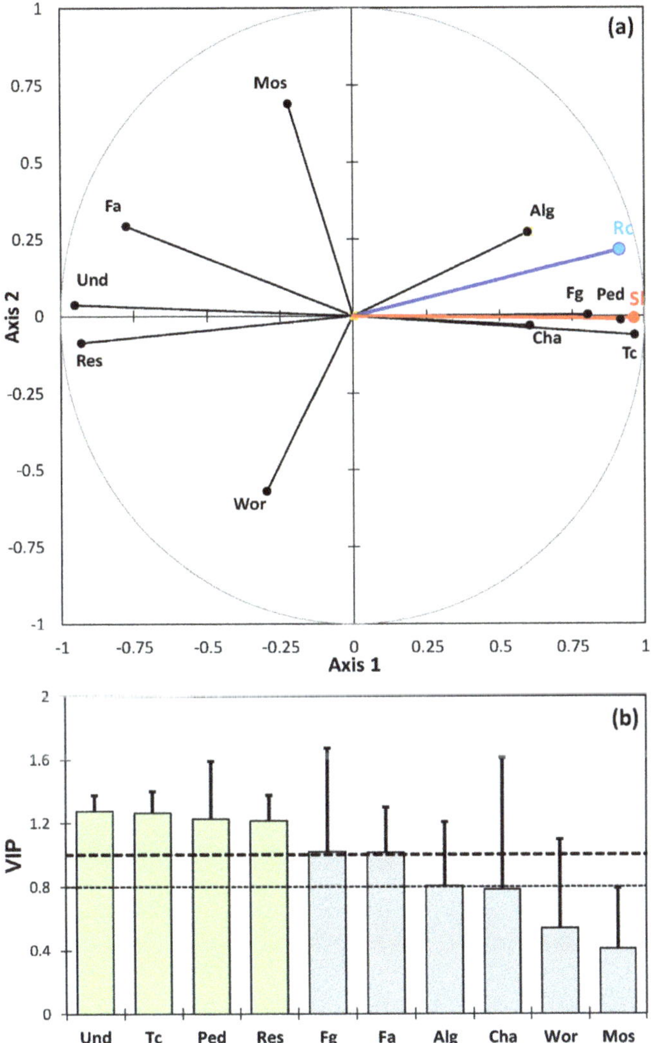

Figure 8. (**a**) Partial least squares regression (PLSR) biplot for seasonal surface runoff coefficient (Rc) and seasonal soil loss (Sl) in relation with soil surface features and understory. (**b**) Variable importance for the projection (VIP) score plot of each variable contributing the most to the models of Rc and Sl. Und: understory; Tc: total crust; Ped: pedestals; Res; residues; Fg: free gravel; Fa: free aggregates; Alg: algae; Cha: charcoals; Wor: worm casts; Mos: mosses. All variables were measured in 2017 in Ban Kokngew, Luang Prabang Province, Lao People's Democratic Republic.

Figure 8b displays the score plot of variable importance in projection (VIP) for each soil surface feature and understory treatment. This plot allows the rapid identification of the variables that contribute the most to the models of runoff coefficient and soil loss (Figure S3). Understory, total crust, pedestal features cover, and residues, were considered the most important variables and therefore the best predictors in the model. The coefficients of each variable contributing to the models' equation are listed in Table S1.

Figure 9 shows the observed and the modelled soil loss proposed by [31]. The soil loss is a function of the runoff coefficient with the equation $\ln(Sl) = -1.30 + 2.36 \ln(Rc)$, where Sl is the soil loss and Rc is the runoff coefficient. The model provided significant statistics ($R^2 = 0.91$, $p < 0.0001$) and is a promising framework for the prediction of soil loss based on the runoff coefficient.

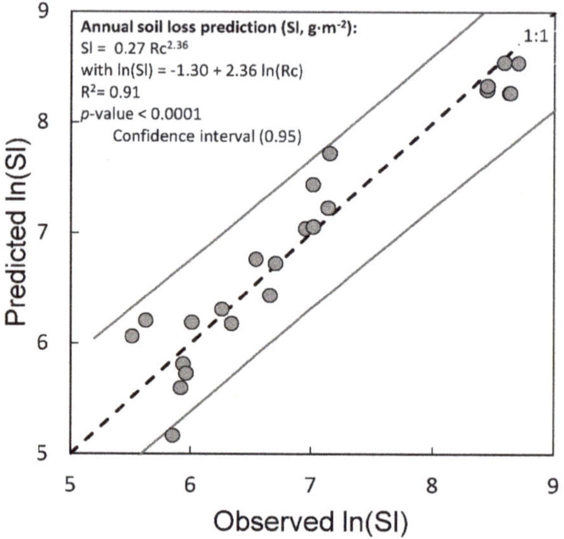

Figure 9. Observed and modelled seasonal soil loss. Each point represents the seasonal soil loss of the 24 microplots. Values (in $g \cdot m^{-2}$) were predicted using the simple formulation $\ln(Sl) = -1.30 + 2.36 \ln(Rc)$, where Rc is the seasonal runoff coefficient and Sl is the seasonal soil loss, as proposed by [31]. Observed Rc and Sl were measured from 4 June to 15 October 2017, in Ban Kokngew, Luang Prabang Province, Lao People's Democratic Republic.

4. Discussion

4.1. Understory Limits Surface Runoff and Soil Erosion

Soil loss is strongly related to surface runoff (Figures 6 and 8, and Table 2). Soil loss increased with increasing surface runoff, especially in TNU, which produced the most surface runoff, over the rainy season (Figure S1). Patin et al. [31] successfully applied the Terrace Erosion and Sediment Transport (TEST) model developed by [43] to estimate soil detachment observed for various land covers in the Houay Pano catchment, northern Lao PDR. They also proposed a simple soil loss model describing the soil loss as a function of the runoff coefficient. We applied this simplified model and also found that seasonal surface runoff coefficient is a reliable predictor of the seasonal soil loss (Figure 9). This result shows that, at the plot scale, surface runoff is statistically the main process responsible for the transfer of solid particles. Our results show that under the studied conditions, sediment production, which is high, is not the limiting factor to soil loss. We found that soil loss had a strong correlation with structural crust and a strong negative correlation with free aggregates and residues. Overall, these findings are in accordance with the findings reported by [30], namely that soil loss increases with an increasing surface runoff coefficient proportionally predicted by structural crust. The percentage of erosion crust is one of the main drivers of soil erosion [30]. At the plot scale, soil detachment firstly depends on splash [5]. The packing of the soil particles, which is due to the compaction by large drops, leads to crust forming [5]. Soil detachment by splash eventually leads to pedestal features [5]. We found that the structural crust represents in average 93–100% of the total crust, similar to the 91% reported by [30] for mature teak trees. Consistently, the percentage of pedestal features cover (Figure 4a) and total crust (82% of the percentage area of soil surface feature; Figure 4b) was relatively high in TNU. The highest surface runoff and soil loss were observed in TNU, due to the highly crusted area together resulting from less residues cover and the absence of understory cover. Hence, runoff generation occurs on the crusted soil area which restricts infiltration [60,61], depending on the presence of plant residues at soil surface and understory [62,63]. Residues do not only prevent crust formation [62], but also favor the biological activity of organisms which decompose organic matter from the residues, especially macro-faunal communities (also referred to as soil engineers) [64]. This biological process contributes to increase the infiltration rate through the development of soil biological porosity and breakdown of soil crusts resulting from previous inappropriate land use [63,65]. The surface runoff coefficient is a function of structural crust [30] and our results confirmed this finding (Table 2). A similar result was obtained by [61,66,67] who found that a soil crusted area enhances surface runoff and soil erosion. Crusted soil has less influence on soil detachment due to its physical protection [33]. This implies that structural crust is not the main cause of soil erosion but the main driver of surface runoff generation [30]. Once generated, surface runoff carries towards downstream the soil particles detached by splash [5].

Our models based on PLSR analysis confirmed the importance of each soil surface feature, including understory cover, in predicting seasonal surface runoff coefficient and seasonal soil loss at the plot scale (Figure S3 and Table S1). High soil detachment and surface runoff could be alleviated by the physical protection of both understory covers, including broom grass, and plant residues (originating from teak leaves and understory). Soil splash depends on understory cover and on the percentage of cover by residues [5], while surface runoff decreases when understory cover increases [68]. The understory and a thick residue layer reduces the rain splash effect and allows a high rate of infiltration, which then reduces surface runoff and soil erosion [69]. The soil is protected by understory, which dissipates the high raindrops' kinetic energy under teak trees [5]. In the absence of understory, the height of trees plays a decisive role with respect to soil detachment (Figure 3). Tall tree canopy in dense tree plantations intercepts the raindrops. Since teak tree leaves are broad, leaves dramatically increase the diameter of raindrops with throughflow, hence their kinetic energy and erosivity [27,30,70]. Plant residue attenuates the effect of rain splash and reduces the percentage area of total crust [30]. A number of authors have shown that the removal of understory by slashing and burning induces increased surface runoff and soil erosion [52,65,71,72].

Runoff coefficients in the four treatments varied in time and increased in the middle of the rainy season (July 2017, Figure S1), which is explained by antecedent rainfall, i.e., the rain falling before any rainfall event of interest [73]. Rainfall events happen more frequently in the middle of the rainy season. Since there is less time between two subsequent rainfall events to dry the soil, soil moisture increases, and soils get saturated or nearly saturated by water. Consequently, surface runoff generation is enhanced, and the runoff coefficients increase in the middle of the rainy season. Surface runoff in TNU plots were much higher than in all other treatments over the rainy season (Figure S1). The runoff coefficient in TNU was higher than in the other treatments since the very beginning of the season despite the low soil moisture and the low hydraulic conductivity of soil. Overall, the runoff coefficient in TNU was the highest with 59.7% (Figure S2), which is comparable to the values found by [30] in mature teak tree plantations with similar conditions of understory. On the contrary, the runoff coefficients of THU and TBG were around 23% (Figure S2), which is comparable to the overall runoff coefficient calculated for the treatments associating young teak trees with a variety of understory in [30].

The least erosive treatment in this study, i.e., TBG (381 g·m^{-2}), produced soil loss about 7.5-times higher than the broom grass treatment in Houay Dou catchment in northern Lao PDR [30]. The difference between our finding and broom grass treatment in [30] can be explained by the lower residue cover and the greater height of teak trees in our study, and by the fact that broom grass canopy was less dense because it was grown under the shadow of the teak trees and frequently harvested. Our values in TBG treatment were also greater than the values reported by [66,74] in northern Vietnam, with similar understory, topography, and climate, but with different vegetation types, namely mixed tree plantation and fallow, respectively. Similarly, the most erosive treatment in this study, i.e., TNU (5455 g·m^{-2}), produced soil loss about 3.5-times higher than the mature teak tree treatment in [30]. The values in TNU were also higher than the values reported by [30,33,66] in the same region with similar understory conditions. Such discrepancies may be related to soil surface conditions or plant and understory characteristics (height, plant density, percentage of cover), which were usually discussed in this study. However values 5.4-times higher were measured by [75] in Costa Rica, where antecedent land use as rangeland most likely greatly and durably reduced soil hydraulic conductivity [72]. Overall, the soil loss in TBG was 14-times less than in TNU. Hence, teak tree plantation owners could divide soil loss by 14 by keeping understory, such as broom grass, within teak tree plantations.

Considering the finding of our study at the plot scale, the farmers would also have saved about 40–50 ton·ha^{-1} of soil over the 2017 rainy season if they had grown understory under teak trees. However, this result must be qualified since the value cannot be extrapolated at scales larger than the plot scale, such as the catchment scale, because erosion processes at catchment scale are different (suspended sediments deposition or resuspension, gully formation on steep slope). At the catchment scale, a supplementary mitigation measure to trap water and eroded soil particles before entering the river network would be riparian buffers and vegetation filter strips [54,76].

4.2. Broom Grass Grown in Teak Tree Plantations: Agronomic Aspects and Ecosystem Services

Tree spacing is a determinant factor of teak tree productivity [65]. In our study, densities of 800–1200 tree·ha^{-1} [47] (Table 1) were relatively high compared to densities reported for other plantation areas (<800 tree·ha^{-1}) [65,77,78]. Planting trees at lower densities in our study site should be seriously considered as not only would it warrant higher productivity [65], possibly higher than in natural forest (40 tree·ha^{-1}) [77], but it would also allow growing intercropped understory adding economic value to the overall plantation yield.

The introduction of agroforestry into agricultural practices ensures an increased food security by a restored soil fertility for food crops [79] and a sustainable production of wood [64]. Broom grass is neither a food nor a feed crop but similar to agroforestry systems, it is grown under trees. Thus, the management practice may possibly increase the overall productivity (biomass, economic yield [45,47]) of the plot, by making an optimal use of the resources (water and nutrients), that would otherwise not be utilized by a single crop, consumed by plant species at different soil layers depending on their root length [80]. Although it seems to be more productive when grown in full light [30], broom grass can be grown under teak trees, similar to other shade-tolerant crops, such as patchouli [78].

Growing broom grass in teak tree plantations may supply a range of ecosystem services that is not limited to the supply of raw material for brooms and to the prevention of high surface runoff and soil loss in teak tree plantations. At the plot scale, the litter from broom grass leaves may increase top soil organic matter content [64], thus improving soil structure, soil nutrient availability [65], soil carbon sequestration [81], and increasing water infiltration and soil moisture retention [44,63]. Diversifying vegetation strata may diversify the habitats for bird species and other forest-dependent species [77,82,83], thus increasing predator biodiversity [84,85] and reducing the need for chemical inputs (insecticides, herbicides, etc.) in e.g., surrounding annual crop plots. At the catchment scale, favoring water infiltration and reducing surface runoff may mitigate natural disasters (floods and droughts) [65,81] and increase the transfer time of contaminants deposited at the soil surface [86]. Decreasing the sediment supply to the stream network would increase the life span of dam reservoirs [87] and avoid dredging costs [88–90], an issue of particular concern along the Mekong River where the number of dams is dramatically increasing [91,92].

4.3. Percentage of Cover of Pedestal Features: An Indicator of Soil Erosion

Both the PLSR model and the simplified model proposed by [31] that we presented in this study provided promising results to predict soil loss. Nevertheless, such models require input data from experimental microplots which are rather difficult and time consuming to be implemented, particularly within the framework of a large-scale approach. It would be more practical to use simple indicators that are known to be proxies of soil loss, such as soil surface features and/or understory characteristics (Table 2). On one hand, understory cover varies throughout the year, which causes difficulties in visual observation and inconsistent estimation. On the other hand, observing soil surface features (such as total crust, free aggregates, free gravel, residues, etc.) requires some expertise. However, among soils surface features, pedestal feature cover is a reliable proxy of soil erosion intensity in the field (Figure 4a) that is easy to assess visually. Hence, lay people could use the proxy of the percentage of cover of pedestal features to identify the impact of the agricultural management of their land on soil degradation through surface runoff and soil loss.

5. Conclusions

We investigated the impact of different types of understory on surface runoff and soil loss in a teak tree plantation of the mountainous region of northern Lao PDR. We analyzed the relationship between understory management, soil surface features (including pedestal features), and surface runoff and soil loss at the plot scale. This paper clearly demonstrates that teak tree plantations, especially in steep sloping lands, can be prone to considerable soil loss if not properly managed. Our main findings, which are graphically synthesized in Figure 10, are that:

- Understory cover acts as an umbrella that protects soil surface from rain splash despite the height of teak trees and the large size of their leaves which contribute to produce raindrops of high kinetic energy. Teak tree plantation owners could divide soil loss by 14 by keeping understory, such as broom grass, within teak tree plantations. Hence, growing understory under teak trees is a mitigation management practice that can be reliably promoted to limit surface runoff and soil erosion.
- Residues from both teak tree leaves and understory not only protect the soil but also enhance the infiltrability of water into the soil. In contrast, the main driver of surface runoff and soil erosion is the percentage of crusted area.
- Understory such as broom grass provides several benefits to the relevant stakeholders in the area, in terms of incomes and ecosystem services. For example, the farmers can sell the brooms made from broom grass.
- The percentage of cover of pedestal features appears as a good indicator of soil erosion that farmers and teak tree plantations owners could easily use to assess the degradation of their land.

In such a context, to minimize surface runoff and soil erosion in steep slope areas such as the montane regions of south–east Asia, decision makers should, if not legally enforce the maintenance of understory strata in teak tree plantation, at least recommend the plantations owners to maintain understory and avoid understory and plant residue layers burning.

At the plot scale, our findings are relevant to farmers concerned about soil loss and soil fertility in their teak tree plots. At catchment scale, they are relevant to decision makers concerned with the management of costs (such as the cost of water treatment, and/or infrastructure rehabilitation such as dam reservoir dredging) resulting from soil loss induced by upslope activities such as tree plantations and improper agricultural land management. In addition to maintaining the understory strata, encouraging the use of e.g., riparian zone buffers along the streams [93,94] could also be recommended to trap soil particles from the cultivated hillslopes and favor runoff infiltration, and thus ensure the sustainability of the system.

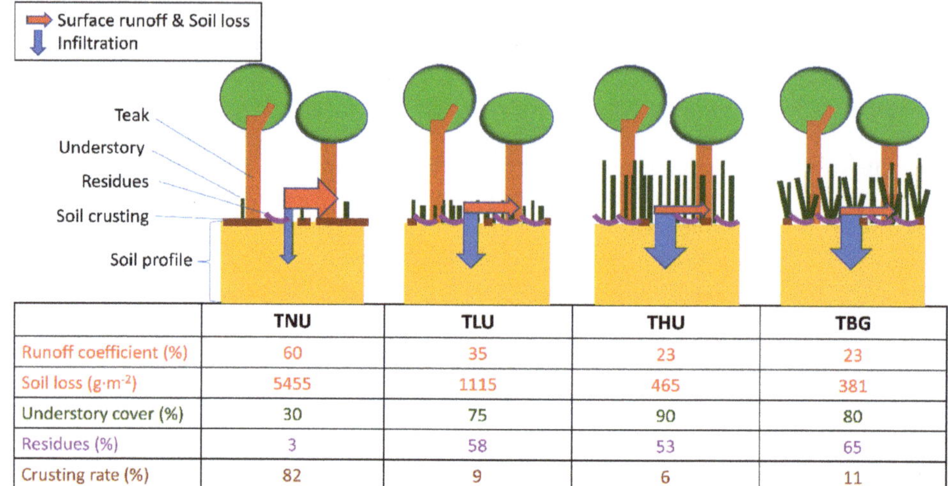

	TNU	TLU	THU	TBG
Runoff coefficient (%)	60	35	23	23
Soil loss (g·m⁻²)	5455	1115	465	381
Understory cover (%)	30	75	90	80
Residues (%)	3	58	53	65
Crusting rate (%)	82	9	6	11

Figure 10. Graphical abstract. TNU: teak with no understory; TLU: teak with low density of understory; THU: teak with high density of understory; TBG: teak with broom grass.

Supplementary Materials: The following are available online at http://www.mdpi.com/2073-4441/12/9/2327/s1, Figure S1: (**a**) Cumulated rainfall (mm). (**b**) boxplots of runoff coefficient (%). (**c**) boxplot of surface runoff (mm). (**d**) boxplot of soil loss (g·m⁻²) in each treatment measured from 4 June to 15 October 2017, in Ban Kokngew, Luang Prabang Province, Lao People's Democratic Republic. TNU: teak with no understory; TLU: teak with low density of understory; THU: teak with high density of understory; TBG: teak with broom grass. Each rainfall bar represents the accumulated rainfall over the period previous to the sampling. Each boxplot contains the extreme of the lower whisker (dashed line), the lower hinge (thin line), the median (bold line), the upper hinge (thin line), and the extreme of the upper whisker (dashed line). The whiskers extend to the most extreme data point, which is no more than 1.5-times the interquartile range from the box. Figure S2: (**a**) Cumulative surface runoff (mm) versus cumulative rainfall and runoff coefficient (%, total surface runoff divided by total rainfall), and (**b**) cumulative soil loss (g·m⁻²) versus cumulative rainfall (mm), measured from 4 June to 15 October 2017, in Ban Kokngew, Luang Prabang Province, Lao People's Democratic Republic. TNU: teak with no understory; TLU: teak with low density of understory; THU: teak with high density of understory; TBG: teak with broom grass. Figure S3: Observed and modelled seasonal runoff coefficient (Rc; in %) and seasonal soil loss (Sl; in g·m−2) from partial least squares (PLS) regression method. Observed Rc and Sl were measured from 4 June to 15 October 2017, in Ban Kokngew, Luang Prabang Province, Lao People's Democratic Republic. Table S1: Coefficients of each variables in the runoff coefficient (%) and soil loss (g·m−2) partial least squares (PLS) models. SD: standard deviation; Fa: free aggregates; Fg: free gravel; Tc: total crust; Cha: charcoals; Res: residues; Wor: worm casts; Alg: algae; Mos: mosses; Ped: pedestals; Und: understory.

Author Contributions: Conceptualization, O.S., B.S. (Bounthan Souksavath), B.L., and O.R.; methodology, A.d.R., C.V., and O.R.; formal analysis, L.S., O.R., and L.B.; data curation, L.S., C.V., P.S., A.d.R., B.S. (Bounsamay Soulileuth), N.S., and O.R.; writing—original draft preparation, L.S.; writing—review and editing, L.B., O.R., A.P., C.V., and A.d.R.; visualization, L.S. and O.R.; supervision, L.B., O.R., and C.O.; funding acquisition, O.R. and A.P. All authors have read and agreed to the published version of the manuscript.

Funding: This research was funded by the French National Research Agency TecItEasy project, grant number ANR-13-AGRO-0007, and by a scholarship of French government and Ministry of Education, Youth and Sports of Cambodia (L.S.).

Acknowledgments: This study was supported by the Luang Prabang Teak Programme (LPTP), the French Institute for Sustainable Development (IRD) through the international joint laboratory LUSES (Dynamic of Land Use Changes and Soil Ecosystem Services), and the Faculty of Agriculture of Lao PDR and Lao students, namely VACHOIMA Sythong and TONGSENG Chong Moua. The authors sincerely thank the M-TROPICS Critical Zone Observatory (https://mtropics.obs-mip.fr/), which belongs to the French Research Infrastructure OZCAR (http://www.ozcar-ri.org/), and the Lao Department of Agricultural Land Management (DALaM) for their support, including granting the permission for field access.

Conflicts of Interest: The authors declare no conflict of interest.

References

1. Gerstengarbe, F.W.; Werner, P.C. Precipitation Pattern. In *Encyclopedia of Ecology*; Jørgensen, S.E., Fath, B.D., Eds.; Elsevier Science: Amsterdam, The Netherlands, 2008; pp. 2916–2923. [CrossRef]
2. Descroix, L.; González Barrios, J.L.; Viramontes, D.; Poulenard, J.; Anaya, E.; Esteves, M.; Estrada, J. Gully and sheet erosion on subtropical mountain slopes: Their respective roles and the scale effect. *Catena* **2008**, *72*, 325–339. [CrossRef]
3. Valentin, C.; Agus, F.; Alamban, R.; Boosaner, A.; Bricquet, J.-P.; Chaplot, V.; De Guzman, T.; De Rouw, A.; Janeau, J.-L.; Orange, D. Runoff and sediment losses from 27 upland catchments in Southeast Asia: Impact of rapid land use changes and conservation practices. *Agric. Ecosyst. Environ.* **2008**, *128*, 225–238. [CrossRef]
4. Nandi, A.; Luffman, I. Erosion related changes to physicochemical properties of Ultisols distributed on calcareous sedimentary rocks. *J. Sustain. Dev.* **2012**, *5*, 52. [CrossRef]
5. Valentin, C.; Rajot, J.L. Erosion and Principles of Soil Conservation. In *Soils as a Key Component of the Critical Zone 5: Degradation and Rehabilitation*; Johns Hopkins University Press: Baltimore, MD, USA, 2018; Volume 5, pp. 39–82.
6. Sartori, M.; Philippidis, G.; Ferrari, E.; Borrelli, P.; Lugato, E.; Montanarella, L.; Panagos, P. A linkage between the biophysical and the economic: Assessing the global market impacts of soil erosion. *Land Use Policy* **2019**, *86*, 299–312. [CrossRef]
7. Panagos, P.; Borrelli, P.; Poesen, J.; Ballabio, C.; Lugato, E.; Meusburger, K.; Montanarella, L.; Alewell, C. The new assessment of soil loss by water erosion in Europe. *Environ. Sci. Policy* **2015**, *54*, 438–447. [CrossRef]
8. Pimentel, D.; Burgess, M. Soil erosion threatens food production. *Agriculture* **2013**, *3*, 443–463. [CrossRef]
9. Pimentel, D. Soil erosion: A food and environmental threat. *Environ. Dev. Sustain.* **2006**, *8*, 119–137. [CrossRef]
10. Owens, P.; Batalla, R.; Collins, A.; Gomez, B.; Hicks, D.; Horowitz, A.; Kondolf, G.; Marden, M.; Page, M.; Peacock, D. Fine-grained sediment in river systems: Environmental significance and management issues. *River. Res. Appl.* **2005**, *21*, 693–717. [CrossRef]
11. Yuan, S.; Tang, H.; Xiao, Y.; Xia, Y.; Melching, C.; Li, Z. Phosphorus Contamination of the Surface Sediment at a River Confluence. *J. Hydrol.* **2019**, *573*, 568–580. [CrossRef]
12. Chartin, C.; Evrard, O.; Onda, Y.; Patin, J.; Lefèvre, I.; Ottlé, C.; Ayrault, S.; Lepage, H.; Bonté, P. Tracking the early dispersion of contaminated sediment along rivers draining the Fukushima radioactive pollution plume. *Anthropocene* **2013**, *1*, 23–34. [CrossRef]
13. Turner, J.N.; Brewer, P.A.; Macklin, M.G. Fluvial-controlled metal and As mobilisation, dispersal and storage in the Rio Guadiamar, SW Spain and its implications for long-term contaminant fluxes to the Donana wetlands. *Sci. Total. Environ.* **2008**, *394*, 144–161. [CrossRef] [PubMed]
14. Domagalski, J.L.; Kuivila, K.M. Distributions of pesticides and organic contaminants between water and suspended sediment, San Francisco Bay, California. *Estuaries* **1993**, *16*, 416–426. [CrossRef]
15. Gateuille, D.; Evrard, O.; Lefevre, I.; Moreau-Guigon, E.; Alliot, F.; Chevreuil, M.; Mouchel, J.M. Mass balance and decontamination times of Polycyclic Aromatic Hydrocarbons in rural nested catchments of an early industrialized region (Seine River basin, France). *Sci. Total. Environ.* **2014**, *470*, 608–617. [CrossRef] [PubMed]
16. Ribolzi, O.; Cuny, J.; Sengsoulichanh, P.; Mousques, C.; Soulileuth, B.; Pierret, A.; Huon, S.; Sengtaheuanghoung, O. Land use and water quality along a Mekong tributary in northern Lao P.D.R. *Environ. Manag.* **2011**, *47*, 291–302. [CrossRef] [PubMed]
17. Boithias, L.; Choisy, M.; Souliyaseng, N.; Jourdren, M.; Quet, F.; Buisson, Y.; Thammahacksa, C.; Silvera, N.; Latsachack, K.; Sengtaheuanghoung, O.; et al. Hydrological Regime and Water Shortage as Drivers of the Seasonal Incidence of Diarrheal Diseases in a Tropical Montane Environment. *PLoS. Negl. Trop. Dis.* **2016**, *10*, e0005195. [CrossRef] [PubMed]
18. Kim, M.; Boithias, L.; Cho, K.H.; Sengtaheuanghoung, O.; Ribolzi, O. Modeling the Impact of Land Use Change on Basin-scale Transfer of Fecal Indicator Bacteria: SWAT Model Performance. *J. Environ. Qual.* **2018**, *47*, 1115–1122. [CrossRef]
19. Thothong, W.; Huon, S.; Janeau, J.-L.; Boonsaner, A.; de Rouw, A.; Planchon, O.; Bardoux, G.; Parkpian, P. Impact of land use change and rainfall on sediment and carbon accumulation in a water reservoir of North Thailand. *Agric. Ecosyst. Environ.* **2011**, *140*, 521–533. [CrossRef]

20. Zarris, D.; Vlastara, M.; Panagoulia, D. Sediment Delivery Assessment for a Transboundary Mediterranean Catchment: The Example of Nestos River Catchment. *Water Resour. Manag.* **2011**, *25*, 3785. [CrossRef]

21. Efthimiou, N.; Lykoudi, E.; Panagoulia, D.; Karavitis, C. Assessment of soil susceptibility to erosion using the EPM and RUSLE Models: The case of Venetikos River Catchment. *Global NEST J.* **2016**, *18*, 164–179.

22. Sidle, R.C.; Ziegler, A.D.; Negishi, J.N.; Nik, A.R.; Siew, R.; Turkelboom, F. Erosion processes in steep terrain—Truths, myths, and uncertainties related to forest management in Southeast Asia. *For. Ecol. Manag.* **2006**, *224*, 199–225. [CrossRef]

23. Patin, J.; Mouche, E.; Ribolzi, O.; Chaplot, V.; Sengtahevanghoung, O.; Latsachak, K.O.; Soulileuth, B.; Valentin, C. Analysis of runoff production at the plot scale during a long-term survey of a small agricultural catchment in Lao PDR. *J. Hydrol.* **2012**, *426*, 79–92. [CrossRef]

24. Ziegler, A.D.; Giambelluca, T.W.; Tran, L.T.; Vana, T.T.; Nullet, M.A.; Fox, J.; Vien, T.D.; Pinthong, J.; Maxwell, J.F.; Evett, S. Hydrological consequences of landscape fragmentation in mountainous northern Vietnam: Evidence of accelerated overland flow generation. *J. Hydrol.* **2004**, *287*, 124–146. [CrossRef]

25. Ziegler, A.D.; Giambelluca, T.W.; Nullet, M.A.; Sutherland, R.A.; Tantasarin, C.; Vogler, J.B.; Negishi, J.N. Throughfall in an evergreen-dominated forest stand in northern Thailand: Comparison of mobile and stationary methods. *Agric. For. Meteorol.* **2009**, *149*, 373–384. [CrossRef]

26. Valentin, C. (Ed.) Soil Surface Crusting of Soil and Water Harvesting. In *Soils as a Key Component of the Critical Zone 5: Degradation and Rehabilitation*; John Wiley & Sons: New York, NY, USA, 2018; Volume 5, pp. 21–38.

27. Goebes, P.; Seitz, S.; Kühn, P.; Li, Y.; Niklaus, P.A.; Oheimb, G.v.; Scholten, T. Throughfall kinetic energy in young subtropical forests: Investigation on tree species richness effects and spatial variability. *Agric. For. Meteorol.* **2015**, *213*, 148–159. [CrossRef]

28. Lacombe, G.; Ribolzi, O.; de Rouw, A.; Pierret, A.; Latsachak, K.; Silvera, N.; Pham Dinh, R.; Orange, D.; Janeau, J.L.; Soulileuth, B.; et al. Contradictory hydrological impacts of afforestation in the humid tropics evidenced by long-term field monitoring and simulation modelling. *Hydrol. Earth Syst. Sci.* **2016**, *20*, 2691–2704. [CrossRef]

29. Ribolzi, O.; Evrard, O.; Huon, S.; de Rouw, A.; Silvera, N.; Latsachack, K.O.; Soulileuth, B.; Lefevre, I.; Pierret, A.; Lacombe, G.; et al. From shifting cultivation to teak plantation: Effect on overland flow and sediment yield in a montane tropical catchment. *Sci. Rep.* **2017**, *7*, 3987. [CrossRef]

30. Lacombe, G.; Valentin, C.; Sounyafong, P.; de Rouw, A.; Soulileuth, B.; Silvera, N.; Pierret, A.; Sengtaheuanghoung, O.; Ribolzi, O. Linking crop structure, throughfall, soil surface conditions, runoff and soil detachment: 10 land uses analyzed in Northern Laos. *Sci. Total. Environ.* **2018**, *616*, 1330–1338. [CrossRef]

31. Patin, J.; Mouche, E.; Ribolzi, O.; Sengtahevanghoung, O.; Latsachak, K.O.; Soulileuth, B.; Chaplot, V.; Valentin, C. Effect of land use on interrill erosion in a montane catchment of Northern Laos: An analysis based on a pluri-annual runoff and soil loss database. *J. Hydrol.* **2018**, *563*, 480–494. [CrossRef]

32. Ribolzi, O.; Patin, J.; Bresson, L.M.; Latsachack, K.O.; Mouche, E.; Sengtaheuanghoung, O.; Silvera, N.; Thiébaux, J.P.; Valentin, C. Impact of slope gradient on soil surface features and infiltration on steep slopes in northern Laos. *Geomorphology* **2011**, *127*, 53–63. [CrossRef]

33. Chaplot, V.; Khampaseuth, X.; Valentin, C.; Bissonnais, Y.L. Interrill erosion in the sloping lands of northern Laos subjected to shifting cultivation. *Earth Surf. Process. Landf.* **2007**, *32*, 415–428. [CrossRef]

34. Vannoppen, W.; Vanmaercke, M.; De Baets, S.; Poesen, J. A review of the mechanical effects of plant roots on concentrated flow erosion rates. *Earth Sci. Rev.* **2015**, *150*, 666–678. [CrossRef]

35. Shinohara, Y.; Otani, S.; Kubota, T.; Otsuki, K.; Nanko, K. Effects of plant roots on the soil erosion rate under simulated rainfall with high kinetic energy. *Hydrol. Sci. J.* **2016**, *61*, 2435–2442. [CrossRef]

36. Roose, E. *Land Husbandry: Components and Strategy*; FAO: Rome, Italy, 1996.

37. Ehigiator, O.A.; Anyata, B.U. Effects of land clearing techniques and tillage systems on runoff and soil erosion in a tropical rain forest in Nigeria. *J. Environ. Manag.* **2011**, *92*, 2875–2880. [CrossRef] [PubMed]

38. Cerdà, A.; Rodrigo-Comino, J.; Giménez-Morera, A.; Novara, A.; Pulido, M.; Kapović-Solomun, M.; Keesstra, S.D. Policies can help to apply successful strategies to control soil and water losses. The case of chipped pruned branches (CPB) in Mediterranean citrus plantations. *Land Use Policy* **2018**, *75*, 734–745. [CrossRef]

39. Li, X.; Niu, J.; Xie, B. The effect of leaf litter cover on surface runoff and soil erosion in Northern China. *PLoS ONE* **2014**, *9*, e107789. [CrossRef]

40. Durán Zuazo, V.H.; Martínez, J.R.F.; Pleguezuelo, C.R.R.; Martínez Raya, A.; Rodríguez, B.C. Soil-erosion and runoff prevention by plant covers in a mountainous area (se spain): Implications for sustainable agriculture. *Environmentalist* **2006**, *26*, 309–319. [CrossRef]

41. Poesen, J. Conditions for gully formation in the Belgian loam belt and some ways to control them. In Proceedings of the of Soil Erosion Protection Measures in Europe. Proc. EC Workshop, Freising, Germany, 24–26 May 1988; pp. 39–52.

42. Valentin, C.; Ruiz Figueroa, J. Effects of kinetic energy and water application rate on the development of crusts in a fine sandy loam soil using sprinkling irrigation and rainfall simulation. *Micromorphol. Sols* **1987**, 401–408.

43. Van Dijk, A.; Bruijnzeel, L.; Eisma, E. A methodology to study rain splash and wash processes under natural rainfall. *Hydrol. Process.* **2003**, *17*, 153–167. [CrossRef]

44. Neyret, M.; Robain, H.; de Rouw, A.; Janeau, J.-L.; Durand, T.; Kaewthip, J.; Trisophon, K.; Valentin, C. Higher runoff and soil detachment in rubber tree plantations compared to annual cultivation is mitigated by ground cover in steep mountainous Thailand. *Catena* **2020**, *189*, 104472. [CrossRef]

45. Pachas, A.N.A.; Newby, J.C.; Siphommachan, P.; Sakanphet, S.; Dieters, M.J. Broom grass in Lao PDR: A market chain analysis in Luang Prabang Province. *For. Trees Livelihoods* **2020**, 1–18. [CrossRef]

46. Olson, D.M.; Dinerstein, E.; Wikramanayake, E.D.; Burgess, N.D.; Powell, G.V.N.; Underwood, E.C.; D'amico, J.A.; Itoua, I.; Strand, H.E.; Morrison, J.C.; et al. Terrestrial Ecoregions of the World: A New Map of Life on Earth: A new global map of terrestrial ecoregions provides an innovative tool for conserving biodiversity. *Bioscience* **2001**, *51*, 933–938. [CrossRef]

47. Pachas, A.; Sakanphet, S.; Midgley, S.; Dieters, M. Teak (Tectona grandis) silviculture and research: Applications for smallholders in Lao PDR. *Aust. For.* **2019**, *82*, 94–105. [CrossRef]

48. Casenave, A.; Valentin, C. A runoff capability classification system based on surface features criteria in semi-arid areas of West Africa. *J. Hydrol.* **1992**, *130*, 231–249. [CrossRef]

49. Ribolzi, O.; Lacombe, G.; Pierret, A.; Robain, H.; Sounyafong, P.; de Rouw, A.; Soulileuth, B.; Mouche, E.; Huon, S.; Silvera, N. Interacting land use and soil surface dynamics control groundwater outflow in a montane catchment of the lower Mekong basin. *Agric. Ecosyst. Environ.* **2018**, *268*, 90–102. [CrossRef]

50. Janeau, J.L.; Bricquet, J.P.; Planchon, O.; Valentin, C. Soil crusting and infiltration on steep slopes in northern Thailand. *Eur. J. Soil. Sci.* **2003**, *54*, 543–554. [CrossRef]

51. Valentin, C.; Casenave, A. Infiltration into sealed soils as influenced by gravel cover. *Soil Sci. Soc. Am. J.* **1992**, *56*, 1667–1673. [CrossRef]

52. Lacombe, G.; Ribolzi, O.; de Rouw, A.; Pierret, A.; Latsachak, K.; Silvera, N.; Pham Dinh, R.; Orange, D.; Janeau, J.L.; Soulileuth, B.; et al. Afforestation by natural regeneration or by tree planting: Examples of opposite hydrological impacts evidenced by long-term field monitoring in the humid tropics. *Hydrol. Earth Syst. Sci.* **2015**, *12*, 12615–12648. [CrossRef]

53. Janeau, J.-L.; Gillard, L.-C.; Grellier, S.; Jouquet, P.; Le, T.P.Q.; Luu, T.N.M.; Ngo, Q.A.; Orange, D.; Pham, D.R.; Tran, D.T.; et al. Soil erosion, dissolved organic carbon and nutrient losses under different land use systems in a small catchment in northern Vietnam. *Agric. Water Manag.* **2014**, *146*, 314–323. [CrossRef]

54. Vigiak, O.; Ribolzi, O.; Pierret, A.; Sengtaheuanghoung, O.; Valentin, C. Trapping efficiencies of cultivated and natural riparian vegetation of northern Laos. *J. Environ. Qual.* **2008**, *37*, 889–897. [CrossRef]

55. Valentin, C.; Bresson, L.-M. Morphology, genesis and classification of surface crusts in loamy and sandy soils. *Geoderma* **1992**, *55*, 225–245. [CrossRef]

56. Quinn, G.; Keough, M. *Experimental Design and Data Analysis for Biologists*; Cambridge University Press: Cambridge, UK, 2002.

57. Abdi, H. Partial least squares regression and projection on latent structure regression (PLS Regression). *Wiley Interdiscip. Rev. Comput. Stat.* **2010**, *2*, 97–106. [CrossRef]

58. Zaldívar Santamaría, E.; Molina Dagá, D.; Palacios García, A.T. Statistical Modelization of the Descriptor "Minerality" Based on the Sensory Properties and Chemical Composition of Wine. *Beverages* **2019**, *5*, 66.

59. Wold, S. PLS for multivariate linear modeling. *Chemom. Methods Mol. Des.* **1995**, 195–218.

60. Bu, C.-F.; Wu, S.-F.; Yang, K.-B. Effects of physical soil crusts on infiltration and splash erosion in three typical Chinese soils. *Int. J. Sediment Res.* **2014**, *29*, 491–501. [CrossRef]

61. Fox, D.; Bryan, R.; Fox, C. Changes in pore characteristics with depth for structural crusts. *Geoderma* **2004**, *120*, 109–120. [CrossRef]

62. Neave, M.; Rayburg, S. A field investigation into the effects of progressive rainfall-induced soil seal and crust development on runoff and erosion rates: The impact of surface cover. *Geomorphology* **2007**, *87*, 378–390. [CrossRef]

63. Rey, F.; Ballais, J.-L.; Marre, A.; Rovéra, G. Rôle de la végétation dans la protection contre l'érosion hydrique de surface. *C. R. Geosci.* **2004**, *336*, 991–998. [CrossRef]

64. Angelsen, A.; Kaimowitz, D. Is agroforestry likely to reduce deforestation. In *Agroforestry and Biodiversity Conservation in Tropical Landscapes*; Schroth, G., Fonseca, G.A.d., Harvey, C.A., Gascon, C., Vasconcelos, H.L., Izac, A.-M.N., Eds.; Island Press: Washington, DC, USA, 2004; pp. 87–106.

65. Blanco, J.A.; Lo, Y.-H. *Forest Ecosystems: More Than Just Trees*; BoD–Books on Demand: Norderstedt, Germany, 2012.

66. Podwojewski, P.; Orange, D.; Jouquet, P.; Valentin, C.; Janeau, J.; Tran, D.T. Land-use impacts on surface runoff and soil detachment within agricultural sloping lands in Northern Vietnam. *Catena* **2008**, *74*, 109–118. [CrossRef]

67. Kosmas, C.; Danalatos, N.; Cammeraat, L.H.; Chabart, M.; Diamantopoulos, J.; Farand, R.; Gutierrez, L.; Jacob, A.; Marques, H.; Martinez-Fernandez, J. The effect of land use on runoff and soil erosion rates under Mediterranean conditions. *Catena* **1997**, *29*, 45–59. [CrossRef]

68. Nouwakpo, S.K.; Weltz, M.A.; Green, C.H.M.; Arslan, A. Combining 3D data and traditional soil erosion assessment techniques to study the effect of a vegetation cover gradient on hillslope runoff and soil erosion in a semi-arid catchment. *Catena* **2018**, *170*, 129–140. [CrossRef]

69. Calder, I.R.; Hall, R.L.; Prasanna, K. Hydrological impact of Eucalyptus plantation in India. *J. Hydrol.* **1993**, *150*, 635–648. [CrossRef]

70. Geißler, C.; Kühn, P.; Böhnke, M.; Bruelheide, H.; Shi, X.; Scholten, T. Splash erosion potential under tree canopies in subtropical SE China. *Catena* **2012**, *91*, 85–93. [CrossRef]

71. Le, H.T.; Rochelle-Newall, E.; Ribolzi, O.; Janeau, J.L.; Huon, S.; Latsachack, K.; Pommier, T. Land use strongly influences soil organic carbon and bacterial community export in runoff in tropical uplands-. *Land. Degrad. Dev.* **2020**, *31*, 118–132. [CrossRef]

72. Fernández-Moya, J.; Alvarado, A.; Forsythe, W.; Ramírez, L.; Algeet-Abarquero, N.; Marchamalo-Sacristán, M. Soil erosion under teak (*Tectona grandis* Lf) plantations: General patterns, assumptions and controversies. *Catena* **2014**, *123*, 236–242. [CrossRef]

73. Jadidoleslam, N.; Mantilla, R.; Krajewski, W.F.; Goska, R. Investigating the role of antecedent SMAP satellite soil moisture, radar rainfall and MODIS vegetation on runoff production in an agricultural region. *J. Hydrol.* **2019**, *579*, 124210. [CrossRef]

74. Phan Ha, H.A.; Huon, S.; Henry des Tureaux, T.; Orange, D.; Jouquet, P.; Valentin, C.; De Rouw, A.; Tran Duc, T. Impact of fodder cover on runoff and soil erosion at plot scale in a cultivated catchment of North Vietnam. *Geoderma* **2012**, *177*, 8–17. [CrossRef]

75. Santamaría Leandro, F. Evaluación de la Pérdida de Suelo en Plantaciones de Teca, Bajo la Aplicación de Sistemas de Conservación de Suelos en Nicoya, Guanacaste. 1992. Available online: http://www.sidalc.net/cgi-bin/wxis.exe/?IsisScript=rednia.xis&method=post&formato=2&cantidad=1&expresion=mfn=005930 (accessed on 2 February 2019).

76. Cao, X.; Song, C.; Xiao, J.; Zhou, Y. The optimal width and mechanism of riparian buffers for storm water nutrient removal in the Chinese eutrophic Lake Chaohu watershed. *Water* **2018**, *10*, 1489. [CrossRef]

77. Koonkhunthod, N.; Sakurai, K.; Tanaka, S. Composition and diversity of woody regeneration in a 37-year-old teak (*Tectona grandis* L.) plantation in Northern Thailand. *For. Ecol. Manag.* **2007**, *247*, 246–254. [CrossRef]

78. Kumar, D.; Bijalwan, A.; Kalra, A.; Dobriyal, M.J. Effect of shade and organic manure on growth and yield of patchouli [Pogostemon cablin (blanco) benth.] under teak (*Tectona grandis* lf) based agroforestry system. *Indian For.* **2016**, *142*, 1121–1129.

79. Schroeder, P. Agroforestry systems: Integrated land use to store and conserve carbon. *Clim. Res.* **1993**, *3*, 53–60. [CrossRef]

80. Chitra-Tarak, R.; Ruiz, L.; Dattaraja, H.S.; Kumar, M.M.; Riotte, J.; Suresh, H.S.; McMahon, S.M.; Sukumar, R. The roots of the drought: Hydrology and water uptake strategies mediate forest-wide demographic response to precipitation. *J. Ecol.* **2018**, *106*, 1495–1507. [CrossRef]

81. Balmford, A.; Whitten, T. Who should pay for tropical conservation, and how could the costs be met? *Oryx* **2003**, *37*, 238–250. [CrossRef]

82. Harvey, C.A.; Villalobos, J.A.G. Agroforestry systems conserve species-rich but modified assemblages of tropical birds and bats. *Biodivers. Conserv.* **2007**, *16*, 2257–2292. [CrossRef]

83. Wunderle, J.M., Jr. The role of animal seed dispersal in accelerating native forest regeneration on degraded tropical lands. *For. Ecol. Manag.* **1997**, *99*, 223–235. [CrossRef]

84. Imron, M.; Tantaryzard, M.; Satria, R.; Maulana, I.; Pudyatmoko, S. Understory Avian Community In A Teak Forest Of Cepu, Central Java. *J. Trop. For. Sci.* **2018**, *30*, 509–518. [CrossRef]

85. Perrin, R. Pest management in multiple cropping systems. *Agro-Ecosystems* **1976**, *3*, 93–118. [CrossRef]

86. Monaghan, R.; Smith, L.; Muirhead, R. Pathways of contaminant transfers to water from an artificially-drained soil under intensive grazing by dairy cows. *Agric. Ecosyst. Environ.* **2016**, *220*, 76–88. [CrossRef]

87. Annandale, G.W. Reservoir sedimentation. In *Encyclopedia of Hydrological Sciences*; Wiley: New York, NY, USA, 2006. [CrossRef]

88. Boithias, L.; Terrado, M.; Corominas, L.; Ziv, G.; Kumar, V.; Marques, M.; Schuhmacher, M.; Acuna, V. Analysis of the uncertainty in the monetary valuation of ecosystem services—A case study at the river basin scale. *Sci. Total. Environ.* **2016**, *543*, 683–690. [CrossRef]

89. Crowder, B.M. Economic costs of reservoir sedimentation: A regional approach to estimating cropland erosion damage. *J. Soil Water Conserv.* **1987**, *42*, 194–197.

90. McHenry, J.R. RESERVOIR SEDIMENTATION 1. *J. Am. Water Resour. Assoc.* **1974**, *10*, 329–337. [CrossRef]

91. Dang, T.D.; Cochrane, T.A.; Arias, M.E. Quantifying suspended sediment dynamics in mega deltas using remote sensing data: A case study of the Mekong floodplains. *Int. J. Appl. Earth Obs. Geoinf.* **2018**, *68*, 105–115. [CrossRef]

92. Arias, M.E.; Cochrane, T.A.; Kummu, M.; Lauri, H.; Holtgrieve, G.W.; Koponen, J.; Piman, T. Impacts of hydropower and climate change on drivers of ecological productivity of Southeast Asia's most important wetland. *Ecol. Model.* **2014**, *272*, 252–263. [CrossRef]

93. Bhat, S.A.; Dar, M.U.D.; Meena, R.S. Soil erosion and management strategies. In *Sustainable Management of Soil and Environment*; Springer: New York, NY, USA, 2019; pp. 73–122.

94. Ahmad, N.S.B.N.; Mustafa, F.B.; Gideon, D. A systematic review of soil erosion control practices on the agricultural land in Asia. *Int. Soil Water Conserv. Res.* **2020**. [CrossRef]

 *water*

Article

Comparison of the MUSLE Model and Two Years of Solid Transport Measurement, in the Bouregreg Basin, and Impact on the Sedimentation in the Sidi Mohamed Ben Abdellah Reservoir, Morocco

Mohamed Abdellah Ezzaouini [1,2,*], Gil Mahé [3], Ilias Kacimi [1] and Abdelaziz Zerouali [2]

[1] Geoscience, Water and Environment Laboratory, Faculty of Sciences, Mohammed V University, Avenue Ibn Batouta, Rabat 10100, Morocco; iliaskacimi@yahoo.fr
[2] Basin of the Bouregreg and Chaouia Agency, Benslimane BP 262, Morocco; azlso@yahoo.fr
[3] Laboratoire HydroSciences, Université Montpellier, IRD, 34090 Montpellier, France; gilmahe@hotmail.com
* Correspondence: ezzaouini70@yahoo.fr; Tel.: +212-661-110-897

Received: 18 May 2020; Accepted: 15 June 2020; Published: 1 July 2020

Abstract: The evaluation and quantification of solids transport in Morocco often uses the Universal Soil Loss Model (USLE) and the revised version RUSLE, which presents a calibration difficulty. In this study, we apply the MUSLE model to predict solid transport, for the first time on a large river basin in the Kingdom, calibrated by two years of solid transport measurements on four main gauging stations at the entrance of the Sidi Mohamed Ben Abdellah dam. The application of the MUSLE on the basin demonstrated relatively small differences between the measured values and those expected for the calibrated version, these differences are, for the non-calibrated version, +5% and +102% for the years 2016/2017 and 2017/2018 respectively, and between −33% and +34% for the calibrated version. Besides, the measured and modeled volumes that do not exceed 1.78×10^6 m^3/year remain well below the dam's siltation rate of 9.49×10^6 m^3/year, which means that only 18% of the dam's sediment comes from upstream. This seems very low because it is calculated from only two years. The main hypothesis that we can formulate is that the sediments of the dam most probably comes from the erosion of its banks.

Keywords: modeling; MUSLE; erosion; solid transport; dam; Bouregreg; Morocco

---

1. Introduction

Erosion is a natural phenomenon that reduces the capacity of dam reservoirs around the world. The natural erosive process is aggravated by anthropogenic activities including pastoral activity [1], deforestation [2–4], and climate change [5] with the advent of periods of heavy rainfall and increasingly frequent dry periods. This phenomenon constitutes a major challenge for water resource management at the scale of the Bouregreg basin [6,7] in northern Morocco.

The Sidi Mohamed Ben Abdellah (SMBA) dam, commissioned in 1974 and raised in 2007, is intended solely to supply drinking water to the coastal area between Rabat and Casablanca, which represents nearly eight million inhabitants. It has a relatively low silting rate compared to other dams in the Kingdom [8].However, it has lost 132 Mm^3 since its commissioning of which 58% of this volume was lost before rising the dam height, with this loss constituting a very significant reduction in its storage capacity. Given the magnitude of this situation, the modeling of soil losses in the basin aims at achieving the following objectives:

- analyzing the biophysical environment;

- describing and evaluating the erosive processes affecting the Bouregreg Basin, and the solid transport by the main tributaries to the SMBA dam;
- and identifying the priority areas contributing to siltation, in order to better guide spatial planning actions.

Universal Soil Loss Equation (USLE) was first established in USA to model erosion in small agricultural catchment [9]. It is based on several parameters linked to climate, soil cover and properties, topography, and human activities. This equation has been modified and adapted several times. The MUSLE model includes the use of water flow rates [10]. Despite the difficulties encountered in calibrating and adapting the Universal Land Loss Model (USLE) to conditions in Morocco [11], most studies on watershed management in Morocco [12–15] and bordering Mediterranean regions, in particular in Algeria and Tunisia [15–17], continue to use USLE, and the revised version RUSLE [18–21], or an event normalized plot soil loss estimated by a modified USLE model—USLE-MM—as in Italy [22], most often for small basins of much less than 5000 km^2. As USLE [23] and RUSLE [24] were developed for the rough assessment of annual land loss at the scale of small plots, their application to large areas leads to rather large errors [25,26]. However, their accuracy increases when coupled with hydrological models [27]. To overcome the difficulties in assessing the accuracy of using a simple erosion equation like USLE, Alewell et al. [28] recommend to strengthen and extend measurement and monitoring programs to build up validation data sets.

Thus, Williams [10] developed a modified version of the USLE (MUSLE) that takes into account the flow load at the outlet by taking into account the biophysical characteristics of the watershed. This model has already been applied to micro-watersheds [29,30] and gave very reliable results compared to measurements. Indeed, in the Sidi Sbaa basin in Morocco, the deviation of the results compared to the MUSLE model was −4% by underestimating the solid inputs [31]. Samaras and Koutitas et al. [32] use MUSLE with SWAT to simulate the potential impact of land cover change on sediment yields to the sea in Greece, but with no observed validation data; while other authors like Fang [33] use the WaTEM/SEDEM model, which includes the RUSLE formula, to estimate erosion, but always without observations to compare. Only a few studies compare the erosion rates with the three USLE formulas. In Maghreb, only one study by Djoukbala et al. [34] compared them on the small basin of 384 km^2 in the north of Algeria, with erosion rates quite similar between the three methods, but they were slightly superior in the case MUSLE. Unfortunately, they could not compare their results with observation data, which does not allow an assessment of the validity of the erosion rates produced in regard to real natural processes.

2. Materials and Methods

2.1. Study Area

The Bouregreg Basin, located in central western Morocco, covers a total area of approximately 10,130 km^2 at its mouth in the north of the city of Rabat (Figure 1). The main rivers of the basin are the Bouregreg, the Grou, and the Korifla and its tributary, the Machraa.

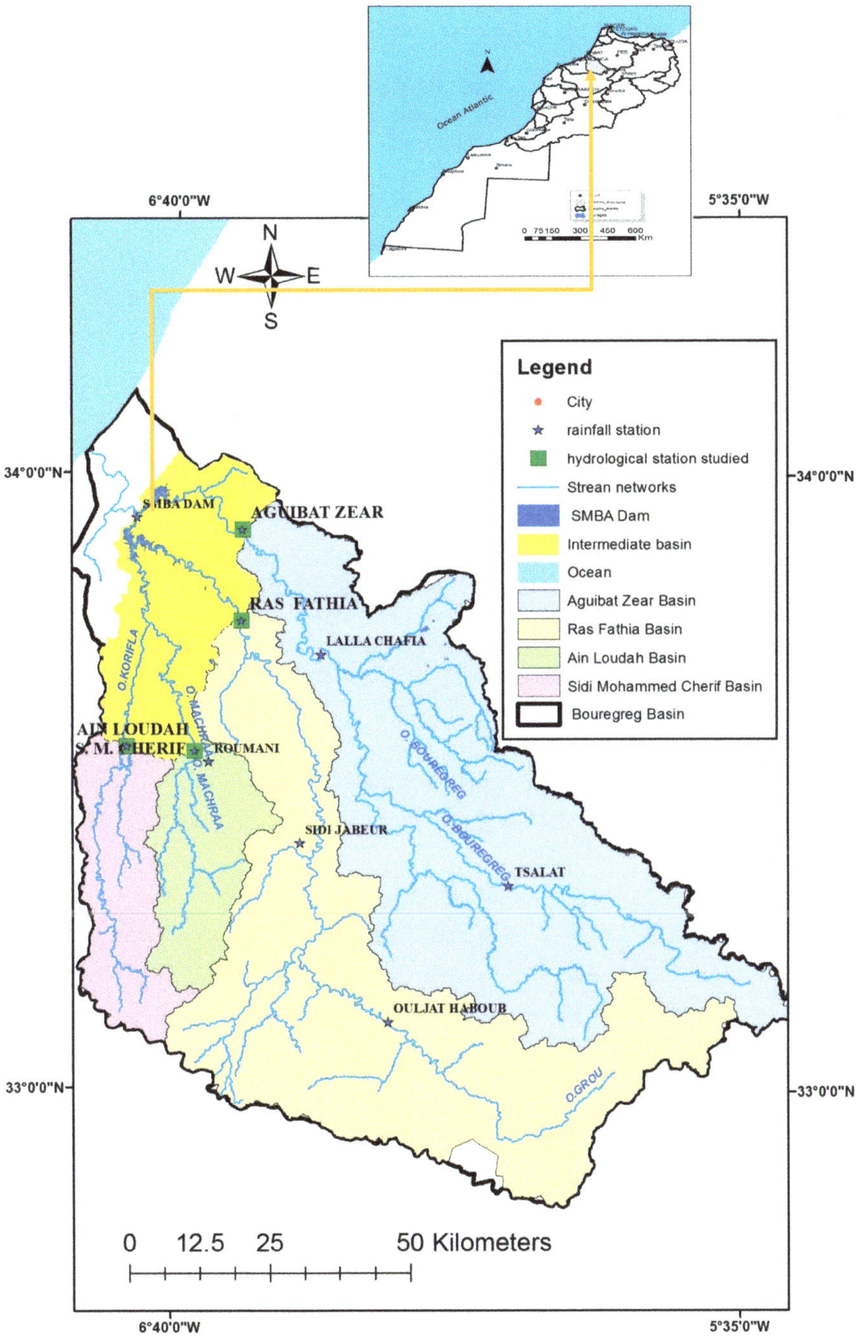

Figure 1. Situation of the Bouregreg Basin on the north Morocco Atlantic coast.

The climate in the study area is Mediterranean with oceanic influence, with an average annual rainfall over the basin varying from 450 mm in Rabat in the north-west, to nearly 750 mm in the mountainous area in the south-east.

Rain events generate water volume of 680×10^6 m^3/year, i.e., an annual mean of 22 m^3/s. It is regulated by the capacity of the SMBA dam, located downstream of the confluence between the Bouregreg, Korifla/Machraa, and Grou rivers. Table 1 summarizes the characteristics of the SMBA dam.

Table 1. Sidi Mohamed Ben Abdellah (SMBA)dam data.

	Watershed Area (km^2)	Initial Dam Capacity (10^6 m^3)	Inter-Annual Water Resources (10^6 m^3)	Opening Date	Number of Bathymetric Measurements
SMBA at its construction	9800	508.6	680	1974	5
SMBA after raising its dike		974.0		2007	3

2.2. Basic Data

2.2.1. Discharges Measurements

The hydrometric network of the Bouregreg basin upstream of the SMBA dam is composed of three major tributaries: Bouregreg, Grou, and Korifla (including its tributary the Machraa). Four hydrological stations located on these tributaries at the entrance of the dam's reservoir were chosen to carry out the measurements of solid transport. These hydrological stations control a basin of about 8521 km^2, i.e., 87% of the catchment area of the SMBA dam. They also have human and material means to ensure the measurement of flows, rainfall, and the concentration of suspended solids. Table 2 gives the characteristics of the four hydrological stations.

Table 2. Characteristics of the four hydrological stations studied.

Hydrological Station	Name of River	Date of Commissioning	Averaging Period	N°IRE ABHBC Code	Watershed Area (km^2)	Lambert Coordinates		
						X	Y	Z
Aguibat Zear	Bouregreg	1975	1975/2018	3118/13	3681	394.500	368.150	90
Ras Fathia	Grou	1975	1975/2018	989/20	3485	394.250	351.800	100
Ain Loudah	Korifla	1971	1971/2018	2673/20	699	373.750	329.150	175
Sidi Mohamed Cherif	Machraa	1971	1971/2018	2674/21	656	385.850	328.200	270

Figure 2 shows that the two years, 2016/2017 and 2017/2018, are dry years which did not record significant water inflows.

Examination of the historical measurement data from the four observation stations shows that 80% of the inflows to the dam are recorded between October and May (Figure 3). The Bouregreg and Grou rivers contribute more than 90% of the inflows generated in the basin.

Table 3 shows the maximum flows, volumes, and number of flood events recorded during the 2016/2017 and 2017/2018 hydrological years at the four hydrological stations studied. The number of recorded events is between six and seventeen, depending on the observation station and the year.

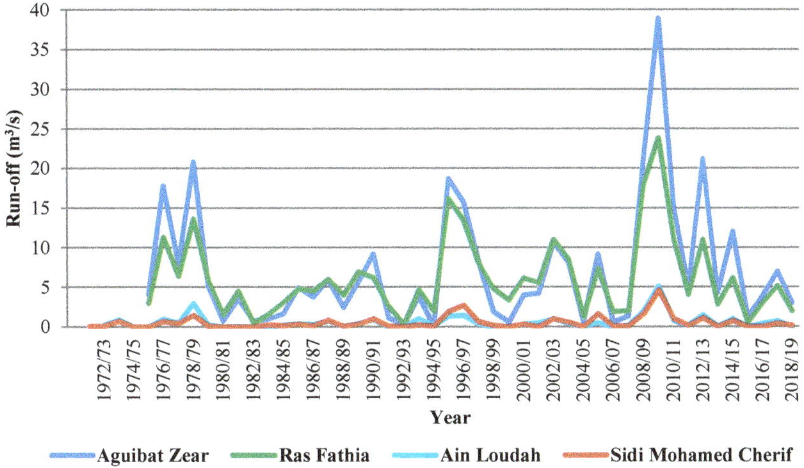

Figure 2. Annual discharges at the 4 hydrological stations studied.

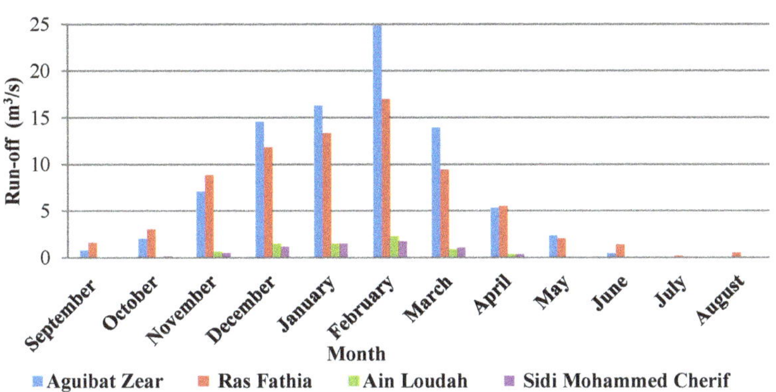

Figure 3. Mean monthly discharges at the hydrological stations studied.

Table 3. Flood events for the four hydrological stations studied.

Hydrological Station	Hydrological Year	Maximum Discharge Recorded Per Year (m^3/s)	Date of Flood Event	Volume (10^6 m^3)	Number of Flood Events
Aguibat Zear	2016/2017	263.4	24/02/2017 at 23h00	52	13
	2017/2018	265.9	07/03/2018 at 14h00	77	14
Ras Fathia	2016/2017	256.4	25/02/2017 at 07h30	33	11
	2017/2018	423.8	07/03/2018 at 09h00	59	10
Ain Loudah	2016/2017	24.5	13/10/2016 at 11h00	0.7	10
	2017/2018	198.7	24/04/2018 at 23h00	7.6	17
Sidi Mohamed Cherif	2016/2017	62.6	13/10/2016 at 07h00	1.8	6
	2017/2018	48.0	11/12/2017 at 17h00	1.28	6

2.2.2. Concentration of Suspended Solids (CSS)

Observers at the four hydrological stations studied, who are contracted by the Hydraulic Basin of the Bouregreg and Chaouia Agency (ABHBC), carry out daily sampling during low-flow periods and hourly sampling during periods of flooding. At each instantaneous sampling, the date, time, and scale rating are noted on the specimen bottles catalogued. The samples are then analyzed in the laboratory and filtered under a vacuum using filtering membranes (0.45μm). The available data for measuring the concentration of suspended solids cover two hydrological years, 2016/2017 and 2017/2018. The number of measurement samples is presented in Table 4.

Table 4. Number of concentration of suspended solids (CSS)samples and floods events per station.

Name of Hydrological Station	River	Watershed Area (km^2)	2016/2017 and 2016/2018	
			Number of CSS Samples	Number of Flood Events
AguibatZear	Bouregreg	3681	727	27
Ras Fathia	Grou	3485	636	21
Ain Loudah	Korifla	699	233	27
Sidi Mohamed Cherif	Machraa	656	789	12

2.2.3. Bathymetric Data

In order to draw up an inventory of the silting of the SMBA dam reservoir, the bathymetric surveys carried out by the Moroccan Directorate of Water Research and Planning were collected and analyzed. A total of seven bathymetric surveys were collected, the oldest dating back to 1974, while the most recent were carried out in 2013. Analysis of the bathymetric data shows that the silting up of the SMBA dam reservoir was of the order of 2.65×10^6 m^3/year, i.e., a specific degradation of 270.4 m^3/km^2/year before raising. After raising, the silting increased to 9.49×10^6 m^3/year, i.e. a specific degradation of 968.37 m^3/km^2/year, which increased the silting rate by 400% [4]. However, the data is subject to doubt for values in the early 2000s [35].The difference between the beginning and the end of the chronicle is of reliable quality. Therefore, the total silting of the SMBA dam reached 132×10^6 m^3, i.e., an average silting rate of 3.7×10^6 m^3/year since its commissioning. Subsequently, the dam has exceeded its dead unit, which is sized for 100×10^6 m^3 and has lost 32×10^6 m^{3}of its useful reserve since its commissioning. Table 5 summarizes the silting status of the dam.

Table 5. Summary of SMBA dam siltation calculation results since its impoundment.

Name of the Dam	Initial Dam Capacity (10^6 m^3)	Total Siltation (10^6 m^3)	Number of Years	Rate of Siltation (10^6 m^3)	Dead-unit Volume at Dam Building (10^{6}m^3)	Lost Volume (%)	Current Capacity (10^6 m^3)
SMBA before raising	508.60	76.88	29	2.65	100.00	15	431.72
SMBA after raising	974.79	55.28	6	9.49		6	919.51

The silting of the dam reservoir represents a real threat to the sustainability of the mobilization of surface water resources in the Bouregreg basin to satisfy required needs. The regulation of the SMBA dam, prior to its raising, was done on a seasonal basis (capacity lower than the annual inflow). In other words, the dam had to discharge the surplus inflows most often from flood spillway or bottom discharge. This technique of management favored the elimination of solid deposits and thus a reduction in the siltation rate of the dam. After the dam was raised and consequently the water capacity of the reservoir increased, the SMBA dam moved to multi-year regulation (capacity greater than annual inflows). By using the restitution devices of the dam, this method of regulation favors storage to the detriment of evacuation, which accentuates the silting rate. Figure 4 illustrates the evolution over time of the normal capacity of the SMBA dam.

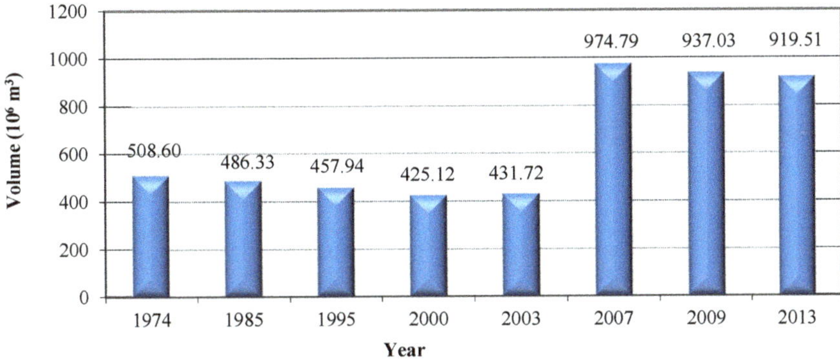

Figure 4. Evolution of the normal capacity of the SMBA dam reservoir, in millions of m^3.

2.3. Methods

2.3.1. Model Selection

In this study we apply the Williams [10] model based on the Modified Universal Soil Loss Equation (MUSLE), integrated in the ArcGIS Geographic Information System for the determination of the soil loss potential at the level of the Bouregreg watershed to the four hydrological stations located immediately upstream of the SMBA dam, as previously done by Khali Issa et al. [36] in another region of Morocco in the North of the country, over a very small basin of 38 km^2.

This model evaluates the average annual rate of erosion at the outlet of the basin. It uses hydrological parameters, measured at the four hydrological stations, taking into account biophysical characteristics. Thus, the model equation is as follows:

$$A = a(Q_{max} \times V_t)^b K \times LS \times C \times P, \tag{1}$$

where A: amount of sediment produced at the outlet in tons, a and b: in this study, we used the scale factor values of the Sidi Sbaa micro-basin [30] (a = 11.8 and b = 0.56), Q_{max}: maximum flow rate in m^3/s, V_t: total volume of runoff water in m^3, K: average soil erodibility (mg MJ-1mm-1), LS: average topographic factor, C: average vegetation cover factor, P: average cultural practices and amenities factor.

2.3.2. Methods Selection

We adopted the suite of methods and operations explained in the flowchart (Figure 5) below to assess siltation rates at hydrological stations upstream of the SMBA dam [37].

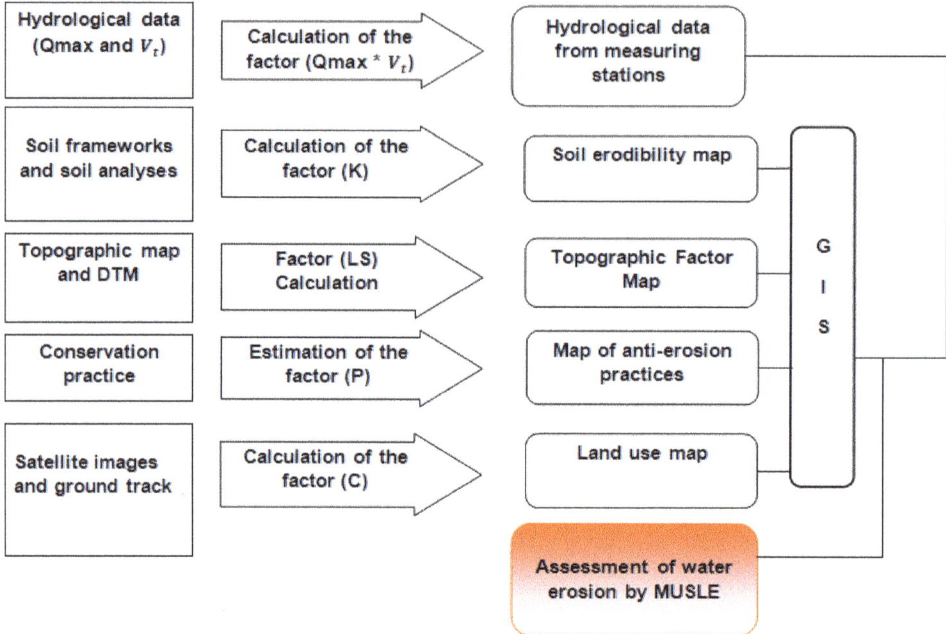

Figure 5. Methodology adopted for the assessment of erosion in the Bouregreg basin.

3. Results

3.1. Rainfall and Hydrometric Analysis

Solid transports are calculated at the four hydrological stations concerned in the hydrological years 2016/2017 and 2017/2018. These two years were respectively dry and wet in terms of rainfall. The average rainfall recorded at all the rainfall stations in the basin reached 354 mm in 2016/2017 and 496 mm in 2017/2018. Thus, the rainfall differences recorded in relation to the arithmetic mean of the data from the rainfall stations located in the Bouregreg basin varied respectively by −10% and +26% since the commissioning dates of the stations, based on data available to study from the ABHBC.

Table 6 summarizes the rainfall variations recorded in relation to the average rainfall of the stations in the Bouregreg basin.

Table 6. Rainfall for the 2016/2017 and 2017/2018 water years.

Rain Gauge Station	Basin	Cumulative Rainfall (mm)		Averaging Period	Mean Year (mm)	Deviation from the Mean (%)	
		2016/2017	2017/2018			2016/2017	2017/2018
LallaChafia	Bouregreg	409	569	1972/2019	486	−16	17
AguibatZear	Bouregreg	482	578	1975/2019	429	12	35
Sidi Jabeur	Grou	260	432	1971/2019	317	−18	36
Tsalat	Bouregreg	454	622	1977/2019	469	−3	33
Roumani	Machraa	261	431	1933/2019	342	−24	26
Ras Fathia	Grou	290	505	1975/2019	387	−25	30
S. M Cherif	Machraa	290	495	1971/2019	367	−21	35
Barrage SMBA	Bouregreg	518	479	1984/2019	469	10	2
OuljatHaboub	Grou	321	350	1972/2019	311	3	12
Ain Loudah	Krofla	258	498	1971/2019	347	−26	43
Total		354	496		392	−10	26

With the exception of the Korifla sub-basin at the right of the Ain Loudah station which recorded a surplus of nearly 24% during 2017/2018, the contributions were in deficit on the rest of the sub-basins during the two hydrological years with more pronounced deficits during the dry year 2016/2017. Table 7 summarizes the inflows recorded at the four stations of the study.

Table 7. Water supplies to the four hydrological stations studied.

Hydrological Station	River	Water Supply (Mm^3)				Deviation from the Mean (%)	
		2016/2017	2017/2018	Averaging Period	Annual Average (mm)	2016/2017	2017/2018
AguibatZear	Bouregreg	122.2	221.5	1975/2019	227.6	−46	−3
Ras Fathia	Grou	95.0	162.8	1975/2019	194.2	−51	−16
Ain Loudah	Korifla	13.3	23.5	1971/2019	18.9	−29	+24
Sidi Mohammed Cherif	Machraa	3.7	12.9	1971/2019	16.7	−77	−23

3.2. Analysis of the Biophysical Environment

The analysis of the biophysical environment consists of determining the average factors used by the model for each sub-basin. These factors, that affect soil erosion, are: soil type, topography, land use and cropping practices, and erosion control facilities [38].

3.2.1. Soil Erodibility Factor (K)

Erodibility is defined as the degree to which soils are resistant to erosion. The factors that have a major influence on the response of soils to erosion, namely the detachment and transport of particles by rain and runoff, are texture, structure, organic matter, and permeability. The methodology for estimating RUSLE K has been applied, and is written as follows [39]:

$$K = \left[2.1 \times 10^{-4}(12 - MO)M^{1.14} + 3.25(S - 2) + 2.5(P - 3)\right]/10, \tag{2}$$

with K: soil erodibility expressed in t.ha.h/ha.MJ.mm (tonne. hectare. hour/hectare. mega joule. millimeter); MO: percentage of organic matter; M: textural term % fine sand + % silt; S: structure class code 1 to 4, with 1 fragmented structure and 4 coarse structure, soil structure affects both landslide susceptibility and infiltration, the profiles described on the Bouregreg have a subangular polyhedral structure and fall under class (3); and P: the permeability code (1 to 6), its value can be inferred indirectly from the organic matter content by calculating the infiltration given by the Equation [40]:

$$Y = 3.53 \times X + 2.08, \tag{3}$$

with: Y = infiltration in cm/h, X = organic matter in %.

Thereafter, the soil erodibility factor K will be calculated using the Harmonized World Soil Database (HWSD), developed by the Food and Agriculture Organization of the United Nations (FAO) (http://www.fao.org/soils-portal/fr/). This database gives the distribution of silt, sand, and clay soil compositions by soil type [41].

Soil is composed of organic and mineral matter. Its texture is determined by the size of the soil particles and their respective quantities. There are three categories of particles that determine soil texture: sand, silt, and clay. They are distinguished by particle diameter: sand 0.05 mm to 2 mm, silt 0.002 mm to 0.05 mm, clay 0.002 mm and less. The modified illustrated triangular graph [42] was used to determine soil texture classification.Soil texture is classified according to the percentage of silt, sand, and clay.

Once the textures have been determined, it is possible to establish the correspondence between the standard texture and the K-factor [43]. These values are given in tons/ha and ton/acre (US system). Although this methodology provides an approximation in the calculation of the K-factor, it has the advantage of lending itself to the constraints imposed by the study area. The rate of organic matter at the level of each watershed for each texture value is calculated by converting organic carbon into organic matter. The conversion factor of 1.724 is commonly used to convert the organic carbon content of a soil sample to organic matter. The conversion factor is old and has survived the test of time and modern analytical methods. According to authors [44], this conventional factor is attributed to the 19th century authors Van Bemmelen [45], Wolff [46], or even Sprengel [47]. It is based on well-established and very old studies showing that soil organic matter contains 58% carbon. Since the C/MO ratio would be equal to 0.58, the MO/C ratio would be equal to 1.724.

Thus, the calculation of the organic matter is carried out by the following formula, which uses the value 1.724 and is widely used in Morocco:

$$MO = CO \times 1.724, \tag{4}$$

the average erodibility factor K per basin is calculated by the formula:

$$K_{average} = \frac{\sum (K \times number)}{\sum number}, \tag{5}$$

with K: K-factor per value, number: counting of pixels with the same K-value in the Arcmap allocation table.

The results of the calculation of the average K-factor for each watershed are in Table 8.

Table 8. Average erodibility factor for each watershed.

Name Watershed	Erodibility Factor $K_{average}$
Bouregreg	0.352
Grou	0.350
Ain Loudah	0.348
Sidi Mohamed Cherif	0.348

The Kmoy factor used is 0.35. This value corresponds to the silt, with the latter being a sedimentary formation whose grains are of intermediate size between clays and sands. The loss of silt leads to a decrease in the water retention capacity of the soil. The result is an increased erodibility and an increased risk of erosion. Because silt is often suspended in water, it is easily transported by floods and can contribute to the siltation of dam reservoirs. Figure 6 shows the distribution of the K factor over the Bouregreg basin for the four hydrological stations located immediately upstream of the SMBA dam.

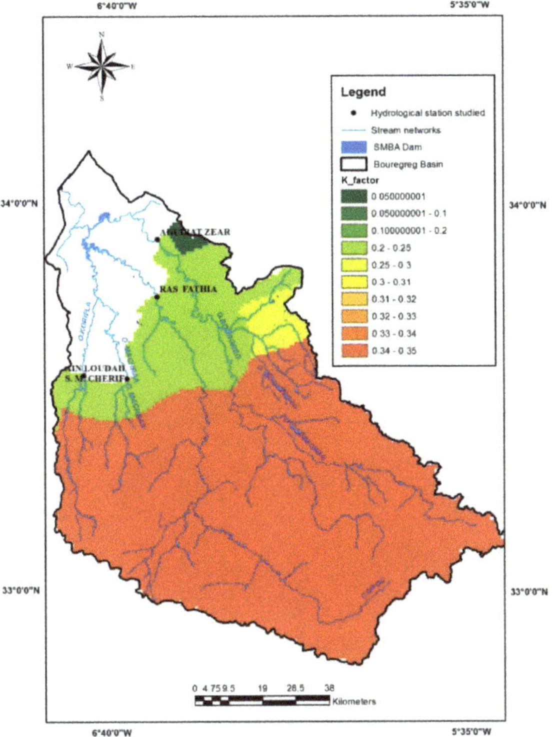

Figure 6. K-factor of the Bouregreg basin at the four hydrological stations.

3.2.2. Topographic Factor (LS)

The topographic factor LS is an essential parameter of the model. It expresses the result of erosion due to the combined effect of the degree of slope and its length. Most recent studies for the determination of the soil loss potential at the watershed level by the MUSLE model utilize the equation established by Wischmeier and Smith [9,23], which is expressed as:

$$LS = \left(\frac{\lambda}{22.13}\right)^{m} \times 65.4 \sin^{2} \beta + 4.56 \sin \beta + 0.0654, \tag{6}$$

with: λ: length of the slope in m, β: slope in degrees, $m = 0.5$ if $\beta \geq 5\%$, $m = 0.3$ if $1 < \beta < 5\%$, $m = 0.2$ if $\beta \leq 1$.

First, the slope map was established by the spatial analyst tool of the ArcGis software using a 90 m resolution DTM. Then, the LS map (Figure 7) by basin was generated using the Wishmeierand Smith formula [9,23]. The mean slopes and mean LS values are given in Table 9.

Table 9. Mean slope and mean topographic factor (LS).

	Mean Slope (%)	Mean Topographic Factor (Ls)
Bouregreg à Lala Chafia	16.6	0.48
Grou à Ras Fathia	14.6	0.49
Korifla à Ain Loudah	9.8	0.27
Machraa à Sidi Mohamed Cherif	12.2	0.58

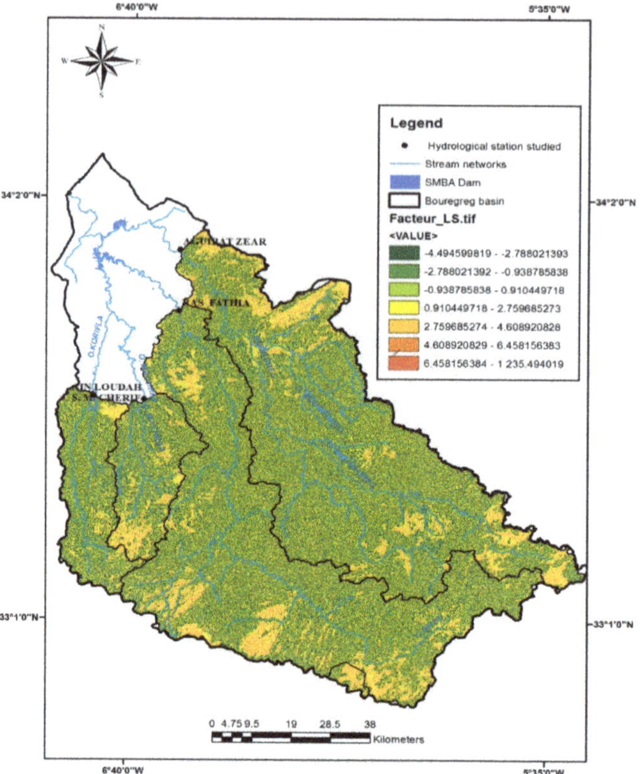

Figure 7. LS factor of the Bouregreg basin at the four hydrological stations.

3.2.3. Land Cover Factor (C)

The soil protection factor (C) indicates the degree of soil protection by vegetation cover. This factor has undergone several changes since the establishment of the universal soil loss equation. The vegetation cover is—after topography—the second most important factor controlling the risk of soil erosion. The value of C depends mainly on the percentage of vegetation cover and the growth phase. The Cfactor map for the Bouregreg catchment area was derived from the land use maps. These were determined from the use of remote sensing data and field observations [48]. The land cover map was extracted from SPOT satellite images at 20 m resolution combined with recent Landsat ETM+ images (2011/2012) using the supervised classification method [41]. Another approach could be to derive the C factor from NDVI maps as practiced on the Wadi Mina in Algeria by Toumi et al. [16].

The distribution of surfaces according to the nature of the vegetation cover is carried out by the ArcGis tool, which subdivides the catchment area into several polygons. Each polygon corresponds to a specific type of vegetation cover. Each vegetation cover corresponds to a factor given by Wischmeir and Smith according to the theme [49]. A color is assigned to each polygon that represents a land cover type. For the calculation of the Cfactor, an independent calculation is required for each sub-basin. Indeed, the sensitivity to erosion of the different classes is determined from the main land use themes, i.e., forest formations, rangelands, agricultural land, arboriculture, water, and bare soil, whose values vary between 0 and 1. The calculation of the average vegetation cover factor depends on the factor given for each land use and surface area by the following formula:

$$C_{average} = \frac{\sum S_i.C_i}{\sum S_i},\tag{7}$$

with S_i: partial polygon area, C_i: value of the C factor according to the theme. Figure 8 shows the C factors for each sub-basin of the Bouregreg River.

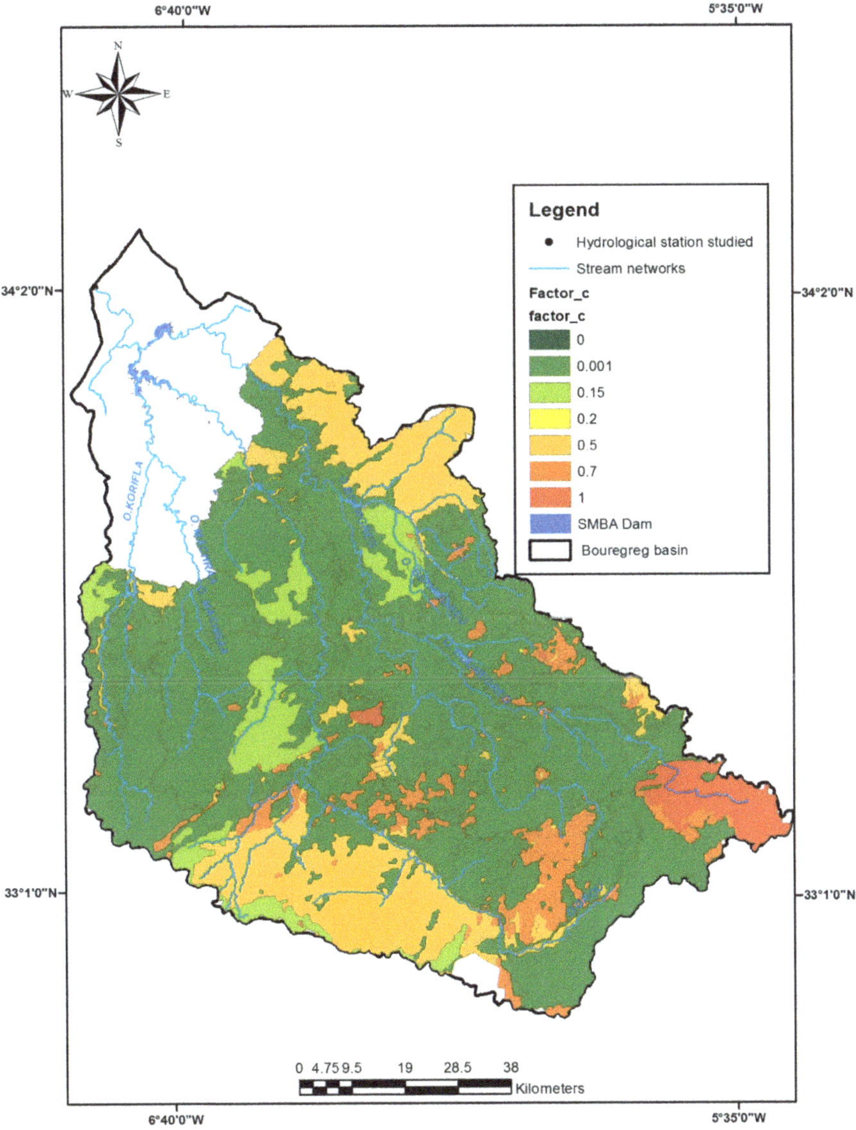

Figure 8. Factor C of the Bouregreg basin at the four hydrological stations.

The application of the formula leads to the results presented in Table 10.

Table 10. Average vegetation cover factor C for each watershed.

Watershed	Land Cover Factor (C) $C_{average}$
Bouregreg	0.26
Grou	0.45
Korifla	0.41
Machraa	0.33

The vegetation cover factor (C) can vary from close to 0 for well-protected soils, to 1 for striated surfaces that are very sensitive to gully erosion. The determination of this factor for the Bouregreg watershed and sub-basins is based on the density of vegetation and the height of the vegetation strata. These data were deduced from the field updating of the land use map by the Royal Centre for Remote Sensing in Space of Morocco (CRTS) available at the ABHBC. The values assigned to the different land use patterns are based on Wischmeier and Smith's tables for forests, matorrals, and pastures. The results show that the values of the C factor range from 0.26 for the sub-basin located in the northeast region, the Bouregreg, the wettest, to 0.45 for the sub-basin that extends furthest south, which is more arid [50,51]. The spatial distribution of the vegetation cover index by class for the Bouregreg watershed shows on 50.45% of the total surface area, i.e., nearly 181,556 ha, an index lower than or equal to 0.2, indicating good protection, while 49.5% of the surface area shows very low protection against erosion, i.e., about 177,986 ha. It can also be noted that the upper northeastern basin, which is both the most humid and the least covered with vegetation, is not, however, subject to high erodibility [52].

3.2.4. Average Cultivation Practices and Amenities Factor P

The P factor is a dimensionless factor expressing soil protection through agricultural practices (P).This factor takes into account purely anti-erosion practices. Contour, strip or terrace cultivation, bench planting, and ridging are the most effective soil conservation practices. These practices proportionally affect erosion by altering the flow pattern or direction of surface runoff and by reducing the amount and speed of runoff. P values are less than or equal to 1. A value of 1 is assigned to lands where none of the practices listed are used. P values vary between 0 and 1 depending on the practice adopted and also on the slope. In the Bouregreg watershed, there are few anti-erosion facilities and farmers do not use anti-erosion cultivation practices. These actions are small-scale and do not have a major impact on reducing erosion due to the size of the basin. As a result, a P value equal to 1 has been assigned to the entire area of the basin.

3.3. *Application and Calibration of the MUSLE Model*

3.3.1. Application of the MUSLE Model

The parameters required for the application of the MUSLE model for the evaluation of solid inputs by the MUSLE model are summarized in Table 11.

Table 11. Averaging factors for the universal soil loss equation modified per sub-basin.

	Mean Topographic Factor (LS)	Mean Erodibility Factor (K)	Mean Vegetation Cover Factor (C)	Mean Development Factor (P)
Bouregreg	0 48	0.32	0.26	1
Grou	0.49	0.32	0.45	1
Korifla	0.27	0.34	0.33	1
Machraa	0.58	0.31	0.41	1

By introducing the various factors of the MUSLE model, the amount of solid inputs produced at the outlet of each watershed is calculated using the flow and volume measurement data for each flood, according to the model equation. The solids transport is then calculated using the measured suspended solids concentrations for the same floods. The concentrations are linearly interpolated over the range of flows in each hydrograph. The bottom solid transport is taken as 10% of the total volume of suspended solids [53], which remains an empirical but widely used value.

3.3.2. Calibration of the MUSLE Model

Calibration of the model at each hydrological station is carried out using the percentage bias method (PBIAS), which consists of minimizing the difference between the observed mean and the mean predicted by the model by acting on the model's scaling parameters a and b:

$$\mathrm{PBIAS} = \left[\frac{\sum_{i=1}^{n}\left(Y_i^{obs} - Y_i^{sim}\right) \times (100)}{\sum_{i=1}^{n}\left(Y_i^{obs}\right)} \right],\tag{8}$$

with Y_i^{obs}: observed solid transport, Y_i^{sim}: solid transport simulated by MUSLE, n: number of observations.

In our case, a and b are respectively equal to 11.80 and 0.56. The calibration period is spread over two years of observations available. The optimal value of PBIAS is 0, where PBIAS has the ability to clearly indicate poor model performance [54]. To ensure model reliability, the model is also calibrated using the KGE (Kling-Gupta Efficiency) Method [55]:

$$\mathrm{KGE} = 1 - \sqrt{(r-1)^2 + \left(\frac{\sigma_{sim}}{\sigma_{Obs}} - 1\right)^2 + \left(\frac{\mu_{sim}}{\mu_{Obs}} - 1\right)^2},\tag{9}$$

with r: simple correlation coefficient between observed and simulated values, σ_{Obs}: standard deviation of observed solid transport values, σ_{sim}: standard deviation of simulated solid transport values, μ_{Obs}: mean of observed solid transport values, μ_{sim}: mean of simulated solid transport values. The optimal value of the KGE factor is 1.

The PBAIS calibration of the model shows that the scaling factor (a) does not change while (b) is more sensitive, but remains close to the initial value. Table 12 shows the results of the PBIAS model calibration.

Table 12. Model calibration results at the four stations studied by percentage bias method (PBIAS).

Name of Basin	Scaling Parameters after Calibration		Calibration Criteria		
	A	B	Bias	KGE	r
Bouregreg	11.799	0.548	0.00	0.48	0.58
Grou	11.799	0.535	0.00	0.11	0.26
Korifla	11.800	0.541	0.00	0.49	0.57
Machraa	11.799	0.502	0.00	0.65	0.70

Calibration of the model by the KGE method shows that the scaling factors a and b are very sensitive, with the terms being very far from the initial values, and are quite or widely and varyingly different from one basin to another. Table 13 shows the results of model calibration using the KGE method.

Table 13. Model calibration results for the four stations studied by KGE.

Name of Basin	Scaling Parameters after Calibration		Calibration Criteria		
	A	b	Bias	KGE	r
Bouregreg	0.039	0.807	0.024	0.601	0.603
Grou	11.799	0.546	−0.236	0.149	0.272
Korifla	1.727	0.643	−0.055	0.537	0.545
Machraaf	1.421	0.620	−0.017	0.689	0.690

3.4. Comparison of Observations with MUSLE Model Results

3.4.1. Aguibat Zear Hydrological Station on the Bouregreg River

During the two hydrological years studied, twenty-seven floods were recorded. The total solid volumes observed were 742,334 tons in 2016/2017 and 439,735 tons in 2017/2018. Table 14 summarizes the results from applying the model to the Aguibat Zear hydrological station.

Table 14. Results of the MUSLE model at Aguibat Zear on the Bouregreg and a comparison with observations.

Hydrologic Year	Observed Solid Transport (tons)	Solid Transport (MUSLE) (tons)			Difference Observation-MUSLE (%)		
		Uncalibrated	Calibration PBAIS	Calibration KGE	Uncalibrated	Calibration PBAIS	Calibration KGE
2016/2017	742 334	619 023	487 494	436 922	−17	−34	−41
2017/2018	439 735	888 011	694 575	716 809	+102	+58	+63

Examination of the results of the MUSLE model calibrated by the two approaches—PBAIS and KGE—shows that calibration by the PBAIS method better simulates solid transport. Figure 9 compares the simulated and observed solid transport. The latter events are less well simulated by MUSLE than the former.

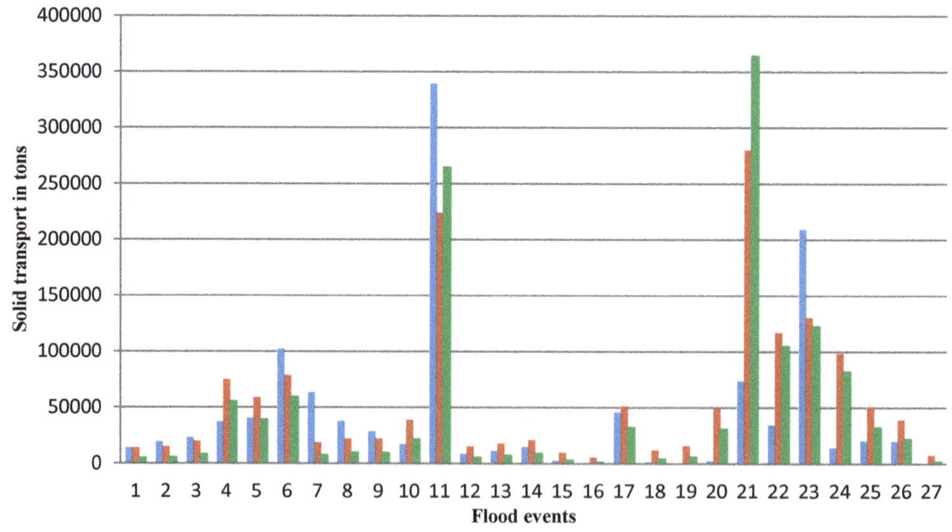

Figure 9. Comparison between MUSLE and observations at Aguibat Zear on the Bouregreg.

3.4.2. Ras Fathia Hydrological Station on the Grou River

During the two hydrological years studied, twenty-one floods were recorded. The total observed solid volumes amount to 517,194 tons in 2016/2017 and 896,438 tons in 2017/2018. Table 15 summarizes the results of applying the model to the Ras Fathia hydrological station for the two versions of the model calibration.

Table 15. Results of the MUSLE model at Ras Fathia on the Grou and comparison with observations.

Hydrologic Year	Observed Solid Transport (tons)	Solid Transport (MUSLE) (tons)			Difference Observation-MUSLE (%)		
		Uncalibrated	Calibration PBAIS	Calibration KGE	Uncalibrated	Calibration PBAIS	Calibration KGE
2016/2017	673 551	875 366	517 194	649 389	+30	−23	−4
2017/2018	740 083	1 446 831	896 438	1 098 213	+95	+21	+48

Examination of the results of the MUSLE model calibrated by the two approaches PBAIS and KGE, shows that calibration by the PBAIS method better simulates solid transport. Figure 10 shows the comparison of simulated and observed solid transport. As for the Bouregreg, the events of the second year are less well represented by MUSLE.

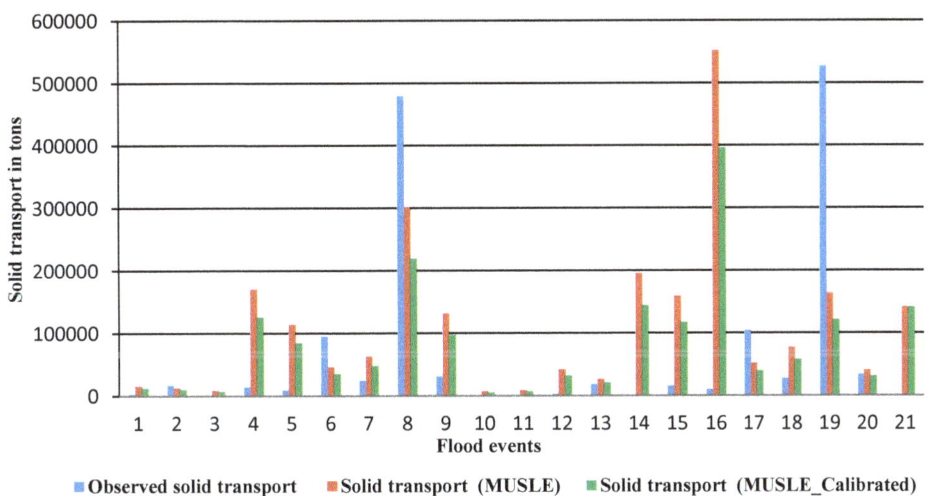

Figure 10. Comparison between MUSLE and observations at Ras Fathia on the Grou.

3.4.3. Sidi Mohamed Cherif Hydrological Station on the Machraa River

During the two hydrological years studied, twelve floods were recorded. The total solid volumes observed were 28,155 tons in 2016/2017 and 48,422 tons in 2017/2018. Table 16 summarizes the results of the application of the model to the hydrological station Sidi Mohamed Cherif.

Table 16. Results of the MUSLE model at Sidi Mohamed Cherif on the Machraa and comparison with observations.

Hydrologic Year	Observed Solid Transport (tons)	Solid Transport (MUSLE) (tons)			Difference Observation-MUSLE (%)		
		Uncalibrated	Calibration PBAIS	Calibration KGE	Uncalibrated	Calibration PBAIS	Calibration KGE
2016/2017	28,155	75,194	27,431	26,225	+167	−3	−7
2017/2018	48,422	141,249	49,146	51,657	+192	+1	+7

Examination of the results of the MUSLE model calibrated by the two approaches—PBAIS and KGE—shows that calibration by the PBAIS method better simulates solid transport. Figure 11 shows the comparison of simulated and observed solid transport. Again, the MUSLE simulations are significantly too high for the events in the recording portion.

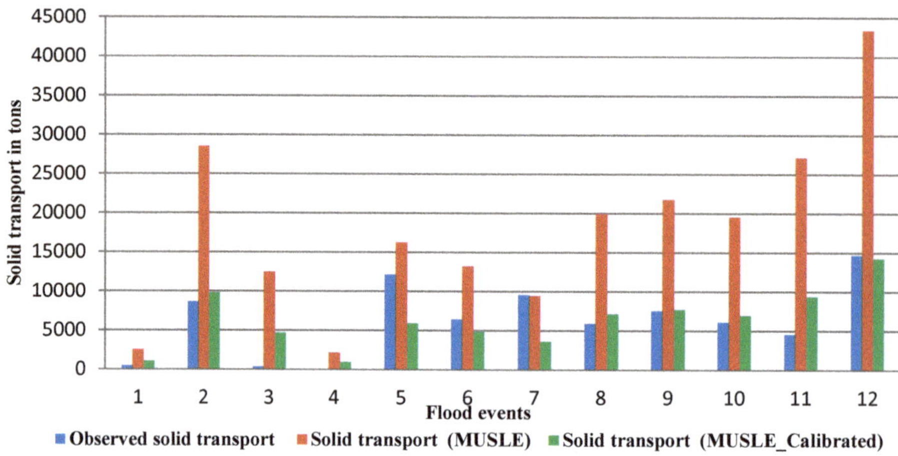

Figure 11. Comparison between MUSLE and observations at Sidi Mohamed Cherif on the Machraa.

3.4.4. Ain Loudah Hydrological Station on the Korifla River

During the two hydrological years studied, twenty-seven floods were recorded. The total solid volumes observed were 68,759 tons in 2016/2017 and 92,324 tons in 2017/2018. Table 17 summarizes the results of the application of the model to the Ain Loudah hydrological station.

Table 17. Results of the MUSLE model at Ain Loudah on the Korifla and comparison with observations.

Hydrologic Year	Observed Solid Transport (tons)	Solid Transport (MUSLE) (tons)			Difference Observation-MUSLE (%)		
		Uncalibrated	Calibration PBAIS	Calibration KGE	Uncalibrated	Calibration PBAIS	Calibration KGE
2016/2017	68,759	41,083	28,881	27,739	−40	−58	−60
2017/2018	92,324	191,859	132,202	142,138	+108	+43	+54

Examination of the results of the MUSLE model calibrated by the two approaches PBAIS and KGE, shows that calibration by the PBAIS method better simulates solid transports. Figure 12 shows the comparison of simulated and observed solid transports. Out of the four stations studied, it is on the Korifla at Ain Loudah that the simulations seem to be the most efficient.

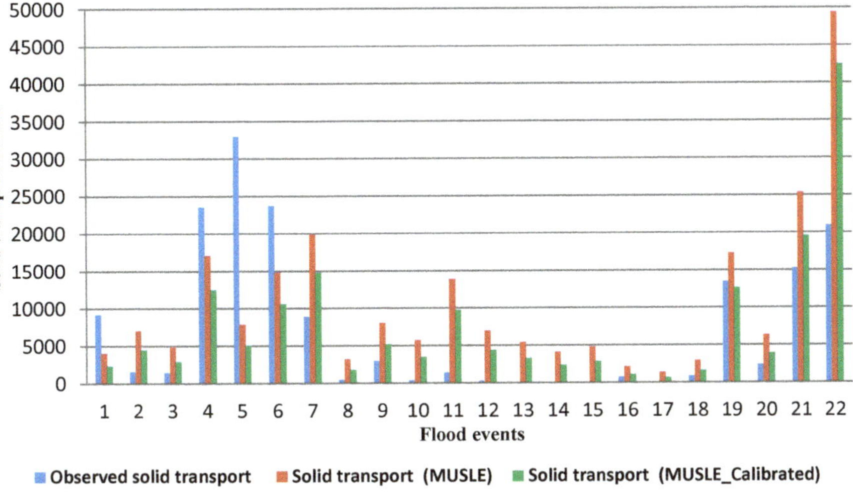

Figure 12. Comparison between MUSLE and observations at Ain Loudah on the Korifla.

From the deviations from observations at the four hydrological stations, it can be deduced that in general, the model calibrated by the PBAIS method gives better results than the model calibrated by the KGE method. Thus, from the observation we note that the differences varied between −58% and +58% for the model calibrated by PBAIS, between −60% and +63% for the model calibrated by the KGE method, and between −40% and +192% for the model not calibrated.

3.5. Comparison of Observations and Results of the MUSLE Model at the Four Hydrological Stations Upstream from the SMBA Dam, with Bathymetric Data

To validate the calibration method, the Nash and Sutcliffe NSE index was used to measure the performance of the model. According to Nash–Sutcliffe [56], NSE is defined as:

$$\text{NSE} = 1 - \frac{\sum_{i=1}^{n} \left(Y_i^{\text{obs}} - Y_i^{\text{Sim}} \right)^2}{\sum_{i=1}^{n} \left(Y_i^{\text{obs}} - \mu_{\text{Obs}} \right)^2},\tag{10}$$

With NSE: Nash coefficient, Y_i^{obs}: observed solid transport, Y_i^{sim}: solid transport simulated by MUSLE, μ_{Obs}: average of the observed solid transport values.

The NSE index varies from $-\infty$ to 1, such that if NSE = 1, then the modeled values match the observations perfectly, while a value above 0 shows a relationship between simulation and reality, and a value below 0 shows that there is no relationship between the two. In other words, the closer the efficiency is to 1, the more the model is observed to be accurate.

Table 18 shows the different calculated values of the Nash–Sutcliffe index, which compares observations to the values calculated by the uncalibrated MUSLE model and observations to the values of the calibrated MUSLE model.

Table 18. Nash–Sutcliffe index (NSE) at study stations.

Name of Basin	Scaling Parameters after Calibration		Nash–Sutcliffe Index (NSE)	
	a	B	MUSLE	MUSLE calibrated
Bouregreg à Aguibat Zear	11.799	0.548	0.38	0.46
Grou à Ras Fathia	11.799	0.535	−0.07	0.02
Korifla à Ain Loudah	11.800	0.541	−0.35	0.30
Machraa à Sidi Mohammed Cherif	11.799	0.502	−9.8	0.47

In general, the uncalibrated MUSLE model has negative NSE values except for those in the Bouregreg basin, which shows that the model values are far from reality, whereas the NSE values of the calibrated model are positive, varying between 0.02 and 0.47, which shows that the calibrated model is superior in reliably representing reality.

Though, for the continuity of the experiment we are continuing the calculation on the four basins in order to complete the comparisons with the volumes of sediment deduced from the bathymetric measurements carried out in the SMBA dam reservoir. For the calculation of the volume of suspended matter, an earthy density of 1.5 t/m^3 is adopted [57]. The siltation of the dam is considered to be equal to 9.49×10^6 m^3/year. Table 19 summarizes the results of the calculations for all the hydrological stations.

Table 19. Summary table of observation results, application of the MUSLE, calibrated MUSLE_ model and bathymetry for the four sub-basins of the Bouregreg.

Name of Basin	Solid Transport 2016/2017			Solid Transport 2017/2018		
	Observation	MUSLE	MUSLE calibrated	Observation	MUSLE	MUSLE calibrated
Bouregreg	742,334	619,023	487,494	439,735	888,011	694,575
Grou	673,551	875,366	517,194	740,083	1,446,831	896,438
Korifla	68,759	41,083	28,881	92,324	191,859	132,202
Machraa	28,155	75,194	27,431	48,422	141,249	49,146
Total (tons)	1,512,799	1,610,666	1,061,000	1,320,564	2,667,950	1,772,361
Total (10^6 m^3)	1.01	1.07	0.71	0.88	1.78	1.18

Thus, the results listed in Tables 19 and 20 show that the calibrated model is closer to the measurement than the non-calibrated model, and that the observed values of sediment transport at the four stations, which vary between 1.01×10^6 m^3 and 0.88×10^6 m^3 for the two hydrological years 2016/2017 and 2017/2018 respectively, and remain largely lower than the quantity of sediments obtained by bathymetric measurements in the SMBA dam, which is approximately 9.49×10^6 m^3.

Table 20. Comparison of observation results, application of the calibrated MUSLE, MUSLE_ model, and bathymetry for the four sub-basins of the Bouregreg during 2016/2017 and 2017/2018.

Hydrologic Year	Bathymetry (10^6 m^3)	Difference MUSLE/ Observation	Difference MUSLE Calibrated/ Observation	Difference Observation/ Bathymetry	Difference MUSLE/ Bathymetry	Difference MUSLE Calibrated/ Bathymetry
2016/2017	9.25	+5%	−30%	−89%	−89%	−93%
2016/2018		+102%	+34%	−91%	−81%	−88%

4. Discussion and Conclusions

The relationships between man, environment, and solid transport are well understood at the spatial scale of catchment areas, and modeling tools using satellite data are particularly suitable [58]. The results obtained from the application of the model (MUSLE), show that the soils of the Bouregreg catchment area are affected by several factors engendering erosion, i.e., steep slopes, low vegetation cover and soil erodibility. Thus, the processing of spatial remote sensing images to determine the biophysical characteristics of the watershed has made it possible to identify areas of high erodibility, low vegetation cover, and steep slopes, which can be used for potential development purposes.

The analysis of solid transports observed during two years at the four hydrological stations that control the main tributaries of the SMBA dam reservoir showed that the Bouregreg and the Grou generated 1,415,885 tons of suspended matter, i.e., 94% and 956,929 tons of suspended matter, i.e., 73% of the total suspended matter generated by the four basins during the 2016/2017 and 2017/2018 hydrological years. Compared to bathymetry, the four sub-basins generate only 12% of the solid transport to the dam in the best case. Hence the relevance of concentrating anti-erosion actions on the dam banks and the intermediate basin which contribute to more than 88% of the solid transport. These actions will have to be carried out in areas of high erodibility and slope to ensure the effectiveness of the intervention.

On the other hand, examination of the figures for MUSLE and calibrated MUSLE values shows that the model in these two versions does not predict peaks in solid transport well. It overestimates low values and underestimates high solid transport values most of the time. Indeed, the deviations from observations range from +5% for the dry year, to +102% for the wet year for the non-calibrated model. These results are similar to several applications of the model in different areas around the world [40]. The deviations recorded for the calibrated model are significantly improved. They vary respectively between −30% and +34% [59]. Thus, the calibrated version of the model gives better results with less than +55% error in sediment prediction [60]. These results are interesting on an interannual average, but do not allow the prediction of quantities transported to the stations during unit floods due to the very high dispersion of the relationship between observation and modeling, which results in average-to-low Nash index values. The large positive differences are due to the underestimation of solid transport peaks by the model. Indeed, the model involves the entire basin without concern for the spatial and temporal distribution of rainfall.

Although the model overestimates values relative to observations, comparison of the model to bathymetric data shows that the differences between mean siltation and model results range from −89% to −81% for the non-calibrated version of the model, and from −93% to −88% for the calibrated version. This underestimation of the inflows to the dam in relation to the mean silting is explained by the fact that the upstream basins do not control the whole SMBA dam basin and that the basins immediately surrounding the banks of the dam reservoir bring significant quantities of sediment to the dam, as assumed by the first results of Maleval [61] on the SMBA dam reservoir, and this is mainly after its elevation, for reasons that are perhaps not only related to the physics of the environment, but also perhaps to the evolution of agro-sylvo-pastoral practices in an environment undergoing rapid socio-economic change due to the proximity of the capital and the development of the "Grand Rabat" and new traffic routes [62,63]. The gaps between the observed solid transport and the silting rate, which are respectively −89% and −91%, confirm this observation, which is also the situation observed by Hallouz et al. [64] on the Wadi Mina basin and the Sidi Mohamed Ben Aouda dam, of a size comparable to the Bouregreg in Western Algeria.

In order to improve the model's output, its scaling parameters should be reviewed and its sensitivity to different physical factors including basin size by adopting a distributed form of sub-basin modeling should be tested [65]. This approach requires a long series of observations thatare currently unavailable, but which are being acquired since the ABHBC, which manages the water resource on the Bouregreg catchment area and the SMBA dam, has set up permanent concentration of suspended solids observation on the stations of the basin, and soon a long time series will be available to refine these first results. Michalec et al. [66] have pointed out the necessity to calibrate the MUSLE model with long time series of observations in order to reduce the uncertainty of simulated erosion provided, even if in some cases very good correlations have been obtained from a calibration with only 10 events like in Nigeria [30]. These measurements will allow a monitoring of the contribution of each sub-basin in order to better direct intervention measurements towards the most productive areas on the one hand, thus reducing the overall cost of the measures to reduce erosion, and to improve modeling and subsequent solid transport forecasting on the other hand.

Eventually, the results of the MUSLE model confirm the relevance of its application in watershed management studies once calibrated, since it integrates all forms of erosion observed at the watershed scale, coupled with hydrological data, which greatly improves accuracy compared to USLE [67] and to RUSLE [68]. Indeed, the average specific degradation at the level of the Bouregreg basin alone at the Aguibat Zear station is of the order of 13.81 t/ha/year according to Moussebbih et al. [69] when calculated by the RUSLE method, whereas those of calibrated MUSLE and observation are of the order of 1.6 t/ha/year. The studies carried out on the wadi Mina basin in Algeria, a basin similar to the Bouregreg basin [64], confirm this finding with deviations of the RUSLE model results from the measurement of suspended solids concentrations, which can reach 79%. In Spain, Ramos-Diez et al. [70] have used the USLE formula to assess the amount of sediment inputs to 25 very small check dams used in an important restoration project over a 9 km^2 area in the north of Spain, showing interesting but still mitigated results and concluding that the size and shape of the dams had an impact on the quality of the USLE assessment.

Kronvang et al. [71] showed that bank erosion was the dominant sediment source (90–94%) in the River Odense catchment in Denmark during three study years. They add that in-channel and overbank sediment sinks and storage dominated the sediment budget, as 79–94% of the sediment input from all sources was not exported from the catchment during the three study years. This is in accordance with our results, as the regular bathymetric survey to monitor the silting up rate of the dam have shown that half of the sediment input to the reservoir comes from the dam banks. Our results are very important for the forthcoming works for erosion mitigation on the Bouregreg catchment; they will enable more efficient orientation of the soil conservation and restoration works that may be carried out in the future, with priority being given to the environment close to the dam, in order to preserve the water capacity of the SMBA dam reservoir while optimizing the financial resources mobilized. Palazon and Navas [72] have shown the same interest to monitor the sediment sources draining to a large reservoir in the Esera River, Ebro basin in Spain. The small basin size (1500 km^2) allowed them to use the SWAT model as an alternative to MUSLE with two gauging stations, each one being covered by different types of soil. This might make it possible to implement the SWAT model on the Bouregreg basin, when the ABHBC will have started the monitoring of sediment transport at all the stations of the basin.

To conclude, it should be pointed out that in the context of climate change, which predicts, in Morocco, an increase in temperatures and a drop in rainfall [73–75], the projections of erosion and sediment transport evolution that are possible from climate model outputs, and which are based on USLE or SWAT type approximation models [76], will need observed data to calibrate and validate their results.

Author Contributions: Conceptualization, M.A.E. and G.M.; methodology, M.A.E.; software, M.A.E.; validation, G.M., I.K. and A.Z.; formal analysis, M.A.E.; investigation, M.A.E.; resources, A.Z.; data curation, A.Z.; writing—original draft preparation, M.A.E. and G.M.; writing—review and editing, M.A.E.; visualization, M.A.E.; supervision, G.M., I.K. and A.Z.; project administration, G.M., I.K. and A.Z. All authors have read and agreed to the published version of the manuscript.

Funding: This research received no external funding.

Acknowledgments: Thanks to ABHBC for the hydrological and rainfall data.

Conflicts of Interest: The authors declare no conflict of interest.

References

1. Mahe, G.; Emran, A.; Brou, Y.T.; Tra Bi, A.Z. Analyse statistique de l'évolution de la couverture végétale à partir d'images MODIS et NOAA sur le bassin versant du Bouregreg (Maroc). *Géo Obs.* **2012**, *20*, 33–44.

2. Laouina, A.; Aderghal, M.; Al Karkouri, J.; Antari, M.; Chaker, M.; Laghazi, Y.; Machmachi, I.; Machouri, N.; Nafaa, R.; Naïmi, K. The efforts for cork oak forest management and their effects on soil conservation. *For. Syst.* **2010**, *19*, 263–277. [CrossRef]

3. Schmidt, S.; Alewell, C.; Meusburger, K. Monthly RUSLE soil erosion risk of Swiss grasslands. *J. Maps* **2019**, *15*, 247–256. [CrossRef]

4. Yahiaoui, S.; Zerouali, A. Etude de l'évolution de l'occupation du sol sur deux grands bassins d'Algérie et du Maroc, et relation avec la sédimentation dans les barrages. In *Considering Hydrological Change in Reservoir Planning and Management*; IAHS Publ 362; Schumann, A., Belyaev, V.B., Gargouri, E., Kucera, G., Mahe, G., Eds.; IAHS Press: Wallingford, UK, 2013; pp. 115–124.

5. Khomsi, K.; Mahe, G.; Sinan, M.; Snoussi, M. Hydro-climatic variability in two Moroccan watersheds: A comparative analysis of temperature, rain and flow regimes. In *Climate and Land Surface Changes in Hydrology*; IAHS Publ. 359; Boegh, E., Blyth, E., Hannah, D.M., Hisdal, H., Kunstmann, H., Su, B., Yilmaz, K.K., Eds.; IAHS Press: Wallingford, UK, 2013; pp. 183–190.

6. Khomsi, K.; Mahe, G.; Tramblay, Y.; Sinan, M.; Snoussi, M. Regional impacts of global change: Seasonal trends in extreme rainfall, runoff and temperature in two contrasted regions of Morocco. *Nat. Hazards Earth Syst. Sci.* **2016**, *16*, 1079–1090. [CrossRef]

7. Mahe, G.; Emran, A.; Brou, Y.T.; Tra Bi, A.Z. Impact de la variabilité climatique sur l'état de surface du bassin versant du Bouregreg (Maroc). *Eur. J. Sci. Res.* **2012**, *84*, 417–425.

8. ABV. *Plan National d'Aménagement des Bassins Versant*; Haut-Commissariat aux Eaux et Forêts et à la Lutte Contre la Désertification: Rabat, Morocco, 1996.

9. Wischmeier, W.H.; Smith, D.D. *Predicting Rainfall-Erosion Losses from Cropland East of the Rocky Mountains: Guide for Selection of Practices for Soil and Water Conservation*; US Department of Agriculture: Washington, DC, USA, 1965.

10. Williams, J.R. *Sediment-Yield Prediction with Universal Equation Using Runoff Energy Factor*; ARS-S. Southern Region, Agricultural Research Service, US Department of Agriculture: Washington, DC, USA, 1975; Volume 40, p. 244.

11. El Bahi, S.; El Wartiti, M.; Yassin, M.; Calle, A.; Casanova, J.L. Applying revised universal loss equation model to forest lands in Central Plateau of Morocco. *Rev. Teledetección* **2005**, *23*, 89–97.

12. Chadli, K. Estimation of soil loss using RUSLE model for Sebou watershed (Morocco). *Model. Earth Syst. Environ.* **2016**, *2*, 51. [CrossRef]

13. Elaloui, A.; Marrakchi, C.; Fekri, A.; Maimouni, S.; Aradi, M. USLE-based assessment of soil erosion by water in the watershed upstream Tessaoute (Central High Atlas, Morocco). *Modeling Earth Syst. Environ.* **2017**, *3*, 873–885. [CrossRef]

14. Hara, F.; Achab, M.; Emran, A.; Mahe, G.; Fhel, B.E. Estimate the Risk of Soil Erosion Using USLE through the Development of an Open Source Desktop Application: DUSLE (Desktop Universal Soil Loss Equation). In Proceedings of the 3rd International Conference on African Large River Basin Hydrology, Alger, Algeria, May 2018; Available online: https://hal.archives-ouvertes.fr/hal-02397686/ (accessed on 17 June 2020).

15. Boufala, M.; El Hmaidf, A.; Chadli, K.; Essahlaoui, A.; El Ouali, A.; Lahjouj, A. Assessment of the Risk of Soil Erosion Using RUSLE Method and SWAT Model at the M'dez Watershed, Middle Atlas, Morocco. In Proceedings of the E3S Web Conf., The Seventh International Congress "Water, Waste and Environment" (EDE7-2019), Salé, Morocco, 20–22 November 2019; Volume 150. [CrossRef]

16. Toumi, S.; Meddi, M.; Mahe, G.; Brou, Y.T. Cartographie de l'érosion dans le bassin versant de l'Oued Mina en Algérie par télédétection et SIG. *Hydrol. Sci. J.* **2013**, *58*, 1–17. [CrossRef]

17. Mihi, A.; Benarfa, N.; Arar, A. Assessing and mapping water erosion-prone areas in northeastern Algeria using analytic hierarchy process, USLE/RUSLE equation, GIS, and remote sensing. *Appl. Geomat.* **2019**. [CrossRef]

18. Gaubi, I.; Chaabani, A.; Ben Mammou, A.; Hamza, M.H. A GIS-based soil erosion prediction using the Revised Universal Soil Loss Equation (RUSLE) (Lebna watershed, Cap Bon, Tunisia). *Nat. Hazards* **2017**, *86*, 219–239. [CrossRef]

19. Toubal, A.K.; Achite, M.; Ouillon, S.; Dehni, A. Soil erodibility mapping using the RUSLE model to prioritize erosion control in the Wadi Sahouat basin. North-West Alger. *Environ. Monit. Assess.* **2018**, *190*, 4. [CrossRef] [PubMed]

20. El Aroussi, O.; Mesrar, L.; El Garouani, A.; Lahrach, A.; Beaabidaate, L.; Akdi, B.; Jabrane, R. Predicting the potential annual soil loss using the revised universal soil loss equation (RUSLE) in the oued El Malleh catchment (Prerif, Morocco). *Present. Environ. Sustain. Dev.* **2011**, *5*, 5–16.

21. Lamyaa, K.; M'bark, A.; Brahim, I.; Hicham, A.; Soraya, M. Mapping Soil Erosion Risk Using RUSLE, GIS, Remote Sensing Methods: A Case of Mountainous Sub-watershed, Ifni Lake and High Valley of Tifnoute (High Moroccan Atlas). *J. Geogr. Environ. Earth Sci. Int.* **2018**, *14*, 1–11. [CrossRef]

22. Bagarello, V.; Ferro, V.; Giordano, G.; Mannocchi, F.; Todisco, F.; Vergni, L. Statistical check of USLE-M and USLE-MM to predict bare plot soil loss in two Italian environments. *Land Degrad. Dev.* **2018**, *29*, 2614–2628. [CrossRef]

23. Wischmeier, W.H.; Smith, D.D. *Predicting Rainfall Erosion Losses: A Guide to Conservation Planning*; Department of Agriculture, Science and Education Administration: Washington, DC, USA, 1978.

24. Renard, K.G. *Predicting Soil Erosion by Water: A Guide to Conservation Planning with the Revised Universal Soil Loss Equation (RUSLE)*; United States Government Printing: Washington, DC, USA, 1997.

25. Kinnell, P.I.A. Why the Universal Soil Loss Equation and the Revised Version of It Do Not Predict Event Erosion Well. *Hydrol. Process.* **2005**, *19*, 851–854. [CrossRef]

26. Haan, C.T.; Barfield, B.J.; Hayes, J.C. Design Hydrology and Sedimentology for Small Catchments. *J. Hydrol. -Amst.* **1963**, *176*, 296–297.

27. Novotny, V. *Water Quality: Prevention, Identification and Management of Diffuse Pollution*; Van Nostrand-Reinhold Publishers: New York, NY, USA, 1994; ISBN 978-0442005597.

28. Alewell, C.; Borrelli, P.; Meusburger, K.; Panagos, P. Using the USLE: Chances, challenges and limitations of soil erosion modelling. *Int. Soil Water Conserv. Res.* **2019**, *7*, 203–225. [CrossRef]

29. Odongo, V.O.; Onyando, J.O.; Mutua, B.M.; van Oel, P.R.; Becht, R. Sensitivity analysis and calibration of the Modified Universal Soil Loss Equation (MUSLE) for the upper Malewa Catchment, Kenya. *Int. J. Sediment Res.* **2013**, *28*, 368–383. [CrossRef]

30. Adegede, A.P.; Mbajiorgu, C.C. Event-based sediment yield modeling for small watersheds using MUSLE in north-central Nigeria. *Agric. Eng. Int.: CIGR J.* **2019**, *21*, 7–17.

31. Yassin, M.; Pépin, Y.; El Bahi, S.; Zante, P. Evaluation de l'érosion au micro-bassin de Sidi Sbaa: Application du modèle MUSLE. *Ann. Rech. For. Maroc* **2013**, *42*, 171–181.

32. Samaras, A.G.; Koutitas, C.G. Modeling the impact of climate change on sediment transport and morphology in coupled watershed-coast systems: A case study using an integrated approach. *Int. J. Sediment Res.* **2014**, *29*, 304–315. [CrossRef]

33. Fang, H. Impact of land use changes on catchment soil erosion and sediment yield in the northeastern China: A panel data model application. *Int. J. Sediment Res.* **2020**. [CrossRef]

34. Djoukbala, O.; Hasbaia, M.; Benselama, O.; Mazour, M. Comparison of the erosion prediction models from USLE, MUSLE and RUSLE in a Mediterranean watershed, case of Wadi Gazouana (N-W of Algeria). *Modeling Earth Syst. Environ.* **2018**. [CrossRef]

35. Mahe, G.; Benabdelfadel, H.; Dieulin, C.; Elbaraka, M.; Ezzaouini, M.; Khomsi, K.; Rouche, N.; Sinan, M.; Snoussi, M.; Tra Bi, A.; et al. Evolution des débits liquides et solides du Bouregreg. In *Gestion Durable des Terres. Proceedings de la Réunion Multi-Acteurs Sur le Bassin du Bouregreg. CERGéo, Faculté des Lettres et Sciences Humaines, Université Mohammed V-Agdal, Rabat, 28 mai 2013*; Laouina, A., Mahe, G., Eds.; Edité par ARGDT: Rabat, Morocco, 2014; pp. 21–36, ISBN 978-9954-33-482-9.

36. Khali Issa, L.; Ben Hamman Lech-Hab, K.; Raissouni, A.; El Arrim, A. Cartographie quantitative du risque d'érosion des sols par approche SIG/USLE au niveau du bassin versant Kalaya (Maroc Nord Occidental). *J. Mater. Environ. Sci.* **2016**, *7*, 2778–2795.

37. El Jazouli, A.; Barakat, A.; Ghafiri, A.; El Moutaki, S.; Ettaqy, A.; Khellouk, R. Soil erosion modeled with USLE, GIS, and remote sensing: A case study of Ikkour watershed in Middle Atlas (Morocco). *Geosci. Lett.* **2017**, *4*, 1. [CrossRef]

38. Sabri, E. Impact des Processus Erosifs sur les Ressources Naturelles des Bassins Hydrographiques et sur L'état D'envasement des Retenues des Barrages: Application au Bassin Versant de L'oued El Abid en Amont du Barrage de Bin El Ouidane- Maroc. Ph.D. Thesis, Univ. Cadi Ayyad, Marrakech, Morocco, 2018.

39. Sadeghi, S.H.R.; Gholami, L.; Khaledi Darvishan, A.; Saeidi, P. A review of the application of the MUSLE model worldwide. *Hydrol. Sci. J.* **2014**, *59*, 365–375. [CrossRef]

40. Aboulabbes, O. Infiltration Characteristics on a Small Watershed. Master's Thesis, Utah State University, Logan Utah, UT, USA, 1984.

41. El Gaatib, R.; Erraji, A.; Larabi, A. Impact des processus érosifs sur les ressources naturelles des bassins hydrographiques et sur l'état d'envasement des retenues de barrages: Application au Bassin Versant de l'Oued Beht en amont du barrage El Kansra, (Maroc). *Géo Obs.* **2014**, *21*, 4.

42. Jamagne, M. *Bases et Techniques d'une Cartographie des Sols'*; Institut National de la Recherche Agronomique: Paris, France, 1967.

43. Stone, R.P.; Hilborn, D. *Universal Soil Loss Equation, USLE*; Factsheet; Government of Ontario, Ministry of Agriculture, Food and Rural Affairs: Canada, 2000. Available online: http://www.omafra.gov.on.ca/english/engineer/facts/12-051.htm (accessed on 17 June 2020).

44. Pribyl, D.W. A critical review of the conventional SOC to SOM conversion factor. *Geoderma* **2010**, *156*, 75–83. [CrossRef]

45. Van Bemmelen, J.M. "Über die Bestimmung des Wassers, des Humus, des Schwefels, der in den colloïdalen Silikaten gebundenen Kieselsäure, des Mangansuswim Ackerboden". *Die Landwirthschaftlichen Vers.-Stn.* **1890**, *37*, 279–290.

46. Wolff, E. Entwurf zur boden analyse. *Z. Anal. Chem.* **1864**, *3*, 85–115. [CrossRef]

47. Sprengel, C. Ueber Pflanzenhumus, Humussaüre und humussaure Salze. *Arch. für die Gesammte* **1826**, *8*, 145–220. [CrossRef]

48. Tra Bi, Z.A.; Emran, A.; Brou, Y.T.; Mahé, G. Cartographie par arbre de décision de la dynamique de l'occupation du sol du bassin versant du Bouregreg, en région semi-aride au centre Nord-Ouest du Maroc, Abidjan, Côte d'Ivoire. *Rev. Sci. Int. Géomatique* **2014**, *1*, 43–54.

49. Sadiki, A.; Bouhlassa, S.; Auajjar, J. Utilisation d'un SIG pour l'évaluation et la cartographie des risques d'érosion par utilisation d'un SIG pour l'évaluation et la cartographie des risques d'érosion par l'équation universelle des pertes en sol dans le Rif oriental (Maroc). *Bull. de l'Inst. Sci.* **2004**, *26*, 69–79.

50. Goussot, E.; Brou, Y.T.; Laouina, A.; Chaker, M.; Emran, A.; Machouri, N.; Mahe, G.; Sfa, M.; Tra Bi, A. Land covers dynamic and agricultural statistics on the Bouregreg watershed in Morocco. *Eur. J. Sci. Res.* **2014**, *126*, 191–205.

51. Tra Bi, A.Z.; Brou, T.; Emran, A.; Mahe, G. Remote sensing and GIS analysis of the dynamic of vegetation in climate variability context on the Bouregreg watershed. In *Climate and Land Surface Changes in Hydrology*; IAHS Publ. 359; Boegh, E., Blyth, D.M., Hannah, H., Hisdal, H., Kunstmann, B., Su, K.K., Yilmaz, Eds.; IAHS Press: Wallingford, UK, 2013; pp. 403–410.

52. El Hadraoui, Y. *Étude Diachronique de L'occupation du sol et de Modélisation des Processus Erosifs du Bassin Versant du Bouregreg (Maroc) à Partir des Données de L'observation de la Terre*; Mémoire d'ingénieur du CNAM, Spécialité Géomètre et Topographe: Le Mans, France, 2013; Available online: https://dumas.ccsd.cnrs.fr/dumas-00920465 (accessed on 17 June 2020).

53. Lahlou, A. *La Dégradation Spécifique des Bassins Versants et son Impact sur L'envasement des Barrages*; IAHS Publ. 137; IAHS Press: Wallingford, UK, 1982; pp. 163–169. Available online: https://pdfs.semanticscholar.org/cddc/4f4b21f52e34043e1b0547d00d0e9dd389b9.pdf (accessed on 17 June 2020).

54. Gupta, H.V.; Sorooshian, S.; Yapo, P.O. Status of automatic calibration for hydrologic models: Comparison with multilevel expert calibration. *J. Hydrol. Eng.* **1999**, *4*, 135–143. [CrossRef]

55. Gupta, H.V.; Kling, H.; Yilmaz, K.K.; Martinez, G.F. Decomposition of the mean squared error and NSE performance criteria: Implications for improving hydrological modelling. *J. Hydrol.* **2009**, *377*, 80–91. [CrossRef]

56. Nash, J.E.; Sutcliffe, J.V. River flow forecasting through conceptual models part I—A discussion of principles. *J. Hydrol.* **1970**, *10*, 282–290. [CrossRef]

57. Lahlou, A. *Etude du transport solide à la station Dar Es Soltane sur l'oued Bouregreg*; Rapport de travaux du service GDE; Gestion des eaux, Ministère des travaux publics et des communications, Direction de l'hydraulique: Rabat, Morocco, 1971.

58. Mahe, G.; Aksoy, H.; Brou, Y.T.; Meddi, M.; Roose, E. Relationships among man, environment and sediment transport: A spatial approach. *Rev. Sci. Eau* **2013**, *26*, 235–244. [CrossRef]

59. De Vente, J.; Verduyn, R.; Verstraeten, G.; Vanmaercke, M.; Poesen, J. Factors controlling sediment yield at the catchment scale in NW Mediterranean geoecosystems. *J. Soils Sediments* **2011**, *11*, 690–707. [CrossRef]

60. Moriasi, D.N.; Arnold, J.G.; Van Liew, M.W.; Bingner, R.L.; Harmel, R.D.; Veith, T.L. Model evaluation guidelines for systematic quantification of accuracy in watershed simulations. Transactions of the ASABE. *Am. Soc. Agric. Biol. Eng.* **2007**, *50*, 885–900.

61. Maleval, V. Premiers résultats des mesures d'érosion ravinaire sur les versants lacustres du barrage Sidi Mohamed ben Abdellah (Maroc) et perspectives de recherche. In *Gestion Durable des Terres. Proceedings de la réunion multi-acteurs sur le bassin du Bouregreg. CERGéo, Faculté des Lettres et Sciences Humaines, Université Mohammed V-Agdal, Rabat, 28 mai 2013*; Laouina, A., Mahe, G., Eds.; Edité par ARGDT: Rabat, Morocco, 2014; pp. 53–62, ISBN 978-9954-33-482-9.

62. Laouina, A.; Chaker, A.; Aderghal, M.; Machouri, N.; Al Karkouri, J. Echelles d'évaluation de la dynamique des terres et de leur gestion, du territoire social au bassin-versant expérimental. In *Gestion Durable des Terres. Proceedings de la réunion multi-acteurs sur le bassin du Bouregreg. CERGéo, Faculté des Lettres et Sciences Humaines, Université Mohammed V-Agdal, Rabat, 28 mai 2013*; Laouina, A., Mahe, G., Eds.; Edité par ARGDT: Rabat, Morocco, 2014; pp. 119–146, ISBN 978-9954-33-482-9.

63. Chaker, M.; Laouina, A.; El Marbouh, M. Changement agropastoral et dégradation des terres dans le plateau Sehoul. In *Gestion Durable des Terres. Proceedings de la Réunion Multi-Acteurs sur le Bassin du Bouregreg. CERGéo, Faculté des Lettres et Sciences Humaines, Université Mohammed V-Agdal, Rabat, 28 mai 2013*; Laouina, A., Mahe, G., Eds.; Edité par ARGDT: Rabat, Morocco, 2014; pp. 85–102, ISBN 978-9954-33-482-9.

64. Hallouz, F.; Mahé, G.; Toumi, S.; Rahmani, S.E.A. Erosion, suspended sediment transport and sedimentation on the Wadi Mina at the Sidi M'Hamed Ben Aouda Dam, Algeria. *Water* **2018**, *10*, 895. [CrossRef]

65. Gwapedza, D.; Hughes, D.A.; Slaughter, A.R. Spatial scale dependency issues in the application of the Modified Universal Soil Loss Equation (MUSLE). *Hydrol. Sci. J.* **2018**, *63*, 1890–1900. [CrossRef]

66. Michalec, B.; Walega, A.; Cupak, A.; Strutyński, M. Verification of the musle to determine the amount of suspended sediment. *Carpathian J. Earth Environ. Sci.* **2017**, *12*, 235–244.

67. Iroumé, A.; Carey, P.; Bronstert, A.; Huber, A.; Palacios, H. GIS application of USLE and MUSLE to estimate erosion and suspended sediment load in experimental catchments, Valdivia, Chile. *Aplicación SIG de USLE y MUSLE para Estimar la Erosión y el Transporte de Sedimentos en Suspensión en Cuencasexperimentales* **2011**, *34*, 119–128.

68. Arekhi, S.; Shabani, A.; Alavipanah, S.K. Evaluation of integrated KW-GIUH and MUSLE models to predict sediment yield using geographic information system (GIS) (Case study: Kengir watershed, Iran). *Afr. J. Agric. Res.* **2011**, *6*, 4185–4198.

69. Moussebbih, A.; Souissi, M.; Larabi, A.; Faouzi, M. Modeling and mapping of the water erosion risk using Gis/Rusle approach in the Bouregreg river watershed. *Int. J. Mech. Prod. Eng. Res. Dev.* **2019**, *9*, 1605–1618. [CrossRef]

70. Ramos-Diez, I.; Navarro-Hevia, J.; San Martín, R.; Díaz-Gutiérrez, V.; Mongil-Manso, J. Evaluating methods to quantify sediment volumes trapped behind check dams, Saldaña badlands (Spain). *Int. J. Sediment Res.* **2016**, *28*, 2446–2456. [CrossRef]

71. Kronvang, B.; Andersen, H.E.; Larsen, S.E.; Audet, J. Importance of bank erosion for sediment input, storage and export at the catchment scale. *J. Soils Sediments* **2013**, *13*, 230–241. [CrossRef]

72. Palazón, L.; Navas, A. Modeling sediment sources and yields in a Pyrenean catchment draining to a large reservoir (Ésera River, Ebro Basin). *J. Soils Sediments* **2014**, *14*, 1612–1625. [CrossRef]

73. Driouech, F.; Mahé, G.; Déqué, M.; Dieulin, C.; El Heirech, T.; Milano, M.; Benabdelfadel, H.; Rouché, N. Evaluation d'impacts potentiels de changements climatiques sur l'hydrologie du bassin versant de la Moulouya au Maroc. In *Global Change: Facing Risks and Threats to Water Resources. Proceedings of the Sixth World FRIEND Conference, Fez, Morocco, October 2010*; IAHS Publ. 340; IAHS Press: Wallingford, UK, 2010; pp. 561–567. Available online: https://horizon.documentation.ird.fr/exl-doc/pleins_textes/divers15-08/010055009.pdf (accessed on 17 June 2020).

74. Filahi, S.; Tramblay, Y.; Mouhir, L.; Diaconescu, E.P. Projected changes in temperature and precipitation indices in Morocco from high-resolution regional climate models. *Int. J. Climatol.* **2017**, *37*, 4846–4863. [CrossRef]

75. Tramblay, Y.; Jarlan, L.; Hanich, L.; Somot, S. Future scenarios of surface water resources availability in North African dams. *Water Resour. Manag.* **2018**, *32*, 1291–1306. [CrossRef]

76. Raclot, D.; Naimi, M.; Chikhaoui, M.; Nunez, J.P.; Huard, F.; Herivaux, C.; Sabir, M.; Pepin, Y. Distinct and combined impacts of climate and land use scenarios on water availability and sediment loads for a water supply reservoir in northern Morocco. *Int. J. Soil Conserv.* **2020**. [CrossRef]

Article

Anthropization and Climate Change: Impact on the Discharges of Forest Watersheds in Central Africa

Valentin Brice Ebodé [1,*], Gil Mahé [2,*], Jean Guy Dzana [1] and Joseph Armathé Amougou [1]

[1] Department of Geography, University of Yaounde 1, P.O. Box 755, Yaounde, Cameroon; dzana1@yahoo.fr (J.G.D.); joearmathe@yahoo.fr (J.A.A.)

[2] International Joint Laboratory DYCOFAC, IRGM-UY1-IRD, BP 1857, Yaoundé (Cameroon) and UMR HSM, IRD-CNRS-University of Montpellier, 34090 Montpellier, France

* Correspondence: ebodebriso@gmail.com (V.B.E.); gil.mahe@ird.fr (G.M.); Tel.: +237-694-426-200 (V.B.E.); Fax: +33-467-144-774 (G.M.)

Received: 6 August 2020; Accepted: 15 September 2020; Published: 29 September 2020

Abstract: Climate change and anthropization are major drivers of river flows variability. However, understanding their simultaneous impact on discharges is limited. As a contribution to address this limitation, the objective of this study is to assess the impact of climate change and anthropization on the discharges of two watersheds of Central Africa (Nyong and Ntem) over a recent period. For this, the hydropluviometric data of the watersheds concerned were analyzed using the Pettitt test. Similarly, the dynamics of the main land use modes (LUM) have been assessed, through classifications obtained from the processing of Landsat satellite images of the watersheds studied on two dates. The results of this study show that in Central Africa, annual discharges have decreased significantly since the 1970s, and yet the decline in annual rainfall does not become significant until the 2000s. The discharges of the rainy seasons (spring and autumn) recorded the most important changes, following variations in the rainfall patterns of the dry seasons (winter and summer) that precede them. Winters experienced a significant decrease in precipitation between the 1970s and 1990s, which caused a drop in spring flows. Their rise, which began in the 2000s, is also accompanied by an increase in spring flows, which nevertheless seems rather slight in the case of the Nyong. Conversely, between the 1970s and 1990s, there was a joint increase in summer rainfall and autumn flows. A decrease of summer rainfall was noted since the 2000s, and is also noticeable in autumn flows. Maximum flows have remained constant on the Nyong despite the slight drop in rainfall. This seems to be the consequence of changes in land use patterns (diminution of forest and increasing of impervious areas). The decrease in maximums flows noted on the Ntem could be linked to the slight drop in precipitation during the rainy seasons that generates it. Factors such as the general decrease in precipitation during the winter and the reduction in the area occupied by water bodies could justify the decrease in minimum flows observed in the two watersheds. These findings would be vital to enhance water management capabilities in the watersheds concerned and in the region. They can also give some new elements to study and understand the seasonal variation and fresh water availability in downstream, estuaries and coastal areas of the regional rivers.

Keywords: climate change; urbanization; impervious area; Cameroon; runoff

1. Introduction

River flows are essentially variable over time. Their evolution is generally considered to be the result of their interactions with climate change and/or anthropization [1–6], although it is recognized that sensitivity of watercourses to these factors depends on the natural predispositions (size, slope system, type of soil, etc.) of their watersheds [7]. Apart from some interesting attempts, such as that of Dzana et al. [1], research studies in which the effects of these two factors are dissociated and evaluated separately are rare. Most of the studies undertaken thus far around the world have generally focused on one or another of these interferences, as given in the results in West Africa, South Africa, Ethiopia and even Uruguay, linking land cover change/or climate change with river discharges [8–11].

In sub-Saharan Africa, the work devoted to the research of the driving factors of the flows variability has been carried out mainly since the 1980s. Those in which the authors correlate the precipitations and the flows are based on the detection of discontinuities in the hydropluviometric series. Their results confirmed, in the case of West Africa, that the 1970s appeared as the main period of discontinuity, marking the beginning of the hydroclimatic drought in this region [12–15]. In Central Africa, the most obvious fluctuations in discharges and precipitation have been observed with seasonal time steps [16–18]. Liénou et al. [19] demonstrated in the case of three equatorial rivers (Ntem, Nyong and Kienke), that the most important climatic variations leading to changes in flows result from variations in precipitation in dry seasons. These authors explain the sensitivity of watersheds to this precipitation variability by the fact that their reduction induces a significant deficit in soil and groundwater storage, resulting in a decrease in the runoff. Conversely, their increase keeps the soil wetter during the dry season, and therefore, improves runoff from the start of the rainy season. It seems, therefore, that the variability of the regimes of equatorial rivers can be better appreciated when these precipitations are taken into account. Regarding the impact of anthropic action (changes in land use modes (LUM)) on runoff, reference works are generally based on supervised classifications of satellite images on at least two dates, with a view to assessing the dynamics of land use and its possible impact on flows [20–26]. Their results generally confirm an increase in flows resulting from an increase in Impervious Area (IA) such as buildings, roads and cultivated areas. Paturel et al. [24] specifies, in the context of a Sahelian basin in West Africa (Nakambe in Burkina-Faso), that this increase in flows is linked to the reduction in the water capacity of the soils resulting from the increase of the impervious surface. However, Dezetter et al. [27] had noted, in a humid tropical climate under forest cover (in the case of the upper Niger River basin in Guinea), an insignificant impact of the increase in agricultural areas, to the detriment of the forest, on flows. According to them, evaporation (due to the increase in temperature) from this region would have a greater impact than that of changes in LUM on the hydrological balance of this region. These conclusions follow those of a former study by Fritsch [28]. This author noted earlier in the context of small Guyanese basins, also under equatorial forest cover, a considerable impact of deforestation on runoff; however this did not last long, as after two years the runoff coefficients returned to nearly past values on surfaces where vegetation was left wild or over plantations, while the runoff remained higher over grass lands (for cattle grazing). This led Fritsch to specify that the magnitude of the hydrological changes observed in the different cases depends on the type and importance of the developments carried out in these basins, together with the impact of the inter-annual variability of natural phenomena.

Few studies have been carried out on this type of basins in forested equatorial area, and there is still a large uncertainty on what are the main drivers of the variability of the hydrological balance in equatorial basins within the context of increasing temperature due to climate change.

The Nyong and Ntem watersheds offer us the opportunity to make possible progress on this issue, as these are two basins that have been observed for decades in this equatorial area, and for which there have already been several former studies on the hydroclimatological variability, which showed some questionable issues about the origin of the observed change in the river flow regime since some decades. These watersheds are forested, but since the early 1970s, climatic fluctuations have been observed there, as in the rest of sub-Saharan Africa [19]. Moreover, their population is increasingly growing [29]. This population growth generally has as a corollary the increase in IA, which can have repercussions on the hydrological functioning of these hydrosystems. However, unlike the West African watersheds, the Nyong and Ntem basins were the subject of a very small number of works after the 2000s, due to the absence of observation data in certain cases, and significant gaps in existing observation series in other cases. Likewise, the difficulty of acquiring satellite images of sufficient quality (without cloud) for processing over large areas would have led the authors to ignore the study of the impact of anthropization (change of LUM) on discharges. Most of the studies cited above relate to West Africa, apart from those of Moffo [25] and Ebodé [26]. Despite these difficulties, it is nevertheless advisable to try to update the information available on these watersheds in the limited number of similar investigations that have focused on them before, but also to take note of certain aspects (land use) having important links with flows, which have been less studied in the region thus far, in connection with the dynamics of IA. Such a study seems fundamental for these rivers, which are among the main sources of water and fishery resources in the country, whose recent decreases are partly attributed to hydroclimatic changes. It is particularly useful in the case of Ntem, where the Memve'ele dam was recently built (in Nyabessan, downstream from Ngoazik station), with the aim of ensuring electricity production, replacing the old dam of Songloulou. It is hoped that the results of this study will reinforce the debate on the issue since the rationing in the electrical distribution of the southern Cameroon network is also generally attributed to climate change. In addition, the data and new inputs from this study would contribute to long-term planning of water demand and use, and improve future simulations of the flows of these rivers.

The main objective of this article is to assess the observed changes in streamflow of Nyong and Ntem watersheds, in response to current climate change and anthropization (land use/land cover change). For this, we propose to study long-term rainfall and discharges time series, and changes in land cover over time from satellite images. The links between rainfall and some sources of large-scale climate variability will also be checked.

2. Materials and Methods

2.1. Study Area

This study focuses on the Nyong (14,438 km^2) and Ntem (19,563 km^2) watersheds, respectively at Mbalmayo and Ngoazik outlets. These watersheds are located in Central Africa between latitudes 1°15′ N and 4°44′ N and longitudes 11°5′ E and 13°30′ E (Figure 1). The first is entirely included in Cameroonian territory, while the second is located between Cameroon, Gabon and Equatorial Guinea. They belong to the equatorial (Ntem) and sub-equatorial (Nyong) domains, with abundant precipitation. The annual precipitation oscillates around 1600 mm over the Nyong and 1800 mm over the Ntem, distributed over four seasons of unequal importance, including two dry and two rainy seasons (Table 1). The rainy seasons (spring and autumn) are generally very wet with record totals of around 900 mm and marked by numerous thunderstorms, which sometimes cause significant floods. The winter dry season is the only real dry season in the region, with average precipitation fluctuating around 90 mm and 180 mm, respectively, in the Nyong and Ntem watersheds (Table 1). The summer dry season tends to become more humid in the region since the 1980s [19]. The Nyong and Ntem watersheds are deeply gullied and cut into hills with convex slopes and wide marshy valleys [30]. On both sides of the main rivers, a topographic landscape in steps is observed, which seems to be the result of brittle tectonics. In the middle part of each watershed, there is a major valley (600 m above sea level on average), which represents the collapse ditch. It is a wide corridor, in the middle of which flows the main river. On the two watersheds, this unit is directly followed on each side of the main river by a large plateau with an average altitude of around 700 m, then large hills (800 m of altitude on average), which are bristling with a few inselbergs located mainly at the limit with neighboring watersheds, and whose altitude can sometimes reach 1200 m [30]. The geological substratum of these watersheds is made up of a granito-gneissic base on which ferralitic (on the tops and slopes) and hydromorphic (in the shallows) soils develop. The vegetation encountered on these basins is a dense semi-deciduous forest with Sterculiaceae and Ulmaceae strongly subjected to anthropic actions [31].

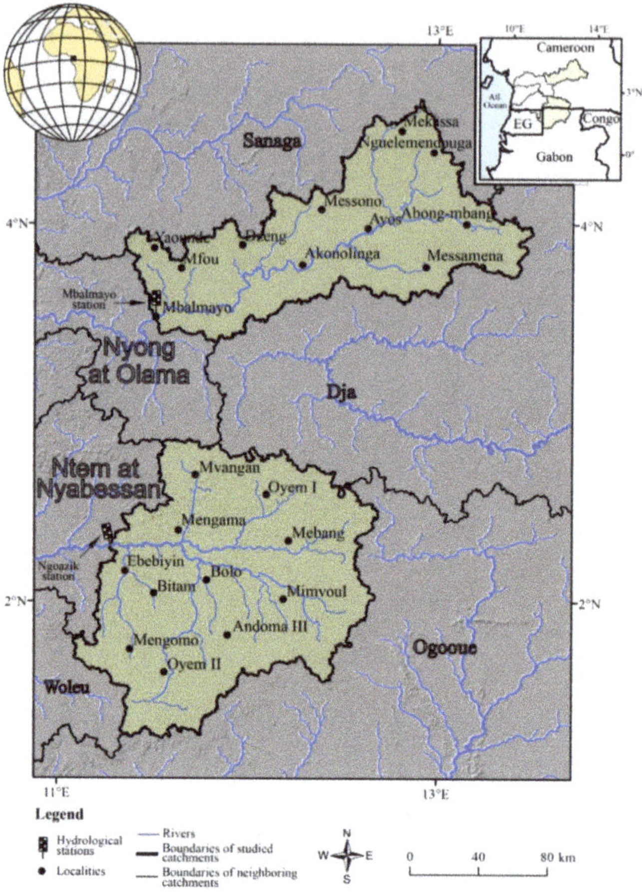

Figure 1. Location map of Nyong and Ntem watersheds.

Table 1. Statistics relating to the evolution of precipitation and average discharges at annual and seasonal time steps over the study period, but also before and after the discontinuity. Cv: coefficient of variation.

Periods	Whole Period		Discontinuity	Mean (m³/s)		Cv (%)	
	Mean (m³/s)	Cv (%)		Before	After	Before	After
Rainfall							
Nyong							
Annual	1627	9	-	-	-	-	-
Spring	634	13	-	-	-	-	-
Summer	274	28	1964–65	220	290	26	26
Autumn	625	11	-	-	-	-	-
Winter	93	43	1975–76	116	78	29	48
Ntem							

Table 1. *Cont.*

Periods	Whole Period		Discontinuity	Mean (m³/s)		Cv (%)	
	Mean (m³/s)	Cv (%)		Before	After	Before	After
Annual	1810	7	-	-	-	-	-
Spring	675	12	-	-	-	-	-
Summer	256	29	1964–65	192	283	29	22
Autumn	693	10	-	-	-	-	-
Winter	184	28	1976–77	208	169	20	32
			Discharges				
			Nyong				
Annual	139	20	1973–74	151	132	19	18
Spring	88	31	1973–74	105	76	28	27
Summer	111	31	-	-	-	-	-
Autumn	243	23	-	-	-	-	-
Winter	123	20	-	-	-	-	-
			Ntem				
Annual	236	23	1970–71	269	215	26	18
Spring	222	34	1970–71	270	193	32	28
Summer	115	49	-	-	-	-	-
Autumn	421	25	1990–91	444	324	22	18
Winter	156	29	1970–71	186	136	25	23

2.2. Data Collection

The data collected for this study are hydroclimatic series and spatial data (satellite images).

2.2.1. Hydroclimatic Data

Nyong discharge data series come from two sources. The gauging station, as well as the methods used for reading water levels and calculating flow rates, however, remain the same from one source to another. The series obtained from the Hydrological Research Center (CRH) cover the period 1951–1987. This center manages a hydrometric database, generally with a daily time step, which contains almost all the measurements carried out on Cameroonian territory since the early 1950s, for most stations. These measurements were carried out first by ORSTOM (Office de la Recherche Scientifique et Technique Outre-Mer/Office for Scientific Research in Overseas Territories, Paris, France) researchers, currently IRD (Institut de Recherche pour le Développement/Research Institute for Development, Marseille, France) and their predecessors from EDF (Electricité De France/France Electricity, Paris, France) and, since 1980, by CRH. No observation data is available on this river between 1988 and 1997. Indeed, after 1987, due to budgetary constraints, the hydrological service no longer guaranteed the continuity of observations. We note then the abandonment of a large number of stations observed, including those of Nyong and Ntem. Over the period 1998–2016, the series of discharges used are those of the ORE-BVET (Observatory for Environmental Research/Experimental Tropical Watersheds). Funded by the Ministry of Research and New Technologies, the National Institute of Universe Sciences, the Research Institute for Development and the Midi-Pyrénées Observatory, this observatory aims to understand the relative influence of climate variability and agriculture on water cycles and biogeochemical cycles in tropical areas. Based on long-term monitoring (>10 years) of meteorological, hydrological and geochemical variables in watersheds, it is implemented both in Cameroon and in India. In Cameroon, the observatory is interested in the Nyong watershed. The complete series of Ntem discharges comes from the CRH, and covers the period 1953–2015, but with gaps during the 1990s and 2000s. There are no observation data on this river between the intervals 1993–2001 and 2006–2009. All of the hydrological data used in this work were collected daily. The monthly, seasonal and annual modules were calculated thereafter.

The rainfall series comes from two sources. Over the period 1950–1999, the SIEREM (Système d'Informations Environnementales pour les Ressources en Eau et leur Modélisation/System of Environmental Informations on Water Resources and their Modeling) rainfall grids were used. SIEREM [32] is a database based on a generic scheme allowing the management of hydroclimatic but also environmental information on Africa. This database was developed by the HydroSciences Montpellier/HSM Laboratory, from a rainfall database (around 7000 gauging stations) from Africa, inherited among others from ORSTOM, and its experience in the management of hydroclimatic databases in the intertropical region [33,34]. Among its recovery products, SIEREM offers monthly rainfall grids over Africa for the period 1940–1999, at the 0.5° square degree, approximately 2750 km^2 of the area considered [35]. The space and time steps chosen seem to be the most appropriate resolutions to be able to appreciate, at the same time, the impacts of climate change and anthropic activities on water resources [36,37]. These grids are freely accessible in ASCII and NetCDF formats on the site http://www.hydrosciences.org/spip.php?article1387. Before use, they had to be validated first by comparison with rain gauges measurements from stations.

From 2000s, the choice was made for TRMM (Tropical Rainfall Measuring Mission) 3B42 V7 rainfall estimates. The TRMM platform was launched in 1997, built jointly by the United States and Japan. It offers several pre-calibrated data sets, including the daily product TRMM 3B42 V7 at a spatial resolution of 0.25° × 0.25°. These data combine syntheses of TRMM images and other satellite data [38]. Thus, microwave sensors from AQUA, NOAA (National Oceanic and Atmospheric Administration), DMSP (Defense Meteorological Satellite Program) and visible/infrared data from geostationary satellites are taken into account in precipitation estimation algorithms. These estimates are then adjusted by incorporating the monthly climate measurements on the ground of the GPCP (Global Precipitation Climatology Project) and CAMS (Climate Assessment and Monitoring System) networks to give the monthly products TRMM 3B43. The adjustment coefficients calculated for this product are finally applied to the daily data to give the final product TRMM 3B42, version 7 [38], downloadable free of charge from the website https://giovanni.gsfc.nasa.gov, in ASCII format. The monthly TRMM rainfall data were first validated through comparisons with the SIEREM rainfall over the years of common availability (1998 and 1999). The correction factors (essentially between 1.17 and 1.4) were subsequently established for the months for which the differences between the two sources were significant (July and August).

Four climatic indices (SOI: Southern Oscillation Index; DMI: Dipole Mode Index; NATL: North Tropical Atlantic sea surface temperature indices; SATL: South Tropical Atlantic sea surface temperature indices) representative of certain sources of large-scale variability (Pacific Ocean, Indian Ocean and Atlantic Ocean), likely to influence the climate of the region [39], were used in this study (Figure 2). These indices were collected at a monthly time step on the NOAA website (https://psl.noaa.gov/data/climateindices/list/). SOI and DMI were collected at the same intervals as the rainfall. The other two indices have only been available since the beginning of the 1980s.

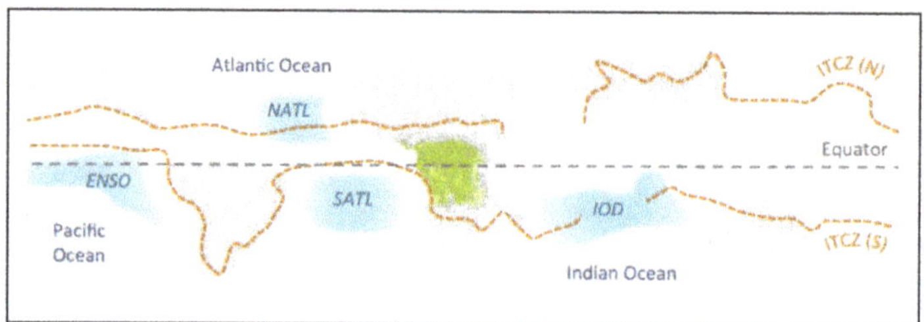

Figure 2. Patterns in oceanic sea surface temperatures (SSTs) known to influence weather in the studied region (blue zones), according to Bush et al. [39]. The Pacific Ocean El Niño Southern Oscillation (ENSO); North and South Tropical Atlantic SSTs (NATL and SATL, respectively) and the Indian Ocean Dipole (IOD).

2.2.2. Spatial Data

The spatial data used for the study of land use in the Nyong watershed are essentially the Landsat 8 satellite images from January 2015 and Landsat MSS from March 1973. Three Landsat scenes cover the entire watershed, corresponding to the path/row 185/57, 185/58, 184/57. The same study was carried out in the Ntem watershed, using four Landsat 8 scenes from January 2015 and Landsat MSS from February 1973, corresponding to path/row 184/58, 185/58, 184/59 and 185/59. All of these images are made available to the general public free of charge by the National Aeronautics and Space Administration (NASA), via the US Geological Survey website (https://earthexplorer.usgs.gov/), in GeoTIFF format. The downloaded images taken during the winter (December to mid-March) were preferred to that of the rainy seasons because they are less affected by cloud disturbances.

2.3. Data Analysis

2.3.1. Hydroclimatic Data

The analysis of rainfall, average discharges and runoff coefficients was carried out using statistical discontinuity detection tests [40] at the significance level 95%. The Pettitt test seems to be the most suitable for the analysis of incomplete series such as ours because it separates the series only into two periods with an overall distinct behavior, which avoids the detection of false discontinuities as can sometimes be observed with other tests such as Hubert segmentation. Its choice at the expense of trend tests (Mann-Kendall test for example) is justified by the fact that it indicates a date from which the change becomes statistically significant (discontinuity), which allows a better appreciation of the variability. Its principle consists in dividing the studied series (of N size) into two sub-samples of sizes m and n, respectively. We then calculate the sum of the ranks of the elements of each sub-sample in the total sample. A statistical study is then carried out based on the two sums thus determined, then it is tested according to the hypothesis that the two subsamples do not belong to the same population. The Pettitt test is non-parametric and derives from that of Mann Whitney. The absence of a discontinuity in the series (Xi) of size N constitutes the null hypothesis. Its implementation supposes that for any instant T between 1 and N, the time series (Xi) 1 to t and t + 1 to N belong to the same population. The variable to be tested is the maximum in the absolute value of the variable Ut, N defined by:

$$Ut, N = \Sigma_{i=1}^{t} \Sigma_{i=t+1}^{N} Dij \tag{1}$$

where Dij = Sign (Xi − Xj) with: sign (x) = 1 if x > 0, 0 if x = 0 and −1 if x < 0.

If the null hypothesis is rejected, an estimate of the date of discontinuity is given by the instant defining the maximum in the absolute value of the variable Ut, N.

The correlation coefficient is the criterion used to assess the links between the rainfall indices and the climatic indices selected. Denoted by "r," the correlation coefficient is the ratio between the covariance (γ) of two variables (X and Y) and the product of their standard deviations. It indicates the strength and direction of the linear relationship between these variables. Varying between −1 and +1, it reflects a strong correlation if it is less than −0.5 or greater than 0.5:

$$r = \frac{\gamma(X, Y)}{\sigma X \sigma Y} \tag{2}$$

where $\gamma(X, Y)$ denotes the covariance of the variables X and Y; σX and σY denote their standard deviations.

To assess the behavior of extreme flows, the Indicators of Hydrologic Alteration (IHA) tool, version 7.1, developed by The Nature Conservancy was used. This tool offers the possibility of comparing the parameters characterizing the flow regimes under different conditions [41]. It uses daily flow values and produces several important statistics. We will only be interested in four of them, considered essential for this study, among which the average and the coefficient of variation (Cv) of extreme flows and the Julian date of the annual minimum and maximum. By dividing the series of values in the period before and after the discontinuity, the tool calculates the change that occurred in the evolution of each of these parameters after the discontinuity. We can thus analyze not only the sign of change between the two periods but also the magnitude of this difference.

2.3.2. Spatial Data

The supervised classification by maximum likelihood of the downloaded Landsat satellite images, using SNAP software (free access), allowed to perform a diachronic analysis of the evolution of land use in the studied watersheds. This operation was preceded by operations of preprocessing and recognition of objects in the field by photography and GPS (Global Positioning System). The preprocessing of satellite images refers to all of the processes applied to raw data to correct the geometric and radiometric errors that characterize certain satellite images. These errors are generally due to the technical problems of the satellites and to the interactions between the outgoing electromagnetic radiation and atmospheric aerosols, also called "atmospheric noise." These atmospheric disturbances are influenced by different factors that are present on the day of acquisition, including weather, fires and other human activities. They affect all the images acquired by passive satellites, including Landsat 4-5-7 and 8. The downloaded Landsat images are ortho-rectified, the preprocessing concerns the atmospheric correction of these images and their re-projection in the local system (WGS_84_UTM_Zone_32N). For this, neo-channels are created, to increase the readability of the data by enhancing certain less obvious properties in the original image, thus showing more clearly the elements of the scene. Three indices are therefore created, namely the Normalized Difference Vegetation Index (NDVI), the Brightness Index (IB) and the Normalized Difference Water Index (NDWI). These indices highlight respectively the vegetated surfaces, the sterile (not-chlorophyllin) elements such as the urban and the water bodies. Their formulas are as follows NDVI = (NIR − R)/(NIR + R); IB = $(R^2 + NIR^2)^{0.5}$ and NDWI = (NIR − MWIR)/(NIR + MWIR), where NIR: ground reflectance of the surface in the near-infrared channel; R: ground reflectance of the surface in the red channel and MWIR: ground reflectance of the surface in the mid-wave infrared channel. Due to the fact that the watersheds studied extend over several scenes, the enhancement operations were followed by the mosaic of the different scenes used on each date. It is an operation of joining two or more adjacent images into a single image. The use of Google Earth, as well as the spaces sampled from the GPS, made it possible to identify with certainty the IA (buildings, savannas, bare soils and crops), water bodies (large rivers, lakes and ponds) and the forests (secondary, degraded, non-degraded and swampy) of each mosaic. Before the classification operation, the separability of the spectral signatures of the sampled objects to avoid interclass confusion was assessed by calculating the "transformed divergence" index. The value of this index is between 0 and 2. A value >1.8 indicates a good separability between two given classes. The different classes used in this study show good separability between them, whatever the image considered, with indices >1.9 (Table 2). The validation of the classifications obtained was carried out using the confusion matrix, which made it possible to obtain treatment details, in order to validate the choice of training plots. After validating the land use/land cover maps produced, the statistical and spatial differences of each class between the two dates were studied.

Table 2. Indices of separability between the different classes considered by watershed and by image. TDI: Transformed Divergence Index and WB: Water Bodies.

Nyong Watershed			
Image from 1973		Image from 2015	
Classes Assessed	TDI	Classes Assessed	TDI
Forest-IA	2	Forest-IA	2
Forest-WB	1.92	Forest-WB	1.98
WB-Forest	1.92	WB-Forest	1.97
WB-IA	2	WB-IA	2
IA-Forest	2	IA-Forest	2
IA-WB	2	IA-WB	2

Table 2. *Cont.*

Ntem watershed			
Image from 1973		Image from 2015	
Classes assessed	TDI	Classes assessed	TDI
Forest-IA	2	Forest-IA	2
Forest-WB	1.97	Forest-WB	1.98
WB-Forest	1.96	WB-Forest	1.98
WB-IA	2	WB-IA	2
IA-Forest	2	IA-Forest	2
IA-WB	2	IA-WB	2

3. Results

3.1. Changes in Discharges

3.1.1. Average Discharges

The average discharges of Nyong and Ntem are analyzed through the annual and seasonal modules. The annual modules of these watercourses evolve statistically downwards (Figure 3). This decrease is accompanied by a reduction in Cv compared to the post-discontinuity period which seems negligible on the Nyong (−1%), but relatively significant on the Ntem (−8%) (Table 1). The Pettitt test shows discontinuities in the series of these rivers in 1973–74 for the Nyong and 1970–71 for the Ntem. The homogeneous sequence following this discontinuity recorded a deficit compared to the long-term average estimated at −5.6% in the first case and −8.9% in the second (Figure 3). However timidly started in the 1970s, the decrease in the annual modules of Nyong and Ntem increased considerably during the 2000s and 2010s (Table 3). These two decades are the driest of the entire observation period. The decadal deviations from the whole period average for each of them are greater than −5% whatever the watershed considered.

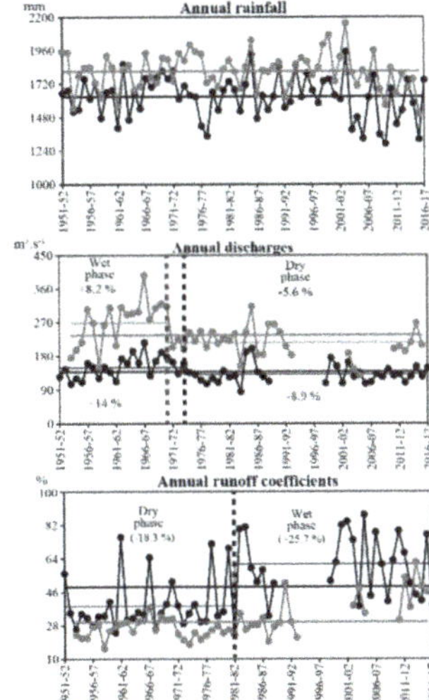

Figure 3. Results of statistical segmentation of Nyong (black) and Ntem (gray) annual rainfall, discharges and runoff coefficients, according to the Pettitt test. The thick horizontal lines indicate the averages of the whole period, while the thin horizontal lines indicate those of the different homogeneous periods. The thick and broken vertical lines indicate the years of discontinuity.

Table 3. Deviations (%) of decadal annual and seasonal averages of rainfall, discharges and runoff coefficients compared to the whole period averages of the studied watersheds.

Basins	Decades	Deviations from the Whole Period Average (%)				
		Annual	Spring	Summer	Autumn	Winter
		Rainfall				
Nyong	1950–59	0.2	3.2	−20.3	3.2	19.9
	1960–69	2.2	0.9	−3.1	1.4	32
	1970–79	−1	−1.8	−2.9	−0.1	3.6
	1980–89	1.5	0.8	10.5	2.1	−24
	1990–99	4.5	2.1	23.8	2.9	−16.1
	2000–09	−5.1	−3.4	−0.3	−8.6	−10.5
	2010–16	−3.1	−3.7	−9.4	0.9	−7.3
Ntem	1950–59	−0.2	3.8	−30.5	3.5	12
	1960–69	−1.7	2.2	−14.9	−5.6	15.9
	1970–79	2.3	0.6	1.9	5.9	−4.6
	1980–89	0.5	2.3	15.9	−0.6	−23.7
	1990–99	2.5	−1.2	30.2	−0.2	−14.5
	2000–09	0.1	−2.8	6.8	−2.1	13.8
	2010–15	−5.9	−7.7	−12.7	−2.2	−4.7
		Discharges				
Nyong	1951–59	−4.8	10.2	1.3	−14	−4.3
	1960–69	17.7	27.3	22.5	13.1	14.9
	1970–79	−2.6	0.1	−12.7	0.9	−6
	1980–88	−0.7	−8.8	−11	8.1	−3.6
	2000–09	−6.5	−14.7	0.8	−4	−7.8
	2010–16	−5.7	−12.3	−8.1	−7.4	7.4
Ntem	1953–59	−4.4	0.2	−27.6	−7.6	3.4
	1960–69	27.8	38.1	34.4	17.1	29.3
	1970–79	−2.2	−3.5	−8.1	−0.4	−5.2
	1980–89	−2.7	−17.4	−0.4	11.5	−12.1
	1990	-	-	-	-	-
	2000	-	-	-	-	-
	2010–15	−9.1	0.2	−12.9	−15.7	−10.4
		Runoff Coefficients				
Nyong	1951–59	−41.3	5.8	24.7	−27.9	−51.2
	1960–69	−34.2	20.7	12.5	12	−48.7
	1970–79	−34.5	−3.9	−15.9	7.8	−44.7
	1980–88	64.5	−8.1	−31.4	8.8	84.1
	2000–09	36.1	−6.2	13.8	9.2	44.4
	2010–16	25.1	−4.2	−1.3	−12.8	34.5
Ntem	1953–59	−13.9	−1.6	8.8	−15.2	−21.7
	1960–69	12.2	33.8	42.9	20.8	−7.3
	1970–79	−11.2	−7.6	−14.1	−1.2	−15.7
	1980–89	2.3	−20.8	−23.4	10.6	11.4
	1990	-	-	-	-	-
	2000	-	-	-	-	-
	2010–15	31.5	24	−14.3	−11.4	93.1

Regarding the discharges of the rainy seasons, a statistically significant decrease was noted in spring in both watersheds that the Pettitt test situated, in the same years as those of the annual modules. As in the annual timescale, the change in Cv of discharges for this season is slight on the Nyong (−1%), but more accentuated on the Ntem (−4%). The deficits of the periods occurring after the discontinuity are −13.6% for the Nyong and −13.1% for the Ntem (Figure 4). The decrease started since the 1970s seems to persist on the Nyong; however, on the Ntem, a considerable increase in the discharges is observed through decadal deviations. These deviations fell from −3.5% and −17.4% in the 1970s and

1980s to 0.2% in the 2010s (Table 3). In autumn, Nyong discharges decrease statistically insignificantly (Figure 4). However, there has been an important decline in them during the 2000s and 2010s, following the increase that began in the 1960s. Ntem discharges statistically decrease during this season since 1990–91 (Figure 4). This decrease is concomitant with a −4% change in Cv (Table 1). The deficit in the sequence following this discontinuity is −23% (Figure 4). The decadal deviations also show a significant decrease during the 2010s, after a long, generally wet period between the 1960s and the 1980s (Table 3).

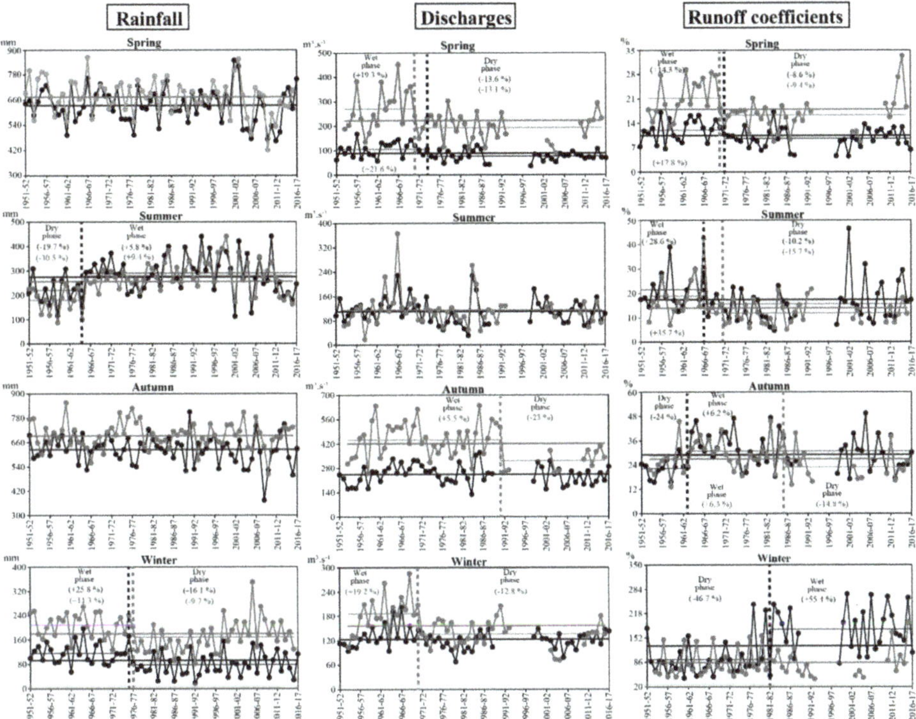

Figure 4. Results of statistical segmentation of Nyong (black) and Ntem (gray) seasonal rainfall, discharges and runoff coefficients, according to the Pettitt test. The thick horizontal lines indicate the averages of the whole period, while the thin horizontal lines indicate those of the different homogeneous periods. The thick and broken vertical lines indicate the years of discontinuity.

In dry seasons, during the summer, the discharges do not change statistically over the two watersheds according to the Pettitt test, although a slight decrease is observed (Figure 4). However, a general decrease since the 1970s can be observed through the decadal deviations, which increased considerably during the 2010s (Table 3). In winter, the Pettitt test does not show any discontinuity in the Nyong discharges, although an increase in them is noted in the 2010s, after the decrease started in the 1970s (Table 3). In the Ntem Watershed, winter flows have decreased statistically and continuously since 1970–71 (Figure 4). The deficit for the post-discontinuity period is −12.8% (Figure 4). Its Cv decreases by 2% (Table 1).

3.1.2. Extreme Discharges

The extreme discharges of the Nyong and the Ntem are analyzed through the maximums and minimums discharges on consecutive days (1, 3, 7, 30 and 90).

As regards the maximums first, it should be noted that in the case of the Nyong, the post-discontinuity period (1973–74 to 2016–17) corresponds to a dry phase. However, despite its dryness, the average maximums have remained essentially constant (Table 4). A slight reduction in variability of the maximums is also noted in this period, which seems more significant in the case of shorter-term ranges (1 day to 7 days), for which the Cv oscillate from 26% before discontinuity to 23% after (Table 4). The maximums appear earlier on the Nyong after the discontinuity. Their average Julian date of observation went from 318 before the discontinuity to 309 after (Table 4).

Table 4. Statistics relating to the maximum and minimum flows of Nyong and Ntem before and after the discontinuity. IHA: Indicators of Hydrologic Alteration.

IHA Statistics	Means (m³/s)		Cv (%)		Change	
	Before Discontinuity	After Discontinuity	Before Discontinuity	After Discontinuity	m³/s	%
Nyong						
Minimum flows						
1-day minimum	25.5	17.9	30	41	−7.5	−29.7
3-day minimum	26.1	18.1	30	41	−8	−30.6
7-day minimum	27.8	19.3	30	38	−8.4	−30.2
30-day minimum	41.5	26.4	28	36	−15	−36.2
90-day minimum	82.9	58.5	31	29	−24.3	−29.3
Maximum flows						
1-day maximum	366	366	26	23	−0.3	−0.08
3-day maximum	363	365	26	23	2	0.55
7-day maximum	361	362	26	23	1.3	0.36
30-day maximum	340	338	26	25	−2.6	−0.76
90-day maximum	274	271	25	25	−2.3	−0.84
Average Julian dates						
Of minimum	61	67				
Of maximum	318	309				
Ntem						
Minimum flows						
1-day minimum	48	34.4	51	61	−13.5	−28.3
3-day minimum	50.24	36.2	51	60	−14	−27.9
7-day minimum	53.1	38.9	51	56	−14.1	−26.5
30-day minimum	68.71	53.8	45	39	−14.8	−21.5
90-day minimum	128.4	107	33	29	−20.8	−16.2
Maximum flows						
1-day maximum	793.6	662	23	29	−131	−16.5
3-day maximum	785.4	656	23	30	−128	−16.3
7-day maximum	767.4	640	24	30	−126	−16.4
30-day maximum	685.7	567	22	32	−118	−17.2
90-day maximum	516.7	426	22	29	−90	−17.4
Average Julian dates						
Of minimum	147	102				
Of maximum	291	275				

As for the Nyong, the post-discontinuity period (1970–71 to 2015–16) is dry on the Ntem. It is characterized by a decrease in the average values of the different ranges of maximum discharges (Table 4). There are no remarkable differences between the rates of change recorded from one range to another after the discontinuity since they are all between −16.37% (for the 3-day maximum discharge) and −17.4% (for the 90-maximum discharge). The variability of these discharges ranges is modest overall but increases slightly after discontinuity. Their Cv oscillate between 22% and 24% before discontinuity, and between 29% and 32% after (Table 4). The Julian date of the maximums is early—16 days on average after the discontinuity (Table 4).

Concerning the minimum discharges, their evolution differs from that of the maximums for the Nyong. All the different ranges taken into account decreased considerably after the discontinuity, regardless of the flow duration. The rates of change recorded after the discontinuity are fairly close from one range to another, varying between −29.3% and −36.2% (Table 4). The 90-day minimum discharge is the only one for which there is a slight reduction in variability after discontinuity. For the other ranges, an increase is noted, which seems more significant for shorter-term minimums (1-day and 3-days). Their Cv went from 30% before the discontinuity to 41% after (Table 4). The minimums have occurred slightly later on the Nyong after the discontinuity. Their average Julian date of appearance increased from 61 before the discontinuity to 67 after (Table 4).

The minimum discharges of the Ntem follow the same trend as the maximums. The means of the five ranges decrease after the discontinuity. The rates of change recorded are more or less close (oscillating between −16.2% and −28.29%). However, a slight decrease in the magnitude of these changes can be noted with increasing flow duration. The decrease is greater for shorter-term minimums (1-day to 7-days). Overall, the variability is greater for shorter-term minimums. Their Cv have increased slightly after the discontinuity, while those of the longer-term minimum have decreased (Table 4). The minimums have appeared on the average 45 days earlier after the discontinuity (Table 4).

The results of this study corroborate those obtained by Liénou et al. [19] in the case of the Ntem, but differ in that of the Nyong. These authors indicated an absence of discontinuity on the Nyong annual modules between 1950s and 1990s, yet this study highlights a statistically significant decrease since 1973–74. This attests to a considerable decrease in discharges from this watershed in recent decades.

3.2. Analysis of the Main Forces Having Important Links with Discharges

3.2.1. Rainfall Variability

Annual Rainfall

The variability of annual rainfall is very moderate in the Nyong and Ntem watersheds (Table 1). A slight downward trend seems to be emerging in both cases, but the Pettitt test does not highlight any major discontinuity in their respective series (Figure 3). However, this absence of discontinuity does not necessarily imply more or fewer long dry and wet sequences in these watersheds. Analysis of the decadal deviations shows that after the slightly dry 1950s and 1960s on the Ntem, there followed three slightly wet decades, most of which had surpluses above the whole period average of around +2% (Table 3). The 1950s and 1960s were wet on the Nyong, then followed by a dry decade and two wet decades, respectively. The beginning of the 2000s marks the return of a dry period on the two watersheds, characterized by a decrease in precipitations amplified during the decade 2000 in the case of Nyong and 2010 in that of Ntem. For these decades, there are negative deviations from their average over the whole period greater than −5% (Table 3). Such deviations had never been recorded before in these watersheds.

Seasonal Rainfall

Rainfall in dry seasons changes in opposite trends (Figure 4), with an increase in summer but a decrease in winter. The Pettitt test highlights discontinuities in the respective time series of summer rainfall in Nyong and Ntem in 1964–65 and 1968–69. The resulting rates of increase from the whole period average are +5.8% and +10.5% (Figure 4). The variability of the rainfall in this season remains intact on the Nyong, while it decreases on the Ntem (Table 1). Although there has been a general increase in precipitation in these watersheds during this season since the 1960s, there is still a stabilization beginning in the 2000s, which was reinforced throughout the 2010s (Table 3).

Concerning winter rainfall, a downward discontinuity is detected in the Nyong and Ntem time series, respectively, in 1976–77 and 1977–78, with significant deficits of −16.1% and −9.7% depending on the watersheds (Figure 4). This decrease is synchronous with a change in Cv of 19% on the Nyong and 12% on the Ntem (Table 1). Though declining since the 1970s, winter precipitations seem to have increased slightly again since the beginning of the 2000s in both watersheds (Table 3).

Unlike the two dry seasons for which there are contrary trends, the rainfall of the rainy seasons (spring and autumn) show similar trends to that of the annual rainfall (Figure 4). The Pettitt test does not indicate discontinuities in their time series. There has been no continuing downward or upward trend for the entire period studied, although some dry sequences have been observed, particularly in the last three decades. The deviations observed are generally small, but slightly more pronounced in the spring, especially during the 2010s (Table 3).

These results are similar to those observed on the Ogooue further south by Mahé et al. [42], which are also found partly in the Congo Watershed [43], and on the rivers of southern Gabon and northern Congo such as Nyanga and Kouilou [19].

Relationships between Rainfall and Some Potential Sources of Variability

The changes in precipitation described above (slight decrease in annual rainfall and rainy seasons, cross-trend evolution in dry season rainfall) motivate us to the examination of the links between the rainfall in these basins and certain relevant sources of large-scale climate variability (SOI, NATL, SATL and DMI), which may influence precipitation in the region [39]. Table 5 presents the correlations between the anomalies of the representative climatic indices of these sources of variability and those of rainfall in the basins studied at annual and seasonal time steps. This analysis shows significant correlations in the two basins between the anomalies of the NATL indices and those of rainfall during spring but also between the anomalies of these same indices and those of rainfall in winter. Important links have also been demonstrated between the rainfall anomalies of these watersheds and those of the SOI in winter.

Table 5. Correlation between rainfall indices and climatic indices at annual and seasonal time steps. Significant correlations are in bold.

Watersheds	Indices	Annual	Spring	Summer	Autumn	Winter
Ntem	DMI	−0.05	−0.15	0.22	−0.1	−0.06
	NATL	−0.03	**−0.52**	−0.04	−0.02	**0.58**
	SATL	−0.06	−0.05	0.06	−0.2	0.43
	SOI	0.03	−0.17	0.12	−0.09	**0.51**
Nyong	DMI	−0.23	−0.09	0.09	−0.23	−0.02
	NATL	−0.17	**−0.54**	−0.09	−0.11	**0.52**
	SATL	0	−0.08	0.14	−0.12	0.36
	SOI	0.21	−0.04	0.1	0.16	**0.53**

NATL indices are negatively correlated with rainfall during spring (Table 5). This implies that warming (cooling) of the Tropical North Atlantic is associated with the low (heavy) precipitation over these watersheds during this season (Figure 5). This configuration was sometimes taken by default on the two watersheds, in particular in 1986–87, 1990–91, 1993–94, 1996–97, 2010–11, etc. Opposite signs of NATL indices and the rainfall anomalies were, however, noted for more than 70% of the years studied in each of these watersheds (Table 6).

Table 6. Frequency of years having the same/opposite signs of ocean indices and rainfall anomalies. NY: Number of years; NYOS: Number of years with opposite signs; NYSS: Number of years with same signs.

Watersheds	NATL&Rainfall (Spring)			NATL&Rainfall (Winter)			SOI&Rainfall (Winter)		
	NY	NYOS	%	NY	NYSS	%	NY	NYSS	%
Nyong	34	24	71	34	23	62	65	35	54
Ntem	33	23	70	33	20	61	64	42	66

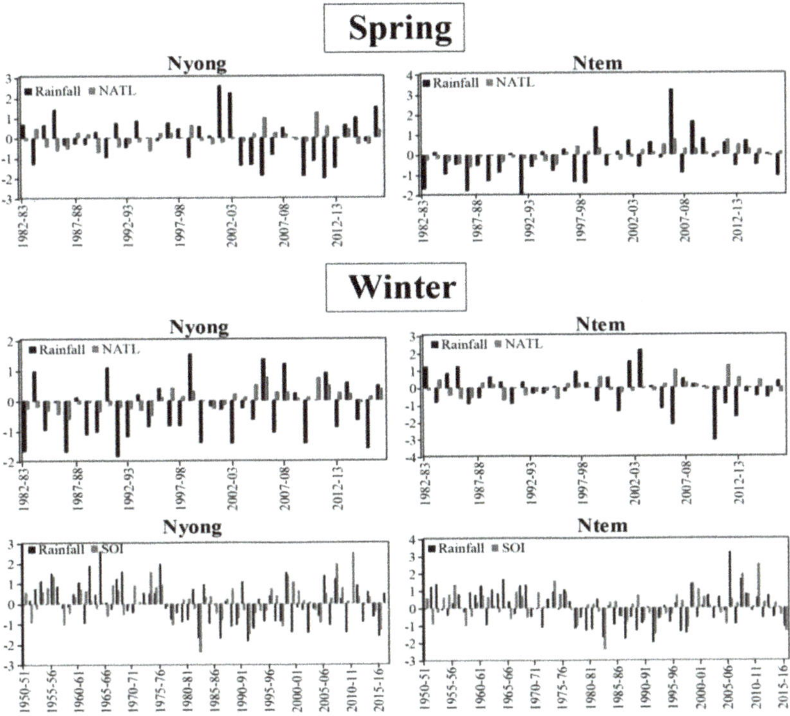

Figure 5. Comparison of climatic indices and normalized rainfall anomalies.

The NATL indices, on the other hand, are positively correlated with winter rainfall (Table 5). An increase in the temperature of the Tropical North Atlantic would be concomitant with abundant rainfall in the watersheds studied during this season and conversely (Figure 5). A different situation was nevertheless observed in these watersheds in 1990–91, 1996–97, 1997–98, 2004–05, 2015–16, etc. The proportion of years for which the same sign of anomalies was observed between these two variables during winter is 62% in the Nyong basin and 61% in that of Ntem (Table 6).

As in the previous case, the SOI is positively correlated with the rainfall in the watersheds studied in winter (Table 5). This implies that a warming of the Pacific Ocean is synchronous with the generally abundant rainfall during this season (Figure 5). This situation was not, however, observed for certain years (1950–51, 1963–64, 1968–69, 1970–71, 1984–85, 1990–91, 1996–97, etc.). The percentage of years having recorded the same sign of anomalies between rainfall and SOI in winter is 54% for the Nyong and 66% for the Ntem. Correlation between rainfall and SOI anomalies is good in both watersheds in winter, but the proportion of years with the same sign of anomalies does not give enough clear information in the case of the Nyong.

The fact that winter rainfall is correlated with both the temperatures of the Tropical North Atlantic and the Pacific Ocean indicates a probable simultaneous control of these two oceans on the precipitation of the whole region during the winter.

Bogning et al. [44] recently demonstrated that an important relationship exists between the March-April-May (MAM) rainfall and the tropical Atlantic Niño. According to them, the negative temperature anomalies in the south-eastern equatorial Atlantic are at the origin of the positive anomalies of the MAM rainfall in the Ogooue watershed. The negative correlation coefficient between these two variables is high (r = −0.74). This study also highlights a relatively strong link (r = −0.52 and −0.54) between the spring rainfall of two equatorial watersheds located further north of the Equator and the temperatures of the Atlantic ocean, with the difference that the link with the rainfall of this season has been established this time with the temperatures of the Tropical North Atlantic of the same season.

3.2.2. Land Use Dynamics in the Studied Watersheds

A diachronic analysis of the classifications carried out on two dates (1973 and 2015) shows for both watersheds changes in land use marked not only by an increase in IA but also by a decrease in forests and water bodies (Figure 6). Between 1973 and 2015, forests and water bodies decreased by 6% and 42%, respectively, in the Nyong watershed. Their decreases on the Ntem between 1973 and 2015 are 1% and 32%. Between these same years, the increase in IA reached 771% in the Nyong watershed and 451% in the Ntem watershed (Table 7). Figure 6 also shows that the disappearance of the forest cover is localized around urban centers, mainly located in the west, near Ngoazik in the case of Ntem. For the Nyong, deforestation is certainly continuous throughout the watershed, but the change is more significant in the west, with the important boom in the city of Yaoundé, not far from Mbalmayo.

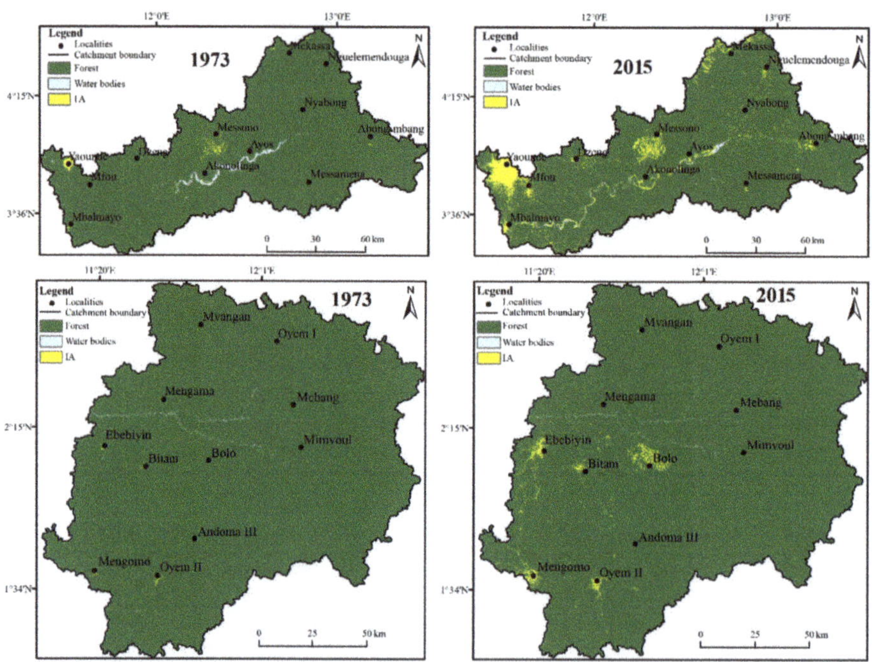

Figure 6. Spatial distribution of the main land use patterns in the studied watersheds at two dates.

Table 7. Evolution of the main land use modes in the Nyong and Ntem watersheds.

Basins	Land Use Modes	Area Occupied in the Basin (km^2)		Change	
		1973	2015	km^2	%
Nyong	Forest	14,174	13,326	−848	−6
	Water bodies	146	84	−62	−42
	IA	118	1028	910	771
Ntem	Forest	19,469	19,301	−168	−1
	Water bodies	53	36	−17	−32
	IA	41	226	185	451

Moffo [25] and Ebodé [26] obtained similar results, respectively, in the Mfoundi (Mefou sub-watershed) and Mefou (Nyong sub-watershed) watersheds.

3.3. The Impacts of Environmental Forces on the Discharges

3.3.1. Rainfall

The impact of rainfall is perceptible in the evolution of flows. In general, the annual precipitation of the two watersheds studied does not decrease significantly, while the flows register downwards discontinuities during the 1970s. The stationarity of the hydrological regime and that of the precipitations of the watersheds at this time step thus departs from the general logic, which admits that hydrological discontinuities occur following rainfall discontinuities. However, for both watersheds, the analysis of the decadal deviations from the long-term average of the two compared variables shows a synchronous evolution between them over the last two decades. For the latter, there is a concomitant decrease in rainfall and flows (Table 3).

During the rainy seasons, primarily in the spring, precipitation has changed moderately, so that no sudden mean inflexion point has been detected in their series. In return, the discharges and the runoff coefficients for this season have decreased significantly since the 1970s. Similarly, the examination of the deviations from the mean of these variables generally shows significant decreases for the discharges and the runoff coefficients, while the rainfall does not decrease or decreases very slightly (Table 3). The decline in winter rainfall since the 1970s appears to be closely related in amplitude to that of spring flows. Although the correlations between them are not significant (r < 0.32 on both catchments), the rhythm of evolution of these two variables over the decades is fairly constant. This is observed during the 2010s, for which a small drop in winter rainfall led to a similar evolution in spring runoff coefficients and discharges (Table 3). This influence of winter precipitation has already been suggested by Liénou et al. [19]. These authors wrote that from the 1970s, the decrease in winter rainfall created a greater water deficit (evaporation and soil water reserve) in the watershed area at the start of the first rainy season in spring. This aridification of winter means that a greater part of the precipitation received during the spring first contributes to filling this water deficit, and therefore, the fraction that generates runoff is reduced. This then translates into lower runoff volumes for the same average depths of precipitation during spring, which explains the drop in the runoff coefficient (Figure 4).

In autumn, rainfall and flows decrease very slightly over the entire study period. However, between the 1970s and 1980s, flows and runoff coefficients increased, while the rainfall did not vary much overall. This increase would be due to the increase in summer rainfall. The impact of Summer rainfall on autumn flows is more noticeable during the 2000s and 2010s, for which there is a joint decrease in summer rainfall, discharges and runoff coefficients of autumn, despite a sometimes-negligible variation in autumn rainfall in the different watersheds (Table 3). In view of these developments, it appears that the increase in summer rainfall during the decades from 1970 to 1990 considerably reduces the deficit of evaporation and soil water reserves at the beginning of the autumn, favoring runoff [19]. The part of the precipitation that actually participates in the runoff increases, hence, there is an increase in the autumn runoff coefficient during these decades (Table 3). The decrease in summer rainfall during the 2000s and 2010s has the opposite effect, which is why there has been a reduction in runoff coefficients during these decades, responsible for a decrease in discharges. This is all the more probable since the analysis of the average monthly flows shows that the variations of the summer rainfall are generally felt more on the flows of the first month (September) of autumn. The model developed for this purpose is significant, with a fairly correct quality of fit (r > 0.6) on the two basins (Figure 7). It could be used to predict fairly accurately the runoff in early autumn (September) based on the rainfall recorded during the summer.

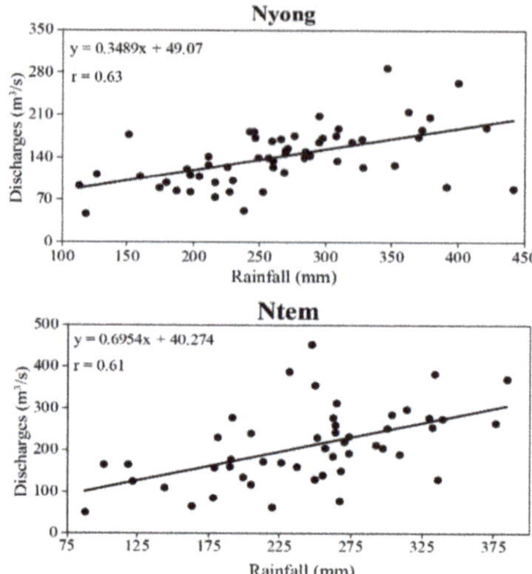

Figure 7. Regression between September (first month of autumn) discharges and summer (July and August) rainfall.

In the summer, we notice that rainfall increases statistically, while flows evolve according to varying magnitudes downwards. These opposite trends could be explained by the fact that the summer flows are partly linked to the volumes precipitated during the spring, which occurs just before, even if the correlation between them is not significant ($r < 0.35$ on both basins). This is why we can note, in the evolution of summer flows, dry decades even when the rainfall was abundant. This was the case in the 1980s, when there was a surplus above average for the summer rainfall, but shortages for both the spring rainfall and the summer flows. However, the variations in summer precipitation themselves seem to also be related to the flows in the season. They participate in the event of a moderate evolution of the spring rainfall, in the maintenance of flows in some cases, or in their reduction in others. This last scenario is observed during the 2010s, where spring rainfall did not vary considerably, but where a decrease in summer rainfall is concomitant with a decrease in flows in the different watersheds (Table 3).

In winter, precipitation and flows decrease according to different magnitudes. The stationarity of these two variables does not reveal an apparent link in the case of Nyong. Additionally, the correlations between them are weak ($r < 0.39$). However, the analysis of the deviations from the average shows clearer links between these variables, since the beginning of the drought started during the 1970s in these two watersheds (Table 3). The succession of deficit rainfall sequences that were observed between the 1970s and 1990s also led to deficits in terms of flows. Similarly, the resumption of rainfall observed during the 2000s caused an upsurge in the flows, which was interrupted again in the two watersheds studied during the 2010s, following the relapse of the rainfall (Table 3).

Several common and divergent points can be noted in the evolution of rainfall and its impact on discharges during the different seasons in the two studied watersheds. Spring discharges have decreased since the beginning of the 1970s, more markedly for the Ntem in the South of Cameroon close to the Equator, while rainfall has not decreased significantly, even though since 2005, several years of slightly weaker spring rainfall have been recorded. In summer there is a net increase in rainfall in the two watersheds from the early 1980s to the late 2000s, then a return to less abundant rainfall. Summer discharges do not seem to follow a variation related to this variability in rainfall. In autumn the rainfall does not drop on the Ntem, except for a few years of less abundance during the most recent decades, while for the Nyong the rainfall has decreased slightly and fairly regularly since the end of the 1980s. Diminution in these watersheds would result from their respective belonging to the equatorial (Ntem) and sub-equatorial (Nyong) domain, which is already at the origin of the significant differences on the duration of each season on these. The discharges seem to have decreased slightly for the Nyong, since the 1990s, in apparent link with the slight decrease of the rainfall, while the discharges of the Ntem seem to decrease also since the middle of the 1990s, without the rainfall having decreased significantly. It should be noted that the period of the possible change in discharges also corresponds to a period of significant gaps in discharge data series, which does not allow great significance for these analyses. In winter the series of rainfall of the two watersheds show a discontinuity in 1976–77, which separates the wet period from the recent drier period. Although during the 2000s the rainfall became close to the totals for the wet period, the discharges of the two rivers followed the same variation, with values generally lower after 1970—even in the 2000s, when the rainfall apparently increased.

The results, obtained in this study from recent data, corroborate the first observations of change in the hydrological regime in the region [19]. These changes attest to the visible and lasting impact of a change in the rainfall regime which began in the late 1970s, and which have major repercussions on hydrological regimes and the seasonal availability of surface water resources.

3.3.2. Changes in Land Use Patterns

The changes in land use patterns seem to have more or less significant links with certain hydrological modifications observed. The slightly early arrival of the maximum noted at the outlets of the two watersheds (Table 4), in a context where rainfall is decreasing slightly, seems to be related to the increase in IA (+771% for the Nyong and +451% for the Ntem) and the diminution in forests (−6% for the Nyong and −1% for the Ntem). These changes could reduce the time it takes for the maximums to appear by increasing the runoff they cause. The stability observed for the Nyong maximums after the discontinuity (Table 4) could also be a consequence of these same developments. Indeed, in a context where the rainy seasons rainfall, which generate the maximum flows, decrease, the most logical would have been to see the maximums decrease, which is, however, not the case. The current modest urbanization rate in this watershed seems to be the most relevant factor to justify this lack of observed trend. In this case, the decrease in precipitation appears to have been offset by the increased runoff. The annual runoff coefficients of this watershed have increased significantly since the 1980s (Table 3), following an accelerated urbanization phase in the western part of this watershed (Yaoundé region) over the 1980s and 1990s, following the subdivision operations undertaken since the end of the 1970s by the municipal authorities [45]. The decrease in the water bodies noted could have contributed to that of the minima in the two watersheds. When there are in large number, water bodies contribute to the supply of rivers during low water periods and contribute to the maintenance of low flows [46]. However, otherwise, their support for the flows decreases, which causes a decrease in minimum flows.

The impact of the increase in the IA on runoff has already been demonstrated in other forest environments (South America), particularly in the cases of certain watercourses such as the Amazon [47], Upper Xingu [48] and Parana [49]. The discharges of these watercourses have indeed increased, following a reduction in the forest area of their watersheds, which seem relatively greater than that observed in the Nyong and Ntem watersheds. The rate of deforestation in the entire Amazon basin, for example, reaches 30% [47].

Certainly, there is reason to wonder whether such small changes in land use could cause changes in extreme flows as visible as those observed in the watersheds investigated. Since the evolution of these extremes does not agree with other phenomena that have a proven impact on flows, in this case the decrease in precipitation highlighted in this study and the rise in temperatures postulated by Sighomnou [50] and Amougou et al. [51] in the region, which act rather in the direction of a reduction in flows and a late observation of the maximums, it seems appropriate to think that the changes in land use, although modest overall, might be the main cause of these evolutions. Their action on flows could have been considerably amplified by their location in the watersheds (to the west part, near the outlets of the watersheds studied), which would favor the rapid routing of rainwater bodies with the least possible losses to the outlets, thereby erasing the size and forest cover effect of these watersheds.

If this situation persists, there is every reason to believe that there will be an increase in maximum flows in the Nyong within a few years, despite the decrease in precipitation.

To be complete, a further study should also analyze the evolution of daily rainfall and dry days within the rainy season, and even the possible changes in the frequency and intensity of rainfall events, which may have significant impact on runoff, as is the case in the Sahel, for instance [52].

4. Discussion: Analysis of Results in the Regional Context

The results of this study have several points in common with similar studies already carried out in Central and West Africa. This study highlights a slight decrease in annual rainfall in the watersheds of Central Africa, rainy seasons and significant changes, but in opposite phases of rainfall in the dry seasons (decrease for winter and increase for summer). The same observation has already been made in other studies carried out in the equatorial region [19]. Discontinuities marking a drop in discharges were highlighted in this study through statistical tests. This corroborates other studies in the region in which the authors have reported discontinuities in the hydrological time series of the Ntem in 1970 [19] and the Ogooue in 1977 [53]. The variability of the flows highlighted for these rivers on a seasonal level reveals a different evolution from the spring and autumn floods. Spring floods decrease significantly compared to those of autumn, which vary little overall. Similar changes have been highlighted by Mahé et al. [42] on Ogooue and Kouilou. In general, this decrease in spring and autumn floods in the equatorial zone is associated with a decrease in rainfall during these seasons which would itself result, not only from anomalies in ocean surface temperatures [54–56], but also local zonal wind anomalies [55,56].

This study reveals changes in land use patterns over time in studied watersheds, which are mainly marked by a decrease in forest and water bodies, and conversely an increase in IA (built, bare soil and agricultural areas). Similar observations had already been made by other authors in Central [26] and West Africa [57–59]. One of the main results of this work reveals that the observed deforestation is greater around urban centers. The same remark was made in Niger by Amogu et al. [23]. They claim in their study that the capital Niamey and its surroundings have experienced more deforestation than the rest of the study area. They add that the causes of this deforestation are essentially linked to the cutting of wood for domestic use.

A few studies in Central Africa [25,26] and several in West Africa [21–24,60] have shown an increase in average discharges, following an increase in IA in the watersheds of certain rivers. This is not the case in this work. Despite the increase in IA, average discharges remain low. This could be linked to the size of the catchments, but also their large forest cover. However, Li et al. [61] stated in their study that IA could not influence the flow rates of a watershed until it reached 50% of its total area. This may be in disagreement with the observation made in this study. Urbanized areas occupy less than 10% of the total surface of the studied watersheds, but their probable effects are still perceptible on extreme flows. The proximity of land-use changes to the outlets of that watersheds could explain the impact of such modest urbanization rates on flows. As ground observations data are sparse and mostly lacking for many rivers in the region, further improvements could come from satellite derived data like for the Ogooue [44] and from a better and exhaustive track of human sources of water abstraction as in the case of the Logone River in Central Africa [62].

5. Conclusions

In this paper we studied the possible impact of climate change and anthropization on river flows of two basins of Central Africa (Nyong and Ntem in South Cameroon), from long-term rainfall and discharges time series and changes in land cover over time from satellite images. Considering the rainfall, the 1970s and 1980s marked the start of a major change in the rainfall pattern during the dry seasons in Central Africa, characterized by an increase in rainfall totals in summer and a decrease in winter. These changes, which prevailed until the end of the 1990s, led to a change in the flows of the rainy seasons that directly follow them. Thus, there has been an increase in autumn flows and a drop in winter flows over the same period. The beginning of the 2000s was marked by a reversal of trends. In the more recent years, there is now a significant drop in rainfall in summer accompanied by a decrease in autumn flows, and conversely, an increase in winter rainfall concomitant with an increase in spring flows, which nevertheless seems rather small on the Nyong. The combined effect of the decrease in rainfall (−16% for Nyong and −8% for Ntem) during the winter and water bodies (−42% in the case of Nyong and −20% in that of Ntem) are the most relevant elements that can be associated to the decreases observed in minimum flows. Considering the survey of the land cover change, the increase of 771% in IA to the detriment of the forest seems to have contributed to the maintenance of maximum flows in the Nyong watersheds, although in a context marked by a deficit in rainfall. Such changes favor an accentuation of the runoff, which modifies the rhythm of the flows, especially those of short duration. This particular aspect could be further investigated by studying the changes in daily rainfall. Eventually, the fluctuations of the average discharges of the rivers of Central Africa depend for the moment mainly on rainfall; however, those of the extreme flows are the result of the combined effect of the two forces analyzed (precipitation and anthropization).

Author Contributions: Conceptualization: V.B.E., G.M., J.G.D. and J.A.A.; methodology: J.G.D., J.A.A. and G.M.; software: V.B.E.; validation: G.M., J.G.D. and J.A.A.; formal analysis: V.B.E. and G.M.; investigation: V.B.E.; data curation V.B.E.; writing—original draft preparation: V.B.E. and G.M.; writing—review and editing: V.B.E. and G.M.; project administration: G.M.; funding acquisition: G.M. All authors have read and agreed to the published version of the manuscript.

Funding: This research received no external funding.

Acknowledgments: The authors wish to thank the Cameroonians authorities and national services of Meteorology and Hydrology, for providing data to the Joint International Laboratory (LMI) DYCOFAC. The authors also thank the National Observing System M-TROPICS (Multiscale TROPIcal CatchmentS) from France, which subsidizes observations in experimental tropical watersheds. The authors warmly thanks the agents of the IRD (Research Institute for Developping countries) Représentation in Cameroon and to the direction of the LMI DYCOFAC, at Yaounde, for their administrative and financial support. The authors also wish to address a memoriam to Ms. Claudine Dieulin, from the IRD, who passed away late January, and who spent years in developing the open-source database HSM-SIEREM, dedicated to the African hydrology and climatology, and which data have been used partly in this study.

Conflicts of Interest: The authors declare no conflict of interest.

References

1. Dzana, J.G.; Ndam, N.J.R.; Tchawa, P. The Sanaga discharge at Edea catchment outlet (Cameroon): An example of tropical rain-fed river system to change in precipitation and groundwater input and to flow regulation. *River Res. Appl.* **2010**, *27*, 754–771. [CrossRef]
2. Chu, M.L.; Knouft, J.H.; Ghulam, A.; Guzman, J.A.; Pan, Z. Impacts of urbanization on river flow frequency: A controlled experimental modeling-based evaluation approach. *J. Hydrol.* **2013**, *495*, 1–12. [CrossRef]
3. Rosburg, T.T.; Nelson, P.A.; Bledsoe, B.P. Effects of urbanization on flow duration and stream flashiness: A case study of Puget sound stream, Western Washington, USA. *JAWRA J. Am. Water Resour. Assoc.* **2017**, *53*, 493–507. [CrossRef]
4. Aulenbach, B.T.; Landers, M.N.; Musser, J.W.; Painter, J.A. Effects of impervious area and BMP implementation and design on storm runoff and water quality on eight small watersheds. *JAWRA J. Am. Water Resour. Assoc.* **2017**, *53*, 382–399. [CrossRef]
5. Diem, J.E.; Hill, T.C.; Milligan, R.A. Diverse multi-decadal changes in streamflow within a rapidly urbanizing region. *J. Hydrol.* **2018**, *556*, 61–71. [CrossRef]
6. Oudin, L.; Salavati, B.; Furusho-Percot, C.; Ribstein, P.; Saadi, M. Hydrological impacts of urbanization at the catchment Scale. *J. Hydrol.* **2018**, *559*, 774–786. [CrossRef]
7. Gibson, C.A.; Meyer, J.L.; Poff, L.E.; Georgakakos, A. Flow regime alterations under changing climate in two river basins: Implications for freshwater ecosystems. *River Res. Appl.* **2005**, *21*, 849–864. [CrossRef]
8. Yira, Y.; Diekkrüger, B.; Steup, G.; Aymar, Y.B. Impact of climate change on hydrological conditions in a tropical West African catchment using an ensemble of climate simulations. *Hydrol. Earth Syst. Sci.* **2017**, *21*, 2143–2161. [CrossRef]

9. Namugize, J.N.; Jewitt, J.; Graham, M. Effects of land use and land cover changes on water quality in the Umngeni river catchment, South Africa. *Phys. Chem. Earth Parts A/B/C* **2018**, *105*, 247–264. [CrossRef]

10. Gorgoglione, A.; Gregorio, J.; Rios, A.; Alonso, J.; Chreties, C.; Fossati, M. Influence of land use/land cover on surface-water quality of Santa Lucìa river, Uruguay. *Sustainability* **2020**, *12*, 4692. [CrossRef]

11. Getahun, Y.S.; Li, M.H.; Chen, P.Y. Assessing impact of climate change on hydrology of Melka Kuntrie Subbasin, Ethiopia with Ar4 and Ar5 projections. *Water* **2020**, *12*, 1308. [CrossRef]

12. Olivry, J.C.; Bricquet, J.P.; Mahé, G. Vers un appauvrissement durable des ressources en eau de l'Afrique humide. In *Hydrology of Warm Humid Regions, 4ème Assemblée IAHS, Yokohama, Japon, 13-15 Juillet 1993*; Gladwell, J.S., Ed.; IAHS Press: Wallingford, UK, 1993; pp. 67–78.

13. Bricquet, J.P.; Bamba, F.; Mahé, G.; Touré, M.; Olivry, J.C. Evolution récente des ressources en eau de l'Afrique atlantique. *Rev. des Sci. L'eau* **2005**, *10*, 321–337. [CrossRef]

14. Servat, E.; Paturel, J.E.; Lubès-Niel, H.; Kouamé, B.; Masson, J.M.; Travaglio, M.; Marieu, B. De différents aspects de la variabilité de la pluviométrie en Afrique de l'Ouest et Centrale. *Rev. des Sci. L'eau* **2005**, *12*, 363–387. [CrossRef]

15. Mahé, G.; L'Hôte, Y.; Olivry, J.C.; Wotling, G. Trends and discontinuities in regional rainfall of West and Central Africa—1951–1989. *Hydrol. Sci. J.* **2001**, *46*, 211–226. [CrossRef]

16. Buisson, A. La grande saison sèche 1985 au Gabon. Situation climatique en Afrique intertropicale. *La Météorologie* **1985**, *15*, 5–13.

17. Mahé, G.; Lerique, J.; Olivry, J.C. L'Ogooué au Gabon. Reconstitution des débits manquants et mise en évidence de variations climatiques à l'équateur. *Hydrol. Cont.* **1990**, *5*, 105–124.

18. Bigot, S.; Moron, V.; Melice, J.L.; Servat, E.; Paturel, J.E. Fluctuations pluviométriques et analyse fréquentielle de la pluviosité en Afrique centrale. In *Water Ressources Variability in Africa during the XXth Century, Abidjan, Côte d'ivoire, Novembre 1998*; Servat, E., Hughes, D., Fritsch, J.M., Hulme, M., Eds.; IAHS Press: Wallingford, UK, 1998; pp. 71–78.

19. Lienou, G.; Mahé, G.; Paturel, J.E.; Servat, E.; Sighomnou, D.; Ekodeck, G.E.; Dezetter, A.; Dieulin, C. Evolution des régimes hydrologiques en région équatoriale camerounaise: Un impact de la variabilité climatique en zone équatoriale? *Hydrol. Sci. J.* **2008**, *53*, 789–801. [CrossRef]

20. Kouassi, A. Caractérisation d'une Modification Eventuelle de la Relation Pluie-Débit et ses Impacts Sur Les Ressources en eau en Afrique de l'Ouest: Cas du Bassin Versant du N'zi (Bandama) en Côte d'Ivoire. Ph.D. Thesis, Université de Cocody, Abidjan, Côte d'ivoire, 2007.

21. Souley, Y.K. L'evolution De L'occupation Des Sols A L'echelle Des Bassins Versants De Wankama Et Tondi Kiboro: Quelles Consequences Sur Les Debits Et L'evapotranspiration Reelle (Etr). Master Thesis, Université Abdou Moumouni, Niamey, Niger, 2008.

22. Leblanc, M.; Favreau, G.; Massuel, S.; Tweed, S.; Loireau, M.; Cappelaere, B. Land clearance and hydrological change in the Sahel: South-west Niger. *Glob. Planet. Chang.* **2008**, *61*, 49–62. [CrossRef]

23. Amogu, O.; Descroix, L.; Yéro, K.S.; Le Breton, É; Mamadou, I.; Ali, A.; Vischel, T.; Bader, J.-C.; Moussa, I.B.; Gautier, E.; et al. Increasing river flows in Sahel? *Water* **2010**, *2*, 170–199. [CrossRef]

24. Paturel, J.E.; Mahé, G.; Diello, P.; Barbier, B.; Dezetter, A.; Dieulin, C.; Karambiri, H.; Yacouba, H.; Maiga, A. Using land cover changes and demographic data to improve hydrological modeling in the Sahel. *Hydrol. Process.* **2017**, *31*, 811–824. [CrossRef]

25. Moffo, Z.M. Contribution Des Systemes D'information Geographiques Pour La Cartographie Des Zones A Risques A Yaounde: Application Au Bassin Du Mfoundi. Master's Thesis, Université de Yaoundé I, Yaounde, Cameroon, 2011.

26. Ebodé, V.B. Etude De La Variabilite Hydroclimatique Dans Un Bassin Versant Forestier En Voie D'urbanisation Acceleree: Le Cas De La Mefou. Master's, Université de Yaoundé I, Yaounde, Cameroun, 2017. Master's Thesis, Université de Yaoundé I, Yaounde, Cameroun, 2017.

27. Dezetter, A.; Paturel, J.E.; Ruelland, D.; Ardoin-Bardin, S.; Ferry, L.; Mahé, G.; Dieulin, C.; Servat, E. Prise en compte des variabilités spatio-temporelles de la pluie et de l'occupation du sol dans la modélisation semi-spatialisée des ressources en eau du haut fleuve Niger. In *Global Change: Facing Risks and Threats to Water Resources, Proceedings of the Sixth World FRIEND Conference, Fez, Morocco, 25–29 October 2010*; IAHS Press: Wallingford UK, 2010; pp. 544–552.

28. Fritsch, J.M. Les Effets Du Defrichement De La Foret Amazonienne Et De La Mise En Culture Sur L'hydrologie De Petits Bassins Versants. Ph.D. Thesis, Université de Montpellier II, Montpellier, France, 1990.

29. BUCREP. *Rapport de présentation des résultats définitifs du recensement de la population en 2005*; BUCREP: Yaoundé, Cameroun, 2011.

30. Olivry, J.C. *Fleuves et Rivieres du Cameroun*; MESRES-ORSTOM: Paris, France, 1986.

31. Letouzey, R. *Notice de la Carte Phytogéographique du Cameroun au 1/500000*; Institut de la carte internationale de la végétation: Toulouse, France, 1985.

32. Boyer, J.F.; Dieulin, C.; Rouché, N.; Crès, A.; Servat, E.; Paturel, J.E.; Mahé, G. SIEREM: An environmental information system for water resources. In *Water Resource Variability–Hydrological Impacts, Proceedings of the Fifth FRIEND World Conference, Havana, Cuba, 26 November 2006*; IAHS Press: Wallingford, UK, 2006; pp. 19–25.

33. Rouché, N.; Mahé, G.; Ardoin-Bardin, S.; Brissaud, B.; Boyer, J.F.; Crès, A.; Dieulin, C.; Bardin, G.; Commelard, G.; Dezetter, A.; et al. Constitution d'une grille de pluies mensuelles pour l'Afrique (période 1900-2000). *Sécheresse* **2010**, *21*, 336–338. [CrossRef]

34. Paturel, J.E.; Boubacar, I.; L'Aour, A.; Mahé, G. Analyses de grilles pluviométriques et principaux traits des changements survenus au 20ème siècle en Afrique de l'Ouest et Centrale. *Hydrol. Sci. J.* **2010**, *55*, 1281–1288. [CrossRef]

35. Dieulin, C.; Mahé, G.; Paturel, J.E.; Ejjiyar, S.; Tramblay, Y.; Rouché, N.; Mansouri, B.E. A new 60-year monthly-gridded rainfall data set for Africa. *Water* **2019**, *11*, 387. [CrossRef]

36. Gleick, P.H. Methods for evaluating the regional hydrologic impacts of global climatic changes. *J. Hydrol.* **1986**, *88*, 97–116. [CrossRef]

37. Arnell, N.W.; Reynards, N.S. The effects of climate change due to global warming on river flows in Great Britain. *J. Hydrol.* **1996**, *183*, 397–424. [CrossRef]

38. Huffman, G.J.; Adler, R.F.; Bolvin, D.T.; Gu, G.; Nelkin, E.J.; Bowman, K.P.; Hong, Y.; Stocker, E.F.; Wolff, D.B. The TRMM Multisatellite Precipitation Analysis (TMPA): Quasi-Global, Multiyear, Combined-Sensor Precipitation Estimates at Fine Scales. *J. Hydrom.* **2007**, *8*, 38–55. [CrossRef]

39. Bush, E.R.; Jeffery, K.; Bunnefeld, N.; Tutin, C.; Musgrave, R.; Moussavou, G.; Mihindou, V.; Malhi, Y.; Lehmann, D.; Edzang, N.J.; et al. Rare ground data confirm significant warming and drying in western equatorial Africa. *PeerJ* **2020**, *8*, e8732. [CrossRef]

40. Lubès, H.; Masson, J.M.; Servat, E.; Paturel, J.E.; Kouame, B.; Boyer, J.F. *Caractérisation des fluctuations dans une série chronologique par applications de tests statistiques. Etudes bibliographiques*; Programme ICCARE, Rapport n3. ORSTOM: Montpellier, France, 1994.

41. Richter, B.D.; Baumgartner, J.V.; Braun, D.P.; Powell, J. A spatial assessment of hydrologic alteration within river network. *Regul. Rivers Res. Mgmt.* **1998**, *39*, 329–340. [CrossRef]

42. Mahé, G.; Lienou, G.; Descroix, L.; Bamba, F.; Paturel, J.E.; Laraque, A.; Meddi, M.; Habaieb, H.; Adegea, O.; Dieulin, C.; et al. The rivers of Africa: Witness of climate change and the human impact on the environment. *Hydrol. Process.* **2013**, *27*, 2105–2114. [CrossRef]

43. Tsalefac, M.; Hiol Hiol, F.; Mahe, G.; Laraque, A.; Sonwa, D.; Sholte, P.; Pokam, W.; Haensler, A.; Beyene, T.; Ludwig, F.; et al. Climate of Central Africa: Past, present and future. In *The Forests of the Congo Basin. Forests and Climate Change*; De Wasseige, C., Tadoum, M., Eba'a, A.R., Doumenge, C., Eds.; Weyrich: Neufchâteau, Belgium, 2015; Volume 2, pp. 37–52.

44. Bogning, S.; Frappart, F.; Paris, A.; Blarel, F.; Ninõ, F.; Picart, S.S.; Lanet, P.; Seyler, F.; Mahé, G.; Onguene, R.; et al. Hydro-climatology study of the Ogooué River basin using hydrological modeling and satellite altimetry. *Adv. Space Res.* **2020**. [CrossRef]

45. Dzana, J.G.; Amougou, J.A.; Onana, V. Modélisation spatiale des facteurs d'aggravation des écoulements liquides à Yaoundé. Application au bassin versant d'Akë. *Mosella* **2004**, *29*, 78–91.

46. Tardif, S. Regionalisation Et Facteurs De La Variabilite Spatiale Des Debits Saisonniers Et Extremes Journaliers Au Quebec Meridional. Master's Thesis, Université du Québec, Québec, Canada, 2005.

47. Coe, M.T.; Costa, M.H.; Soares-Filho, B.S. The influence of historical and potential future deforestation on the streamflow of the Amazon river—Land surface processes and atmospheric feedbacks. *J. Hydrol.* **2009**, *369*, 165–1774. [CrossRef]

48. Dias, L.C.; Macedo, M.N.; Costa, M.H.; Coe, M.T.; Neil, C. Effects of land cover change on evapotranspiration and streamflow of small catchments in the Upper Xingu river basin, Central Brazil. *J. Hydrol.* **2015**, *4*, 108–122. [CrossRef]

49. Lee, E.; Livino, A.; Han, S.C.; Zhang, K.; Briscoe, J.; Kelman, J.; Moorcroft, P. Land cover change explains the increasing discharge of the Paraná river. *Reg. Env. Chang.* **2018**, *18*, 1871–1881. [CrossRef]

50. Sighomnou, D. Analyse Et Redefinition Des Regimes Climatiques Et Hydrologiques Du Cameroun: Perspectives D'evolution Des Ressources En Eau. Ph.D. Thesis, Université de Yaoundé I, Yaoundé, Cameroun, 2004.

51. Amougou, J.A.; Ndam, N.J.R.; Djocgoue, P.F.; Bessoh, B.S. Variabilité climatique et régime hydrologique dans un milieu bioclimatique de transition: Cas du bassin fluvial de la Sanaga. *Afr. Sci.* **2015**, *11*. Available online: https://www.semanticscholar.org/paper/Variabilit%C3%A9-climatique-et-r%C3%A9gime-hydrologique-dans-Amougou-Ngoupayou/614a4602790ca1b9c87683c0bb1234ffdbfd7551 (accessed on 15 September 2020).

52. Panthou, G.; Lebel, T.; Vishel, T.; Quantin, G.; Sane, Y.; Ba, A.; Ndiaye, O.; Diongue-Niang, A.; Diopkane, M. Rainfall intensification in tropical semi-arid regions: The Sahelian case. *Environ. Res. Lett.* **2018**, *13*. [CrossRef]

53. Conway, D.P.; Persechino, A.; Ardoin–Bardin, S.; Hamandawana, H.; Dieulin, C.; Mahé, G. Rainfall and river flow variability in sub-saharan Africa during the 20th century. *J. Hydrom.* **2009**, *10*, 41–59. [CrossRef]

54. Mahé, G.; Citeau, J. Interactions between the Ocean, Atmosphere and Continent in Africa, Related to the Atlantic Monsoon Flow: General Pattern and the 1984 Case Study. *Veille Clim. Satell.* **1993**, *44*, 34–54.

55. Nicholson, E.N.; Dezfuli, A.K. The relationship of rainfall variability in western equatorial Africa to the tropical oceans and atmospheric circulation. Part I: The boreal Spring. *J. Clim.* **2013**, *26*, 45–65. [CrossRef]

56. Nicholson, E.N.; Dezfuli, A.K. The relationship of rainfall variability in western equatorial Africa to the tropical oceans and atmospheric circulation. Part II: The boreal Autumn. *J. Clim.* **2013**, *26*, 66–84. [CrossRef]

57. Mahé, G.; Leduc, C.; Amani, A.; Paturel, J.E.; Girard, S.; Servat, E.; Dezetter, A. Augmentation récente du ruissellement de surface en région soudano-sahélienne et impact sur les ressources en eau. In *Hydrology of the Mediterranean and Semiarid Regions*; Servat, E., Najem, W., Leduc, C., Shakeel, A., Eds.; IAHS Press: Wallingford, UK, 2003; pp. 215–222.

58. Kergoat, L.; Hiernaux, P.; Baup, F.; Boulain, N.; Cappelaere, B.; Cohard, J.M.; Descroix, L.; Galle, S.; Guilbert, S.; Guichard, F.; et al. Land surface in AMMA Extending Ecosystem, Energy and Water Balance Studies in Space and Time is Some-Times Surprising. In Proceedings of the 2nd International Conference of AMMA Program, Karlsruhe, Germany, 26–30 November 2007.

59. D'Orgeval, T.; Polcher, J. Impacts of precipitation events and land-use changes on West African river discharges during the years 1951–2000. *Clim. Dyn.* **2008**, *31*, 249–262. [CrossRef]

60. Mahé, G.; Olivry, J.C.; Servat, E. Sensibilité des cours d'eau ouest-africains aux changements climatiques et environnementaux: Extrêmes et paradoxes. In *Regional Hydrological Impacts of Climatic Change—Hydroclimatic Variability, Proceedings of symposium S6 held during the Seventh IAHS Scientific Assembly, Foz do Iguaçu, Brazil, 3–9 April 2005*; IAHS Press: Wallingford, UK, 2005; pp. 167–177.

61. Li, K.Y.; Coe, M.T.; Ramankutty, N.; De jong, R. Modeling the hydrological impact of land–use change in West Africa. *J. Hydrol.* **2007**, *337*, 258–268. [CrossRef]

62. Tellro Wai, N.; Ngatcha, B.N.; Mahe, G.; Doumnang, J.C.; Delclaux, F.; Ngolona, G.; Genthon, P. Influence des activités anthropiques sur le régime hydrologique du fleuve Logone de 1960 à 2000. In *Hydrology in a Changing World: Environmental and Human Dimensions, Proceedings of FRIEND-Water 2014, Montpellier, France, 7–10 October 2014*; Daniell, T., van Lanen, H.A.J., Demuth, S., Laaha, G., Servat, E., Mahe, G., Boyer, J.F., Paturel, J.E., Dezetter, A., Ruelland, D., Eds.; IAHS Press: Wallingford, UK, 2014; pp. 438–442.

Article

Expanding Rubber Plantations in Southern China: Evidence for Hydrological Impacts

Xing Ma [1,2]**, Guillaume Lacombe** [3,]*****, **Rhett Harrison** [4]**, Jianchu Xu** [2] **and Meine van Noordwijk** [5]

[1] Yunnan Institute of Environmental Sciences, Kunming 650034, China; maxing@yies.org.cn
[2] World Agroforestry Centre (ICRAF), East Asia, Kunming 650204, China; j.c.xu@cgiar.org
[3] International Water Management Institute (IWMI), Vientiane PO BOX 4199, Laos
[4] World Agroforestry Centre (ICRAF), East & Southern Africa Region, 13 Elm Road, Woodlands, Lusaka 999134, Zambia; r.harrison@cgiar.org
[5] World Agroforestry Centre (ICRAF), Southeast Asia, Bogor 16001, Indonesia; m.vannoordwijk@cgiar.org
* Correspondence: guillaume.lacombe@cirad.fr

Received: 21 January 2019; Accepted: 23 March 2019; Published: 29 March 2019

Abstract: While there is increasing evidence concerning the detrimental effects of expanding rubber plantations on biodiversity and local water balances, their implications on regional hydrology remain uncertain. We studied a mesoscale watershed (100 km^2) in the Xishuangbanna prefecture, Yunnan Province, China. The influence of land-cover change on streamflow recorded since 1992 was isolated from that of rainfall variability using cross-simulation matrices produced with the monthly lumped conceptual water balance model GR2M. Our results indicate a statistically significant reduction in wet and dry season streamflow from 1992 to 2002, followed by an insignificant increase until 2006. Analysis of satellite images from 1992, 2002, 2007, and 2010 shows a gradual increase in the areal percentage of rubber tree plantations at the watershed scale. However, there were marked heterogeneities in land conversions (between forest, farmland, grassland, and rubber tree plantations), and in their distribution across elevations and slopes, among the studied periods. Possible effects of this heterogeneity on hydrological processes, controlled mainly by infiltration and evapotranspiration, are discussed in light of the hydrological changes observed over the study period. We suggest pathways to improve the eco-hydrological functionalities of rubber tree plantations, particularly those enhancing dry-season base flow, and recommend how to monitor them.

Keywords: agroforestry; catchment hydrology; humid tropics; hydrological modeling; impact assessment; land-cover change; Montane Southeast Asia; rubber; trend detection; water balance

1. Introduction

Over recent decades, commercial plantations of rubber trees (*Hevea brasiliensis*) have expanded rapidly across tropical areas around the world, especially in the uplands of Southeast Asia. These plantations have largely encroached on lands that were covered by forests, including land used for annual cropping as part of rainfed rice-based shifting cultivation systems [1,2]. Between 2003 and 2010, about 15,000 km^2 of land were converted to rubber plantations in Cambodia, Southern China, Thailand, and Vietnam [3]. This massive land-cover conversion is threatening biodiversity [4–7], resulting in the loss of ecosystem services [1,8–12]. One important environmental effect of land conversion on rubber plantations is the modification of local and regional water balances [13–18]. *Hevea brasiliensis* is an equatorial species indigenous to the Amazon rain forest. Rubber trees are bigger water consumers than native forest species in Southeast Asia, due in part to their large xylem vessels [19] and an extended root system allowing a wide part of the soil to be explored for water uptake [20–22]. Depending on the season and the zoning of soil moisture, the tree is able to shift from the shallow to the deep soil layer to extract water from where it is most abundant at that point in time, indicating significant plasticity in sources of water uptake [18–23]. These key features enable rubber trees to thrive through the period of greatest water demand. As opposed to the original primary or secondary forest, which exhibit diverse timing in phenology, rubber trees show strong synchronicity for flushing and shedding [17].

The effect of rubber plantations on water balance has been assessed at the plot level [13]. Using measured soil moisture profiles and above-canopy radiations, the authors showed that the rates and timing of water consumption by rubber trees and traditional land-cover types (tea, secondary forest, and grassland) are drastically different. Deep root water uptake of rubber during the dry season is controlled by leaf phenology (leaf loss and renewal), which is determined by day-length, whereas the increased water demand of native vegetation starts with the arrival of the first monsoon rainfall. At the basin level, simulations with a rubber evapotranspiration (ET) model suggested greater water losses through ET from rubber-dominated landscapes than from traditional vegetation cover [14]. Fifteen years of paired watershed rainfall and runoff measurements and one year of paired eddy covariance water flux data from a primary tropical rain forest (51.1 ha) and a tropical rubber plantation (19.3 ha) in the Xishuangbanna prefecture in Southern China helped assessing how rubber plantations affect local water resources [15]. The mean annual ET from the rubber plantation (about 1130 mm) surpassed that of the indigenous forest (about 950 mm), despite a similar rainfall input. A drying up of the streamflow was observed downstream of the rubber plantation in the dry season, while the stream coming from the tropical forest remained perennial. When soil water is limited, rubber trees can regulate stomatal conductance to prevent excessive xylem cavitation [24]. This adaptive capacity to cope with a temporary water shortage enhances the aptitude of rubber trees to deplete soil water resources when they become available again, at the beginning of the wet season. However, the replacement of native forest vegetation with rubber can increase run-off variability and the magnitude of storm streamflow. This situation is observed in tree plantations subject to soil compaction that reduces the soil permeability, despite a greater potential of root water uptake [16]. This analysis indicates that the hydrologic impact of tropical forest conversion to rubber plantation is controlled not only by the different phenology of the trees, but also by the land management, which can strongly influence soil hydraulic properties and lead to greatly modified soil infiltration capacity, as has been observed elsewhere in Southeast Asia [25].

The effects of land-cover change on watershed hydrology are often compounded by climate variability [26]. Several methods exist to separate the relative contributions of climate variability and land-use/land-cover change to watershed hydrological changes, with the most robust ones combining modeling and statistical approaches [27]. Two main techniques are usually applied in small- to medium-size watersheds (i.e., <1000 km^2). Paired watershed studies establish the statistical relationships for outflow variables between two neighboring watersheds, ideally similar in geomorphology, area, land cover, and climate. Following calibration, land-cover treatments are applied to one watershed, and changes in the statistical relationships are indicative of the land treatment effect on hydrology. Important limitations of this approach are the relatively few samples used for model development, and the spatial variability of rainfall events between the two watersheds [28]. A second approach involves the calibration of a rainfall-runoff model in one single watershed. The model is first calibrated before a land-cover treatment occurs. The model is then used as a virtual control watershed, along with the rainfall observed after the land-cover treatment, in order to reconstitute runoff as if no change in the watershed had occurred. An underlying assumption for this approach is that the watershed behaviour is stationary in each of the pre-treatment and post-treatment periods. This assumption is seldom tested. In addition, very few studies have tested the statistical significance of observed hydrological changes [28].

Our objectives were twofold: (a) verify if the conversion of forest to rubber tree plantations in a second-order tributary of the Mekong River in Southern China actually reduced seasonal streamflow, as can be expected from local water balances and larger-scale predictive modeling studies, and (b) assess how this relationship is influenced by land use and geomorphological heterogeneities. For these aims, the following steps were undertaken:

1. Select a watershed (about 100 km^2) which has been subject to intense land conversion to rubber tree plantations during a long period, with available rainfall and streamflow records,
2. Map land-cover types and their changes over time and according to elevations and slopes,
3. Quantify the spread of rubber trees and the recession of other land-cover types, and see how this is influenced by geomorphology,
4. Model the rainfall–runoff relationship, assess the statistical significance of its temporal variations, and verify if they are consistent with observed changes in land cover, particularly those related to the spread of rubber tree plantations,
5. Interpret the correlations between observed land-use changes and hydrological changes.

2. Materials and Methods

2.1. Study Site

In the Xishuangbanna prefecture, about 80% of rubber plantations are located at elevations below 1000 m, on slopes varying between 8 and 25 degrees [29]. More recently, rubber trees were planted in the less suitable remaining areas, with steeper slopes at a higher elevation [30].

The watershed of the Nan'a River (95.1 km^2) in the Upper Mekong Basin, is located in the southwestern part of the Xishuangbanna prefecture, Yunnan Province, China, about 11 km from Myanmar. The elevation and annual average temperature range from 590 to 2174 m [31] (Figure 1) and from 30.4 to 22.2 °C, respectively. According to the hydrometeorological data recorded by the Xishuangbanna Water Resources and Hydrological Bureau, mean annual rainfall (1992–2005) varied from 1373 mm in the lower part of the basin to 1767 mm in the upper part; over the same period, mean annual streamflow yielded 754 mm. About 85% of annual rainfall occurs in the wet season (May–October). The mean lowest flow (26.1 mm·mo^{-1}) and highest flow (132.3 mm·mo^{-1}) are generally observed in April and August, respectively. In 2013, the watershed included six villages inhabited by 2860 persons whose main income originated from rubber. Before rubber plantations started expanding in the watershed in the late 1980s, farmland in the lower part of the watershed included paddy rice and vegetables. Farmland in the upper part of the watershed included rainfed rice and corn. Most of the farmland in the lowland was converted to banana plantations in the 2010s, following a drop in rubber price. No major water infrastructures, such as dams, existed in the watershed during the study period.

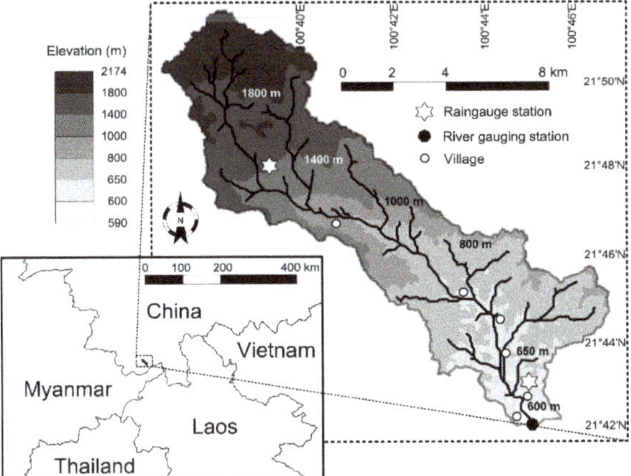

Figure 1. Location of the Nan'a watershed and measurement devices. Contour lines were derived from the ASTER Global Digital Elevation Model [31].

2.2. Hydrometeorological Data

The two rain gauges installed by the Xishuangbanna Water Resources and Hydrological Bureau are located in the downstream and upstream parts of the watershed (Figure 1) and have been recording daily rainfall since 1992. The watershed-scale daily areal rainfall was derived from the point measurements using the Thiessen polygons method. The daily stream water level has been measured at the outlet of the watershed since 1992 with a water level recorder at 10-minute time intervals (Figure 1). Due to the stability of the river cross section at the watershed outlet, the rating curve (relationship between water level and discharge) was determined only once, when measurements were initiated in 1992. The streamflow gauging station was abandoned in 2006. These hydrometeorological data underwent quality control by the Yunnan Water Resources and Hydrological Bureau. Inter-annual means of areal monthly reference evapotranspiration (ET_0) were derived from the Climate Research Unit database [32].

2.3. Land-Cover Mapping

A visual interpretation of three Landsat products (TM1992, ETM2002, and TM2007) with 30 m resolution [33], and one RapidEye product from 2010 with 5 m resolution [34] resampled to 30 m resolution allowed the mapping of six major land-cover types inside the watershed: forest, rubber, grassland, farmland, settlement, and open water bodies. The overall accuracy of this classification was assessed by calculating an error matrix applied to 59 ground control points for the 2010 land-cover map and by computing the Kappa coefficient. Land-cover transitions between 1992 and 2002 and between 2002 and 2007 were quantified using the intersection method available in ArcGIS [35,36]. The spatial distribution of rubber plantations across elevations and slopes in 1992, 2002, and 2007 was analyzed using a 30 m digital elevation model [31] and standard algorithms available in ArcMap 10.2.

2.4. Assessment of Hydrological Changes

Inter-annual variation in the rainfall–runoff relationship was quantified by producing cross-simulation matrices using the model GR2M [37]. GR2M is a monthly lumped conceptual water balance model that was empirically developed using a sample of 410 basins under a wide range of climate conditions. GR2M requires areal rainfall and ET_0 to estimate streamflow. The model includes a soil moisture accounting production store and a routine store that empties following a quadratic function. The two parameters of the model determine the capacity of the production store and the value of an underground water exchange coefficient applied to the routine store. Compared with several widely used water balance models, GR2M ranks amongst the most reliable and robust [37]. For this analysis, like in most hydrological analyses performed in the Mekong Basin, each hydrological year starts in April and ends in March of the following year, so as to group rainfall and related streamflow in the same year. The model was repeatedly calibrated over each hydrological year of the study period. The first year (April 1992–March 1993) was dedicated to model warming for the adjustment of the initial water levels in the model reservoirs. Thirteen model calibrations were produced from April 1993 to March 2006. Over each hydrological year, the model was calibrated by maximizing the Nash-Sutcliffe efficiency criteria (NSEq) applied to streamflow q ($mm \cdot mo^{-1}$) for the evaluation of hydrological changes during the wet season. The Nash-Sutcliffe efficiency criteria calculated on the logarithm of q (NSElnq) was used for the evaluation of hydrological changes during the dry season [38]. While each of these two efficiency criteria are calculated with the 12 monthly flow values of each 1-year calibration period (including wet- and dry-season streamflow), NSEq and NSElnq give more weight to high- and low-flow values, respectively. The constraint of a less than 10% bias on cumulated streamflow over each sub-period was applied to the calibrations. Each model Mj calibrated over the hydrological year j ($1 < j < 13$) was run in simulation mode over the whole study period, April 1993–March 2006, to produce a cross-simulation matrix (Figure 2). This process was performed twice to produce two matrices: one matrix of wet season streamflow (cumulated from May to October) simulated with models optimized with NSEq, and one matrix of dry season streamflow (cumulated from November to April) simulated with models optimized with NSElnq.

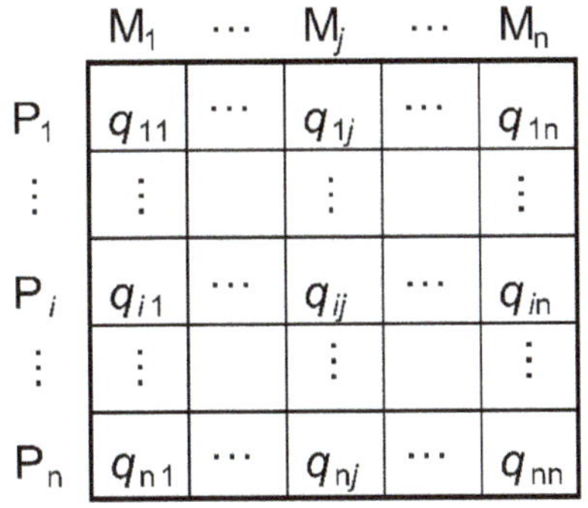

$$
\begin{array}{c|c|c|c|c|c}
 & M_1 & \cdots & M_j & \cdots & M_n \\
\hline
P_1 & q_{11} & \cdots & q_{1j} & \cdots & q_{1n} \\
\vdots & \vdots & & \vdots & & \vdots \\
P_i & q_{i1} & \cdots & q_{ij} & \cdots & q_{in} \\
\vdots & \vdots & & \vdots & & \vdots \\
P_n & q_{n1} & \cdots & q_{nj} & \cdots & q_{nn}
\end{array}
$$

Figure 2. Cross-simulation matrix. M_j ($j \in [1 - n]$) defines the set of model parameters calibrated over the hydrological year j. P_i ($i \in [1 - n]$) defines the rainfall that occurred over the hydrological year i. Streamflow q_{ij} is simulated by model M_j with rainfall P_i.

Streamflow q_{ij} is simulated with the model M_j capturing the watershed behaviour of hydrological year j, using rainfall from the hydrological year i ($i \in [1 - n]$). In a given row i, streamflow variations between columns are due to non-climatic changes (e.g., land-cover changes). In a given column j, streamflow variations between rows reflect inter-annual rainfall variability under stable watershed conditions. Variations in simulated streamflow between the columns of the matrices were plotted against time to illustrate temporal changes in the hydrological behaviour of the watershed. In these simulations, the monthly rainfall input to the model is used in loop each year and corresponds to the rainfall of the hydrological year April 1996–March 1997, exhibiting median annual depth over the studied period. Similar modeling framework was successfully performed in a previous study [25].

To assess the statistical significance of changes in the rainfall–runoff relationship over any multi-year sub-period, the non-parametric test proposed by [39] was applied to the corresponding cross-simulation matrix. Following the prescribed methodology, the cross-simulations matrices (one for wet season streamflow and one for dry season streamflow) were re-sampled by permuting columns. The statistic S was calculated for the original matrix and for each re-sampled one using Equation (1):

$$
S = \sum_{i=1}^{n} \left[\sum_{j=1}^{i-1} (q_{ii} - q_{ij}) + \sum_{j=i+1}^{n} (q_{ij} - q_{ii}) \right]. \tag{1}
$$

Under the null hypothesis H_0 of an absence of unidirectional trend in the hydrological behavior of the watershed over the studied period, the value of S associated to the original matrix should be close to zero. A negative (resp., positive) value denotes a decreasing (resp., increasing) trend in the basin water yield. The p-value of a negative (resp., positive) trend is assessed by calculating the non-exceedance (resp., exceedance) frequency of the original S values compared to the range of S values derived from the permuted matrices.

3. Results

3.1. Land-Cover Change

The land-cover classification was found to be reliable, with an overall accuracy of 89.1% and a Kappa coefficient of 0.85. Figure 3 shows spatial and temporal changes in the spatial distribution of land-cover types between 1992 and 2010. Temporal changes in the cumulative areal percentages of each land-cover type are displayed in Figure 4. Rubber plantations appeared first in the downstream part of the watershed nearby the six main villages and gradually spread upstream, replacing farmland and forest. From 1992 to 2010, the areal percentage of rubber tree plantations in the watershed increased from 10% to 44%, while forest area declined from 47% to 21% and farmland declined from 26% to 12%. Changes in grassland area (from 16% to 21%) and settlement area (from 0.7% to 1.2%) were moderate.

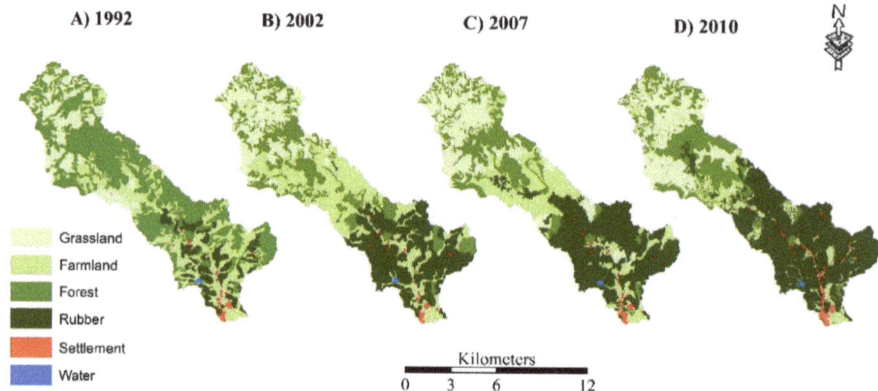

Figure 3. Land-cover change in the Nan'a watershed, based on Landsat (years 1992, 2002, and 2007) and RapidEye (year 2010) imagery [33,34].

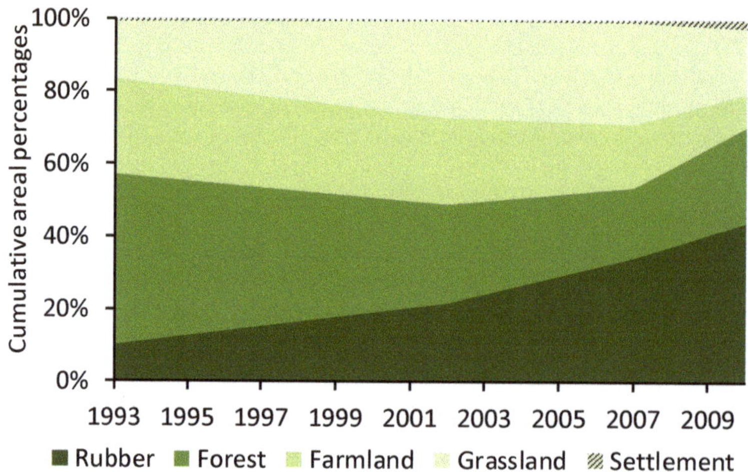

Figure 4. Areal percentages of each land-cover type in the Nan'a watershed.

Figure 5 quantifies land-cover transitions. From 1992 to 2002, the forested area declined drastically from 4517 ha to 2132 ha. About 42%, 30%, and 28% of the deforested area was converted to farmland, rubber plantations, and grassland, respectively. While this land conversion increased the farmland area by 1000 ha of fertile land, another 880 ha of more degraded farmland were planted with rubber trees as a strategy to maintain the soil productivity. Over the same period, the grassland areas gained 665 ha from deforestation and lost 240 ha allocated to farmland. Land-cover dynamics were different from 2002 to 2007. Rubber tree plantations continued expanding at the expense of other areas, primarily farmland, and secondarily forest. About 300 ha of farmland reverted to grassland as a result of a land-use policy targeting slopping land for vegetation regeneration [40]. Over the whole period 1992–2007, rubber tree plantations and forest constantly expanded and declined, respectively.

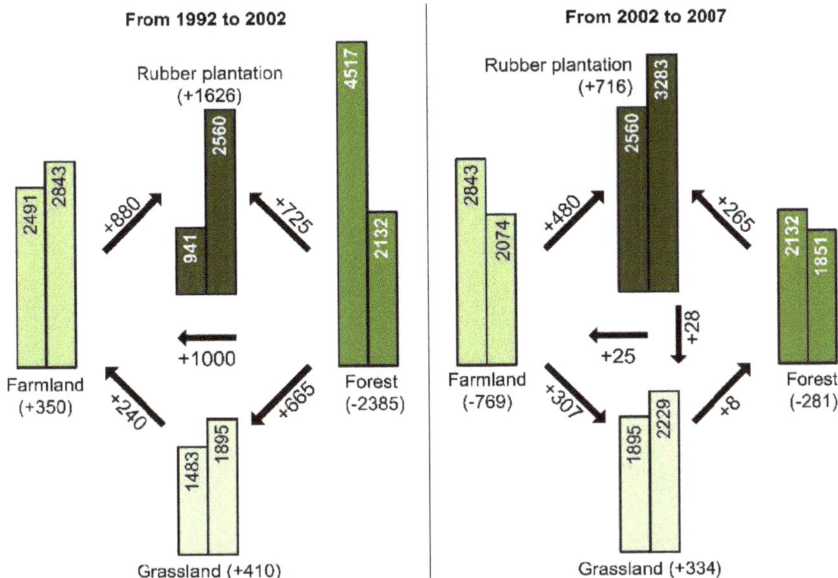

Figure 5. Land-cover conversions in the Nan'a watershed. Numbers refer to areas in hectares. "+" = gain. "−" = loss. Histograms quantify the areas in the first year (left bar) and last year (right bar) of each period.

Figure 6 shows the temporal changes in the spatial distribution of rubber tree plantations in the watershed, according to elevation and slope. From 1992 to 2007, rubber plantations first occupied the most suitable and convenient land near the villages and roads in the lower part of the watershed, on gentle slopes and flat land. As the plantations continued expanding, the most suitable land became rarer and new rubber trees were planted on higher and steeper land. Between 1992 and 2007, the average elevation of rubber tree plantations increased from 657 m to 740 m (Figure 6a). Over the same period, the mean slope of rubber tree plantations increased from 12 to 17 degrees (Figure 6b).

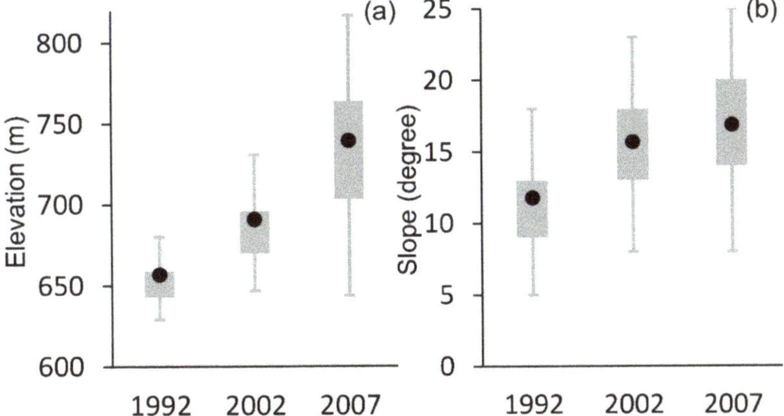

Figure 6. Distribution of rubber trees according to elevation (**a**) and slope (**b**) in the Nan'a watershed. Boxes: 40–60% quantiles. Whiskers: 20–80% quantiles. Black dots: averages.

3.2. Hydrological Change

Figure 7 shows the temporal variation in annual rainfall, streamflow, and the runoff coefficient in the Nan'a watershed from 1993 to 2005. Inter-annual variability of rainfall exhibits 3- to 6-year cycles of gradual increase, with maximums in 1995 (1893 mm) and 2001 (2114 mm, the wettest year), and minimums in 1993 (1250 mm), 1997 (1285 mm), and 2003 (1063 mm, the driest year). While annual streamflow correlated with annual rainfall ($r^2 = 0.85$, p = 0.00) (Figure 8), no correlation ($r^2 = 0.00$) was observed between annual rainfall and annual runoff coefficient, yielding an inter-annual mean of 0.49.

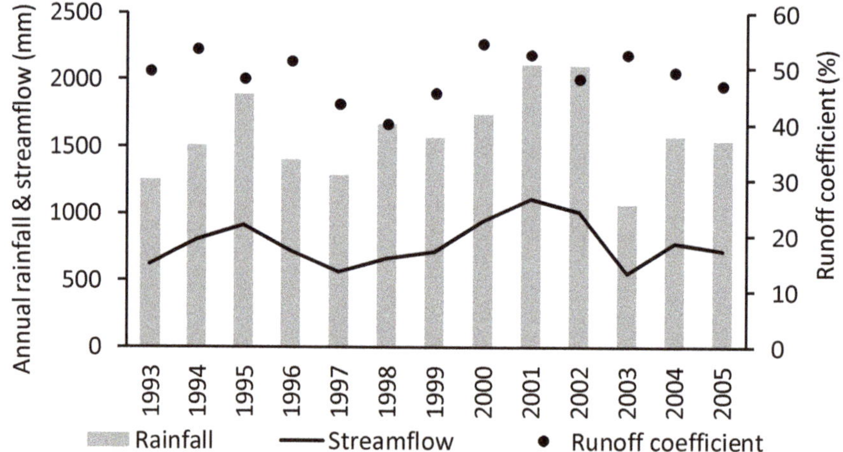

Figure 7. Annual rainfall, streamflow, and runoff coefficient in the Nan'a watershed.

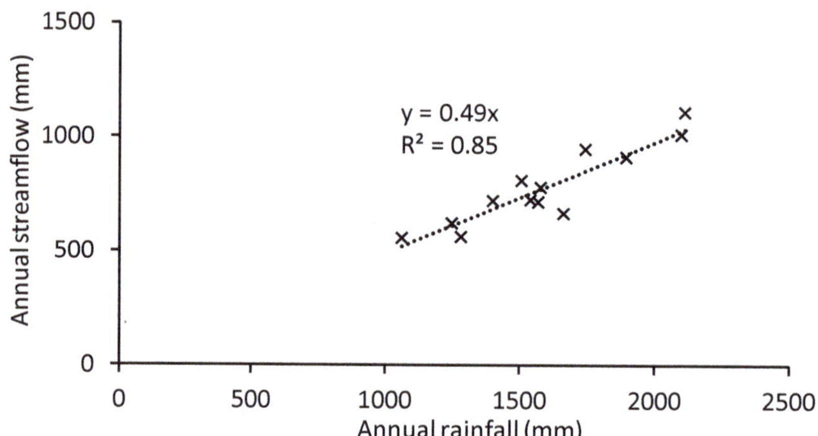

Figure 8. Relationship between annual rainfall and annual streamflow in the Nan' a watershed. Slope of the regression line is equivalent to the inter-annual mean of the annual runoff coefficient.

The annual values of NSEQ and NSElnQ averaged over the whole study period are high: 88.8% and 90.2%, respectively. The lowest values were obtained for hydrological years 2004 (NSElnQ = 81.7%) and 1994 (NSEQ = 80.9%). All these values exceed 0.75, confirming that the performance of GR2M is very good in the studied watershed [41].

Figure 9 depicts the temporal variation in streamflow, simulated under stable rainfall conditions. Unlike Figure 7, where streamflow variation is mainly caused by inter-annual rainfall variability, the variation shown in Figure 9 is caused by non-climatic factors, such as land-cover changes. Wet season and dry season streamflow follow the same patterns, with a reduction from 1993 to 2002 (−196 mm and −65 mm, respectively) followed by an increase from 2002 to 2005 (+80 mm and +44 mm, respectively). The cross-simulation test [39] applied to seasonal streamflow over the whole study period revealed no significant hydrological change at the 90% confidence level. However, a significant decline was observed from 1993 to 2002, with p-values of 0.09 and 0.01 for wet season and dry season streamflow respectively. The same test applied to any other sub-periods yielded higher (i.e., less significant) p-values except for the period 1993–1998, which exhibits the most significant trend of decline (p-values of 0.06 and 0.03 for wet season and dry season streamflow, respectively). In contrast, no significant change was observed from 2002 to 2005.

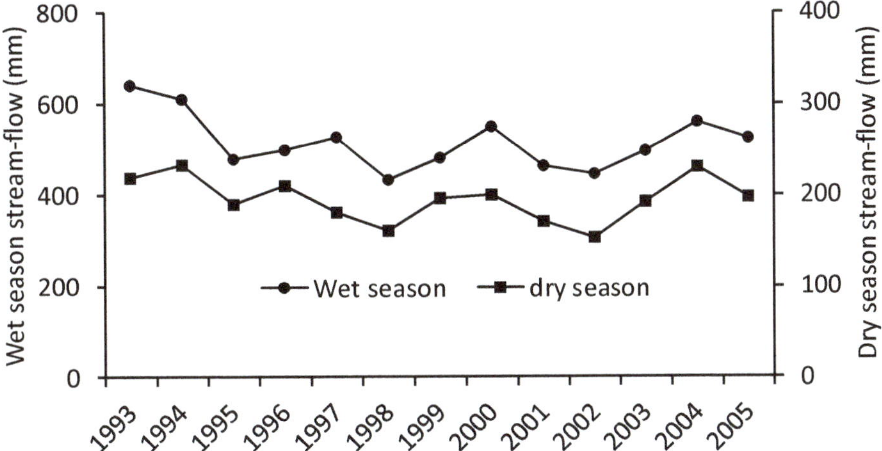

Figure 9. Wet season and dry season streamflow simulated with GR2M calibrated over each hydrological year (indicated on x-axis) and run with the same rainfall input (hydrological year April 1996–March 1997).

4. Discussion

4.1. Relationships between Hydrological Changes and Land-Cover Changes

Between 1993 and 2002, the basin water yield significantly decreased by 30%. The wet season and dry season streamflow simulated under stable rainfall conditions reduced by 196 mm and 65 mm, respectively (Figure 9). Over the same period, the areal percentage of the rubber tree plantations increased by 1626 ha—equivalent to 17.1% of the watershed area—at the expense of forest and farmland, mainly in the lower part of the watershed (Figures 3–6). As in the rest of the region [42], the resulting loss of food-productive land was compensated by converting forest to farmland at elevations higher than 800 m, where rubber trees are less productive and more prone to frost damage. The 30% streamflow reduction likely reflects an increase in ET, caused by greater root water abstraction in relation to the replacement of natural forest and farmland by rubber trees [22]. Indeed, compared to natural forest and farmland, due to a denser and deeper root system, rubber trees have the ability to absorb a larger amount of water in the soil [15,18,23], ultimately reducing watershed streamflow [14]. The transpiration of rubber trees is greater in the wet season than in the dry season [43]. However, the decreased dry season streamflow quantified in the present study could be caused by the ability of rubber trees to shift from the shallow to deep soil layer to extract water where it is the most abundant [18,22,23]. The lower part of the watershed, where rubber trees were planted between 1992 and 2002, is where the water table is likely the shallowest. This increase in ET may have been partially compensated by the concurrent conversion of forest to farmland in the upper part of the basin. This conversion to a sparser vegetation cover, limiting root water uptake, likely increased streamflow production. However, this counteracting change could not fully compensate the hydrological effect of rubber expansion for the following two reasons: (1) Forest conversion to farmland occurred at a higher elevation in the watershed (Figure 3) where the water table is deeper, thereby limiting root water uptake. This means that any change in the vegetation water demand for evapotranspiration would have a smaller effect on the local water balance than the land-cover changes happening in the lower part of the basin. (2) The conversion of forest to farmland involved an area half that of the conversion of forest to rubber tree plantations (Figure 5).

Between 2002 and 2006, the simulated wet and dry season streamflow increased by 80 mm and 44 mm, respectively (Figure 9). These changes were found to be statistically insignificant at the 90% confidence level. Over the same period, the area of rubber tree plantations increased by 716 ha (about 7.5% of the watershed area) at the expense of farmland and forest (Figure 5). On average, these new rubber trees were planted at elevations about 83 m higher and on slopes steeper (by about 5 degrees) than existing rubber trees. Concurrently, about 300 ha of farmland were converted to grassland. These two types of land conversion correspond to a theoretical increase in ET, resulting from a land conversion to a denser vegetation cover with a deeper root system. However, both wet and dry season streamflow simulated under stable rainfall conditions increased during this period, although insignificantly. This hydrological change is likely to happen when soil permeability is reduced, despite a greater evapotranspiration water demand. This situation typically happens on steep slopes exposed to the erosive power of incident rainfall [44]. This is the case where soils under the canopy of tree plantations are cleared and not protected from throughfall by any understory [25,45]. In this environment, enhanced throughfall accelerates soil surface crusting, soil erosion, and surface runoff. Greater rates for these processes were observed under rubber trees planted on steeper slopes in an experimental site nearby the study area [45]. Similarly, a lower soil infiltration capacity is observed where grassland forms on eroded soils [46], as observed in the study area. Although terracing practiced on steep hillslopes under rubber plantations aims to promote runoff infiltration, in some instances such interventions led to soil compaction, reducing soil permeability and ultimately leading to increased runoff [16]. Further field investigations in the studied watershed are required to verify these hypotheses.

4.2. Limitations of the Approach and Recommendations

This study relied on two types of data: continuous hydro-meteorological records collected from one river water level gauging station and two rain gauges, and four land-cover maps at different dates, partly overlapping the period of streamflow records. While plausible explanations could be proposed to explain the succession of two opposite hydrological changes, they could not be ascertained for several reasons. The first one is intrinsically linked to the nature of observed land-cover changes and to their variations according to the watershed geomorphology. The attribution of hydrological trends was complicated by concurrent counteracting changes in vegetation cover. In addition, successive land-cover changes occurred in different parts of the watershed, where surface–groundwater interactions were likely influenced by variations in slope, elevation, and other soil and geological features not reported in this study. The effects of this heterogeneity on the watershed hydrology could not be captured by the limited available data, which provided snapshots of land-cover patterns for a few years only. However, the figures and literature review suggest a continuous increase and decrease in areas of rubber tree plantations and forest, respectively [1–3], with a gradual shift of the land-cover changes toward areas at greater elevations and with steeper slopes [30], supporting our interpretations.

In this study, hydrological modeling was performed at a monthly time step. It could be argued that a daily time step would have provided more precise results. As shown in previous modeling studies performed with a conceptual-lumped rainfall-runoff model in the Mekong Basin [25,47,48], GR2M has several advantages compared to daily models. Aggregating daily rainfall and streamflow in monthly steps avoids the risk of shifting errors between daily rainfall and streamflow time series. This temporal aggregation results in Nash coefficients of GR2M calibrations (>80%) greater than those obtained with models working at a daily time step [47]. This high performance enables GR2M to capture inter-annual variations of the catchment behavior for comparison with long-term changes in land cover, which was the main aim of this study and others [25,47,48].

Future investigations of the hydrological impact of rubber tree plantations at watershed scales should pay particular attention to their effect on streamflow during the dry season. Indeed, when the monsoon recedes, streamflow often becomes the only available water resource that sustains biodiversity and enhances ecosystem services, such as water supply for domestic uses. Rubber plantations may lead to either a reduction or an increase in dry season streamflow, depending on the previous land cover, and the effects of the land conversion on soil permeability and root water uptake. If the gain in infiltration surpasses the increase in ET, groundwater recharge and the related base-flow will increase [49,50]. The effectiveness of rubber plantation in promoting water infiltration will vary according to the management of its understory. Enhanced levels of soil organic matter from litter inputs improve soil structure, enhance aggregate stability, and promote faunal activity, leading to higher macro porosity, thereby creating preferential pathways for infiltration [51]. Permitting understory vegetation to grow increases litter input and reduces over-land flow rates. Roots also create preferential pathways for water infiltration. In contrast, a systematic clearing of vegetation under the tree canopy reduces litter input and increases soil exposure to rainfall intensity, promoting soil crusting that will, in turn, reduce infiltration and increase runoff during the rainy season [45], ultimately leading to a reduction of dry season streamflow [25]. Moreover, increased runoff accelerates soil detachment, erosion, and soil organic carbon depletion, which compromises crop yield over the long term [52]. Another factor influencing the infiltration trade-off is tree density. At a low to intermediate tree cover density, each new tree can improve soil hydraulic properties beyond its canopy edge, which means that the infiltration gains can be proportionally higher than the additional losses from increased transpiration and interception [53]. This process is upper bounded by a tree density, over which any additional tree will further increase ET, while infiltration has already reached a maximum [54]. Optimal tree density is therefore not only a question of yield maximization, but also an ecological consideration.

To improve infiltration under the canopy of monoculture rubber tree plantations, intercropping with understory species is another solution with potential economic return [55]. Before the fourth year of the rubber plantation, young trees are intercropped with fruit species (e.g., pineapples, mangosteen, mangos, and durian), which contribute to maintaining soil organic matter, faunal activity, and permeability. After the closure of the rubber tree canopy, intercrops are replaced by shade-tolerant varieties, such as tea, coffee, cacao, or slow-growing timber species [56,57]. Over time, timber-based systems could become highly diverse and are analogous to advanced secondary forests [58].

Future research should pay particular attention to the trade-off between infiltration and ET, and its relationship to plantation management. Proper measurement devices should be put in place accordingly. They should include microplots to survey soil surface structures and permeability [59], piezometers, river water level gauging stations, and rain gauges. Most importantly, the size of the gauged watershed should be small enough to capture homogeneous land-cover changes. Ideally, it should be positioned downstream of a small headwater watershed likely to experience homogenous land conversion to rubber tree plantation.

5. Conclusions

The capacity of rubber trees to deplete groundwater resources was previously evidenced locally. But the regional hydrological effects of the rapid expansion of rubber tree plantations across southern montane China have not been quantified. In an attempt to address this shortfall, our investigations focused on a 100 km^2 gauged watershed where rubber tree plantations increased from 10% to 44% between 1992 and 2010. While it was possible to isolate hydrological changes from those caused by rainfall variability, their attribution to land-cover changes, and specifically to rubber tree plantations, was compounded by the heterogeneity of land-cover transitions occurring in the watershed, and their variations across a contrasted geomorphology. The processes that explain the lack of unequivocal linkage between tree rubber planting and watershed hydrological change were reviewed, and we formulated practical recommendations in terms of monitoring devices and data collection to improve the strength of trend attribution in future studies. While the spread of rubber tree plantations is likely to continue in the humid tropics—although at a pace moderated by market fluctuations—it is important that the largest number of tree growers adopt sustainable practices. Such techniques should aim at maximizing rainwater infiltration on planted hillslopes, while maintaining biodiversity and diversifying crop productions. Related ecosystem services include the provision of dry season water resources in the form of stream base-flow. Promoting intercropping in rubber tree plantations appears to be a suitable technique to reach these objectives.

Author Contributions: Conceptualization, X.M. and G.L.; methodology, X.M. and G.L.; validation, X.M., G.L. and R.H.; formal analysis, X.M.; investigation, X.M.; resources, X.M.; data curation, X.M. and G.L.; writing—original draft preparation, G.L. and X.M.; writing—review and editing, G.L., R.H., M.v.N., and X.M.; visualization, G.L. and X.M.; supervision, G.L., J.X., M.v.N. and R.H.; project administration, R.H.; funding acquisition, R.H and J.X.

Funding: This research was funded by the CGIAR research program on Integrated Systems for the Humid Tropics, the ICRAF Post-doctoral Program (2012–2014) and the Bundesministerium für Wirtschaftliche Zusammenarbeit und Entwicklung/Deutsche Gesellschaft für Internationale Zusammenarbeit research program that supported the Green Rubber project No. 13.1432.7-001.00.

Acknowledgments: We are grateful to the Xishuangbanna Water Resources and Hydrological Bureau for the provision of Hydro-Meteorological data.

Conflicts of Interest: The authors declare no conflict of interest.

References

1. Ziegler, A.D.; Fox, J.M.; Xu, J. The Rubber Juggernaut. *Science* **2009**, *324*, 1024–1025. [CrossRef] [PubMed]
2. Fox, J.M.; Castella, J.C. Expansion of rubber (Hevea brasiliensis) in Mainland Southeast Asia: What are the prospects for smallholders? *J. Peasant Stud.* **2013**, *40*, 155–170. [CrossRef]
3. Association of Natural Rubber Producing Countries Home Page. Natural Rubber Trends and Statistics Archive. Available online: http://www.anrpc.org/ (accessed on 17 January 2019).
4. Yi, Z.F.; Cannon, C.H.; Chen, J.; Ye, C.X.; Swetnam, R.D. Developing indicators of economic value and biodiversity loss for rubber plantations in Xishuangbanna, Southwest China: A case study from Menglun township. *Ecol. Indic.* **2014**, *36*, 788–797. [CrossRef]
5. Warren-Thomas, E.; Dolman, P.M.; Edwards, D.P. Increasing demand for natural rubber necessitates a robust sustainability initiative to mitigate impacts on tropical biodiversity. *Conserv. Lett.* **2015**, *8*, 230–241. [CrossRef]
6. Ahrends, A.; Hollingsworth, P.M.; Ziegler, A.D.; Fox, J.M.; Chen, H.; Su, Y.; Xu, J. Current trends of rubber plantation expansion may threaten biodiversity and livelihoods. *Glob. Environ. Chang.* **2015**, *34*, 48–58. [CrossRef]
7. Sreekar, R.; Huang, G.; Yasuda, M.; Quan, R.C.; Goodale, E.; Corlett, R.T.; Tomlinson, K.W. Effects of forests, roads and mistletoe on bird diversity in monoculture rubber plantations. *Sci. Rep.* **2016**, *6*. [CrossRef] [PubMed]
8. Wu, Z.L.; Lia, H.M.; Liu, L.Y. Rubber cultivation and sustainable development in Xishuangbanna, China. *Int. J. Sustain. Dev. World Ecol.* **2001**, *8*, 337–345. [CrossRef]
9. Hu, H.B.; Liu, W.J.; Cao, M. Impact of land use and land cover changes on ecosystem services in Menglun, Xishuangbannan, Southwest China. *J. Environ. Monit. Assess.* **2008**, *146*, 147–156. [CrossRef] [PubMed]
10. Mann, C.C. Addicted to rubber. *Science* **2009**, *325*, 564–566. [CrossRef] [PubMed]
11. Qiu, J. Where the rubber meets the garden. *Nature* **2009**, *457*, 246–247. [CrossRef]
12. Fox, J.M.; Castella, J.C.; Ziegler, A.D.; Westley, S.B. *Rubber Plantations Expand in Mountainous Southeast Asia: What Are the Consequences for the Environment?* East-West Center: Honolulu, HI, USA, 2014; 8p.
13. Guardiola-Claramonte, M.; Troch, P.A.; Ziegler, A.D.; Giambelluca, T.W.; Vogler, J.B.; Nullet, M.A. Local hydrologic effects of introducing non-native vegetation in a tropical catchment. *Ecohydrology* **2008**, *1*, 13–22. [CrossRef]
14. Guardiola-Claramonte, M.; Troch, P.A.; Ziegler, A.D.; Giambelluca, T.W.; Durcik, M.; Vogler, J.B.; Nullet, M.A. Hydrologic effects of the expansion of rubber (Heveabrasiliensis) in a tropical catchment. *Ecohydrology* **2010**, *3*, 306–314. [CrossRef]
15. Tan, Z.H.; Zhang, Y.P.; Song, Q.H.; Liu, W.J.; Deng, X.B.; Tang, J.W.; Deng, Y.; Zhou, W.J.; Yang, L.Y.; Yu, G.R.; et al. Rubber plantations act as water pumps in tropical China. *Geophys. Res. Lett.* **2011**, *38*, L24406. [CrossRef]

16. Liu, W.J.; Liu, W.Y.; Lu, H.J.; Duan, W.P.; Li, H.M. Runoff generation in small catchments under a native rain forest and a rubber plantation in Xishuangbanna, Southwestern China. *Water Environ. J.* **2011**, *25*, 138–147. [CrossRef]

17. Carr, M.K.V. The water relations of rubber (*Hevea Brasiliensis*): A review. *Exp. Agric.* **2012**, *48*, 176–193. [CrossRef]

18. Liu, W.J.; Li, J.T.; Lu, H.J.; Wang, P.Y.; Luo, Q.P.; Liu, W.Y.; Li, H.M. Vertical patterns of soil water acquisition by non-native rubber trees (*Hevea brasiliensis*) in Xishuangbanna, southwest China. *Ecohydrology* **2013**, *7*, 1234–1244. [CrossRef]

19. Ayutthaya, S.I.N.; Do, F.C.; Pannangpetch, K.; Junjittakarn, J.; Maeght, J.L.; Rocheteau, A.; Cochard, H. Water loss regulation in mature Hevea brasiliensis: Effects of intermittent drought in the rainy season and hydraulic regulation. *Tree Phys.* **2011**, *31*, 751–762. [CrossRef]

20. Srinivasan, K.; Kunhamu, T.K.; Mohan Kumar, B. Root excavation studies in a mature rubber (*Hevea brasiliensis* Muell. Arg.) plantation. *Nat. Rubber Res.* **2004**, *17*, 18–22.

21. van Noordwijk, M.; Rahayu, S.; Williams, S.E.; Hairiah, K.; Khasanah, N.; Schroth, G. Crop and tree root-system dynamics. In *Belowground Interactions in Tropical Agroecosystems*; van Noordwijk, M., Cadisch, G., Ong, C.K., Eds.; CAB International: Wallingford, UK, 2004; pp. 83–107.

22. Giambelluca, T.W.; Mudd, R.G.; Liu, W.; Ziegler, A.D.; Kobayashi, N.; Kumagai, T.; Miyazawa, Y.; Lim, T.K.; Huang, M.; Fox, J.; et al. Evapotranspiration of rubber (*Hevea brasiliensis*) cultivated at two plantation sites in Southeast Asia. *Water Resour. Res.* **2016**, *52*, 1–20. [CrossRef]

23. Maeght, J.C.; Gonkhamdee, S.; Clément, C.; Isarangkool Na Ayutthaya, S.; Stokes, A.; Pierret, A. Seasonal patterns of fine root production and turn over in a mature rubber tree (*Hevea brasiliensis* Müll. Arg.) Stand differentiation with soil depth and implications for soil carbon stocks. *Front. Plant Sci.* **2015**, *6*, 1022. [CrossRef] [PubMed]

24. Kumagai, T.; Mudd, R.G.; Giambelluca, T.W.; Kobayashi, N.; Miyazawa, Y.; Lim, T.K.; Liu, W.; Huang, M.; Fox, J.M.; Ziegler, A.D.; et al. How do rubber (*Hevea brasiliensis*) plantations behave under seasonal water stress in northeastern Thailand and central Cambodia. *Agric. For. Meteorol.* **2015**, *213*, 10–22. [CrossRef]

25. Lacombe, G.; Ribolzi, O.; de Rouw, A.; Pierret, A.; Latsachak, K.; Silvera, N.; Pham Dinh, R.; Orange, D.; Janeau, J.L.; Soulileuth, B.; et al. Contradictory hydrological impacts of afforestation in the humid tropics evidenced by long-term field monitoring and simulation modelling. *Hydrol. Earth Syst. Sci.* **2016**, *20*, 2691–2704. [CrossRef]

26. Zhang, M.; Liu, N.; Harper, R.; Li, Q.; Liu, K.; Wei, X.; Ning, D.; Hou, Y.; Liu, S. A global review on hydrological responses to forest change across multiple spatial scales: Importance of scale, climate, forest type and hydrological regime. *J. Hydrol.* **2017**, *546*, 44–59. [CrossRef]

27. Wei, X.; Liu, W.; Zhou, P. Quantifying the relative contributions of forest change and climatic variability to hydrology in large watersheds: A critical review of research methods. *Water* **2013**, *5*, 728–746. [CrossRef]

28. Zégre, N.; Skaugset, A.E.; Som, N.A.; McDonnell, J.J.; Ganio, L.M. In lieu of the paired catchment approach: Hydrologic model change detection at the catchment scale. *Water Resour. Res.* **2010**, *46*, W11544. [CrossRef]

29. Liu, X.N.; Feng, Z.M.; Jiang, L.G.; Li, P.; Liao, C.H.; Yang, Y.Z.; You, Z. Rubber plantation and its relationship with topographical factors in the border region of China, Laos and Myanmar. *J. Geogr. Sci.* **2013**, *23*, 1019–1040. [CrossRef]

30. Chen, H.; Yi, Z.F.; Schmidt-Vogt, D.; Ahrends, A.; Beckschäfer, P.; Kleinn, C.; Ranjitkar, S.; Xu, J. Pushing the limits: The pattern and dynamics of rubber monoculture expansion in Xishuangbanna, SW China. *PLoS ONE* **2016**, *11*, e0150062. [CrossRef] [PubMed]

31. ASTER GDEM Validation Team. *ASTER Global Digital Elevation Model Validation Summary Report*; METI & NASA: Washington, DC, USA, 2009; 28p.

32. Harris, I.; Jones, P.D.; Osborn, T.J.; Lister, D.H. Updated high-resolution grids of monthly climatic observations—The CRUTS3.10 Dataset. *Int. J. Climatol.* **2014**, *34*, 623–642. [CrossRef]

33. USGS (U.S. Geological Survey). Earth Explorer. 2019. Available online: https://earthexplorer.usgs.gov/ (accessed on 26 March 2019).

34. DLR (German Aerospace Center). The RapidEye Earth observation system. 2019. Available online: https://www.dlr.de/ (accessed on 26 March 2019).

35. Long, H.; Tang, G.; Li, X.; Heilig, G.K. Socio-economic driving forces of land-use change in Kunshan, the Yangtze River Delta economic area of China. *J. Environ. Manag.* **2007**, *83*, 351–364. [CrossRef]

36. Ma, X.; Xu, J.C.; Luo, Y.; Prasad Aggarwal, S.; Li, J.T. Response of hydrological processes to land-cover and climate changes in Kejie watershed, South-West China. *Hydrol. Process.* **2009**, *23*, 1179–1191. [CrossRef]

37. Mouelhi, S.; Michel, C.; Perrin, C.; Andréassian, V. Stepwise development of a two-parameter monthly water balance model. *J. Hydrol.* **2006**, *318*, 200–214. [CrossRef]

38. Pushpalatha, R.; Perrin, C.; Le Moine, N.; Andréassian, V. A review of efficiency criteria for evaluating low-flow simulations. *J. Hydrol.* **2012**, *420–421*, 171–182. [CrossRef]

39. Andréassian, V.; Parent, E.; Michel, C. A distribution-free test to detect gradual changes in watershed behavior. *Water Resour. Res.* **2003**, *39*. [CrossRef]

40. Wang, G.; Innes, J.L.; Lei, J.; Dai, S.; Wu, S.W. China's Forestry Reforms. *Science* **2007**, *318*, 1556–1557. [CrossRef]

41. Moriasi, D.N.; Arnold, J.G.; Van Liew, M.W.; Bingner, R.L.; Harmel, R.D.; Veith, T.L. Model evaluation guidelines for systematic quantification of accuracy in watershed simulations. *Trans. ASABE* **2007**, *50*, 885–900. [CrossRef]

42. Li, H.; Mitchell Aide, T.; Ma, Y.; Liu, W.; Cao, M. Demand for rubber is causing the loss of high diversity rain forest in SW China. *Biodivers. Conserv.* **2007**, *16*, 1731–1745. [CrossRef]

43. Kobayashi, N.; Kumagai, T.; Miyazawa, Y.; Matsumoto, K.; Taterishi, M.; Lim, T.K.; Mudd, R.G.; Ziegler, A.D.; Giambelluca, T.W.; Yin, S. Transpiration characteristics of a rubber plantation in central Cambodia. *Tree Phys.* **2014**, *34*, 285–301. [CrossRef] [PubMed]

44. Liu, W.J.; Zhu, C.J.; Wu, J.; Chen, C.F. Are rubber-based agroforestry systems effective in controlling rain splash erosion? *Catena* **2016**, *147*, 16–47. [CrossRef]

45. Lacombe, G.; Valentin, C.; Sounyafong, P.; de Rouw, A.; Soulileuth, B.; Silvera, N.; Pierret, A.; Sengtaheuanghoung, O.; Ribolzi, O. Linking crop structure, throughfall, soil surface conditions, runoff and soil detachment: 10 land uses analyzed in Northern Laos. *Sci. Total Environ.* **2018**, *616–617*, 1330–1338. [CrossRef]

46. Hill, R.D.; Peart, M.R. Land use, runoff, erosion and their control: A review for Southern China. *Hydrol. Process.* **1999**, *12*, 2029–2042. [CrossRef]

47. Lacombe, G.; Pierret, A.; Hoanh, C.T.; Sengtaheuanghoung, O.; Noble, A. Conflict, migration and land-cover changes in Indochina: A hydrological assessment. *Ecohydrology* **2010**, *3*, 382–391. [CrossRef]

48. Lyon, S.W.; King, K.; Polpanich, O.; Lacombe, G. Assessing hydrologic changes across the Lower Mekong Basin. *J. Hydrol. Reg. Stud.* **2017**, *12*, 303–314. [CrossRef]

49. Bruijnzeel, L.A. (de)forestation and dry season flow in the tropics: A closer look. *J. Trop. For. Sci.* **1989**, *1*, 229–243.

50. Bruijnzeel, L.A. Hydrological functions of tropical forests: Not seeing the soil for the trees? *Agric. Ecosyst. Environ.* **2004**, *104*, 185–228. [CrossRef]

51. Bargués Tobella, A.; Rees, H.; Almaw, A.; Bayala, J.; Malmer, A.; Laudon, H.; Ilstedt, U. The effect of trees on preferential flow and soil infiltrability in an agroforestry parkland in semiarid Burkina Faso. *Water Resour. Res.* **2014**, *50*, 3342–3354. [CrossRef]

52. Ribolzi, O.; Evrard, O.; Huon, S.; de Rouw, A.; Silvera, N.; Latsachack, K.O.; Soulileuth, B.; Lefèvre, I.; Pierret, A.; Lacombe, G.; et al. From shifting cultivation to teak plantation: Effect on overland flow and sediment yield in a montane tropical catchment. *Sci. Rep.* **2017**, *7*. [CrossRef] [PubMed]

53. Beven, K.; Germann, P. Macropores and water flow in soils. *Water Resour. Res.* **1982**, *18*, 1311–1325. [CrossRef]

54. Ilstedt, U.; Bargués Tobella, A.; Bazié, H.R.; Bayala, J.; Verbeeten, E.; Nyberg, G.; Sanou, J.; Benegas, L.; Murdiyarso, D.; Laudon, H.; et al. Intermediate tree cover can maximize groundwater recharge in the seasonally dry tropics. *Sci. Rep.* **2016**, *6*, 21930. [CrossRef] [PubMed]

55. Penot, E.; Courbet, P.; Chambon, B.; Ilahang, K. Improved rubber agroforests in Indonesia: Myth of reality? *Plant. Rech. Dév.* **1999**, *6*, 400–414.

56. Min, S.; Huang, J.; Bai, J.; Waibel, H. Adoption of intercropping among smallholder rubber farmers in Xishuangbanna, China. *Int. J. Agric. Sustain.* **2017**, *15*, 223–237. [CrossRef]

57. Langenberger, G.; Cadisch, G.; Martin, K.; Min, S.; Waibel, H. Rubber intercropping: A viable concept for the 21st century? *Agrofor. Syst.* **2017**, *91*, 577–596. [CrossRef]

58. Lacombe, G.; Bolliger, A.M.; Harrison, R.D.; Ha, T.T.T. Integrated tree, crop and livestock technologies to conserve soil and water, and sustain smallholder farmers' livelihoods in Southeast Asian uplands. In *Integrated Systems Research for Sustainable Smallholder Agriculture in the Central Mekong: Achievements and Challenges*; Hiwasaki, L., Bolliger, A.M., Lacombe, G., Raneri, J., Schut, M., Staal, S., Eds.; World Agroforestry Center: Hanoi, Vietnam, 2016; pp. 41–63. ISBN 978-604-943-434-1.
59. Casenave, A.; Valentin, C. A runoff capability classification system based on surface features criteria in semi-arid areas of West Africa. *J. Hydrol.* **1992**, *130*, 231–249. [CrossRef]

Article

Recent Trend in Hydroclimatic Conditions in the Senegal River Basin

Ansoumana Bodian [1,*], Lamine Diop [2], Geremy Panthou [3], Honoré Dacosta [4], Abdoulaye Deme [5], Alain Dezetter [6], Pape Malick Ndiaye [1], Ibrahima Diouf [7] and Théo Vischel [3]

[1] Laboratoire Leïdi "Dynamique des territoires et développement", Université Gaston Berger (UGB), BP 234-Saint-Louis, Senegal; ndiaye.papa-malick@ugb.edu.sn

[2] UFR S2ATA Unité de Formation et de Recherche des "Sciences Agronomiques de l'Aquaculture et des Technologies Alimentaires", Université Gaston Berger (UGB), BP 234-Saint-Louis, Senegal; lamine.diop@ugb.edu.sn

[3] IGE Institut des Géosciences de l'Environnement, Université Grenoble Alpes, 38000 Grenoble, France; geremy.panthou@univ-grenoble-alpes.fr (G.P.); theo.vischel@univ-grenoble-alpes.fr (T.V.)

[4] Département de Géographie, Université Cheikh Anta Diop, BP 5005 Dakar, Senegal; dacosta.honore@gmail.com

[5] Laboratoire LSAO "Laboratoire des Sciences de l'Atmosphère et de l'Océan", Université Gaston Berger (UGB), BP 234-Saint-Louis, Senegal; abdoulaye.deme@ugb.edu.sn

[6] HydroSciences Montpellier, Univ Montpellier, IRD, CNRS, CC 057, 163 rue Auguste Broussonnet, 34090 Montpellier, France; Alain.Dezetter@ird.fr

[7] NOAA Center for Weather and Climate Prediction, 5830 University Research Court, College Park, MD 20740, USA; ibrahima.diouf@noaa.gov

[*] Correspondence: bodianansoumana@gmail.com or ansoumana.bodian@ugb.edu.sn; Tel.: +221-77-811-7553

Received: 21 December 2019; Accepted: 20 January 2020; Published: 6 February 2020

Abstract: Analyzing trends of annual rainfall and assessing the impacts of these trends on the hydrological regime are crucial in the context of climate change and increasing water use. This research investigates the recent trend of hydroclimatic variables in the Senegal River basin based on 36 rain gauge stations and three hydrometric stations not influenced by hydraulic structures. The Man Kendall and Pettitt's tests were applied for the annual rainfall time series from 1940 to 2013 to detect the shift and the general trend of the annual rainfall. In addition, trends of average annual flow rate (AAFR), maximum daily flow (MADF), and low flow rate (LFR) were evaluated before and after annual rainfall shift. The results show that the first shift is situated on average at 1969 whereas the second one is at 1994. While the first shift is very consistent between stations (between 1966 and 1972), there is a significant dispersion of the second change-point between 1984 and 2002. After the second shift (1994), an increase of annual rainfall is noticed compared to the previous period (1969–1994) which indicates a not significant, partial rainfall recovery at the basin level. The relative changes of hydrologic variables differ based on the variables and the sub-basin. Relative changes before and after first change-point are significantly negative for all variables. The highest relative changes are observed for the AAFR. Considering the periods before and second shifts, the relative changes are mainly significantly positive except for the LFR.

Keywords: trends; Senegal River Basin; rainfall shift; hydroclimatic variables; streamflow; climate change

---

1. Introduction

Rainfall is a major factor that conditions food production, is mainly derived from rain-fed agriculture, and plays a central role in the availability of surface and groundwater water resources [1].

Therefore, several studies have focused on how climate variability affects rainfall regime in West Africa over the last century [2,3]. A common feature of these studies is the identification of a succession of drought episodes [4]. Since the beginning of observations in 1854, three main dry periods have been recognized [2]: 1911–1913, 1940–1943 and 1968 to recent decades. Due to its duration and intensity, the latter has been recognized as the greatest drought of the last century [5]. The significant decline in annual rainfall generally led to the depletion of water resources [6,7], the amplification of water deficits [8] and the modification of natural ecosystems and socio-economic systems [4]. It had particularly negative effects on human activities and environment. The drought and its hydrological impacts have been well investigated until the 1990s [3,9–19]. However, the number of studies on hydroclimatic evolution in West Africa over the last three decades has decreased significantly because of the decline of in-situ observation networks and the difficulty for academic structures to access data from national meteorological and/or hydrological services.

Recently, some authors [18,20,21] have questioned whether a few years of excess rainfall of the 1990s could be the sign of a gradual recovery of rainfall leading to more favorable hydroclimatic conditions. Similarly, Diello [22] showed that the area affected by drought has shrunk since 1994 in the Sahelian area. However, other studies have suggested that severe drought conditions were persisting in the Sahel at the end of the 1990s [12,23]. Thus, the question of the persistence or end of drought was discussed in the scientific community in early 2000.Vischel et al. [24] investigated the scientific literature on recent rainfall trends in West Africa. They pointed out that after the severe drought in the Sahel in the late 1960s, the last decade of the 20th century saw a partial recovery of rainfall, particularly in the central part of the Sahel, but without returning to the wet conditions of the 1950s and 1960s. In addition, recent studies have shown that the relative increase in rainfall accumulation was mainly linked to an increase in the intensity of rainfall events, while the occurrence of rainfall events remains at an average level close to that of periods of extreme drought [25,26]. Therefore, the terms "recovery" or "return to wet conditions" often used to characterize rainfall over the past two decades has to be considered with caution.

In this respect, Descroix et al. [27] argued that the end dates of the drought of the 1970s may also vary region to another. Lebel and Ali [16] then Panthou et al. [25] have shown that there are contrasts between the Western Sahel which seems to start a later recovery in annual precipitation than the central Sahel. Nkrumah et al. [26] showed that contrast also exists between sub-regions with in the Southern West Africa with a more marked increase in coastal than inland regions. It is also what Diop et al. [28] have shown in a recent study in Senegal where rainfall recovery in recent decades is significant in the coastal strip of Senegal. All these recent studies push for further analyze recent rainfall trends in West Africa at the sub-regional scale. While climatic zones or country-based approaches dominate the literature, there is a lack of analyses at the scale of regional basins in the West African region. The basin scale is a key scale, not only for analyzing climatic regional contrasts, but also to understand their impacts on water resources. In the context of climate and global changes and increasing water demand, it is a major challenge to jointly analyze rainfall and hydrological changes to improve water resource management [29].

2. Materials and Methods

2.1. Study Area

The Senegal River basin extends from the Sahara in the North to the humid tropical region of Guinea in the South [30]. It drains an area of 300,000 km^2 [31] with four riparian states, from upstream to downstream, Guinea, Mali, Senegal and Mauritania (Figure 1). The Senegal River basin has experienced climate change effects since the 1970s [31]. To remedy the effects of adverse climatic conditions, the Senegal River Basin Development Organization (in French, Organisation pour la Mise en Valeur du Fleuve Sénégal, OMVS) built three dams (Figure 1) in 1986 (Diama), 1988 (Manantali) and 2013 (Felou). The main functions of the Diama dam are to improve irrigation in the Senegal River valley and delta and to facilitate water supply. Manantali is a multi-usage dam with a storage capacity of 11 billion m^3 of water, an energy production of 800 GWh/year and an irrigation capacity of 255,000 ha. Felou is a run-of-river dam with a production capacity of 350 GWh/year. These various hydraulic structures have artificialized the Senegal River regime in some places. From the climate stand point, rainfall in the basin is related to the displacement of the Intertropical Convergence Zone (ITCZ) from the south to north, inducing the penetration of the West African monsoon governed by the thermal contrast between the sea and the continent [32]. Based on the latitudinal distribution of precipitation, four climatic zones have been defined in the Senegal River basin by Dione [30]: Guinea (mean annual precipitation, P > 1500 mm); Southern Sudan (1000 < P < 1500 mm); Northern Sudan (500 < P < 1000 mm); and the Sahel (P < 500 mm).

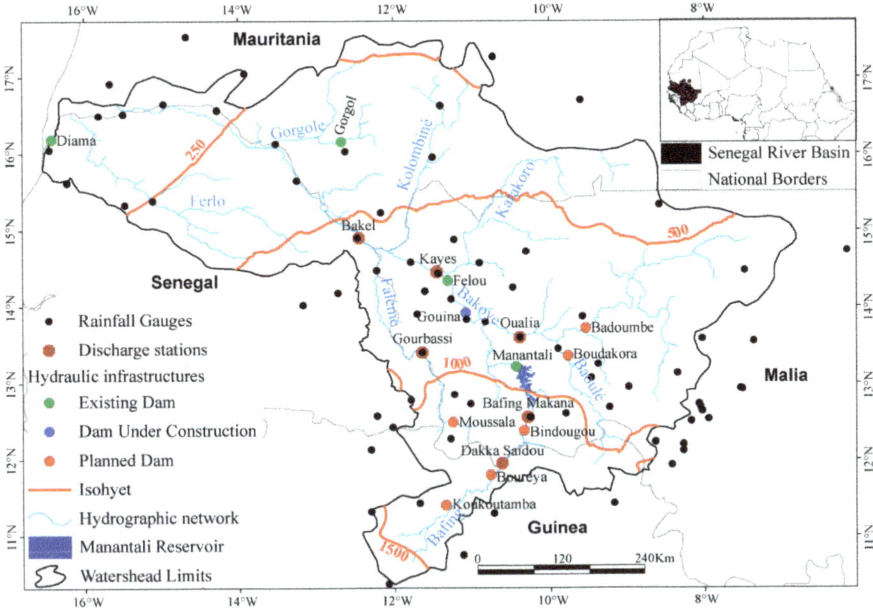

Figure 1. Senegal River Basin, hydrographic network, river flow stations, rainfall gauges, and main hydraulic infrastructures.

2.2. Data

In this study, in-situ annual rainfall and daily observed stream-flows data are used.

2.2.1. Annual Rainfall data, Selected Gauges and Periods of Study

The rainfall data used in this study come from the OMVS database that contains 80 rainfall stations (Figure 1) covering the period 1940 to 2013. For each station, the percentage of gaps in the data series has been calculated. Only stations with less than 50% of missing years were first selected, leading to retain 48 stations. Then, two criteria were used to select the final stations for the study: (i) the coefficient of determination (Figure 2a) between the observed annual rainfall and the calculated one by the Regional Vector Method [33–35] and (ii) stations with no missing values (Figure 2b). Thus, for this study, stations (36) with zero gaps or a coefficient of determination of 0.5, which is an acceptable value in linear regression [36], were selected (Figure 2c). Figure 3 shows the available annual rainfall data over the period 1940–2013. For the stations with gaps, the regional vector method was used to fill the gaps. The regional vector method has already been evaluated by Bodian [37] and has shown to be robust enough to fill missing values in annual rainfall series.

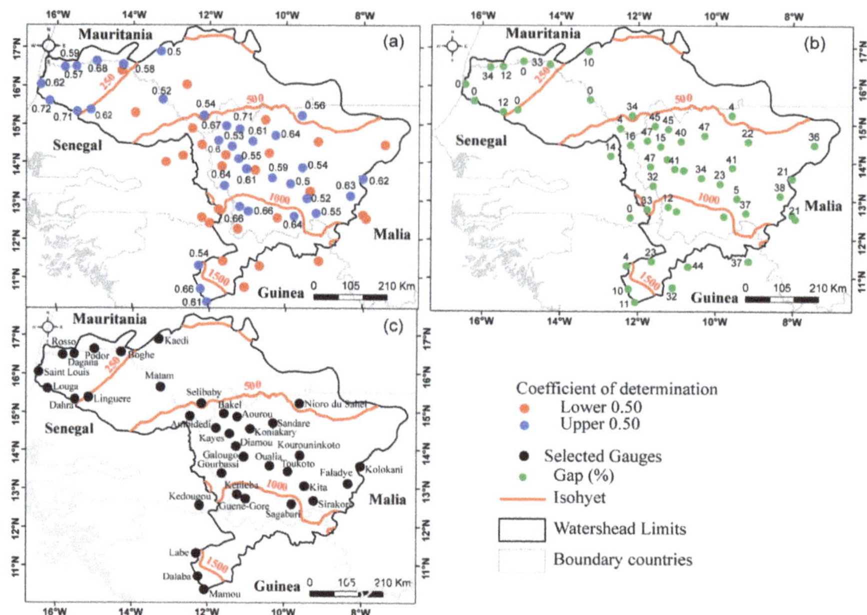

Figure 2. (**a**) Coefficient of determination between the observed annual rainfall and calculated by the regional vector method, (**b**) percentage of gap in the data series by station, (**c**) stations selected for the study.

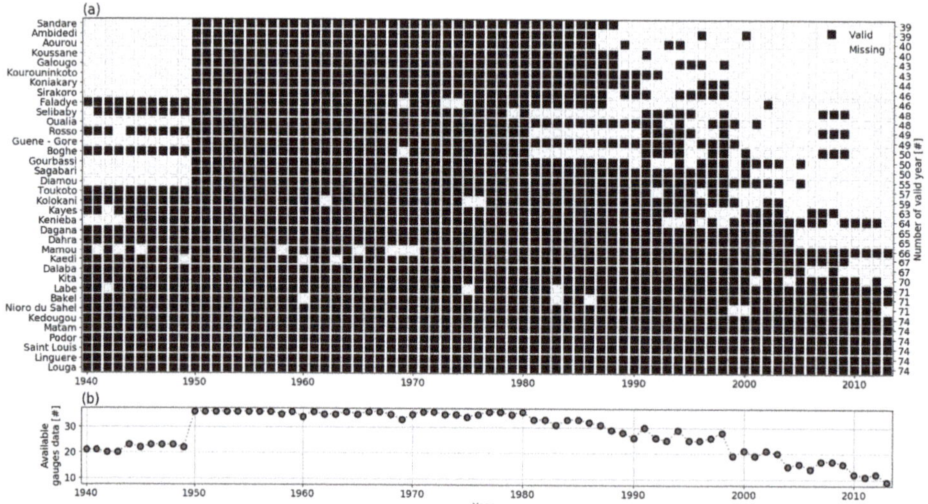

Figure 3. Available annual rainfall data over the period 1940–2013l. (**a**) missing values by station; (**b**) number of available gauges data per year.

2.2.2. Hydrological Data

For this study, three streamflow stations not influenced by hydraulic structures were chosen (Figure 1). Table 1 provides the available hydrometric data from the OMVS database.

Table 1. Available discharge data.

River Basin	Stations	Surface (km^2)	Start of Record (dd/mm/yyyy)	End of Record (dd/mm/yyyy)	% Gaps
Bafing	Dakka Saidou	15660	27/05/1952	21/04/2016	1.02
Faleme	Gourbassi	16264	02/01/1954	24/03/2016	0.17
Bakoye	Oualia	104479	01/06/1954	24/03/2016	1.90

2.3. Methods

2.3.1. Hydroclimatical Variables Trend Analysis

Two methods, namely the Mann-Kendall and Pettitt's tests, were used to detect the annual rainfall trends and shift over the entire Senegal River basin. The Mann-Kendall test [38,39] is a non-parametric test often used to detect the presence of a monotonic trend in a chronological series. The Pettitt's test [40] was used to detect the change-point in the considered series. It is based on the Mann-Whitney two-sample test and allows the detection of a single shift at an unknown time. By combining the Mann Kendall and Pettitt tests, a several-steps approach was applied. First, the Man Kendall and Pettitt's tests were applied for the annual rainfall series from 1940 to 2013 to detect the shift and the general trend of the annual rainfall. Once a change-point was detected, a new annual rainfall series was considered starting after the change point. For this new series, we applied the same tests as previously. Furthermore, trends of average annual flow rate (AAFR), maximum daily flow (MADF), and low flow rate (LFR) were evaluated before and after annual rainfall shifts.

2.3.2. Standardized Precipitation Index (SPI)

The SPI [9,41] is a widely used index to characterize meteorological drought. It reflects a rainfall surplus or deficit for the year compared to a whole period under investigation. A positive value of SPI indicates rainfall surplus whereas a negative value shows rainfall deficit.

3. Results

3.1. Temporal Variability of Annual Rainfall

Figure 4 presents the temporal variability of the annual rainfall in the Senegal River basin. The results show that the stations exhibited different shifts. However, two regime shifts statistically significant at the 5% level are exhibited. The first change-point is situated on average at 1969 whereas the second one is at 1994 (Figure 4). While the first shift is very consistent between stations (comprised between 1966 and 1972), there is a significant dispersion of the second rupture between 1984 and 2002. The two identified shifts lead to distinguish three different periods as depicted by the SPI (Figure 5). From 1940 to 1968, the mean SPI is positive and reveals annual rainfall well above the interannual mean characterizing a very wet period. By contrast, the inter-shift period displays very negative values of SPI showing that the Senegal River basin clearly witnessed the well-documented Sahelian great drought. The dispersion of the break date shows that the identification of an end date for the drought period is much less consistent with some stations remaining very dry while others seem to be starting to recover more quickly.

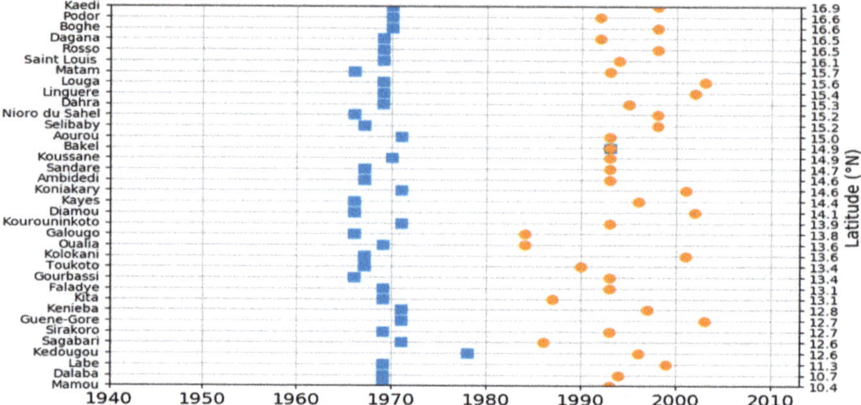

Figure 4. Breaks identified in the annual rainfall series by the Pettitt test at 5% level. The blue dots represent the year of occurrence of the first shift and the orange dots the year of occurrence of the second shift.

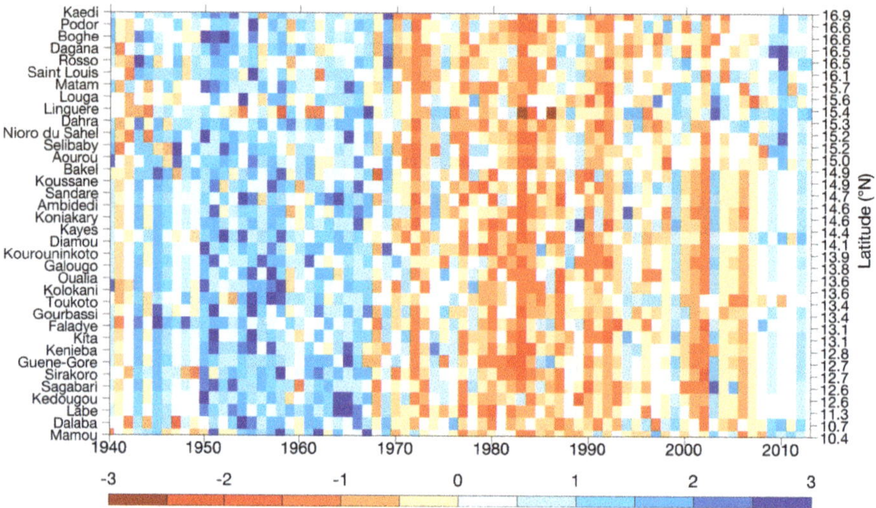

Figure 5. Variation of SPI over the period 1940–2013.

3.2. Annual Rainfall Trend

The trend in annual rainfall is investigated from 1940 to 2013 for both the whole annual rainfall series and the periods after the first shift to 2013. For the whole period (1940–2013), Figure 6a shows that the spatial pattern is dominated by a downward trend except Bakel which exhibits an upward trend but not significant at a 5% significance level. The case of Bakel was mentioned by previous studies [28,37]. However, the period after the first shift to 2013 (Figure 6b) shows a significant positive linear trend (0.05) at the majority of the stations in the Senegal River basin. There are few stations mainly at the upper part of the river basin which exhibit a non-significant positive trend. These results indicate that the Sahelian part of the river basin exhibits a significant upward trend of annual rainfall, whereas the Soudanian one shows an upward trend but is not significant.

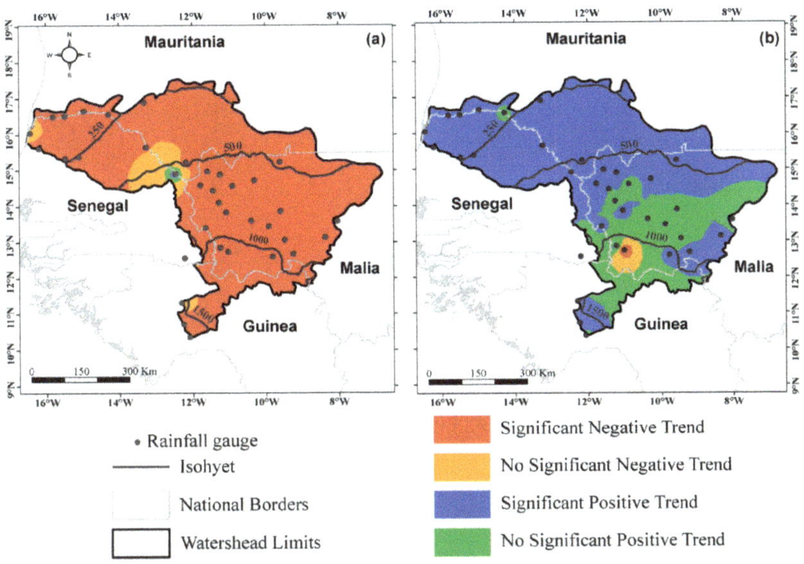

Figure 6. Spatial distribution of the statistics Z of the Mann–Kendall test for the annual rainfall (**a**) period 1940–2013; (**b**) period after shift 1 to 2013.

3.3. Evolution of Hydrological Variables

An analysis was performed in order to evaluate the variability of the hydrologic variables across the three sub-basins of the area: Gourbassi, Oualia and Dakka Saidou for the period from 1940 to 2013 (Table 2 and Figure 7). Both rainfall and hydrologic variables: AAFR, MADF, LFR seem to have similar trends. For all basins, the period before the first break was the wettest with average rainfall varying from 877 mm (Oualia) up to 1239 mm (Gourbassi). Table 2 presents the relative change computed before and after breaks for all variables and for each sub-basin. The table shows that the relative changes differ based on the variables and the sub-basin. Relative changes before (1940–1969) and after first break (1970–2013) are significantly negative (0.05) for all variables. The highest relative changes are observed for the AAFR reaching −54% in Gourbassi sub-basin. When considering the periods before 1970–1994 and second shifts (1995–2013), the relative changes are mainly significantly positive (0.05) except for the LFR. These results are further supported by the tabulated SPI presented above.

Table 2. Mean, standard deviation and changes in hydroclimatic variables.

| Basin | Parameters | Long Term Period | | Period 1 | | Period 2 | | Period 3 | | Change | |
| | | 1940–2013 | | 1940–1969 | | 1970–1994 | | 1995–2013 | | Break 1 | Break 2 |
		$\bar{X}$	σ	$\bar{X}$	σ	$\bar{X}$	σ	$\bar{X}$	σ	$\Delta 1$	$\Delta 2$
Gourbassi	Rainfall	1125	171	1239	97	990	89	1111	86	−0.15 *	0.12
	AAFR	101	52	168	38	67	33	92	29	−0.54 *	0.37 *
	MADF	922	429	1359	356	682	358	899	318	−0.43 *	0.32 *
	LFR	26	11	31	8	20	9	31	13	−0.18 *	0.58
Oualia	Rainfall	822	117	887	147	715	127	855	120	−0.12 *	0.20 *
	AAFR	117	67	202	39	69	40	99	42	−0.58 *	0.44 *
	MADF	1041	605	1578	408	666	429	1066	614	−0.46 *	0.60 *
	LFR	39	18	62	12	31	10	29	11	−0.51 *	−0.04
DakKa Saidou	Rainfall	1472	198	1629	164	1282	125	1461	123	−0.16 *	0.14 *
	AAFR	228	64	309	53	180	34	223	26	−0.35 *	0.24 *
	MADF	1328	472	1796	540	1053	228	1311	376	−0.35 *	0.24 *
	LFR	40	11	48	18	38	6	36	5	−0.23 *	−0.05 *

LFR, low flow rate (m³/s); MADF, maximum daily flow (m³/s); AAFR, average annual flow rate (m³/s); Δ1, relative change between period before and after shift 1 (%); Δ2, relative change between period before and after shift 2 (%). * Significant changes at the 5% threshold.

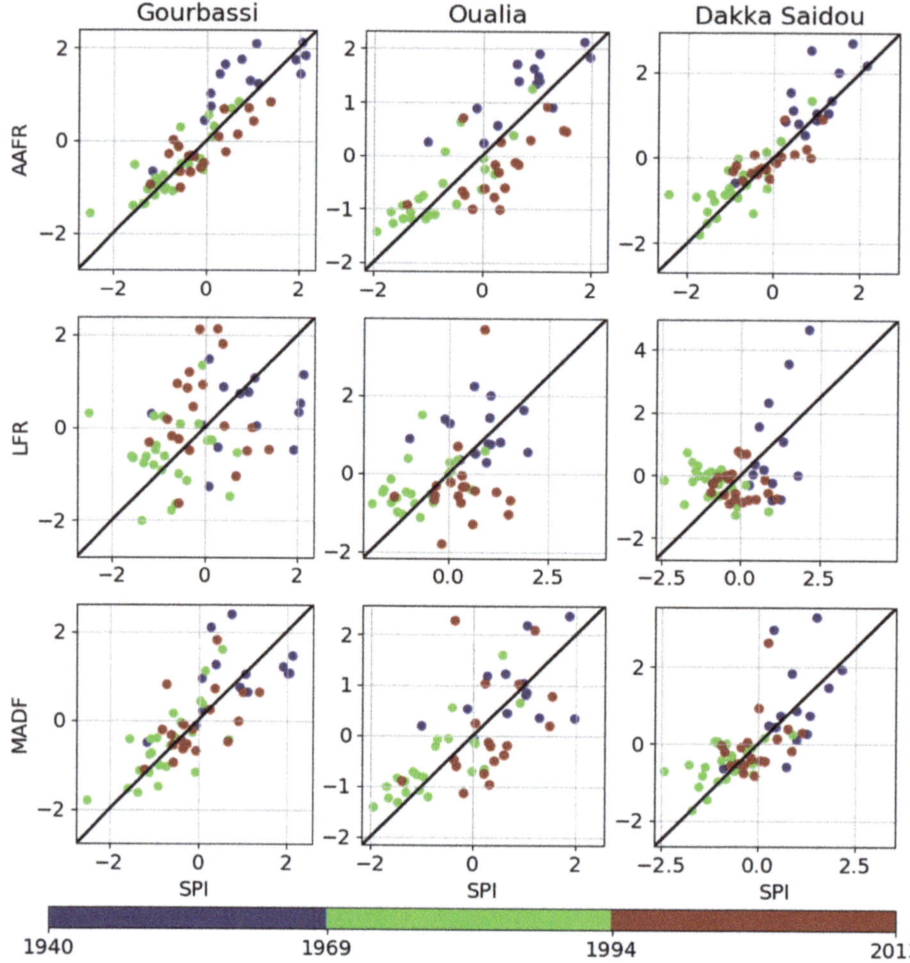

Figure 7. Relationship between SPIs and hydrological variables.

4. Discussion

The results of this paper show two main shifts on annual rainfall in 1969 and 1994 which corroborates the results of Nicholson et al. [19] who showed that a major change in the rainfall regime occurred in both the Sahel and Guinea Coast around 1968/1969. The annual rainfall exhibited a decreasing trend if we considered the whole series. This decreasing trend of the annual rainfall has the same consequence on the hydroclimatic variables: AAFR, MADF, LFR. This is in accordance with Olivry [42] who found a net relationship between annual rainfall and hydoclimatic variables in terms of trend. In addition, Diop et al. [43] by investigating the annual streamflow in the upper Senegal River basin found a significant decreasing trend from 1961 to 2014.

The first shift (1969) marks the starting point of the drought as indicated by the values of the SPI. Our results suggest an onset of the drought similar to that found by Nichloson et al. [19]. After the second shift (1994), there is an increase of annual rainfall compared to the previous period (1969–1994) which indicated a partial rainfall recovery not significant at a basin level.

Hence, the paper confirms the findings suggesting a return of wet period [18,28,44,45]. This leads to an improvement in water availability. The relative changes between period before and after second shift range from 0.24 (Dakka Saidou) to 0.60 (Oualia) for MADFs, from 0.24 (Dakka Saidou) to 0.44 (Oualia) for AAFRs and from 0.58 (Gourbassi) to −0.04 (Oualia and Dakka Saidou). With the exception of the low water flows of Oualia and Dakka Saidou, the various hydrological variables analysed show an uptrend. Flows were sustained for a few years at the beginning of the drought by emptying natural reservoirs; it now takes a few years to restore them. However, the relative change of 0.58 for the Gourbassi low water flow does not seem realistic and may be implied by associated uncertainty with the calculation of low water flows. Indeed, the calculation of low water flows requires the separation of flows. This separation of flows was carried out based on the decrease in the annual hydrograph, which corresponds to a regular decrease in flows or drying phase. After rapid flows transfer, the part of the curve with the steep slope corresponds to the recession and the part that shows a gentle slope to drying up [46] that begins in late November or early December in our study area. This method permits to calculate the annual drying coefficients and corresponding low water flows. However, flow separation is hardly possible for a large river [47] and this can affect the calculated low water flows and lead to discrepancies in the results between stations. In addition, the calibration curves of the different stations used in the study back to the 1950s (1950 for Gourbassi, 1952 for Dakka Saidou and 1954 for Oualia). The existence of probable unearthing can persist at the level of the various hydrometric stations with a greater influence on low water flows than on the other hydrological variables. The reasons for this failure to update the calibration curves are various. Indeed, in recent decades, countries which share the Senegal River basin (Guinea-Conakry, Mali, Senegal and Mauritania) have difficulties in providing hydrological monitoring of the river and its tributaries correctly because of a lack of human and financial resources [48]. Over the last few years, African hydrological services have difficulty to ensure water resource monitoring, resulting in significant gaps in streamflow series. The reasons for the discontinuity of data are part of budgetary austerity measures imposed by international financial institutions.

5. Conclusions

Analyzing trends of annual rainfall and assessing the impacts of these trends on the hydrological regime are crucial in a context of climate and global change and increasing water use. This paper investigates the recent trend of hydroclimatic variables in the Senegal River basin based on 36 rain gauge stations and 3 hydrometric stations not influenced by hydraulic structures. Results show two main shifts on annual rainfall in 1969 and 1994. The first shift (1969) marks the starting point of the drought as indicated by the values of the SPI. After the second shift (1994), there is an increase of annual rainfall compared to the previous period (1969–1994) which indicated a partial rainfall recovery not significant at a basin level.

Overall, these findings demonstrate that there is a recovery of annual rainfall in the Senegal River basin which is leading to the improvement of surface water availability. However, we cannot affirm with precision if this recovery is significant and persistent unless we have a longer data series. Therefore, it is crucial and necessary to have access to recent hydroclimatic data in order to further investigate the so-called recovery period.

Author Contributions: Conceptualization, A.B. and L.D.; methodology, A.B. and L.D.; software, A.B., L.D., G.P., P.M.N.; validation, A.D. (Alain Dezetter), T.V., H.D., and A.D. (Abdoulaye Deme); formal analysis, A.B., L.D., G.P., A.D. (Alain Dezetter), T.V.; data curation, A.B., H.D., I.D.; writing—original draft preparation, A.B. and L.D.; writing—review and editing, A.B., L.D., G.P., A.D. (Alain Dezetter), T.V., I.D., P.M.N.; All authors have read and agreed to the published version of the manuscript.

Funding: This research received no external funding.

Acknowledgments: The authors express their gratitude to the hydrological and meteorological services of the member countries of the Senegal River Basin Development Organization. We would like to thank the IRD and IGE for there technical and financial supports. The authors also thank the three reviewers and the editors.

Conflicts of Interest: The authors declare no conflict of interest.

References

1. Sarr, M.A.; Zorome, M.; Seidou, O.; Bryant, C.R.; Gachon, P. Recent trends in selected extreme precipitation indices in Senegal-A changepoint approach. *J. Hydrol.* **2013**, *505*, 326–334. [CrossRef]
2. Sircoulon, J. Les données hydropluviométriques de la sécheresse récente en Afrique intertropicale. Comparaison avec les sécheresses "1913" et "1940". *Cah. Orstom Ser. Hydrol.* **1976**, *13*, 75–174.
3. Servat, E.; Paturel, J.E.; Lubes-Niel, H.; Kouamé, B.; Masson, J.M.; Travaglo, M.; Marieu, B. De différents aspects de la variabilité de la pluviométrie en Afrique de l'Ouest et Centrale. *J. Water Sci.* **1999**, *12*, 363–387. [CrossRef]
4. Liénou, G. Impacts de la Variabilité Climatique sur les Ressources en eau et les Transports de Matières en Suspension de Quelques Bassins Versants Représentatifs au Cameroun. Ph.D. Thesis, Yaoundé I University, Yaoundé, Cameroon, 2007.
5. Hulme, M.; Doherty, R.; Ngara, T.; New, M.; Lister, D. African climate change: 1900–2100. *Clim. Res.* **2001**, *17*, 145–168. [CrossRef]
6. Hubert, P.; Bader, J.C.; Bendjoudi, H. Un siècle de débits annuels du fleuve Sénégal. *J. Sci. Hydrol.* **2007**, *52*, 68–73. [CrossRef]
7. Abrate, T.; Hubert, P.; Sighomnou, D. A study on hydrological series of the Niger River. *Hydrol. Sci. J.* **2013**, *58*, 1–9. [CrossRef]
8. Olivry, J.C. Les conséquences durables de la sécheresse actuelle sur l'écoulement du fleuve Sénégal et l'hypersalinisation de la Basse-Casamance. In *The influence of Climate Change and Variability on the Hydrologic Regime and Water Resources, Proceedings of the Vancouver Symposium, Vancouver, BC, Canada, 9–22 August 1987 Canada*; IAHS Publication: Wallingford, UK, 1987.
9. Lamb, P.J. Persistence of sub-Saharan drought. *Nature* **1982**, *299*, 46–47. [CrossRef]
10. Carbonnel, J.P.; Hubert, P. Sur la sécheresse au Sahel d'Afrique de l'Ouest. Une rupture climatique dans les séries pluviométriques du Burkina-Faso (ex Haute-Volta). *C. R. Acad. Sci.* **1985**, *301*, 941–944.
11. Nicholson, S.E. The spatial coherence of African rainfall anomalies: Interhemispheric teleconnections. *J. Clim. Appl. Meteorol.* **1986**, *25*, 1365–1381. [CrossRef]
12. L'hôte, Y.; Mahé, G.; Somé, B.; Triboulet, J.P. Analysis of a Sahelian annual rainfall index from 1896 to 2000; the drought continues. *Hydrol. Sci. J.* **2002**, *47*, 563–572. [CrossRef]
13. Dacosta, H.; Konaté, Y.K.; Malou, R. La variabilité spatio-temporelle des précipitations au Sénégal depuis un siècle. In *Regional Hydrology: Bringing the Gap between Reseach and Pratice (FRIEND Conference, Le Cap, Afrique du Sud)*; IAHS Publication: Wallingford, UK, 2002; pp. 499–506.
14. Niel, H.; Leduc, C.; Dieulin, C. Caractérisation de la variabilité spatiale et temporelle des précipitations annuelles sur le bassin du Lac Tchad au cours du 20ème siècle. *Hydrol. Sci. J.* **2005**, *50*, 223–243. [CrossRef]
15. Ali, A.; Lebel, T. The Sahelian standardized rainfall index revisited. *Int. J. Climatol.* **2009**, *29*, 1705–1714. [CrossRef]
16. Lebel, T.; Ali, A. Recent trends in the Central and Western Sahel rainfall regime (1990–2007). *J. Hydrol.* **2009**, *375*, 52–64. [CrossRef]
17. Mahé, G.; Paturel, J.E. 1896–2006 Sahelian annual rainfall variability and runoff increase of Sahelian Rivers. *C. R. Geosci.* **2009**, *341*, 538–546. [CrossRef]
18. Sanogo, S.; Fink, A.H.; Omotosho, J.A.; Ba, A.; Redl, R.; Ermert, V. Spatio-temporal characteristics of the recent rainfall recovery in West Africa. *Int. J. Clim.* **2015**, *35*, 4589–4605. [CrossRef]
19. Nicholson, S.E.; Fink, A.H.; Funk, A. Assessing recovery and change in West Africa's rainfall regime from a 161-year record. *Int. J. Clim.* **2018**, *38*, 3770–3786. [CrossRef]
20. Sene, S.; Ozer, P. Évolution pluviométrique et relation inondations-événements pluvieux au Sénégal. *Bull. Soc. Geogr. Liège* **2002**, *42*, 27–33.
21. Ozer, P.; Erpicum, M.; Demarée, G.; Vandiepenbeck, M. The Sahelian drought may have ended during the 1990s. *Hydrol. Sci. J.* **2003**, *48*, 489–492. [CrossRef]
22. Diello, P. Interrelations Climat—Homme—Environnement dans le Sahel Burkinabé: Impacts sur les états de surface et la modélisation hydrologique. Ph.D. Thesis, University of Montpellier, Montpellier, France, 2007.
23. Ardoin, S.; Lubés-Niel, H.; Servat, E.; Dezetter, A.; Paturel, J.E.; Mahé, G.; Boyer, J.F. Analyse de la persistance de la sécheresse en Afrique de l'Ouest: Caractérisation de la situation de la décennie 90. In *Hydrology of the Mediterranean and Semiarid Regions (Colloque IAHS, Montpellier)*; Servat, É., Najem, W., Leduc, C., Ahmed, S., Eds.; IAHS Publication: Wallingford, UK, 2003; pp. 223–228.

24. Vischel, T.; Lebel, T.; Panthou, G.; Quantin, G.; Rossi, A.; Martineti, M. Le retour d'une période humide au Sahel ? Observations et perspectives. In *Les Sociétés Rurales Face aux Changements Climatiques et Environnementaux en Afrique de l'Ouest*; Sultan, B., Lalou, R., Sanni, M.A., Oumarou, A., Soumaré, M.A., Eds.; Collection Synthèses; IRD Editions: Marseille, France, 2015; pp. 43–60.

25. Panthou, G.; Vischel, T.; Lebel, T.; Quantin, G. Rainfall intensification in tropical semi-arid regions: The Sahelian case. *Environ. Res. Lett.* **2018**, *13*, 064013. [CrossRef]

26. Nkrumah, F.; Vischel, T.; Panthou, G.; Klutse, N.A.B.; Adukpo, D.C.; Diedhiou, A. Recent Trends in the Daily Rainfall Regime in Southern West Africa. *Atmosphere* **2019**, *10*, 741. [CrossRef]

27. Descroix, L.; Guichard, F.; Grippa, M.; Lambert, L.A.; Panthou, G.; Mahe, G.; Gal, L.; Dardel, C.; Quantin, G.; Kergoat, L.; et al. Evolution of surface hydrology in the sahelo-sudanian strip: An updated review. *Water* **2018**, *10*, 748. [CrossRef]

28. Diop, L.; Bodian, A.; Diallo, D. Spatiotemporal Trend Analysis of the Mean Annual Rainfall in Senegal. *Eur. Sci. J.* **2016**, *12*, 231–245. [CrossRef]

29. Kuentz, A. Un Siècle de Variabilité Hydro-Climatique sur le Bassin de la Durance: Recherches Historiques et Reconstitutions. Ph.D. Thesis, Agro Paris Tech, Paris, France, 2013.

30. Dione, O. Evolution Climatique Récente et Dynamique Fluviale Dans les Hauts Bassins des Fleuves Sénégal et Gambie. Ph.D. Thesis, Université Jean Moulin Lyon 3, Lyon, France, 1996. Available online: http://horizon.documentation.ird.fr/exl-doc/pleins_textes/pleins_textes_7/TDM_7/010012551.pdf (accessed on 9 December 2019).

31. Bodian, A. Approche par Modélisation Pluie–Débit de la Connaissance Régionale de la Ressource en eau: Application au Haut Bassin du Fleuve Sénégal. Ph.D. Thesis, Université Cheikh Anta Diop de Dakar, Dakar, Sénégal, 2011. Available online: http://hydrologie.org/THE/BODIAN.pdf (accessed on 9 December 2019).

32. Bodian, A.; Dezetter, A.; Dacosta, H. Apport de la modélisation hydrologique pour la connaissance de la ressource en eau: Application au haut bassin du fleuve Sénégal. *Climatologie* **2012**, *9*, 109–125.

33. Brunet-Moret, Y. Test d'homogénéité. *Cah. Orstom Ser. Hydrol.* **1977**, *14*, 119–129.

34. Hiez, G. Bases théoriques du "Vecteur Régional". Les premières applications et leur mise en œuvre informatique. In *Deuxièmes Journées Hydrologiques de l'ORSTOM (Montpellier)*; Editions de l'ORSTOM: Ballan-Miré, France, 1986; pp. 1–35.

35. Hiez, G.; Cochonneau, G.; Sechet, P.; Fernandes, U.M. Application de la méthode du vecteur régional à l'analyse de la pluviométrie annuelle du bassin versant amazonien. *Veill. Clim. Satellitaire* **1992**, *43*, 39–52.

36. Henseler, J.; Ringle, C.; Sinkovics, R. The use of partial least squares path modeling in international marketing. *Adv. Int. Mark.* **2009**, *20*, 277–320.

37. Bodian, A. Caractérisation de la variabilité temporelle récente des précipitations annuelles au Sénégal (Afrique de l'Ouest). *Physio Géo* **2014**, *8*, 297–312. [CrossRef]

38. Mann, H.B. Nonparametric tests against trend. *Econ. J. Econ. Soc.* **1945**, *13*, 245–259. [CrossRef]

39. Kendall, M.G. Further contributions to the theory of paired comparisons. *Biometrics* **1955**, *11*, 43–62. [CrossRef]

40. Pettitt, A.N. A non-parametric approach to the change-point problem. *J. R. Stat. Soc. Ser. C Appl. Stat.* **1979**, *28*, 126–135. [CrossRef]

41. Ali, A.; Lebel, T.; Amani, A. Signification et usage de l'indice pluviométrique au Sahel. *Sécheresse* **2008**, *19*, 227–235.

42. Olivry, J.C. De l'évolution de la puissance des crues des grands cours d'eau intertropicaux d'Afrique depuis deux décennies. In *Potamologie d'Hier et d'Aujourd'hui, Aménagements et Cours d'eau. Actes des Journées Hydrologiques—Centenaire Maurice Pardé Grenoble*; Vivian, H., Ed.; Institut de Géographie Alpine: Grenoble, France, 1994; pp. 101–108.

43. Diop, L.; Yaseen, Z.M.; Bodian, A.; Djaman, K.; Brown, L. Trend analysis of streamflow with different time scales: A case study of the upper Senegal River. *ISH J. Hydraul. Eng.* **2017**. [CrossRef]

44. Giannini, A.; Salack, S.; Lodoun, T.; Ali, A.; Gaye, A.T.; Ndiaye, O. A unifying view of climate change in the Sahel linking intra-seasonal, interannual and longer time scales. *Environ. Res. Lett.* **2013**, *8*, 024010. [CrossRef]

45. Panthou, G.; Vischel, T.; Lebel, T. Recent trends in the regime of extreme rainfall in the Central Sahel. *Int. J. Clim.* **2014**, *34*, 3998–4006. [CrossRef]

46. Malanda, E.N. Etude Statistique et Modélisation Pluie-Débit à la l'Aide des Modèles Conceptuels Globaux GR4J et GR2M: Applications sur le Bassin Versant du Fleuve Gambie à la Station de Kédougou. Ph.D. Thesis, Université Cheikh Anta Diop, Dakar, Sénégal, 2009.
47. Olivry, J.C. Etude régionale sur les basses eaux; les effets durables du déficit des précipitations sur les étiages et les tarissements en Afrique de l'ouest et du centre. In *XII^{émes} Journée Hydrologiques de l'Orstom (Montpellier)*; Editions de l'ORSTOM: Ballan-Miré, France, 1996; p. 14.
48. Bodian, A.; Dezetter, A.; Dacosta, H. Rainfall-Runoff Modelling of Water Resources in the Upper Senegal River Basin. *Int. J. Water Resour. Dev.* **2015**, *32*, 89–101. [CrossRef]

Article

Spatial Drought Characterization for Seyhan River Basin in the Mediterranean Region of Turkey

Yonca Cavus * and Hafzullah Aksoy

Department of Civil Engineering, Istanbul Technical University, Maslak, Istanbul 34469, Turkey
* Correspondence: cavus17@itu.edu.tr

Received: 22 May 2019; Accepted: 24 June 2019; Published: 27 June 2019

Abstract: Drought is a natural phenomenon that has great impacts on the economy, society and environment. Therefore, the determination, monitoring and characterization of droughts are of great significance in water resources planning and management. The purpose of this study is to investigate the spatial drought characterizations of Seyhan River basin in the Eastern Mediterranean region of Turkey. The standardized precipitation index (SPI) was calculated from monthly precipitation data at 12-month time scale for 19 meteorological stations scattered over the river basin. Drought with the largest severity in each year is defined as the critical drought of the year. Frequency analysis was applied on the critical drought to determine the best-fit probability distribution function by utilizing the total probability theorem. The sole frequency analysis is insufficient in drought studies unless it is numerically related to other factors such as the severity, duration and intensity. Also, SPI is a technical tool and thus difficult to understand at first glance by end-users and decision-makers. Precipitation deficit defined as the difference between precipitation threshold at SPI = 0 and critical precipitation is therefore more preferable due to its usefulness and for being physically more meaningful to the users. Precipitation deficit is calculated and mapped for 1-, 3-, 6- and 12-month drought durations and 2-, 5-, 10-, 25-, 50- and 100-year return periods at 12-month time scale from the frequency analysis of the critical drought severity. The inverse distance weighted (IDW) interpolation technique is used for the spatial distribution of precipitation deficit over the Seyhan River basin. The spatial and temporal characteristics of drought suggest that the Seyhan River Basin in the Eastern Mediterranean region of Turkey experiences quite mild and severe droughts in terms of precipitation deficit. The spatial distribution would alter greatly with increasing return period and drought duration. While the coastal part of the basin is vulnerable to droughts at all return periods and drought durations, the northern part of the basin would be expected to be less affected by the drought. Another result reached in this study is that it could be common for one point in the basin to suffer dry conditions, whilst surrounding points in the same basin experience normal or even humid conditions. This reinforces the importance of spatial analysis over the basin under investigation instead of the point-scale temporal analysis made in each of the meteorological stations. With the use of spatial mapping of drought, it is expected that the destructive and irreversible effects of hydrological droughts can be realized in a more physical sense.

Keywords: critical drought; frequency analysis; Mediterranean region; precipitation deficit; Seyhan River basin; spatial drought analysis; standardized precipitation index (SPI)

---

1. Introduction

Drought is a stochastic natural event which emerges from a remarkable deficiency in precipitation. It has an impact on a large number of sectors since water is the source of life. Drought seriously affects the majority of the population in many ways such as economically, socially and environmentally. The fact that the lack of water affects different sectors makes it more difficult to create one single definition of drought. A certain definition does not exist for drought because of the complicated stochastic nature of hydrometeorological variables and water demand in different regions around the world [1] and the variability in the climate of the region under investigation. Owing to the increase in water demand, drought hydrology has been receiving much attention. As a result, extensive research studies have been performed on drought and numerous review papers have been published [1–6].

Drought typically has probabilistic characteristics which are severity, duration, intensity, frequency and interarrival time [3,7–9]. In the literature, there have been limitless studies on drought characterization [10–16]. A different approximation of frequency analysis was used to derive precipitation deficit from the drought Severity-Duration-Frequency (SDF) curves for drought characterization in different hydrological basins in Turkey [17–19]. As the most recent study on drought in Turkey, Cavus [20] developed a methodology based on precipitation deficit, precipitation threshold and critical precipitation concepts, all newly introduced and detailed below. Also used is a regression equation established between the SPI and the corresponding precipitation. As the outputs of the methodology, drought SDF and intensity–duration–frequency (IDF) curves were plotted with which it is possible to determine the severity and intensity of a drought with a given duration and return period.

Turkey is situated in the East Mediterranean. Annual average precipitation in Turkey is 630 mm, 67% of which falls during the winter and spring months with the influence of Mediterranean depression [21–23]. Climate models predict that, by the end of the 21st century, Europe will face droughts extending over larger areas in the Mediterranean region with increasing intensity and duration [24,25]. Therefore, monitoring of drought and management plans are vital to determine the impact of drought on the Mediterranean region. Drought action plans should be more efficient such that the use of economic resources is optimized. As examples, Vicente-Serrano and Begueria [26] evaluated the impact of drought using remote sensing in the semi-arid Mediterranean region in Spain. Caloiero et al. [27] analyzed drought to calculate different return-period droughts by using the SPI in the northern hemisphere including the European continent, Ireland, UK and the Mediterranean basins. Vicente-Serrano et al. [28] studied drought patterns in eastern Spain of the Mediterranean region and found the frequency, duration and intensity of drought for each region considered. Also, Spinoni et al. [29] classified drought episodes by frequency, duration and severity.

A particular meteorological station is used when the point-scale temporal analysis of drought is concerned. On the other hand, spatial analysis of drought is as important as its temporal analysis, as it could be common for one point in an area to suffer dry conditions, whilst surrounding points in the same area experience normal or even humid conditions. Thus, information related to not only one particular station but also to neighboring stations is needed in making decision for drought mitigation or preparedness at the basin scale. Spatial analysis is performed using data from all available meteorological stations in the basin. This provides spatial drought characterization that utilizes the severity, intensity, duration and return period. It can be presented in the form of spatial patterns of drought intensity contours for a given drought duration and return period. Spatial variability of drought events in the literature has been approached from different perspectives [21–23,30,31]. He et al. [32] investigated drought hazard and spatial characteristic analysis in China using a GIS-based drought hazard assessment model.

Spatial IDF mapping of the precipitation deficit has been the purpose of this study to provide a comprehensive characterization of droughts for meteorological stations in the Seyhan River basin of the Eastern Mediterranean region, Turkey. In the IDF maps, contours based on the precipitation deficit are plotted for a given drought duration and return period. The maps are given in this study for drought durations of 1, 3, 6 and 12 months and return periods of 2, 5, 10, 25, 50, and 100 years at 12-month time scale. They are expected to provide useful information to use in taking actions against drought such that it becomes less destructive and does not create irreversible effects on the economy, society and ecology. This study is also expected to provide measures for reducing the negative effects of the drought as a step in the drought management plan.

2. Seyhan River Basin and Data

The Seyhan River Basin is located in the southern part of Turkey (Figure 1). The climate in the Seyhan River basin is strongly influenced by the topography. The northern part of the basin is characterized by a mountainous, steep, harsh topography while lowlands prevail in the southern part of the basin (Figure 1). The basin extends from the Mediterranean coast to the Central Anatolia and shows three different characteristics in terms of climate. The northern part of the basin exhibits the characteristics of the Central Anatolian climate; thus, it is colder than the southern part of the basin; the highest precipitation is observed at highlands in this part of the basin. In the coastal areas of the Çukurova plain and the surrounding areas, the summer season is hot and dry while the winter is warm and rainy. That part of the basin between the coastal zone in the south and the Tarsus mountains in the north has a semi-arid Mediterranean climate with dry and hot summers and rainy and warm winters.

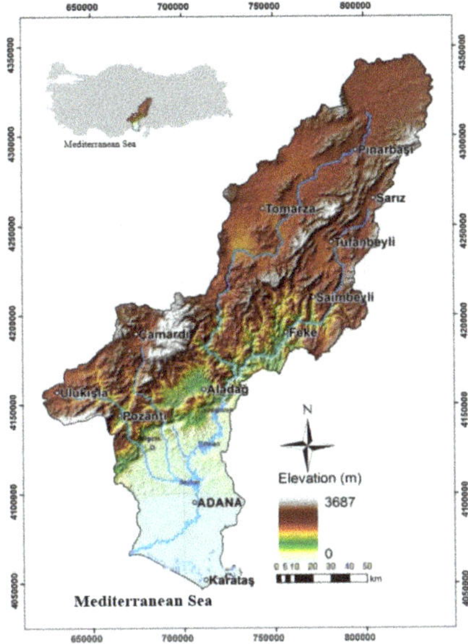

Figure 1. Topography of the Seyhan River basin.

Monthly precipitation data were obtained from 19 meteorological stations operated by the State Meteorological Service (MGM with its Turkish acronym) and from the General Directorate of State Hydraulic Works (DSI). The meteorological stations are listed in Table 1 in which the number and name of the stations are given together with the observation period of each station and the total number of missing data filled (in months). Statistical characteristics calculated from the monthly precipitation time series of each meteorological station are also given in Table 1. They are the minimum, maximum and mean values of monthly precipitation data in each meteorological station. It is seen, for example, that station 17981 (Karataş) among all meteorological stations has the highest percentage of no-rainy months (15.31%) and also the lowest altitude (22 m above mean sea level) compared to other stations. The altitude of the meteorological stations varies greatly within the basin owing to the topography. Data quality is an important issue in meteorological applications [33,34]. Any meteorological station with a minimum of 10 years of observation is taken for the analysis. Any gap in the data not exceeding 12 consecutive months was filled; longer gaps, if any, were not filled. In the case that a gap longer than 12 months exists in the time series, data before or after the gap were used provided that either portion is at least 10 years long. From the data before or after such a gap, the longer one was taken. The long-term monthly mean was taken to replace for the missing value of the month in the gap.

Precipitation time series of the 19 meteorological stations satisfied the above criteria; i.e., all-time series have 10-year minimum length with a missing data gap, if any, of the 12-month uninterrupted length at maximum. The layout of the meteorological stations in the Seyhan River basin is shown in Figure 2 from which it is seen that the stations are scattered over the basin almost homogenously.

Table 1. Meteorological stations in the Seyhan River basin.

Code	Station Name	Latitude	Longitude	Altitude (m)	Institution	Observation Period	Missing Data (month)	No-Rainy Months (%)	Mean (mm)	Min (mm)	Max (mm)
6204	Tufanbeyli	38.2600	36.2195	1415	MGM	1998–2012	3	6.67	545	312	706
6560	Saimbeyli	37.9811	36.0853	1050	MGM	1986–1995	4	4.17	923	625	1233
6893	Çamardı	37.8358	34.9975	1603	MGM	1969–1982	1	7.74	412	316	546
6902	Feke	37.7764	35.9000	583	MGM	1970–1993	1	4.86	910	598	1352
17351	Adana Bölge	37.0041	35.3443	23	MGM	1960–2016	0	11.40	663	317	1265
17802	Kayseri Pınarbaşı	38.7251	36.3904	1542	MGM	1963–2009	5	4.08	423	267	597
17837	Tomarza	38.4522	35.7912	1402	MGM	1965–2010	8	4.53	408	269	585
17840	Sarız	38.4781	36.5035	1599	MGM	1968–2011	0	5.30	524	354	748
17906	Ulukışla	37.5480	34.4867	1453	MGM	1962–2011	11	6.33	322	182	428
17934	Pozantı	37.4758	34.9022	1080	MGM	1963–1992	24	6.11	719	380	1299
17936	Karaisalı	37.2505	35.0628	240	MGM	1965–2011	0	4.61	881	437	1451
17981	Karataş	36.5683	35.3894	22	MGM	1963–2011	5	15.31	777	366	1365
D18M003	Uzunpınar	38.9710	36.8990	1740	DSI	1959–2005	11	7.45	303	161	493
D18M004	Seyhan Baraj	37.7000	35.0830	55	DSI	1974–2015	4	14.48	657	314	1117
D18M011	Kazancık	39.0670	36.7330	1585	DSI	1965–2003	8	7.69	274	176	433
D18M012	Hasan Çavuşlar	37.8330	35.5830	1400	DSI	1990–2005	0	3.65	1006	713	1539
D18M013	Kamışlı	37.5670	34.9500	1225	DSI	1963–2002	2	9.17	628	328	1123
D18M018	Gıcak	37.5670	35.2000	975	DSI	1988–2006	0	10.53	843	520	1173
D18M019	Çeralan	38.1000	36.0000	1600	DSI	1991–2005	4	6.67	970	622	1342

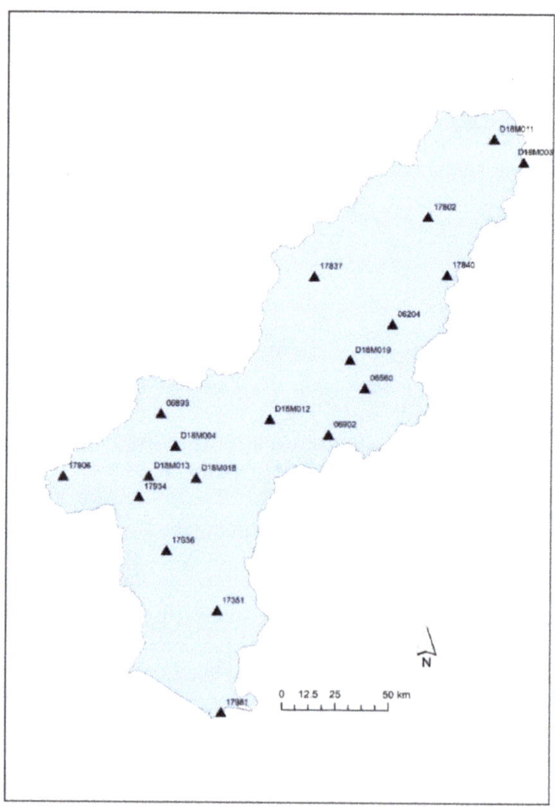

Figure 2. Layout of the meteorological stations in the Seyhan River basin.

The average annual total precipitation is 641 mm in the Seyhan River basin; it changes from 274 mm in Kazancik (D18M011) in the north to 1006 mm in Hasan Cavuslar (D18M012) in the central part of the basin. A general decreasing tendency in the precipitation is observed from the coastal south to the north in the basin (Figure 3). As for the temporal change in the precipitation of the Seyhan River basin, three meteorological stations were taken: Adana (17351) from the southern, Feke (6902) from the central and Uzunpinar (D18M003) from the northern part (Figure 4). The annual total precipitation time series of the meteorological stations depict a visual stationarity along the observation period, meaning that no significant trend is observed. Also, a parallel tendency is noticed from the temporal fluctuations in the precipitation of the selected meteorological stations. This is simply connected to the size of the basin which is not so large as to create a difference.

The drainage area of the basin is 20,731 km^2 which is composed of 2.82% of the surface area of Turkey. The annual mean flow at the outlet of the basin to the Mediterranean Sea is 211.07 m^3/s. The most important river in the basin is the Seyhan River, which gives its name to the basin. It is formed by the confluence of Zamanti and Goksu rivers and flows into the Mediterranean Sea. It has a length of 560 km, making it one of the largest rivers in Turkey [35,36].

The Seyhan River basin has been studied widely due to its importance for irrigation, energy and flood control as well as water resources management and planning. For example, *the Impact of Climate Change on Water Resources Project*, *the Seyhan Basin Pollution Prevention Action Plan* and *the Seyhan Basin Sectoral Water Allocation Plan* are among the studies completed by the General Directorate of Water Management [35,36] and *the Water Management and Preparation of Basin Protection Action Plan* is one performed by the Scientific and Technological Research Council of Turkey [37]. Further examples to be mentioned are given by Dikici et al. [38] who studied drought analysis with the Palmer drought index, Selek and Tuncok [39] who examined the effect of climate change on surface water management, and Topaloğlu [40] who determined the best-fit probability distribution functions for flow and precipitation in the basin. It should be emphasized that these are only a very short list of the studies conducted in the Seyhan River basin. On the other hand, making observation and mapping the formation of drought by using precipitation, soil moisture and plant as indicators the European Drought Observatory (EDO) has found that the most severe drought and precipitation deficit in the Seyhan River basin was recorded in 1990.

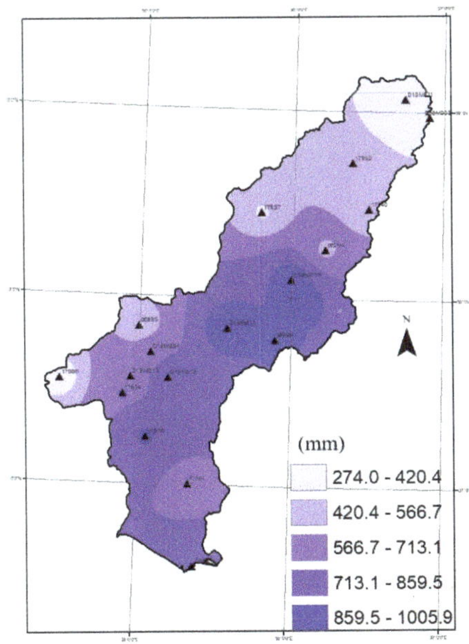

Figure 3. Spatial change in the annual precipitation over the Seyhan River basin.

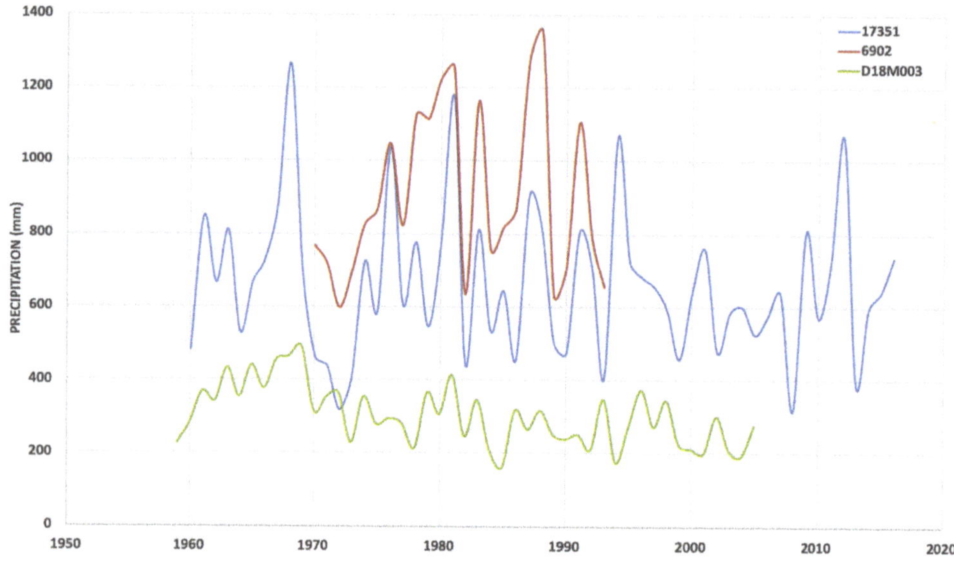

Figure 4. Temporal variability in precipitation.

3. Methodology

3.1. Standardized Precipitation Index (SPI)

With the help of SPI [41], dry periods of the monthly precipitation can be identified and monitored as well as wet periods. The SPI is obtained simply as the standard normal variable transformed from the gamma distribution [42] by (Equation (1))

$$SPI_{ij} = \frac{x_{ij} - \mu_j}{\sigma_j}$$

(1)

where x_{ij} is the precipitation (in mm) in month j ($j = 1, 2, 3, \ldots \ldots , 12$) of year i ($i = 1, 2, \ldots , n$); μ_j and σ_j, are the average and standard deviation of precipitation (in mm) in month j, respectively. The SPI values are calculated for different time scales such as k = 1, 3, 6, 9, 12, 24 months. These arbitrarily selected time scales are used to represent the three types of drought: meteorological, agricultural, and ydrological [41].

3.2. Frequency Analysis

Frequency analysis is performed to characterize the drought and to determine the best-fit probability distribution function. In the frequency analysis of drought, the 2- and 3-parameter Gamma (G2, G3), the Generalized Extreme Value (GEV), the 2- and 3-parameter log-normal (LN2, LN3), Log-Pearson Type 3 (LP3) and the 2- and 3-parameter Weibull (W2, W3) probability distribution functions are often used in the literature [1,3]. For the sake of consistency with the literature, the above probability distribution functions were considered for the frequency analysis of drought in this study.

For months, when no precipitation is observed in certain years, the frequency analysis is applied on the non-zero values only to distinguish zero values from non-zero values, because the frequency analysis would otherwise not be meaningful. Any process that is composed of zero and non-zero values is called a censored or intermittent process. The total probability theorem is available to use for such processes to examine them in two parts: the zero- and non-zero parts.

According to the total probability theorem (Equation (2)) [43]

$$P(X \geq x) = P(X \geq x|X = 0)P(X = 0) + P(X \geq x|X \neq 0)P(X \neq 0) \tag{2}$$

is used. Here (Equation (3)),

$$P(X \geq x|X = 0) = 0 \tag{3}$$

Therefore, Equation (4)

$$P(X \geq x) = P(X \geq x|X \neq 0)P(X \neq 0) \tag{4}$$

is obtained.

$P(X \neq 0)$ is the rate of years with non-zero values in the SPI_k (k = 1, 3, 6, 9, 12, 24 months) time series. Equation (4) can also be written in terms of the cumulative probability distribution as

$$1 - F(x) = (1 - p)[1 - F^*(x)] \tag{5}$$

In Equation (5), p is the probability of zero values. $F(x)$ is the cumulative probability distribution function of all X including zeros which is expressed as $P(X \leq x|X \geq 0)$ and $F^*(x)$ is the cumulative probability distribution function of the non-zero values of X which is expressed as $P(X \leq x|X \neq 0)$. The rate of the non-zero values, $1 - p$ in Equation (5), can be expressed in terms of the probability as (Equation (6))

$$1 - p = P(X \neq 0) \tag{6}$$

The magnitude of an event with return period T can be predicted by solving Equation (5) for $F^*(x)$ and then by using the inverse transformation of $F^*(x)$ to get the value of X. From Equation (5), Equation (7)

$$\frac{1 - F(x)}{1 - p} = 1 - F^*(x) \tag{7}$$

and Equation (8)

$$F^*(x) = \frac{F(x) - p}{1 - p} \tag{8}$$

can be written. Considering that the return period of a given severity for a particular drought duration can be predicted by Equation (9)

$$F(x) = 1 - \frac{1}{T} \tag{9}$$

Equation (8) turns into Equation (10)

$$F^*(x) = \frac{1 - \frac{1}{T} - p}{1 - p} \tag{10}$$

3.3. New Concepts and Definitions

The already known concepts related to the drought process are the dry period length, drought duration, drought severity and drought intensity. Also, the frequency or return period is used to characterize the drought. These concepts are defined as follows:

(a) *Dry period length (L)*: The cluster which consists of consecutive negative values of SPI is referred to as the dry period length (Figure 5). It begins in a month with a negative SPI and continues until a positive SPI value is obtained in the time series. A dry period is shown mathematically as in Equation (11)

$$A = \{SPI|SPI < 0\} \tag{11}$$

where $s(A)$ is the number of elements of set A that shows the length of the dry period as (Equation (12))

$$L = s(A) \tag{12}$$

(b) *Drought duration (D)*: Duration of droughts in an L-month long dry period is $1 \leq D \leq L$.

(c) *Drought severity (S)*: The accumulation of negative SPI values preceded and followed by positive SPI clusters is called severity. The severity of a drought D month-long is calculated by Equation (13)

$$S = \sum_{i=1}^{D} SPI_i, \quad SPI_i \in A \tag{13}$$

In other words, it is the largest absolute value of the cumulative drought index (SPI in this study) in the dry period considered (Equation (14)):

$$S = \sum_{i=1}^{D} |SPI_i| \tag{14}$$

(d) *Drought intensity (I)*: The intensity is obtained by dividing the severity of the drought by its duration (Equation (15)):

$$I = S/D \tag{15}$$

(e) *Return period (or frequency)*: The return period of a drought is defined as the average time between two consecutive drought events. The drought frequency decreases with the increasing return period.

In this study, the following definitions of Cavus [20] are also considered:

(f) *Critical drought severity*: When more than one drought is recorded for any year, drought with the maximum severity is taken as the critical drought. No critical drought is assigned to a year in which drought is not observed.

(g) *Within-period drought*: Any drought with a duration shorter than the dry period length is called within-period drought. For example, in a dry period of 3 months, there are three 1-month droughts and two 2-month droughts. Similarly, there are two 1-month droughts in a dry period of 2 months (Figure 5).

(h) *Singular drought*: Drought that extends over the dry period length is called a singular drought. For example, there exists a 1-month singular drought in a dry period of 1 month; a 2-month singular drought in a dry period of 2 months; a 3-month singular drought in a dry period of 3 months and so on. The length of dry period becomes the same as the drought duration for singular droughts while the former is larger than the latter for within-period droughts.

(i) *No drought year*: Any year with no negative run of SPI is considered a year with no drought. Thus, the critical drought severity is not calculated for such years; a zero value is assigned to the critical drought severity instead.

It should be emphasized that drought is a process which is different than a dry period. It occurs any time when the value of the drought index (SPI in this study) takes a negative value. The drought can be as short as 1 month and as long as the dry period length. However, the critical drought concept introduced in this study considers the most severe drought of the year and eliminates all other milder droughts observed in the same year. The critical drought concept is meteorological station-based and therefore, area under the drought episodes are not considered [44].

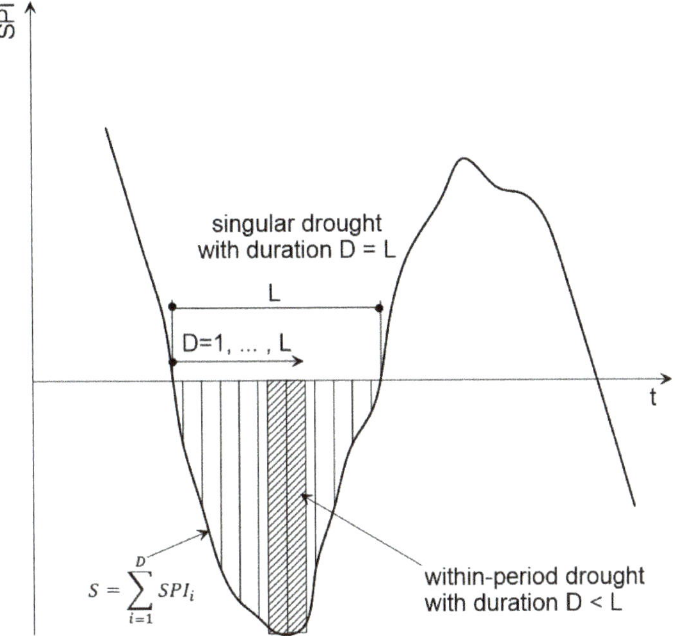

$$S = \sum_{i=1}^{D} SPI_i$$

Figure 5. Dry period length (L), drought duration (D), and drought severity (S). Dry period length (L) is determined as a fixed value for each dry period; drought duration (D) changes, however, from 1 month to L. Drought with duration L is called a singular drought while droughts with duration shorter than the dry period length are called within-period drought.

3.4. Precipitation Deficit

A drought is defined as a period in which the SPI is continuously negative [41,45,46]. In other words, it begins when the SPI first falls below zero and ends with a positive value of SPI [41]. Thus, the retrospective analysis of drought events by using runs of SPI values may be useful to derive tangible information for precipitation required to overcome the drought [19].

Instead of the direct use of drought indices to develop SDF curves as in previous studies [10], the precipitation deficit is calculated in this study. This approach helps to better understand as the accumulated precipitation is used to define the drought event so that it can be easily identified by end-users such as farmers and decision-makers.

As part of the methodology in this study, the relationship between the precipitation and SPI is detected by regression analysis. In the drought analysis, when the drought duration (D month) and return period (T years) increase, precipitation is expected to decrease, and therefore, the precipitation deficit is expected to increase. Any function to be applied on the relationship between precipitation and the corresponding SPI should satisfy this physical expectation. Some types of functions such as the second and the third order polynomials were omitted as they might produce negative precipitation deficit values against this physical reality. On the other hand, some functions such as Gompertz were discarded, because it was discovered after trials that they could not properly fit SPI_k time series of months with high number of zero values (such as SPI_1 in August–July). As a result, it was seen that the logistic function could be appropriate to choose among the functions tested due to the above expectation of the physical realization. The logistic regression equation was fitted to the relation between precipitation and the corresponding SPI values and therefore the logistic regression equation was used to analyze data clusters. It describes a family of sigmoidal curves. The simple logistic function has the form of (Equation (16))

$$f(x) = \frac{a}{1 + be^{-cx}}$$ (16)

in which a, b and c are parameters to be estimated through the use of data scatter.

The regression equation can be used to calculate the precipitation threshold value. For time scales 1, 3, 6 and 9 months, the relation between precipitation and SPI changes from month to month. That is, a particular function should be used for each month of the year. However, one curve exists for time scales of 12 and 24 months. Drought is classified depending on the value that the drought index takes [47]. Referring to Figure 6, the precipitation threshold (P_{TH}) was taken as precipitation at SPI = 0 for all time scales. Precipitation values at the boundary of drought classes ($P_{B,Extreme}$, $P_{B,Severe}$, $P_{B,Moderate}$, $P_{B,Mild}$) are determined based on the classification of McKee et al. [41] as shown in Figure 6. Also shown is the critical precipitation (P_c) which is expected to occur under a critical drought. The difference between the precipitation threshold and the critical precipitation is defined as the precipitation deficit and it is calculated by Equation (17)

$$P_D = P_{TH} - Pc$$ (17)

for each drought of a given duration and return period. A flowchart of the steps that will be implemented for calculating the precipitation deficit is given in Figure 7.

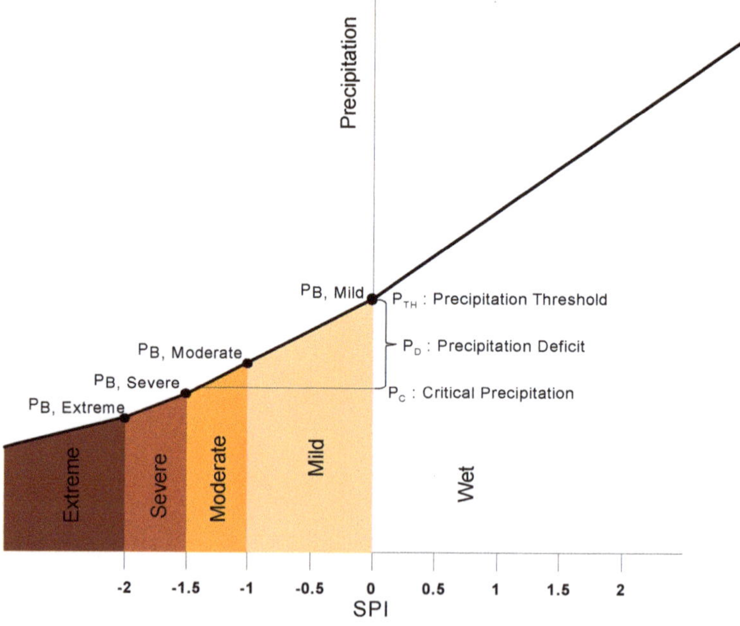

Figure 6. Definition of precipitation threshold, boundary precipitation and critical precipitation.

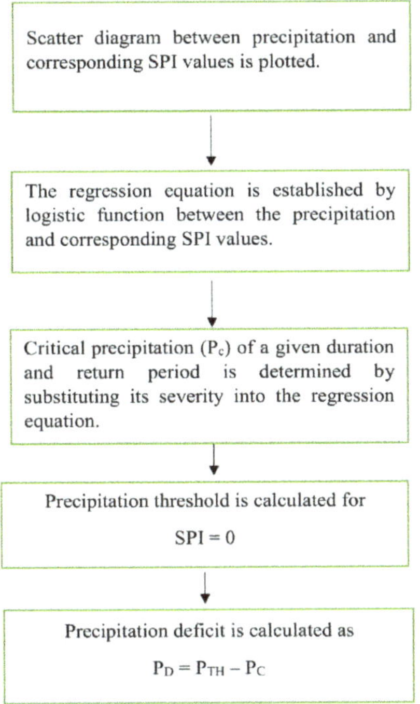

Figure 7. Step-by-step calculation of the precipitation deficit.

4. Spatial Mapping

In this study, the Seyhan River basin in the Mediterranean region of Turkey was investigated for its drought characteristics based on the precipitation deficit by considering 19 meteorological stations. SPI was applied to monthly precipitation data at the stations at k = 12-month time scale. Critical drought severity was calculated from the SPI time series which were first implemented by frequency analysis after which critical drought severities were calculated for return periods of 2, 5, 10, 25, 50 and 100 years. From the critical drought severity, the precipitation deficit of 1-, 3-, 6- and 12-month drought durations and 2-, 5-, 10-, 25-, 50- and 100-year return periods were determined at k = 12-month time scale. The drought intensity values were also obtained as the ratio of the drought severity to its duration. Examples to describe the above analysis are given in Table 2 from which it is clearly seen that no drought exists in a few meteorological stations for longer drought duration and return periods as the likelihood of any drought decreases as its duration and return period increase. Also, it is seen from Table 2 that no drought was determined in station 6560 at k = 12-month time scale although the station has experienced drought at lower time scales. This is due to the short length of the precipitation time series that do not allow one to make a proper frequency analysis and to quantify the drought.

Spatial mapping becomes useful in obtaining information for stations with smaller size observation that prevents making a proper frequency analysis or no observation at all [40]. The resulting precipitation deficit corresponding to D month-drought duration and T year-return period at k = 12-month time scale was mapped using the inverse distance weighted (IDW) interpolation technique (Figure 8). Only the precipitation deficit calculated from the SPI was considered in the interpolation process to derive the spatial mapping of the drought over the basin.

Figure 8 presents boundary values of precipitation deficit changes for each D-month drought. As given in Table 2, the D = 1-month drought has its own boundary value that changes between 42.6 mm (the 2-year return period drought in meteorological station 17802) and 567.9 mm (the 100-year return period drought in meteorological station D18M019). Similarly, for the drought of D = 3 month-duration, boundary values are between 31.2 mm (the 2-year return period drought in meteorological station D18M011) and 305.6 mm (the 100-year return period drought in meteorological station D18M012). This is also applied to D = 6 and 12-month drought durations for which the precipitation deficit intervals are 16.7 mm–231.2 mm and 22.1 mm–74.9 mm, respectively (Table 2). Therefore, it is important to emphasize that the maps in Figure 8 are only comparable for each of the drought durations, because the iso-contours of the precipitation deficit have been fixed for each of the drought durations separately. Darker colors in the maps of the same drought duration imply more severe drought. For example, it is seen that droughts become more severe with moving from shorter return periods to longer return periods for D = 1-month drought. This statement is correct for D = 3-, 6- and 12-month drought as well. However, when droughts with different durations are compared, it should be emphasized also that a darker color in a longer drought duration map does not necessarily mean that the drought is more severe than a drought with a lighter color in the shorter drought duration. Similarly, a lighter color does not necessarily mean the opposite; i.e., the drought is milder. Therefore, the discussion of results should be made through the joint use of Table 2 with Figure 8 to arrive at a conclusive discussion about the severity of the drought. It is clearly seen from Table 2 and Figure 8 that the drought severity becomes milder with the increasing longer return periods. This is an expected result of the fact that drought severity is absorbed along longer drought durations. This is a phenomenon quite similar to or even the same as the less intense precipitation of longer durations in the hydrological practice of precipitation intensity–duration–frequency curves.

The spatial distribution of D = 1-month drought duration indicates that more severe precipitation deficits tend to occur in the coastal parts of the basin for all return periods while the north-eastern part of the basin is prone to a lower precipitation deficit at the same return periods. In other words, the majority of the precipitation deficit that occurred in the coastal part was severe in D = 1-month drought duration. As the return period increases from 2 years to 100 years, the drought intensity increases and more severe intensities move towards the northern part of the basin. However, it is always lower in the north compared to the south. The intensity of precipitation deficit exhibits a more variable behavior over the basin when the return period increases.

At the D = 3-month drought duration, again more severe droughts are typically observed at the coastal part of the basin. Especially, the northern part of the Seyhan River basin exhibits a lower precipitation deficit. Nevertheless, as the return period increases, more severe droughts shift from south towards the north, as was the case for D = 1-month drought. It should be noted from Table 2 that for three meteorological stations (6204, 6560 and D18M019), no drought intensity is calculated. This is because the number of critical drought severities is less than 10 and the frequency analysis was therefore not applied on these particular meteorological stations. The number of such meteorological stations increases to 5 and 6 for D = 6 and 12 months, respectively. These stations are not indicated in the corresponding maps in Figure 8.

Another issue to discuss is the D = 12-month duration drought with 2-year return period (See blank column of T = 2-year return period in D = 12-month drought in Table 2 that corresponds to the blank lower-left cell of Figure 8). The mildest drought of D = 12-month has a return period longer than 2 years. In other words, once a 12-month drought is observed, it is as severe as a drought with a return period longer than 2 years. This is the case for 19 meteorological stations used in this study. Therefore, no map was created for D = 12-month drought at T = 2-year return period. Also, it is noticeable from Table 2 that this has been the case for five meteorological stations (6902, 17802, 17840, 17934, and 17936). Maps in Figure 8 have been created by using meteorological stations for which the precipitation deficit is calculated for a given drought duration and return period.

Table 2. Drought intensity based on precipitation deficit corresponding to 2, 5, 10, 25, 50 and 100-year return periods at k = 12-month time scale.

Station no	D = 1 Month						D = 3 Months						D = 6 Months						D = 12 Months					
Return Period	2	5	10	25	50	100	2	5	10	25	50	100	2	5	10	25	50	100	2	5	10	25	50	100
6204	50.4	152.7	214.8	281.1	321.0	353.5																		
6560																								
6893	81.3	108.7	123.0	138.6	148.7	158.1	65.3	79.3	86.1	93.3	97.8	101.8												
6902	174.3	280.5	320.4	354.2	371.5	384.3	96.4	139.7	218.4	230.4	235.4	238.6		131.3	137.6	141.1	142.5	143.4		73.4	74.4	74.7	74.8	74.9
17351	142.0	230.9	273.7	316.4	342.1	363.7	79.9	150.2	172.4	187.7	194.3	198.6	44.1	96.9	102.4	105.3	106.2	106.7		52.6	53.4	53.6	53.6	53.7
17802	42.6	113.4	156.4	199.8	224.6	243.7	31.4	81.4	99.4	114.8	122.5	127.9		56.2	63.8	67.8	69.0	69.5		32.7	34.4	34.8	34.9	34.9
17837	66.6	113.9	136.9	158.7	170.8	180.4	48.1	76.9	90.7	103.8	111.0	116.5	34.0	52.5	58.8	63.3	65.1	66.1		29.9	32.3	33.2	33.4	33.5
17840	69.7	136.4	170.9	207.8	231.4	252.2	38.6	94.3	113.4	130.4	139.6	146.7		66.6	75.4	81.3	83.7	85.2		40.2	42.3	43.2	43.5	43.6
17906	42.7	96.5	125.2	151.9	165.9	176.4	32.0	66.1	78.4	88.1	92.7	95.9	22.1	45.9	49.7	51.5	52.1	52.4		25.6	26.3	26.4	26.5	26.5
17934	146.1	242.9	294.1	348.3	382.5	412.4	98.4	154.5	178.1	199.0	209.8	217.5		98.9	107.6	113.7	115.7	116.6						
17936	150.8	296.0	374.1	452.4	498.3	535.5	105.8	204.3	228.8	246.5	254.7	260.4		129.6	136.4	140.0	141.2	141.9		70.5	71.3	71.5	71.5	71.6
17981	149.7	270.4	325.1	377.9	408.8	434.0	80.4	179.0	201.9	218.4	226.0	231.2	80.4	179.0	201.9	218.4	226.0	231.2		62.3	62.9	63.1	63.1	63.1
D18M003	57.9	109.1	134.8	159.3	173.0	183.9	45.7	72.1	80.9	87.7	90.8	93.0	31.0	45.3	47.6	48.6	48.9	49.1		24.3	24.5	24.6	24.6	24.6
D18M004	129.2	208.8	254.0	305.7	340.6	372.8	89.4	141.5	161.9	180.1	189.7	196.8	58.9	95.0	100.8	104.3	105.7	106.5		52.2	53.4	53.8	53.9	54.0
D18M011	43.0	85.6	108.3	130.9	143.9	154.3	31.2	60.7	69.8	76.9	80.3	82.8	16.7	39.6	42.8	44.4	44.8	45.0		22.1	22.5	22.6	22.6	22.6
D18M012	152.6	247.5	312.4	396.4	459.6	521.3	70.3	165.4	216.2	264.3	288.9	305.6												
D18M013	146.7	234.8	276.7	317.9	342.2	362.1	96.9	150.2	166.8	179.4	185.3	189.5	61.2	94.0	98.8	101.1	101.8	102.2		50.3	51.1	51.3	51.3	51.3
D18M018	149.2	286.5	350.0	411.9	448.4	478.1	72.7	189.1	220.0	242.3	252.3	259.1												
D18M019	109.9	270.7	356.5	451.3	512.8	567.9																		

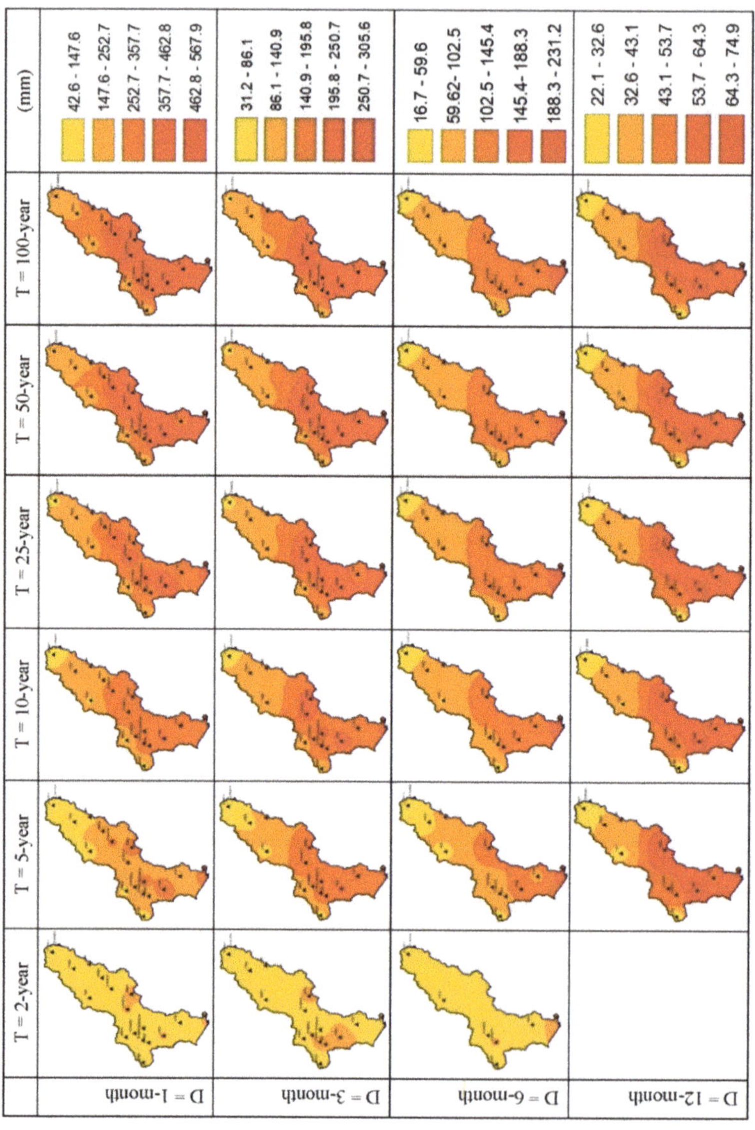

Figure 8. Intensity maps for droughts of 1, 3, 6 and 12-month durations and 2, 5, 10, 25, 50 and 100-year return periods over the Seyhan River basin.

Except for the northern part, the majority of the basin has more severe droughts for all return periods. A conclusive result is that the coastal part of the basin with higher precipitation (Figure 3) experiences more severe drought at all return periods while the northern part of the basin with lower precipitation is characterized by a milder drought. This can be explained by the fact that higher temperature in the southern coastal lowlands increases evapotranspiration that reduces the available precipitation substantially and gives an increase in the precipitation deficit to end up with more severe droughts. In the northern part of the basin with higher altitudes and lower temperature, droughts are milder compared to the southern part due to the net precipitation being reduced by the lower evapotranspiration. It shows also that the coastal parts of the basin are more likely to be affected from hydrological drought with a consequent loss in water resources.

5. Conclusions

This study was assessed as a presentation of a framework of methodologies for the analysis of the spatial characteristics of drought by utilizing frequency analysis in Seyhan River basin in the eastern part of the Mediterranean region in Turkey. The SPI calculated at the 12-month time scale was used as an indicator of drought. Precipitation deficit was newly introduced and calculated using regression analysis between SPI and the corresponding precipitation.

The precipitation deficit was calculated based on the SPI. Instead of the direct use of SPI, the severity/intensity is calculated using the precipitation deficit to characterize the drought to make the procedure more physically based for a better and simpler interpretation. This is because SPI is a technical tool and it is thus difficult to understand, at first glance, by end-users and decision-makers. Precipitation deficit in different durations and return periods is more useful and physically meaningful to the users. The direct information provided by the precipitation deficit allows planning and measures against drought to be taken in advance. Spatial analysis of drought might help decision makers to achieve better natural resources planning by considering the spatial drought vulnerability.

The spatial analysis indicates that the Seyhan River basin in the Mediterranean region of Turkey experiences droughts with quite different severities simultaneously. The spatial distribution would alter greatly with increasing return period and drought duration. While the coastal part of the basin is vulnerable to droughts at all return periods and drought durations, the northern part of the basin would be expected to be less affected by the drought. On the other hand, as the drought duration and return period increase, drought intensity based on precipitation deficit is expected to decrease. Another result reached in this study is that it could be common for one point in the basin to suffer from the drought, whilst surrounding points in the same basin experience normal or even humid conditions. This reinforces the importance of spatial analysis over the basin instead of the point-scale temporal analysis made in each meteorological station in the basin.

The drought characterization based on precipitation deficit completely changes at longer drought duration and return periods. The river basin experiences severe prolonged droughts, which means that, at $k = 12$-month time scale, the coastal part will suffer from severe hydrological drought.

Author Contributions: Conceptualization, H.A. and Y.C.; Data Curation, H.A.; Formal Analysis, H.A. and Y.C.; Investigation, H.A. and Y.C.; Methodology, H.A. and Y.C.; Software, Y.C.; Visualization, Y.C. and H.A.; Writing-Original Draft Preparation, Y.C.; Writing-Review & Editing, H.A. and Y.C.

Funding: This research was supported by Research Fund of the Istanbul Technical University. Project Number: 42141.

Conflicts of Interest: The authors declare no conflict of interest.

References

1. Mishra, A.K.; Singh, V.P. A review of drought concepts. *J. Hydrol.* **2010**, *391*, 202–216. [CrossRef]
2. Heim, R. A review of twentieth-century drought indices used in the United States. *Bull. Am. Meteorol. Soc.* **2002**, *83*, 1149–1165. [CrossRef]
3. Mishra, A.K.; Singh, V.P. Drought modeling—A review. *J. Hydrol.* **2011**, *403*, 157–175. [CrossRef]

4. Zargar, A.; Sadiq, R.; Naser, B.; Khan, F.I. A review of drought indices. *Environ. Rev.* **2011**, *19*, 333–349. [CrossRef]

5. Hao, Z.; Singh, V.P. Drought characterization from a multivariate perspective: A review. *J. Hydrol.* **2015**, *527*, 668–678. [CrossRef]

6. Eslamian, S.; Ostad-Ali-Askari, K.; Singh, V.P.; Dalezios, N.R.; Ghane, M.; Yihdego, Y.; Matouq, M. A review of drought indices. *Int. J. Constr. Res. Civ. Eng.* **2017**, *3*, 48–66.

7. Dracup, J.A.; Lee, K.S.; Paulson, E.G. On the statistical characteristics of drought events. *Water Resour. Res.* **1980**, *16*, 289–296. [CrossRef]

8. Rossi, G.; Benedini, M.; Tsakiris, G.; Giakoumakis, S. On regional drought estimation and analysis. *Water Resour. Manag.* **1992**, *6*, 249–277. [CrossRef]

9. Loaiciga, H.A.; Leipnik, R.B. Stochastic renewal model of Iow-fow streamflow sequences. *Stoch. Hydrol. Hydraul.* **1996**, *10*, 65–85. [CrossRef]

10. Dalezios, N.R.; Loukas, A.; Vasiliades, L.; Liakopoulos, E. Severity-duration-frequency analysis of droughts and wet periods in Greece. *Hydrol. Sci. J.* **2001**, *45*, 751–769. [CrossRef]

11. Tsakiris, G.; Vangelis, H. Establishing a drought index incorporating evapotranspiration. *Eur. Water* **2005**, *9*, 3–11.

12. Nalbantis, I.; Tsakiris, G. Assessment of hydrological drought revisited. *Water Resour. Manag.* **2009**, *23*, 881–897. [CrossRef]

13. Eris, E.; Aksoy, H. Persistency in wet and dry periods in Goztepe meteorological station in Istanbul, Turkey. In Proceedings of the AMHY-FRIEND International Workshop on Hydrological Extremes, Analyses and Images of Hydrological Extremes in Mediterranean Environments, Cosenza, Italy, 10–12 July 2008; pp. 93–99.

14. Vicente-Serrano, S.M.; Begueria, S.; Moreno, J.I.L. A multiscalar drought index sensitive to global warming: The standardized precipitation evapotranspiration index. *J. Clim.* **2010**, *23*, 1696–1718. [CrossRef]

15. Santos, J.; Portela, M.; Pulido-Calvo, I. Regional frequency analysis of droughts in portugal. *Water Resour. Manag.* **2011**, *25*, 3537–3558. [CrossRef]

16. Tigkas, D.; Vangelis, H.; Tsakiris, G. The drought indices calculator (DrinC). *Earth Sci. Inform.* **2015**, *8*, 697–709. [CrossRef]

17. Aksoy, H.; Cetin, M.; Onoz, B.; Yuce, M.I.; Eris, E.; Selek, B.; Aksu, H.; Burgan, H.I.; Esit, M.; Orta, S.; et al. *Hidrolojik havzalarda düşük akımlar ve kuraklık analizi*; Rapor No. TUJJB-TUMEHAP-2015-01; TUJJB Sonuç Raporu: Istanbul, Turkey, 2018.

18. Aksoy, H.; Onoz, B.; Cetin, M.; Yuce, M.I.; Eris, E.; Selek, B.; Aksu, H.; Burgan, H.I.; Esit, M.; Orta, S.; et al. SPI-based drought severity-duration-frequency analysis. In Proceedings of the 13th International Congress on Advances in Civil Engineering, Izmır, Turkey, 12–14 September 2018.

19. Cetin, M.; Aksoy, H.; Onoz, B.M.; Yuce, M.I.; Eris, E.; Selek, B.; Aksu, H.; Burgan, H.I.; Esit, M.; Cavus, Y.; et al. Deriving accumulated precipitation deficits from drought severity duration-frequency curves: A case study in Adana province, Turkey. In Proceedings of the 1st International, 14th National Congress on Agricultural Structures and Irrigation, Antalya, Turkey, 26–28 September 2018.

20. Cavus, Y. Critical Drought Severity-Duration-Frequency Curves Based on Precipitation Deficit. Master's Thesis, Hydraulics and Water Resources Programme, Department of Civil Engineering, Graduate School of Science, Engineering and Technology, Istanbul Technical University, İstanbul, Turkey, May 2019; 160p.

21. Turkes, M. Spatial and temporal analysis of annual rainfall variations in Turkey. *Int. J. Climatol.* **1996**, *16*, 1057–1076. [CrossRef]

22. Komuscu, A.U. An analysis of recent drought conditions in Turkey in relation to circulation patterns. Drought Network News, Summer–Fall 2001, University of Nebraska Lincoln. *Drought-Natl. Drought Mitig. Cent.* **2001**, *13*, 5–6.

23. Sonmez, F.K.; Komuscu, A.U.; Erkan, A.; Turgu, E. An analysis of spatial and temporal dimension of drought vulnerability in Turkey using the standardized precipitation index. *Nat. Hazard.* **2005**, *35*, 243–264. [CrossRef]

24. Jones, P.D.; Hulme, M.; Briffa, K.R.; Jones, C.G. Summer moisture availability over Europe in the Hadley centre general circulation model based on the Palmer drought severity index. *Int. J. Climatol.* **1996**, *16*, 155–172. [CrossRef]

25. Vicente-Serrano, S.M.; Begueria, S. Differences in spatial patterns of drought on different time scales: An analysis of the Iberian penisula. *Water Resour. Manag.* **2006**, *20*, 37–60. [CrossRef]

26. Vicente-Serrano, S.M.; Begueria, S. Evaluating the impact of drought using remote sensing in a Mediterranean, semi-arid region. *Nat. Hazard.* **2007**, *40*, 173–208. [CrossRef]

27. Caloiero, T.; Veltri, S.; Caloiero, P.; Frustaci, F. Drought analysis in Europe and in the Mediterranean basin using the standardized precipitation index. *Water* **2018**, *10*, 1043. [CrossRef]

28. Vicente-Serrano, S.M.; Gonzalez-Hidalgo, J.C.; Luis, M.D.; Raventos, J. Drought pattern in the Mediterranean area: The Valencia region (eastern Spain). *Clim. Res.* **2004**, *26*, 5–15. [CrossRef]

29. Spinoni, J.; Antofie, T.; Barbosa, P.; Bihari, Z.; Lakatos, M.; Szalai, S.; Szentimrey, T.; Vogt, J. An overview of drought events in the Carpathian Region in 1961–2010. *Adv. Sci. Res.* **2013**, *10*, 21–32. [CrossRef]

30. Bonaccorso, B.; Bordi, I.; Cancelliere, A.; Rossi, G.; Sutera, A. Spatial variability of drought: An analysis of the SPI in Sicily. *Water Resour. Manag.* **2003**, *17*, 273–296. [CrossRef]

31. Loukas, A.; Vasiliades, L. Probabilistic analysis of drought spatiotemporal characteristics in Thessaly region, Greece. *Nat. Hazards Earth Syst. Sci.* **2004**, *4*, 719–731. [CrossRef]

32. He, B.; Lv, A.F.; Wang, J.J.; Zhao, L.; Liu, M. Drought hazard assessment and spatial characteristics analysis in Chine. *J. Geogr. Sci.* **2011**, *21*, 235–249. [CrossRef]

33. Murara, P.; Acquaotta, F.; Garzena, D.; Fratianni, S. Daily precipitation extremes and their variations in the Itajaí River Basin, Brazil. *Meteorol. Atmos. Phys.* **2018**. [CrossRef]

34. Baronetti, A.; Acquaotta, F.; Fratianni, S. Rainfall variability from a dense rain gauge network in north-western Italy. *Clim. Res.* **2018**, *75*, 201–213. [CrossRef]

35. Su Yönetimi Genel Müdürlüğü. *Seyhan Havzası Sektörel Su Tahsis Planı Hazırlanması Projesi*; Impact of Climate Change on Water Resources Project; Ankara, Turkey, 2016; (In Turkish). Available online: https://www.tarimorman.gov.tr/SYGM/Belgeler/Seyhan%20Havzas%C4%B1/Sekt%C3%B6rel%20Su%20Tahsis%20Plan%C4%B1.pdf (accessed on 15 May 2019).

36. Su Yönetimi Genel Müdürlüğü. *İklim Değişikliğinin Su Kaynaklarına Etkisi Projesi, Proje Nihai Raporu, Ek 20-Seyhan Havzası*; Change on Water Resources Project; Ankara, Turkey, 2017; (In Turkish). Available online: http://iklim.ormansu.gov.tr/ckfinder/userfiles/files/Iklim_Nihai_Rapor_Seyhan_Ek_20_REV_nihai.pdf (accessed on 15 May 2019).

37. The Scientific and Technological Research Council of Turkey (TUBITAK) Marmara Research Center (MAM) (TUBITAK). *Havza Koruma Eylem Planları-Seyhan Havzası*; Water Management and Preparation of Basin Protection Action Plans; Ankara, Turkey, 2010; (In Turkish). Available online: https://www.tarimorman.gov.tr/SYGM/Belgeler/havza%20koruma%20eylem%20planlar%C4%B1/Seyhan_Havzasi.pdf (accessed on 15 May 2019).

38. Dikici, M.; Ipek, C.; Topcu, I. Seyhan havzasında Palmer indeksleri ile kuraklık analizi. In Proceedings of the 6th International Symposium on Innovative Technologies in Engineering and Science, Antalya, Turkey, 9–11 November 2018.

39. Selek, B.; Tuncok, I.K. Effects of climate change on surface water management of Seyhan basin, Turkey. *Environ. Ecol. Stat.* **2014**, *21*, 391–409. [CrossRef]

40. Topaloğlu, F. Determining suitable probability distribution models for flow and precipitation series of the Seyhan river basin. *Turk. J. Agric. For.* **2002**, *26*, 187–194.

41. McKee, T.B.; Doesken, N.J.; Kleist, J. The relationship of drought frequency and duration to time scales. In Proceedings of the Eighth Conference on Applied Climatology, American Meteorological Society, Anaheim, CA, USA, 17–23 January 1993.

42. Aksoy, H. Use of gamma distribution in hydrological analysis. *Turk. J. Eng. Environ. Tubitak* **2000**, *24*, 419–428.

43. Haan, C.T. *Statistical Methods in Hydrology*; Iowa State University Press: Ames, IA, USA, 1977.

44. González-Hidalgo, J.C.; Vicente-Serrano, S.M.; Peña-Angulo, D.; Salinas, C.; Tomas-Burguera, M.; Beguería, S. High-resolution spatio-temporal analyses of drought episodes in the western Mediterranean basin (Spanish mainland, Iberian Peninsula). *Acta Geophys.* **2018**, *66*, 381–392. [CrossRef]

45. Paula, A.; Pereira, L.S. Drought concepts and characterization: Comparing drought indices applied at local and regional scales. *Int. Water Resour. Assoc. Water Int.* **2006**, *31*, 37–49. [CrossRef]

46. Rahmat, S.N.; Jayasuriya, N.; Bhuiyan, M. Development of drought severity-duration-frequency curves in Victoria, Australia. *Aust. J. Water Resour.* **2015**, *19*, 156–160. [CrossRef]
47. Spinoni, J.; Naumann, G.; Carro, H.; Barbosa, P.; Vogt, J. World drought frequency, duration, and severity for 1951–2010. *Int. J. Climatol.* **2014**, *34*, 2792–2804. [CrossRef]

Article

A New 60-Year 1940/1999 Monthly-Gridded Rainfall Data Set for Africa

Claudine Dieulin [1,*], **Gil Mahé** [1,*], **Jean-Emmanuel Paturel** [1], **Soundouss Ejjiyar** [2], **Yves Tramblay** [1], **Nathalie Rouché** [1] and **Bouabid EL Mansouri** [2]

[1] HydroSciences Montpellier, Université de Montpellier–CC 57, 163 rue Auguste Broussonnet, 34090 Montpellier, France; jean-emmanuel.paturel@ird.fr (J.-E.P.); yves.tramblay@ird.fr (Y.T.); nathalie.rouche@ird.fr (N.R.)

[2] Faculty of Sciences, Hydro-Informatic Section, Campus Maamora, Ibn Tofail University, Kenitra BP 133–14000, Morocco; ejjiyar-soundouss@hotmail.com (S.E.); b_elmansouri@yahoo.fr (B.E.M.)

[*] Correspondence: claudine.dieulin@umontpellier.fr (C.D.); gil.mahe@ird.fr (G.M.); Tel.: +33-467-149-025 (C.D.); Fax.: +33-467-144-774 (C.D. & G.M.)

Received: 23 November 2018; Accepted: 13 February 2019; Published: 22 February 2019

Abstract: The African continent has a very low density of rain gauge stations, and long time-series for recent years are often limited and poorly available. In the context of global change, it is very important to be able to characterize the spatio-temporal variability of past rainfall, on the basis of datasets issued from observations, to correctly validate simulations. The quality of the rainfall data is for instance of very high importance to improve the ef ficiency of the hydrological modeling, through calibration/validation experiments. The HydroSciences Montpellier Laboratory (HSM) has a long experience in collecting and managing hydro-climatological data. Thus, HSM had initiated a program to elaborate a reference dataset, in order to build monthly rainfall grids over the African continent, over a period of 60 years (1940/1999). The large quantity of data collected (about 7000 measurement points were used in this project) allowed for interpolation using only observed data, with no statistical use of a reference period. Compared to other databases that are used to build the grids of the Global Historical Climatology Network (GHCN) or the Climatic Research Unit of University of East Anglia, UK (CRU), the number of available observational stations was significantly much higher, including the end of the century when the number of measurement stations dropped dramatically, everywhere. Inverse distance weighed (IDW) was the chosen method to build the 720 monthly grids and a mean annual grid, from rain gauges. The mean annual grid was compared to the CRU grid. The grids were significantly different in many places, especially in North Africa, Sahel, the horn of Africa, and the South Western coast of Africa, with HSM_SIEREM data (database HydroSciences Montpellier_Système d'Information Environnementales pour les Ressources en Eau et leur Modélisation) being closer to the observed rain gauge values. The quality of the grids computed was checked, following two approaches—cross-validation of the two interpolation methods, ordinary kriging and inverse distance weighting, which gave a comparable reliability, with regards to the observed data, long time-series analysis, and analysis of long-term signals over the continent, compared to previous studies. The statistical tests, computed on the observed and gridded data, detected a rupture in the rainfall regime around 1979/1980, on the scale of the whole continent; this was congruent with the results in the literature. At the monthly time-scale, the most widely observed signal over the period of 1940/1999, was a significant decrease of the austral rainy season between March and May, which has not earlier been well-documented. Thus, this would lead to a further detailed climatological study from this HSM_SIEREM database.

Keywords: Africa; rainfall; monthly grids; database; inverse distance weighted

1. Introduction

On a global scale, the first climate observations began during the second part of the 19th century. Among these climate variables, rainfall data sets have been the most complete, since the beginning of the 1950s [1]. In Africa, the most well-known change in rainfall regime occurred in Sahel, but all of Western and Central Africa experienced an abrupt decrease in mean annual rainfall, at the end of the 1960s or towards the beginning of the 1970s [2–4]. To study rainfall changes at the scale of the whole African continent, rainfall grids can be downloaded from several institutions, but it is well-known that the African continent is less documented than other parts of the world [5,6]. This has led to some discrepancies between rainfall calculated from different databases, as has been shown for Western Africa, by Mahé et al. [7]. Among the available data sets, Climatic Research Unit of University of East Anglia, UK (CRU) [8,9] is used very often and seems to be the most accurate, compared to other sources. However, for instance, for Western and Central Africa, rainfall data are not very well-documented [10,11].

The first objective of the present work is to elaborate monthly grids of the best possible quality, over the African continent. For this, it was necessary to build the most exhaustive rainfall database to help improve understanding of the climatic processes and other applications, such as hydrological modeling. Hydrologists at the Institut de Recherche pour le Développement (IRD) or the French National Research Institute for Sustainable Development, previously ORSTOM (Office de Recherche Scientifique Outre-Mer), had historically participated in the acquisition of climatological data and built up a digital database, as early as the end of the 1960s. However, by the end of the 1980s, the IRD stopped collecting and managing rainfall data for Western and Central Africa. The researchers of the HSM decided to valorize this unique set of data for Africa, first, in the search of additional data to cover the whole continent, and to update the database up to 1999. This project was an opportunity to create a reference database, called HSM_SIEREM database, which is still managed by the HSM. The HSM_SIEREM [12] database has been used for sub-continental studies of climate change, like those by Paturel et al. [10,11], many regional studies of river-runoff relationships [13,14], and studies of the impact of climate change on hydrological regimes in Western Africa [15]. These data were first used at a continental scale, for a study of rainfall-runoff variability over the past decades [16].

In this article, we have presented, for the first time, the methods through which the data were selected and evaluated, the process leading to the creation of the monthly-gridded dataset over Africa (at a 0.5 × 0.5 degree resolution), and a first assessment of the content of this new database. In the second part, we have compared this new HSM_SIEREM grid, with the CRU grid and have pointed out the similarities and the differences to help improve the future uses of both bases. In the third part, we have presented a first study of rainfall variability, over the 1940/1999 time period, for the whole continent, to check the general quality of the dataset, by searching if the main rainfall signals were visible in large-scale regional averaged time-series. This first study will be followed by further detailed studies on the spatio-temporal variability of rainfall over the African continent. The monthly grids we created have been provided, for free, on the website of the HydroSciences Montpellier SIEREM project (http://www.hydrosciences.org/spip.php?article1387), as zipped ASCII grids and in the NetCDF format.

2. Materials and Methods

2.1. History of the HSM_SIEREM Database

Created in 1943, ORSTOM/IRD's mission was to promote scientific cooperation and education in Western and Central Africa. In this framework, hydrologists installed and managed most of the hydro-meteorological stations in French speaking countries—the former colonies. The records were noted in paper booklets and a long process of data entry began, as early as 1967, on punched cards, first, and were then transferred onto magnetic tapes of the CNRS (Centre National de la Recherche Scientifique) computers in Orsay/Paris. This transfer was taken into account to run a first control on the data quality check for duplicate cards, removal of data for non-existent days (such as, 31st of April and 30th of February), before building up the first database. These cards were individually controlled and were modified, kept or removed. During this process, every measurement point was linked to only one data set. Due to the state of technology at the time the data was collected, some measurement stations had no geographical referencing, and it was added later, when possible.

A tripartite agreement between the Comité Interafricain d'Etudes Hydrauliques (CIEH) or the Interafrican Hydraulic studies Committee, the Agence pour la SECurité de la Navigation Aérienne (ASECNA) or Agency for the Safety of Aerial Navigation in Africa and Madagascar, and the ORSTOM, began in 1973, for the collection of data from thirteen countries of Western and Central Africa (Benin, Burkina Faso, Cameroon, Congo, Ivory Coast, Gabon, Mali, Mauritania, Niger, Central Africa, Senegal, Chad, and Togo) and ended in 1989, with the edition of the two rainfall data books, for each country, covering the years up to 1965 and 1966, to 1980 [5]. This agreement allowed to update the database at a daily time-step, up to 1980. Figure 1 [17] shows the benefits of this agreement in the evolution of the rainfall database. From this set of data, a map of the mean annual rainfall over Western and Central Africa was drawn by L'Hote & Mahé [18].

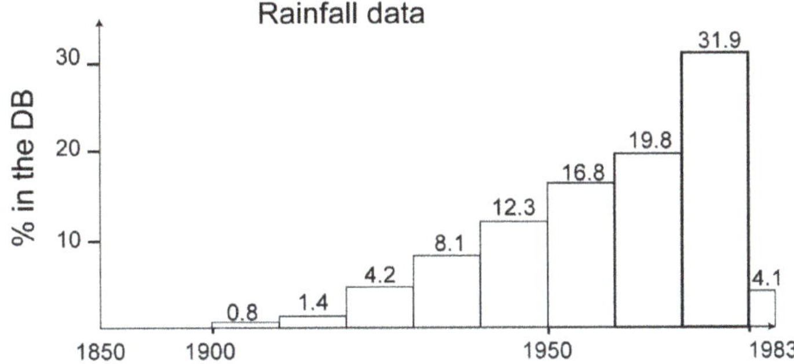

Figure 1. Distribution of data in the ORSTOM rainfall database from the origin of the observation to 1983 [17].

The Water Assessment [19] financed by the World Bank, the UNDP (United Nations Development Program), the African Bank of Development, and the French ministry of Cooperation, initiated a program to collect monthly rainfall data over the whole African continent, for the period of 1981/1990. This program was carried out by Mott MacDonald International, the BCEOM (Bureau Central d'études pour les Equipements Outre-Mer) and the SOGREAH (Société Grenobloise d'Etudes et d'Applications Hydrauliques) offices for the Western and Central part of Africa, and the hydrologists of ORSTOM, who participated to the gathering of data in the Sub-Saharan area.

Up to 1999, the IRD Hydrology Laboratory in Montpellier gathered and managed all the climatological data archived in every country where ORSTOM hydrologists were present, mainly in Africa, as well as the rainfall data from all past programs. Since 2000, the HSM_SIEREM database was only used and enriched by the teams of HSM in Montpellier, with no institutional mandate for commitment to the management or the dissemination of this base. For Eastern and Southern Africa, HSM gathered data through its commitment in the IHP (International Hydrological Program) of UNESCO and mainly in the FRIEND programs (Flow Regimes from International Experimental and Network Data), as HSM managed the implementation and hosting of the databases of the different African FRIEND groups. Many other international programs that the HSM collaborated on in Africa, allowed the IRD hydrologists access to more recent data.

The CRU is specifically acknowledged, as they had developed a collaboration with the hydrologists of HSM which let them access the rainfall data in the areas where HSM did not have any access. However, both CRU and HSM databases where kept independent.

The HSM_SIEREM database, thus, contains rainfall data for more than 6000 stations across Africa.

2.1.1. The Reference Database

The development of a continuous, quality-checked and reliable long-term database of monthly rainfall across Africa, was a key component of this project. The database contained numerous records, due to the fact that any incoming rainfall datum was always stored. According to Rouché et al. [20], the result was that, for some stations, up to ten different data sets were stored (collected by different people, at different time-steps, data corrected for specific purposes, etc.). Therefore, it was essential to eliminate redundant sets and choose the better series.

Constitution of the Reference Data Set

The dataset was built for every African country. Records with missing geographical coordinates were removed, daily, and ten-day data were aggregated at the monthly time-step. Then, if several series for the same station still remained in the database, a "quality label" was given to the series (depending on the origin of the data, the reliability of the provider, and the length of the observation period), to select which one was to be kept. When the retained series had missing periods, it was filled with data from a different station of the same measurement point, when available.

The aim of this process was to keep only one station per measurement point. The graph in Figure 2 shows the number of sets before and after the creation of the reference database, for every country.

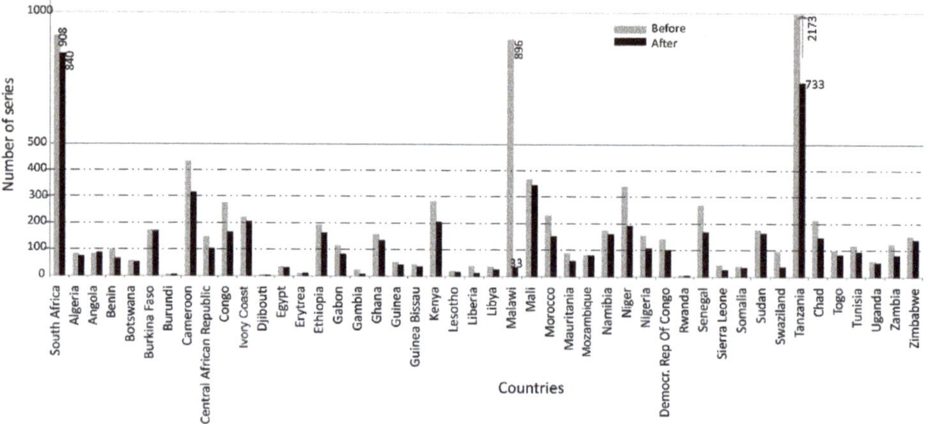

Figure 2. Result of the selection process between redundant time-series—in grey, the number of series at the beginning, and in black, the final number of reference series for the period 1900/1999, where each series corresponds to a unique station. Note: Due to insufficient data availability, Madagascar has not been processed in this project.

Criticism of the Reference Data Set

For every country, patterns were established to define:

- the period of the rainy and dry seasons; and
- a range of monthly rainfall values (minimum and maximum) that were not to be crossed.

During this process, a series of automatic tests were run, and a flag was inserted if any value:

- exceeded the previously defined values for the country;
- was identical to the value of the same month of the previous or next year at the station; or
- was identical to the value of the previous month of the same year.

The flagged months were printed and checked by the HSM hydrologists, who determined if the flag led to the removal of the data or of part of the time-series. This step was the longest but the most important for the quality of the database. Thus, the time-series of each station was both automatically and manually checked, to reduce the risk of erroneous data in the reference database.

2.1.2. The Period

The HSM_SIEREM base was exported in the form of one file per country and per year, over the period 1900/1999. The first goal was to create monthly grids for the whole of the 20th century. However, when plotting the measurement points, the maps showed that before 1940 (Figure 3), the data were too sparse and too heterogeneously distributed to be spatialized over the whole continent. The decision was then made to limit the interpolation process to the period 1940/1999. Contrary to CRU, we decided not to create a reference period like the ones the CRU used to establish patterns to create interpolated rainfall measurement points [9]—even for the years where the distribution of data was heterogeneous or density of stations was low—enabling them to start the interpolated time-series in 1900. This limited our possibility to extend the grid to a period prior to 1940. However, this is also what makes a fundamental difference between the HSM_SIEREM and the CRU grids, i.e., in the HSM_SIEREM database, each month, the grid was calculated only from the available observed data.

Figure 3. Evolution of the number of measurement stations in 1900, 1920, and 1940, in the HSM_SIEREM rainfall database.

All African National Meteorological services were then contacted to update the data series at a monthly time-step, especially from the last 20 years of the 20th century, but only a few of them sent recent monthly data. The number of months (Figure 4) with available data, therefore, varies from 4061 stations in 1975 to 1464 stations in 1999. This shows a decrease of the number of observed rain gauges; this trend started as soon as the 1980s and has been observed worldwide, but it affects Africa more severely. However, the HSM_SIEREM database still has a sufficient number of points, compared to the Global Historical Climatology Network (GHCN) values—2500 gauges, worldwide, for the year 2000 [5], compared to a little less than 1,500 for Africa alone, in our database, for December 1999.

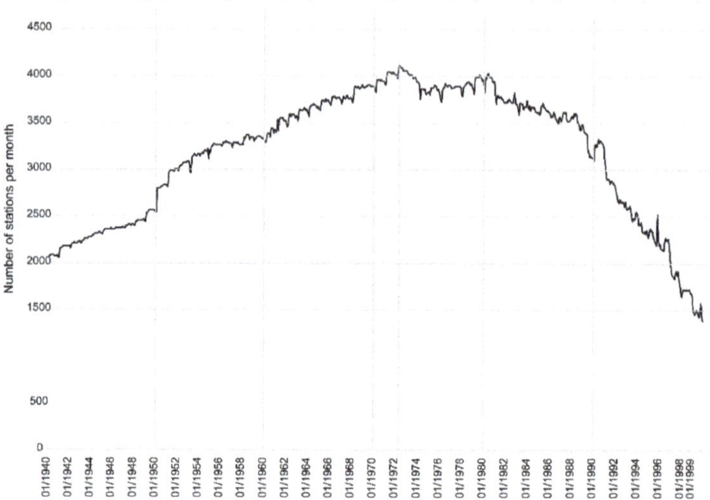

Figure 4. Number of monthly data observed in the HSM_SIEREM base, per year, from January 1940 to December 1999.

The scarcity of rainfall data was a recurrent problem, whatever the purpose [5,9,21]. Due to its history, the HSM laboratory collected a large amount of data. However, it is not allowed, either by the National Meteorological Services or the WMO (World Meteorological Organization) to disseminate raw rainfall data, as they are kept as the property of these Services. It was then decided that interpolated grids would be created, which would be disseminated for free to the community as the HSM_SIEREM database. This database has already been compared to the CRU for hydrological modeling in Western and Central Africa, and had given better results [7].

2.2. Geographical Processing

2.2.1. Reference of the CRU Grid Constitution

Our references for rainfall grids over the African continent were the CRU grids. These were built at a resolution of 0.5 × 0.5 degree. They concern seven climatological variables, namely precipitation, wet-day frequency, mean temperature, diurnal temperature range, vapor pressure, cloud cover, and ground frost frequency. For all data types, the construction of a monthly surface climate over global land areas, excluding Antarctica, covers the 1901/1996 time period (period of the first version, later update extended this period up to the 21st century). An "anomaly" approach was used, this technique attempted to maximize the available station data in space and time. "In this technique, grids of monthly anomalies relative to a standard normal period (1961/1990) (. . .) were first derived. The anomaly grids were then combined with a high-resolution mean monthly climatology to arrive at fields of estimated monthly surface climate." [8]. Rainfall anomalies are expressed in percentage. The standard period was chosen because of the good quantity of data available for this period.

The consistency checks were somewhere similar to the ones we applied on our data sets, defining monthly minimum, mean, and maximum values, and checking the values out of these ranges for confirmation or removal.

The anomalies over the standard normal period of rainfall were calculated, the CRU used tiles; 29 tiles were defined worldwide, with 4 tiles being defined for the African continent: tile 17 (20°W to 20°E, 38°N to 0°), tile 18 (20°E to 52°E, 38°N to 0°), tile 19 (0° to 52°E, 0° to 20°S), and tile 20 (10° to 52°E, 20°S to 36°S).

Three global station datasets were compiled by the CRU, from the basis of the construction of the gridded anomalies of primary variables. The precipitations and mean temperatures were compiled by the CRU, the diurnal temperature range dataset was based on the GHCN maximum and minimum temperature data.

The interpolation method adopted to process the tiles was a thin-plate spline, a function of latitude, longitude, and elevation. "The technique is robust in areas with sparse or irregularly spaced data points. The main advantage of splines over many other geostatistical methods is that prior estimation of the spatial autocovariance structure is not required" [8].

The process was computed per tile at a step of 0.5 × 0.5 degrees, with the tiles overlapping by at least 5 × 5 degrees (the number of stations varied from 200 to 1000 per tile). On the overlapping areas, the grid values were calculated as a weighted average of the contributing tiles; the weights were simply grid points between a particular point and the edge of its tile. The grids were constrained to avoid unrealistic values for rainfall, and the negative values were converted to zero.

"A GCV and its square root (RTGCV) provide an estimate of the mean predictive error (and hence power) of the fitted surface and as such permit an assessment of the relative accuracy of a fitted surface" [8].

The second step was to collect mean temperature, diurnal temperature, and precipitation over the 1901/1996 period. Any series with less than 20 years of data during the 1961/1990 period were excluded from the analysis. Rainfall data were expressed in percentages. Each station time-series was converted to anomalies, relative to the 1961/1990 mean.

The interpolation of the monthly anomaly method chosen for the 1901/1996 series was ADW (Angular Distance-Weighted), using the eight nearest stations, except when there were more than eight stations within a single 0.5° grid cell (only three cells were used in this case). The ADW gridding employed in this study did not permit the inclusion of elevation as a predictor, it was, thus, not included in a tri-variate interpolation technique [9]. "To prevent extrapolation to unrealistic values, the interpolated anomaly fields were forced towards zero at grid points beyond the influence of any stations. This was accomplished by creating synthetic stations with anomaly values of zero in regions where there were no stations within a predefined distance chosen to be equal to the global-mean CDD"; (450 km for precipitation) [9].

2.2.2. Choice of the Interpolation Process Options

Compared to the CRU standard normal reference period, we had a sufficient number of observations for a longer period and could interpolate rainfall from the observed values at each monthly time-step (Figure 5).

One file per month was created in the period 1940/1999 (i.e., 720 files). All files of monthly data were imported in the ArcMap and a point shapefile was created. The following processes were computed, using the ArcInfo Workstation macro-command language (AML). This ensured a homogeneous process.

Considering the size of the region to process and the heterogeneity of the distribution of rain gauge stations, and taking advantage of the CRU experience, the IDW (Inverse Distance Weighted) method was chosen.

With the ArcInfo Workstation, the IDW interpolation method does not allow to take into account the altitude of the measurement points. A few things can be discussed about this point. Due to the very different geographical situations across Africa, the question of altitude influence on precipitation cannot be dealt with one simple approach for the whole continent. For instance, altitude has no influence on rainfall over Western Africa (except in few specific locations, like near the Atakora mountains and close to the Guinea Mounts) and in the Sahel, covering several thousands of kilometers, therefore, the spline method gives the same results as the kriging method [22]. The coast of Cameroon is particularly wet, and the highest rainfall amounts in Africa are recorded at the foot of the Mount Cameroon; the HSM_SIEREM database gives much higher rainfall in this area than the CRU one—an example of the kind of differences between the two grids. In this area the rainfall decreases with altitude, from 10,000 mm per year on the coast of Debundsha (altitude 26 m), to 2000 mm near the top of Mount Cameroon (4040 m). The situation was roughly identical on the coast of the Guinean Mounts, with very high annual amounts on the coast, and decreasing values with elevation. Contrarily, the rainfall amount increased in other hilly areas, over the continent, like the Atlas Mountains in Maghreb, or in the hills of East Africa. In this context, it could be seen that the spatial gradient of precipitation with altitude was not homogeneous over the continent, and to consider altitude in the interpolation framework would require a more regional approach.

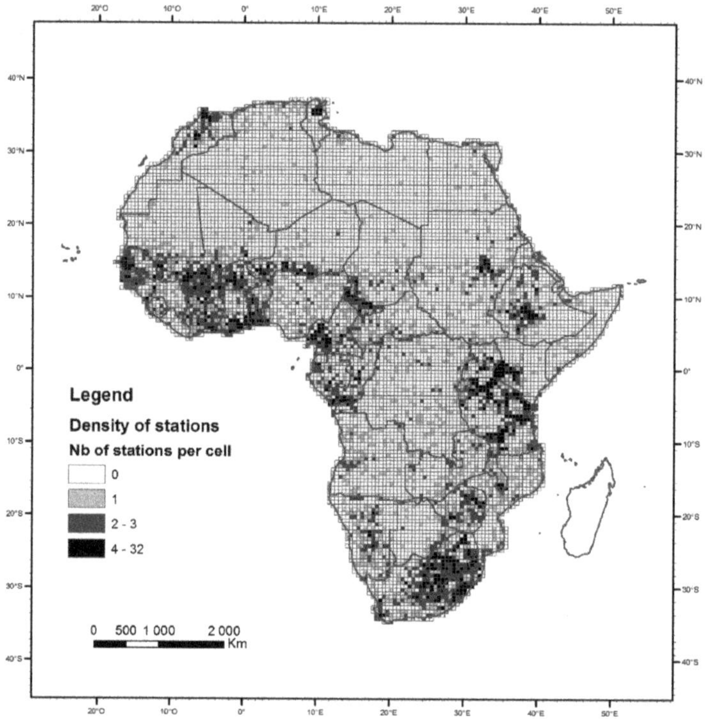

Figure 5. Density of the stations in the HSM_SIEREM rainfall database.

2.2.3. Options for Spatialization

The IDW process with ArcInfo lets the user choose between different options of interpolation. Among these, the options used for this study were:

- The default power value was used: 2.
- Priority was given to the radius around the interpolated point, with a value of six times the cell size of the output grid (in this case, each cell was half a square degree; the radius is then three decimal degrees). All the points within the radius were used for the interpolation.
- A minimum of four observation points were required to spatialize. If this minimum was not filled, the radius was extended until the program found four points.
- The output grid cell size was half a square degree. This was the common resolution of the grids used for regional modeling.
- The crest line of the Atlas Mountains is known to stop the humid Atlantic winds. Therefore, a barrier-line was created along this crest line, which modified the spatialization so that the interpolation did not use the values of points that were beyond the barrier line, on both sides.

- The extent of the spatialization process was 18° west, 51.5° east, 37.5° north, and 35° south. If this option was not present, the spatialization stopped at the stations located at the four edges—west, east, north, and south.

The raster grids at a step of 0.5 degree were converted into vector and then intersected to a regular grid at the step of half a square degree. The ArcInfo covers were converted into shape files and the raster grids (the first step of the process) were exported as ASCII files. Both the shape files and the ASCII grids being the products that would be available as free downloads as well as NetCDF arrays.

2.2.4. Validation of the IDW Interpolation

To evaluate the robustness of the interpolation method considered here, the interpolation methods of inverse distance weighting (IDW) and ordinary kriging (OK) were validated, using a cross-validation, as in [23]. Each rain gauge was in turn removed and the monthly precipitation was estimated with the remaining stations using the different methods. Only stations with at least 30 years of monthly precipitation data, representing 2912 rain gauges, were considered to validate the methods, to ensure robust estimates of the validation metrics; the Pearson correlation coefficient (r) and the relative bias. For both methods, IDW and OK, a search neighborhood with a 300 km radius was considered to perform the validation. The variograms required for the OK method were fitted with a spherical variogram model for each time-step, when rain was measured for at least four stations, otherwise, the IDW interpolation was performed. The spherical variogram model is convenient for precipitation, as it is not a spatially continuous field like temperature, since it provides a value of the decorrelation distance [24].

The validation results indicated very similar performances between the IDW and the OK, with an average correlation coefficient of validation of 0.86 for the IDW and 0.85 for the OK. The relative bias over all stations was equal to 7.8% with IDW and 7.1% with OK. The spatial patterns of the validation results are very similar, as shown in Figure 6, for the correlation coefficients of the validation samples being obtained with IDW or OK. The areas with the lowest density of stations, such as East Africa or south of the Maghreb countries, had the lowest performances in validation. Therefore, the interpolated precipitation in these areas must be interpreted with care.

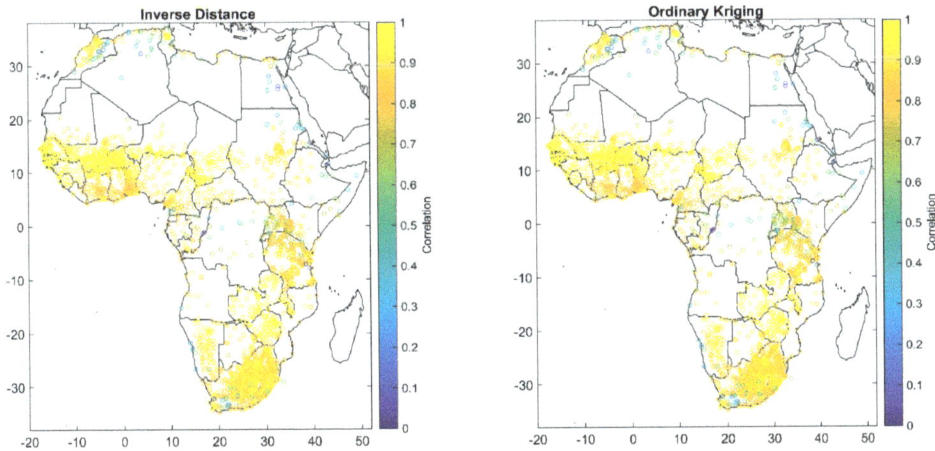

Figure 6. Validation results with independent station data for Inverse Distance (**left**) and Ordinary Kriging (**right**).

This validation confirmed that in the presence of a dense network, different interpolation methods (such as IDW or OK) performed with a similar efficiency.

2.2.5. Comparison between the CRU and HSM_SIEREM Grids

The grids from the CRU are widely used in models. Paturel et al. [11] compared the results of the GR2M hydrological model, using the rainfall grids from CRU and the ones from the previous versions of the HSM_SIEREM, over Western and Central Africa. The HSM_SIEREM database gave better results. Two mean annual grids over the 1940/1999 period were built, one with the HSM_SIEREM database and one with the CRU grids. The CRU monthly historical climate database, converted to the ESRI ASCII raster format by Jawoo Koo (HarvestChoice/IFPRI—Raw data), was downloaded from http://badc.nerc.ac.uk/data/cru). Figure 7 shows that the main features of the African rainfall distribution were similar.

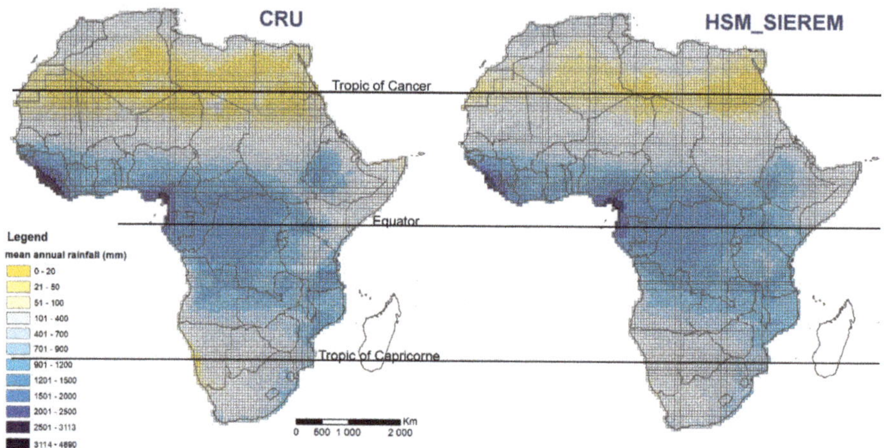

Figure 7. Mean annual Climatic Research Unit of University of East Anglia, UK (CRU) and HSM_SIEREM data base over the period of 1940/1999.

We checked the differences between the two mean inter-annual grids of the African continent, over the period of 1940 to 1999 (Figure 8).

The complete grid contained 10,380 cells, among which:

- A total of 4402 cells (in brown and yellow), i.e., 43%, had values ranging from −10% to −15,640%, compared to the HSM_SIEREM grid. They were mostly located in the Sahara and overlapped with the Sahel. In this area, the CRU grid had wide areas with values of mean annual rainfall between 0 and 20 mm, while the HSM_SIEREM ranged from 10 to 50 mm.
- A total of 4708 cells (in white), i.e., 45%, had values between −10% and +10%, compared to the HSM_SIEREM grid, which is very close.
- A total of 1270 cells (in green), i.e., 12%, had higher values, between +10% and +92%, compared to the HSM_SIEREM grid cells, the highest values being located on the southern slope of the Atlas Mountains (due to the barrier-line created in the HSM_SIEREM spatialization) and in Eastern Egypt (Dubief [22] drew a 5 mm isohyet in this area for the 50 first years of the 20th century).

The map of differences in percent showed two main features—in sub-Saharan Africa, the difference between the two grids mainly ranged between −9% and +10%, which could seem quite low, while the difference was much greater over Sahara and most of Northern Africa. In sub-Saharan Africa, some areas of low rainfall also showed a greater difference, as in the South Western coast of Africa and most of the horn of Africa. The CRU grid showed a higher rainfall over the Guinean mounts, the central part of the Congo basin, and the South of Angola. The HSM_SIEREM grid showed a higher rainfall in the South of Ghana, most of Nigeria, Cameroon and Gabon, along the South Western coast of

Africa, and over most of East and South Africa. About the Sahara region, the CRU grid showed values very different from that of the HSM_SIEREM grid. In most of the Sahara the CRU grid showed near 0 values, while the HSM_SIEREM grid showed a significant amount of rainfall, which were coherent with the Dubief rainfall map [25]. The CRU grid also showed too high a rainfall over the Saharan areas, in Egypt, North Chad, South-East Algeria and the Saharan border of the Atlas Mountains.

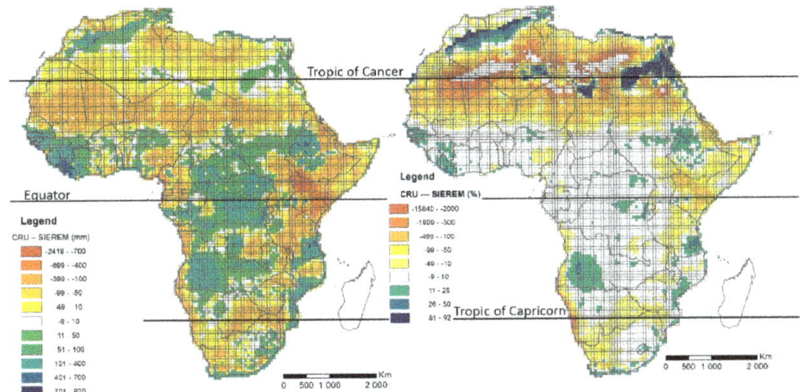

Figure 8. Differences of the CRU grid minus the HSM_SIEREM grid, in millimeters on the left and in percent on the right, over the period 1940/1999.

Even in the regions where the difference in percent was the lowest between the two grids (−9% to +10%), there were many areas where the difference was over 50 or 100 millimeters per year, which could have a significant impact on water-balance issues.

The differences between both grids could be due to the difference in the statistical approach of data processing and due to the density of observed stations. According to the figures given in Eklund et al. [26], the number of stations in Africa in the CRU data set was twice lower than that in the HSM_SIEREM data set, between 1940 and 1985, and three times lower after 1985. The maximum number of stations was over 4,000 for the HSM_SIEREM and less than 2000 for the CRU in the mid-seventies. It is quite likely that a greater number of stations will undergo a better precision in rainfall interpolation, at a grid scale.

3. Grid Production and Quality Assessment

The interpolation process was run over 720 months; every month was exported as an ASCII grid format file, containing one rainfall value at every 0.5 degree over the African continent; 6 zipped files, each containing 10 years of grids. Additionally, the NetCDF format array containing the 720 months of interpolated values, can be freely downloaded from the HSM website.

This first assessment of the HSM_SIEREM grid content by comparison with the CRU grid gives interesting details about where the use of HSM_SIEREM grid led to substantial differences with the CRU grids. This comparison would worth being completed by a first look at the rainfall variability over Africa, to check whether the main climatic signals are well-depicted by the HSM_SIEREM database. To assess the overall quality of this new dataset, we performed a first climatological analysis of rainfall over the whole continent, over the years 1940/1999.

3.1. Analysis of Rainfall Inter-Annual Variability

The standardized precipitation index (SPI) (Figure 9) was calculated, both, with the observed data (Figure 9a) and gridded data (Figure 9b), over the 60-year 1940/1999 period. For the observed data, the SPI was calculated with all stations having at least one complete year of observations. The SPI was calculated as SPI = (Xi-X)/S; where: Xi is the mean annual rainfall of year i, X is the mean annual rainfall over the reference period, and S is the standard deviation of the rainfall set over the reference period.

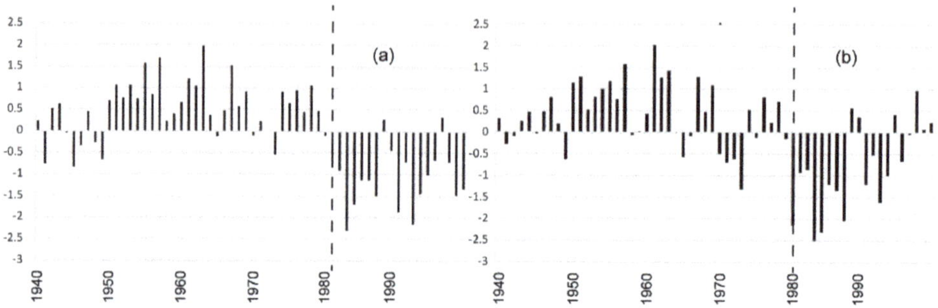

Figure 9. Mean annual rainfall for Africa, standardized precipitation index (SPI) on the observed data (**a**) and on the HSM_SIEREM gridded values (**b**).

SPI calculated with both observed and gridded data sets gave very similar time-series variations. It showed that, for the whole period, 1940/1999, and the whole continent, the SPI was almost permanently negative after 1981. This was coherent with most previous results [2,27]. There were some minor discrepancies between the two signals during the 1940s and at the end of the 1990s, which could, first, be linked to the lower number of stations in Africa during these periods, and, second, to their uneven distribution across the continent.

3.2. Statistical Tests on Mean Annual Rainfall

These tests, checking the random character of the series, were gathered using the Khronostat software [28], which can be downloaded, free of charge, at http://www.hydrosciences.org/spip.php?article1000. The most widely used tests deal with the stationarity of the mean of the series, throughout its observation period [29]. This test applied on the observed data confirmed a rupture in the series at three confidence levels, 99%, 95%, and 90% (Table 1).

Table 1. Results of the Mann–Whitney test, modified by Pettitt [29].

Observed Data	Gridded Data
Null hypothesis rejected at confidence levels of 99%, 95%, and 90%	Null hypothesis rejected at confidence levels of 99%, 95%, and 90%
Probability of exceeding the critical value 1.93×10^{-6} in 1979	Probability of exceeding the critical value 5.44×10^{-3} in 1969

The results of the statistical analysis of the inter-annual observed rainfall time-series over Africa was slightly surprising, as the 1970s rupture in the rainfall time-series is well known and has been largely described in Western and Central Africa [2,30,31] This meant that what happened in Western Africa and pro parte in the Western part of Central Africa, was more or less specific to this area and not to the whole continent. This should inspire further research on regional variables to explain these climatic features. The fact that the rupture date was found to be in 1969, with the gridded, data might be due to two causes—the first being that the rupture in 1969 might not have been the most prominent signal everywhere in Africa, even if it was present in many places. The second points out the impact of the interpolation method in regions with a high heterogeneity of the distribution of stations, and also the impact of the difference of length of a time-series. Indeed, this rupture in 1969 on the gridded data might be mostly driven by the extension of this signal that was most prominent in Western and Central Africa, due to the lack of data in many other parts of sub-tropical Africa, after 1990. This has already been discussed by Singla et al. [32], for Northern Africa.

3.3. Monthly Time-Series

The previous results led to the construction of two new data sets of monthly rainfall for two distinct periods: 1940/1979 and 1980/1999. Then a graph was drawn for the three-monthly time-series (before and after the rupture of 1979, and the entire 60-year period) (Figure 10), for observed and gridded data.

For observed data (in the left, in Figure 10), the graph showed two rainy seasons, February/ay and June–October. The first rainy season occurred in the austral hemisphere and the second occurred in the boreal hemisphere. The second rainy season remained the rainiest and the differences between the three periods were limited. Concerning the first rainy season during the month of March, the monthly values reached during the years 1940/1979 were clearly much higher than those during the years 1980/1999—the mean monthly rainfall value decreased from 90 mm to 64 mm, by almost 30% for the observed values. For the gridded data set (in the right, in Figure 10), values showed the same interdecadal variability, but with lower monthly values and a lower difference between the periods. The difference for March was still significant, with a reduction of 12% after 1980, from 62 mm to 55 mm.

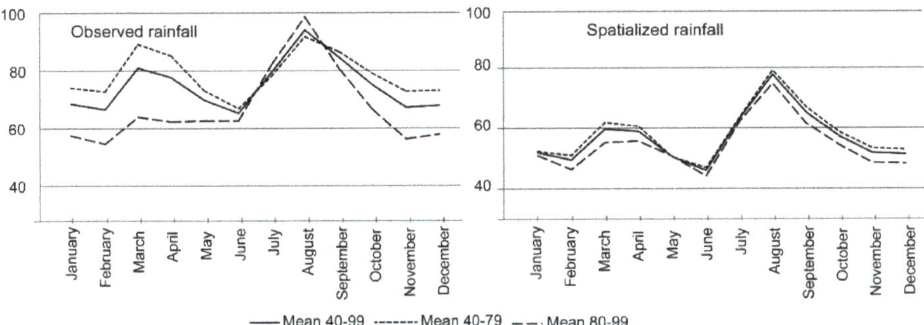

Figure 10. Seasonal variation of the rainfall for the periods 1940/1999, 1940/1979, and 1980/1999, for observed and gridded data.

Several authors [21,33,34] have already noticed a substantial decrease in rainfall over the southern part of Africa, between the equator and 10° South, but none were found to be as significant as the one shown here.

This first result showed that rainfall had decreased on a continental scale, for Africa, for several decades, and that the monthly rainfall regime did not change much between June and September, i.e., it showed a boreal summer tropical rainy season but changed much more during November to April/May, with a strong decrease that took place during the austral summer and autumn tropical rainfall period. This meant that it was mostly the Southern Hemisphere that seemed to have been affected by this decrease. One must note, however, that due to the rupture being detected in 1979/1980 for the whole continent, the decrease of rainfall that occurred in the Western and in part of Central Africa, since 1969/1970, did not appear clearly at the scale of the continent average. It was nonetheless one of the strongest climatic signals ever recorded in Africa.

4. Discussion and Conclusion

This paper is the result of a long duration of work on the collection, criticism, and assessment of a very rich and high-quality database of rainfall. This study could be carried out due to the quantity of data collected during a long period, mainly in Western and Central Africa, by the ORSTOM and the IRD, and due to the long and thorough criticism process that was applied to the data sets. It is especially precious, as the African continent has a particularly poor observation network, compared to the rest of the world.

The HSM_SIEREM rainfall gridded data set showed a very good correlation with the observed data, whatever the interpolation method used. It was compared to the widely used CRU rainfall grid, and their differences have been mapped, to help determine their best use for future studies. The HSM_SIEREM grid was built from twice the number of stations than the CRU grid, the number of stations was even three times higher for the period of 1985/1999, and the distribution of stations better covered the Sahara and some other parts of Africa. This gave the HSM_SIEREM grid more confidence in describing rainfall over Africa, with some regions showing very important differences, especially in the low-rainfall regions, in the Sahara and the Sahel, the Southwestern coast of Africa, and the horn of Africa.

The results confirmed that the African continent was seriously affected by a rainfall deficit and a change in the rainfall regime at the end of the 20th century. The first assessment showed that the two main periods of rupture in the time-series was between 1969/1970 and 1979/1980, which are depicted in the global African rainfall times-series issued from the whole data set, in both, the gridded and observed time-series. It seems that it was the first part of the year, from February to May, which registered the highest rainfall reduction, especially in the austral part of Africa (Figure 11). This reduction was several times higher than the rainfall reduction in the boreal part of Africa, at the same period. Even if part of these figures might be linked to the interpolation of unevenly distributed observed rainfall stations, this could have had strong incidences on the global climate of the area and requires further studies to analyze the regional variability of rainfall over the continent, between different databases.

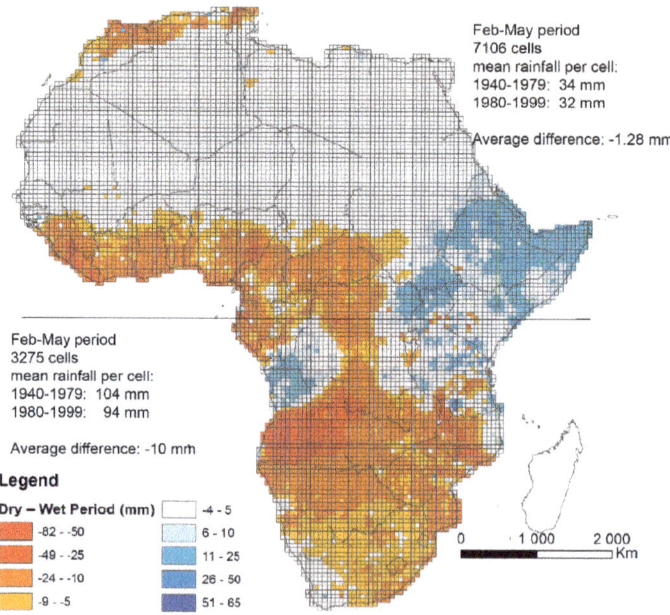

Figure 11. Map of the anomalies, Globally Dry Period (1980/1999)–Globally Wet Period (1940/1979). Average period for the months of February to May.

This study also showed that the use of direct observed data for interpolation gave different results and spatial representations than the use of rainfall patterns and anomalies to generate annual maps, which were used by the CRU and many other data centers. The comparison between the spatially-averaged rainfall data and observed data, over several regions of Morocco [32], showed that averaging th data over large surfaces led to a reduced visibility of climatic variability and the main climatic signals, but the main signals remained visible whatever the size of the region used for averaging. This supports the idea that the results visible with, both, observed and gridded data sets are quite consistent.

All monthly grids built during this study are available for free on the HydroSciences Montpellier website and our aim is to contribute to the maintenance of an accurate database for climatic scenarios (http://www.hydrosciences.org/spip.php?rubrique1387/). These grids are already available on the Researchgate website: https://www.researchgate.net/project/Monthly-rainfall-gridded-data-set-for-Africa.

As it is very difficult to gather observed data covering the whole continent since 2000, it is improbable that an update of this study might be performed, knowing that, since then, a number of rainfall gridded product have been released on the basis of satellite-derived rainfall estimations [35], which help fill the gap between the needs of researchers for recent data and the unavailability of a free international observed rainfall database. The length of the time-series was, thus, adapted for a comparison of historical time-series and rainfall time-series from the GCM/RCM reanalysis, or for study of time-series for regions where data are difficult to obtain, even for historical periods.

Author Contributions: Conceptualization: C.D., G.M.; Methodology: C.D., J.-E.P., Y.T.; Software: C.D., Y.T., S.E.; Validation: G.M., Y.T., J.-E.P., Y.T.; Formal analysis: C.D., N.R.; Investigation: B.E.M.; Data curation: J.-E.P., S.E., N.R.; Witing—original draft preparation: C.D., G.M.; Writing—review and editing: C.D., G.M., Y.T., J.-E.P.; Project administration: G.M.; Funding acquisition: G.M.

Acknowledgments: We thank the CRU of the University of East Anglia for the exchange of raw data at the early stage of this study. We thank the IRD and the HSM Laboratory of Montpellier for the financial and technical support; and the FRIEND program (UNESCO/IHP) for their collaboration through the managers of the databases of the different regional programs in Africa (MEDFRIEND, FRIEND AOC, FRIEND Nile, FRIEND Austral Africa). We thank all the national meteorological services of African countries for their continuous work of data collection, although, we regret that only a few of them accepted to update the data series for the 1990s.

Conflicts of Interest: The authors declare no conflict of interest.

References

1. IPCC. Summary for Policymakers. In *Climate Change 2013: The Physical Science Basis. Contribution of Working*; Group I to the Fifth Assessment Report of the Intergovernmental Panel on Climate Change; Stocker, T.F., Qin, D., Plattner, G.-K., Tignor, M., Allen, S.K., Boschung, J., Nauels, A., Xia, Y., Bex, V., Midgley, P.M., Eds.; Cambridge University Press: Cambridge, UK; New York, NY, USA, 2013.

2. Mahé, G.; Olivry, J.C. Variations des précipitations et des écoulements en Afrique de l'Ouest et Centrale de 1951 à 1989. *Sécheresse* **1995**, *6*, 109–117.

3. Servat, E.; Paturel, J.-E.; Kouamé, B.; Travaglio, M.; Ouedraogo, M.; Boyer, J.F.; Lubès-Niel, H.; Fritsch, J.M.; Masson, J.M.; Marieu, B. *Identification, Caractérisation et Conséquences d'une Variabilité Hydrologique en Afrique de l'Ouest et Centrale*. 'Abidjan'98', Abidjan, Côte d'Ivoire, 16–19 Novembre 1998; IAHS-AISH Publication: Wallingford, UK, 1998; Volume 252, pp. 323–337.

4. Mahé, G.; L'Hote, Y.; Olivry, J.C.; Wotling, G. Trends and discontinuities in regional rainfall of West and Central Africa—1951–1989. *Hydrol. Sci. J.* **2001**, *46*, 211–226. [CrossRef]

5. New, M.; Todd, M.; Hulme, M.; Jones, P. Precipitation measurements and trends in the twentieth century. *Int. J. Climatol.* **2001**, *21*, 1899–1922. [CrossRef]

6. Washington, R.; Harrison, M.; Conway, D.; Black, E.; Challinor, A.; Grimes, D.; Jones, R.; Morse, A.; Kay, G.; Todd, M. African Climate Change. *Bull. Am. Meteorol. Soc.* **2006**, *87*, 1355–1366. [CrossRef]

7. Mahé, G.; Girard, S.; New, M.; Paturel, J.E.; Cres, A.; Dezetter, A.; Dieulin, C.; Boyer, J.F.; Rouché, N.; Servat, E. Comparing available rainfall gridded datasets for West Africa and the impact on rainfall-runoff modelling results, the case of Burkina-Faso. *Water SA* **2008**, *34*, 529–536.

8. New, M.; Hulme, M.; Jones, P. Representing Twentieth-Century Space-Time Climate Variability. Part I: Development of a 1961–90. Mean Monthly Terrestrial Climatology. *J. Clim.* **2000**, *12*, 829–856.

9. New, M.; Hulme, M.; Jones, P. Representing Twentieth-Century Space-Time Climate Variability. Part II: Development of a 1901–96. Monthly grids of Terrestrial Surface Climate. *J. Clim.* **2000**, *13*, 2217–2238.

10. Paturel, J.E.; Boubacar, I.; L'Aour-Cres, A.; Mahé, G. Monthly rainfall grids in West and Central Africa. *Revue des Sciences de l'Eau* **2010**, *23*, 325–333. [CrossRef]

11. Paturel, J.E.; Ouedraogo, M.; Mahé, G.; Servat, E.; Dezetter, A.; Ardoin, S. The influence of distributed input data on the hydrological modelling of monthly river flow regimes in West Africa. *Hydrol. Sci. J.* **2003**, *48*, 881–890. [CrossRef]

12. Boyer, J.F.; Dieulin, C.; Rouché, N.; Cres, A.; Servat, E.; Paturel, J.E.; Mahé, G. SIEREM: An environmental information system for water resources. In *Climate Variability and Change—Hydrological Impacts, Proceedings of the Fifth FRIEND World Conference, Havana, Cuba, November, 26th 2006*; IAHS Publication: Wallingford, UK, 2006; pp. 19–25.

13. Paturel, J.E.; Boubacar, A.; L'Aour, A.; Mahé, G. Analyses de grilles pluviométriques et principaux traits des changements survenus au 20ème siècle en Afrique de l'Ouest et Centrale. *Hydrol. Sci. J.* **2010**, *55*, 1281–1288. [CrossRef]

14. Paturel, J.E.; Ouedraogo, M.; Servat, E.; Mahé, G.; Dezetter, A.; Boyer, J.F. The concept of rainfall and streamflow normals in West and central Africa in a context of climatic variability. *Hydrol. Sci. J.* **2003**, *48*, 125–137. [CrossRef]

15. Ardoin-Bardin, S.; Dezetter, A.; Servat, E.; Paturel, J.E.; Mahé, G.; Niel, H.; Dieulin, C. Using general circulation model outputs to assess impacts of climate change on runoff for large hydrological catchments in West Africa. *Hydrol. Sci. J.* **2009**, *54*, 77–89. [CrossRef]

16. Conway, D.P.; Persechino, A.; Ardoin-Bardin, S.; Hamandawana, H.; Dieulin, C.; Mahé, G. Rainfall and river flow variability in sub-Saharan Africa during the 20th century. *J. Hydrometeorol.* **2009**, *10*, 41–59. [CrossRef]

17. Jaccon, G.; Travaglio, M. La banque de données hydropluviométriques du laboratoire d'hydrologie. In *Systèmes d'information pour l'environnement*; Colloques et Séminaires; ORSTOM: Bondy, France, 1990; pp. 49–59.

18. L'hote, Y.; Mahé, G. *Afrique de l'Ouest et Centrale, Précipitations Moyennes Annuelles (Période 1951–1989)*; ORSTOM/IRD: Bondy, France, 1996.

19. Water Assessment. In *Evaluation Hydrologique de l'Afrique Sub-Saharienne. Pays de l'Afrique de l'Ouest*; Mott MacDonald International, BCEOM, SOGREAH, ORSTOM, Eds.; Sci. Financement Banque Mondiale, PNUD, BAD, MFC; ORSTOM Editions: Montpellier, France, 1992.

20. Rouché, N.; Mahé, G.; Ardoin-Bardin, S.; Brissaud, B.; Boyer, J.F.; Cres, A.; Dieulin, C.; Bardin, G.; Commelard, G.; Paturel, J.E.; et al. Constitution d'une grille de pluies mensuelles pour l'Afrique, période 1900–2000. *Secheresse* **2010**, *21*, 336–338.

21. Nicholson, S.E. The nature of rainfall fluctuations in subtropical West Africa. *J. Am. Meteorol. Soc.* **1980**, *108*, 473–487. [CrossRef]

22. Wotling, G.; Mahé, G.; L'Hôte, Y.; Le Barbé, L. Analysis of the space-time variability of annual rainfall linked to the African monsoon using regional vectors. *Veill. Clim. Satellitaire* **1995**, *52*, 58–73.

23. Tramblay, Y.; Thiemig, V.; Dezetter, A.; Hanich, L. Evaluation of satellite-based rainfall products for hydrological modelling in Morocco. *Hydrol. Sci. J.* **2016**, *61*, 2509–2519. [CrossRef]

24. Lebel, T.; Bastin, G.; Obled, C.; Creutin, J.D. On the accuracy of areal rainfall estimation. *Water Resour. Res.* **1987**, *23*, 2123–2134. [CrossRef]

25. Dubief, J. *Le Climat du Sahara. Mémoire de Recherche de l'Institut de Météorologie et de Physique du Globe de l'Algérie*; Université d'Alger: Alger, Algérie, 1963.

26. Eklund, L.; Romankiewicz, C.; Brandt, M.S.; Doevenspeck, M.; Samimi, C. Data and methods in the environment-migration nexus: A scale perspective. *Die Erde J. Geogr. Soc. Berl.* **2016**, *147*, 139–152. [CrossRef]

27. Nicholson, S.E.; Fink, A.H.; Funk, C. Rainfall over the African continent from the 19th through the 21st century. *Glob. Planet. Chang.* **2018**, *165*, 114–127. [CrossRef]

28. Lubes-Niel, H.; Masson, J.M.; Servat, E.; Paturel, J.E.; Kouame, B.; Boyer, J.F. *Caractérisation de Fluctuations dans une série Chronologique par Application de Tests Statistiques. Etude Bibliographique*; Programme ICCARE. Rapport n°3; ICCARE; ORSTOM: Montpellier, France, 1994.

29. Pettitt, A.N. A non-parametric approach to the change-point problem. *Appl. Stat.* **1979**, *28*, 126–135. [CrossRef]

30. Hulme, M.; Doherty, R.; Ngara, T.; New, M.; Lister, D. African climate change: 1900–2100. *Clim. Res.* **2001**, *17*, 145–168. [CrossRef]

31. Mahé, G.; Olivry, J.C. Changements climatiques et variations des écoulements en Afrique occidentale et centrale, du mensuel à l'interannuel. In Proceedings of the Hydrology for the Water Management of Large River Basins Congrès AISH, Vienne, Autriche, 13–15 août 1991; van de Ven, F.H.M., Gutknecht, D., Loucks, D.P., Salewicz, K.A., Eds.; IAHS Publications: Wallingford, UK, 1991; Volume 201, pp. 163–172.

32. Singla, S.; Mahé, G.; Dieulin, C.; Driouech, F.; Milano, M.; El Guelai, F.Z.; Ardoin-Bardin, S. Evolution des relations pluie-débit sur des bassins versants du Maroc. In *Global Change: Facing Risks and Threats to Water Resources, Proceedings of the Sixth World FRIEND Conference, Fez, Morocco, October 2010*; IAHS Publications: Wallingford, UK, 2010; Volume 340, pp. 679–687.

33. Fauchereau, N.; Trzaska, S.; Rouault, M.; Richard, Y. Rainfall Variability and Changes in Southern Africa during the 20th Century in the Global Warming Context. *Nat. Hazards* **2003**, *29*, 139–154. [CrossRef]

34. Morishima, W.; Akasaka, W. Seasonal trends of Rainfall and Surface Temperature over Southern Africa. *Afr. Study Monogr.* **2010**, *40*, 67–76.

35. Adeyewa, Z.D.; Nakamura, K. Validation of TRMM rainfall data over major climatic regions in Africa. *J. Appl. Meteorol.* **2003**, *42*, 331–347. [CrossRef]

Review

Recent Budget of Hydroclimatology and Hydrosedimentology of the Congo River in Central Africa

Alain Laraque [1,*], Guy D. Moukandi N'kaya [2], Didier Orange [3], Raphael Tshimanga [4], Jean Marie Tshitenge [4], Gil Mahé [5], Cyriaque R. Nguimalet [6], Mark A. Trigg [7], Santiago Yepez [8] and Georges Gulemvuga [9]

[1] Géosciences Environnement Toulouse, Université de Toulouse, CNES, CNRS, IRD, UPS, 31400 Toulouse, France; alain.laraque@ird.fr

[2] LMEI/CUSI/ENSP/Marien N'gouabi University, Brazzaville B.P. 69, Republic of the Congo; guymoukandi@yahoo.fr or guy.moukandi@umng.cg

[3] University of Montpellier, Eco&Sols-UMR IRD/INRA/Cirad/Supagro, UMR210 IRD, Place Viala (Bt. 12), CEDEX 2, 34060 Montpellier, France; didier.orange@ird.fr

[4] Department of Physics, Faculty of Sciences (JMT), University of Kinshasa, B.P. 190 (JMT), Kinshasa XI, Democratic Republic of the Congo; raphtm@yahoo.fr (R.T.); jeanmarie12@gmail.com or Jeanmarie.tshitenge@unikin.ac.cd (J.M.T.)

[5] HydroSciences Montpellier, UMR IRD-CNRS-Université de Montpellier, Maisons des Sciences de l'Eau, 300 avenue Emile Jeanbrau, 34090 Montpellier, France; gil.mahe@ird.fr

[6] Geography Department, University of Bangui, Bangui, Central African Republic; cyrunguimalet@gmail.com

[7] School of Civil Engineering, University of Leeds, Leeds LS2 9JT, UK; m.trigg@leeds.ac.uk

[8] Departamento Manejo de Bosques y Medio Ambiente, Facultad de Ciencias Forestales, Universidad de Concepción—UdeC, Concepción 407374, Chile; syepez@udec.cl

[9] International Commission for Congo-Ubangi-Sangha basin (CICOS), Kinshasa B.P. 190, Democratic Republic of Congo; georges_gul@yahoo.fr or georges.gulemvuga@cicos.info

* Correspondence: alain.laraque@ird.fr; Tel.: +33-476-415-118

Received: 15 July 2020; Accepted: 10 September 2020; Published: 18 September 2020

Abstract: Although the Congo Basin is still one of the least studied river basins in the world, this paper attempts to provide a multidisciplinary but non-exhaustive synthesis on the general hydrology of the Congo River by highlighting some points of interest and some particular results obtained over a century of surveys and scientific studies. The Congo River is especially marked by its hydrological regularity only interrupted by the wet decade of 1960, which is its major anomaly over nearly 120 years of daily observations. Its interannual flow is 40,500 m^3 s^{-1}. This great flow regularity should not hide important spatial variations. As an example, we can cite the Ubangi basin, which is the most northern and the most affected by a reduction in flow, which has been a cause for concern since 1970 and constitutes a serious hindrance for river navigation. With regard to material fluxes, nearly 88×10^6 tonnes of material are exported annually from the Congo Basin to the Atlantic Ocean, composed of 33.6×10^6 tonnes of TSS, 38.1×10^6 tonnes of TDS and 16.2×10^6 tonnes of DOC. In this ancient flat basin, the absence of mountains chains and the extent of its coverage by dense rainforest explains that chemical weathering (10.6 t km^{-2} $year^{-1}$ of TDS) slightly predominates physical erosion (9.3 t km^{-2} $year^{-1}$ of TSS), followed by organic production (4.5 t km^{-2} $year^{-1}$ of DOC). As the interannual mean discharges are similar, it can be assumed that these interannual averages of material fluxes, calculated over the longest period (2006–2017) of monthly monitoring of its sedimentology and bio-physical-chemistry, are therefore representative of the flow record available since 1902 (with the exception of the wet decade of 1960). Spatial heterogeneity within the Congo Basin has made it possible to establish an original hydrological classification of right bank tributaries, which takes into account vegetation cover and lithology to explain their hydrological regimes. Those of the Batéké plateau present a hydroclimatic paradox with hydrological regimes that are among the most stable on the planet, but also with some of the most pristine waters as a result of the intense drainage of an immense sandy-sandstone aquifer. This aquifer contributes to the regularity of the Congo River flows, as does the buffer role of the mysterious "Cuvette Centrale". As the study of this last one sector can only be done indirectly, this paper presents its first hydrological regime calculated by inter-gauging station water balance. Without neglecting the indispensable in situ work, the contributions of remote sensing and numerical modelling should be increasingly used to try to circumvent the dramatic lack of field data that persists in this basin.

Keywords: hydroclimatology; hydrosedimentology; hydrogeochemical; Congo River Basin

---

1. Introduction

In terms of area and discharge, the Equatorial Congo River Basin is the second largest river in the world and the largest in Africa, but paradoxically remains one of the least known. This basin can still be considered as almost pristine because, given its size, it is relatively less anthropogenicized. This is at least partly due to the lack of transport infrastructure related to the inaccessibility of its large central swampy 'Cuvette' named 'Cuvette Centrale' covered by dense rainforest that occupies the heart of the basin. Shem and Dickinson (2006) [1] observed that, despite its crucial position as the third largest deep convection center in the world, the Congo Basin has not yet received adequate attention in the field of climate and hydrological research.

Considering the importance of the Congo Basin to global fluxes of water, energy, carbon, various suspended or dissolved elements to the ocean and to the atmosphere, and the recent renewed interest of the international scientific community, the purpose of this paper is to present a state-of-the-art review of existing studies and a synthesis of new results, especially on hydroclimatology and hydrosedimentology. Significantly, some passages that are useful for the international scientific community have been translated from previous publications, which unfortunately remain little consulted, despite their importance, mainly because they are written in French, the main spoken language in this basin.

Specifically, it presents and discusses: (i) an overview of the Congo Basin; (ii) the history of hydropluviometric networks in the Congo Basin; (iii) the rainfall and hydrologic features of the whole Congo Basin and four important sub-catchments, which have the longest multi-decadal chronology of data, and represent its mean hydro-ecosystems; (iv) the first indirect estimation of the hydrological regime of the mysterious 'Cuvette Centrale'; (v) water quality sediment transport dynamics on the Congo River from in situ data at its main Brazzaville-Kinshasa Station; and (vi) some hydrological and hydrogeochemical singularities in the Congo catchment. It ends by illuminating the modeling and remote sensing "studies" and their main results and it concludes with the current state of human use and exploitation and the need for a coherent policy for its future protection and management. In this review paper, the methodologies that were used for each parameter studied can be found in the references that are cited as the origin of the results presented.

2. Geographical Presentation of the Congo River Basin (CRB)

The Congo is the largest river in the African continent with a basin area of about 3.7×10^6 km^2 and a mean annual discharge of 40,500 m^3 s^{-1}, calculated using 117 years of data, from 1902 to 2019 at its main hydrological station of Brazzaville-Kinshasa. This station represents 97% of the total basin area, is situated at an altitude of 277 m a.s.l., and is 498 km upstream of the river's mouth to the Atlantic Ocean (Figure 1a,e). It is the second-largest river in the world in terms of discharge and catchment area. The 4700-km long Congo River provides half of all the river waters discharged from the African continent to the Atlantic Ocean [2]. The source of the Congo is located at 1420 m a.s.l. in the village of Musofi, south of the Katanga region (southeast of the Democratic Republic of Congo).

In the Northern hemisphere, on the right bank, the Ubangi River is the second-most important tributary in terms of discharge [3], after the major tributary, the Kasaï, located further downstream on the left bank (Figure 1). Just upstream of the major hydrometric stations of Brazzaville-Kinshasa, the Congo flows through Malebo Pool (ex. Stanley Pool), which corresponds to a wide channel reach about 20–25 km long, characterized by numerous sand bars, which emerge during low flows. Between Malebo Pool and its mouth, the Congo falls around 280 m over a distance of about 500 km. Here the many narrow channels create the rapids known as the Livingstone Falls. It is within this section of the river that the only important hydropower plant (Inga) was constructed on the Congo River. The Congo Basin spans nine different countries (Angola, Burundi, Cameroon, Democratic Republic of the Congo (DRC), Central Africa Republic (CAR), Republic of Congo, Rwanda, Tanzania and the Zambia). It comprises several thousands of navigable waterways, depending on the hydrologic period and on the draft of the ships and their tonnage.

The form, relief, geology, climate, and vegetation cover of the Congo Basin (Figure 1a–d) are generally concentrically distributed. Surrounded by savannahs on smoothly rounded hills from deeply weathered Mesozoic Precambrian formations [4], less than 44% of the catchment area is covered by rainforest, which promotes the capacity to recycle the basin's moisture [5]. The basin can be divided into six main hydrologic units (Figure 1c), which are its main distinctive physiographic regions. Their areas and hydroclimatic features are listed in Table 1. Five of the six units are monitored by gauging stations.

The two main overlapping 60-year rainfall study periods are 1940–1999 and 1952–2012, and their data comes from the SIEREM database website (www.hydrosciences.fr/sierem) [6] and BRLi (2016) [7], respectively. This quite concentric shape of Congo watershed presents the following main physiographic characteristics.

Its northern and southern margins are dominated by a set of shield plateaus covered by shrub and tree savannah vegetation and are characterized by a humid tropical transition climate and annual rainfall between 1400 to 1800 mm year^{-1}. The more northerly part of the Congo Basin, which is drained by the Ubangi tributary, is a less humid region, well described by Runge and Nguimalet (2005) [8]. With a total rainfall of around 1500 mm year^{-1}, its watershed benefits from a transitional humid tropical climate with the Sudanese-Sahelian regions farthest north. In the south of the Congo Basin, the sub-basin of the Kasaï River receives rainfall of about 1460 mm year^{-1}.

In the lowlands center of the basin, under an equatorial climate marked by rainfall of 1600 to 1800 mm year^{-1}, the "Cuvette Centrale" is a topographic depression of about 1,176,000 km^2 [7]. Its floodplain (Figure 1d) covers about 30%, or 360,000 km^2 [9] of the total basin area. This area is characterized by very small topographical slopes (2 cm km^{-1}) on unconsolidated Cenozoic sediments [4], constituted of alluvial deposits. These formations are covered by rainforests, permanently or periodically flooded, according to the hydro-rainfall cycle. There are also extensive but shallow lakes such as Lac Tumba, Lac Maï Ndombe, etc.

In its northwestern part, the Sangha basin is mainly covered by dense rainforest, and under an equatorial climate receives a mean rainfall ranging of 1625 mm year^{-1}.

In the west, the Batéké plateau constitutes a high-altitude sandstone aquifer recharged by rainfall (1800 to 2000 mm year^{-1}), which controls the hydrological regime of the watercourses draining this aquifer. Both the runoff coefficient (50%) and the specific discharge (close to 35 L s^{-1} km^{-2}) are the highest in this region [10].

In the east is the highest point of the basin, the Karisimbi volcano with 4507 m a.s.l situated in the Virunga mountain chain of the East-African rift. This region is located in the Lualaba basin, which receives between 1110 and 1680 mm year^{-1} depending on the location [7].

Straddling the Equator, the Congo River shows a poorly-contrasted equatorial regime, with low seasonal discharge variability of 3.3, measured as the ratio between the highest monthly mean discharge of 75,500 m^3 s^{-1} and the lowest of 23,000 m^3 s^{-1}. The Congo River also has a relatively low inter-annual variability of 1.66, measured as the ratio between the lowest and highest known annual discharges. The hydrological year, begins in August and is marked by the alternation of two periods of high flows (October–January; April–May) and two periods of low flows (February–March; June–September) and presents an amplitude range of 3.65 m, which can arise 6.54 m at Brazzaville-Kinshasa.

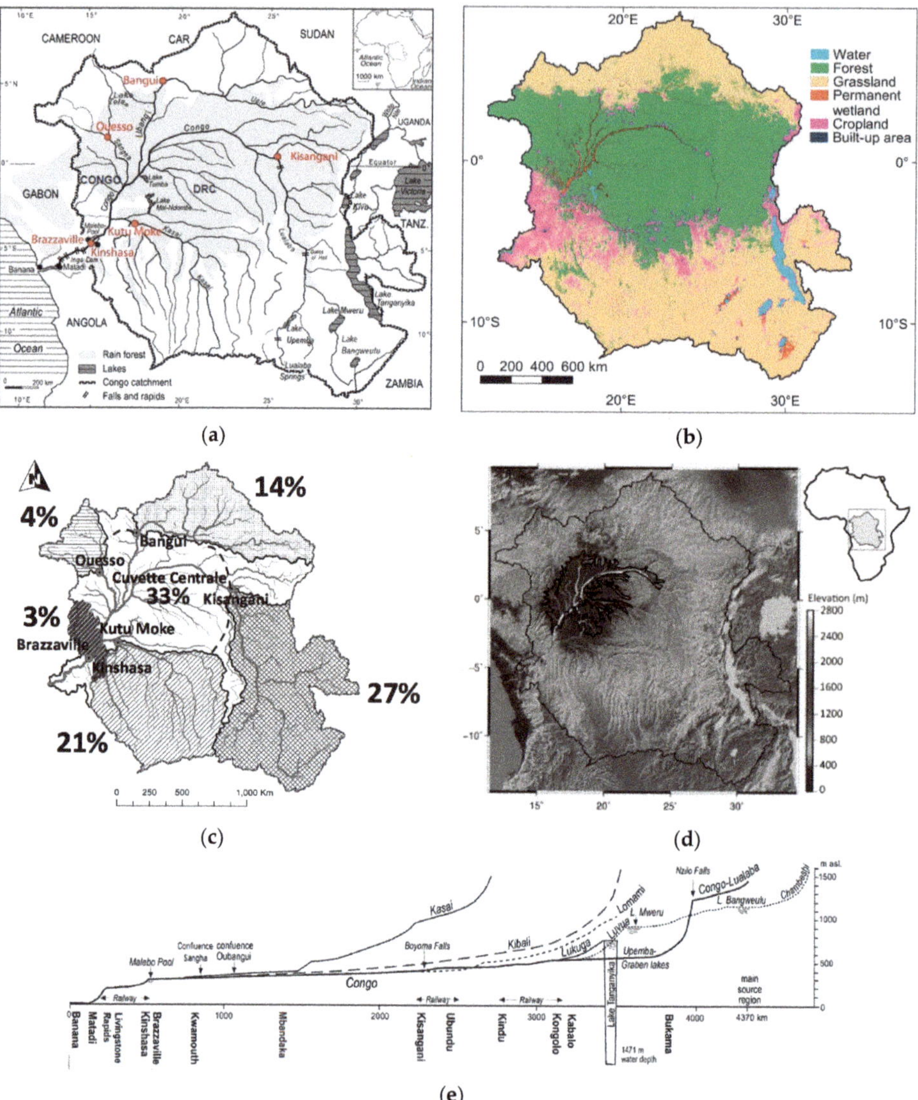

(**a**)

(**b**)

(**c**)

(**d**)

(**e**)

Figure 1. (**a**) Overview of Congo basin (adapted from Runge, 2007) [4], (**b**) Land cover on Congo basin [11], (**c**) Main drainage units studied in Congo basin: Lualaba at Kisangani St., Kasaï at Kutu Moke St., Sangha at Ouesso St., Ubangi at Bangui St., Batéké plateau and 'Cuvette Centrale', with their percentage of area of Congo basin represented by the BZV/KIN station. Legend: the dashed line encircles the general area of the 'Cuvette Centrale'. The area with inclined dark hatching is that of the Batéké plateau, (**d**) Topography of the Congo Basin and of the flooded portion of the Central Basin. The topography comes from the SRTM [12] and the flooded portion of the Central Cuvette (the black polygon) is generalized by Bwangoy et al. (2010) [9], (**e**) Longitudinal profiles of the Congo River and its main tributaries (adapted from Runge, 2007) [4].

Table 1. The hydro-climatic characteristics in the Congo basin represented at its main hydrological stations [13].

		Geographical Coordinates			Basin Area km²			Hydro-Climatic Characteristics				
		Latitude	Longitude E	Altitude m	Station	Outlet	Ratio st./ex.	% CRB at BZV/KIN	Rainfall mm	Interannual Average Discharge m³ s⁻¹	Specific Discharge L s⁻¹ km⁻²	Seasonal Variation Flow
Lualaba [1]	Kisangani	00°30′11″ N	25°11′30″	373	974,140	974,140	1.00	26.62	1307	7640	7.8	1.9
Kasaï [2]	Kutu-Moke	03°11′50″ S	17°20′45″	303	750,000	897,540	0.84	20.49	1456	8070	10.8	1.9
Sangha [3]	Ouesso	01°37′00″ N	16°03′00″	326	159,480	213,670	0.75	4.36	1625	1550	9.7	2.3
Ubangi [4]	Bangui	04°22′00″ N	18°35′00″	345	494,090	650,480	0.76	13.50	1499	3660	7.4	2.9
Batéké Plateau [5]	-	-	-	305	42,570	90,000	0.47	1.16	1900	1330	31.24	1.1–1.5
'Cuvette Centrale' [6]	-	-	-	305–335		1,192,190		32.57	1700	-		-
Congo [7]	BZV/KIN	04°16′21.5″ S	15°17′37.2″	314	3,659,900	3,730,740	0.98	100	1447	40,500	11.07	1.7

The interannual modules are calculated at the different periods studied: [1] 1951 to 2012; [2] 1948 to 2012; [3] 1947 to 2018; [4] 1936 to 2018; [5&6] 1947 to 1994, and [7] 1903 to 2018. For the 1940–1999 period, the rainfall comes from the SIEREM database website [6], except for [5] & [6] from Laraque et al. (1998) [10]. Legend: st. = hydrometric station, ex. = outlet. The hydro-climatic characteristics of the Forest Central Basin and the Bateké plateau are derived from Laraque et Maziezoula (1995) [14] and Laraque et al. (1998 and 2009) [10,15], about right bank tributaries of Congo River represented by hydrologic stations. The coordinates X, Y, Z are those of the hydrological stations.

3. Brief History of Hydropluviometric Networks in the Congo Basin

The hydropluviometric networks originate from the colonial era, when Belgium occupied the left bank of the river, which is now the Democratic Republic of Congo (DRC) and France occupied the right bank, now known as the Central African Republic (CAR), the Republic of Congo, and part of Cameroon.

3.1. Rainfall Network

ORSTOM (Office de Recherche Scientifique et Technique d'Outre Mer), known today as IRD (Institut de Recherche pour le Développement), participated in the management of the network of rainfall measurements on the right bank of the Congo river since the 1940s until the independence of the Republic of Congo, the Central African Republic, and Cameroon.

After the independence of these countries at the beginning of the 1960s, climate surveys were continued by ASECNA (Agency for Aerial Navigation Safety in Africa and Madagascar) and by the National Meteorological Departments.

In the DRC, due to the importance of the rainforest vegetation and the very few developed road networks, the density of the rainfall network is scarce in most of the central part of the basin, except on the Ubangi sub-basin area (Figure 2a). Observed rainfall data are also scarce in Angola. Observation data are a little more numerous in Zambia and Tanzania, but rainfall data are poorly available during and after the last years of the 20th century in most of these countries, due to a general decline in support of the networks. In CAR, the pluviometric database was completed through the PIRAT and PEGI programs [16] and then Callède et al. (2009) [17] produced a monograph on the Ubangi basin. In this basin, Nguimalet and Orange (2019) [3] studied a rainfall series from 1940 to 2015.

In the entire Congo Basin, a compilation of available and accessible rainfall data could only be carried out for the period 1940–1999, from the free access SIEREM database website [6] and the annual rainfall map of Africa (Figure 2b) [18] is today the last reference for this basin in addition to the older reference of Bultot (1971) [19].

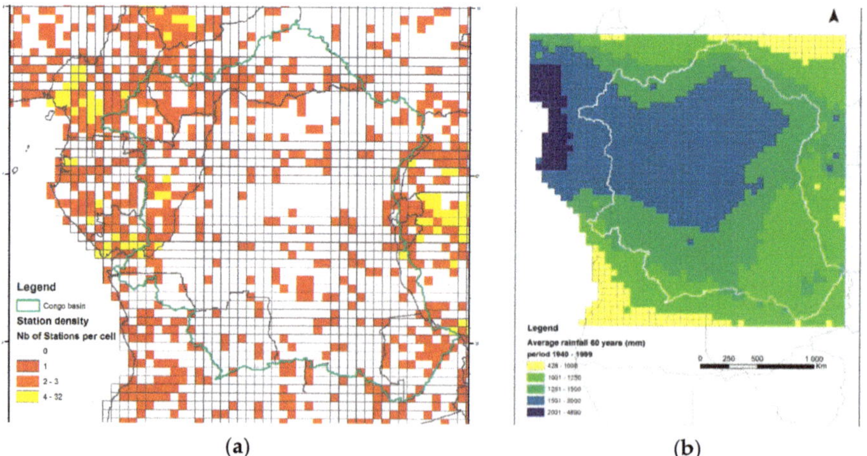

(a)　　　　　　　　　　　　　　(b)

Figure 2. (**a**) Density of observed rainfall stations over the Congo River Basin area [18], (**b**) Inter-annual averaged rainfall over the Congo River Basin, period 1940–1999, from the SIEREM database [6,20].

3.2. Discharges Network

In the entire Congo Basin, during the first half of the 20th century, there were more than 400 functional hydrometric stations (Figure 3a). Since the basin countries' independence, only fifteen are still operational today (Figure 3b). On the left bank, the hydrometric network was more or less abandoned. It was maintained on the right bank tributaries until the end of the 20th century by ORSTOM (ex. IRD), which had created hydro-pluviometric networks during the 1940s. However, the National Inland Navigation Services, RVF (Régie des Voies Fluviales) on the left bank, and GIE-SCEVN (Groupement d'Intérêt Economique—Service Commun d'Entretien des Voies Navigables) on the right bank, have continued to this day the hydrometric surveys of some stations located on the navigable routes. The CICOS (Commission Internationale du bassin Congo-Oubangui-Sangha) is responsible for the IWRM (Integrated Water Resources Management) within the Basin. Cameroon is the only country to have created a real operational NHS (National Hydrological Service).

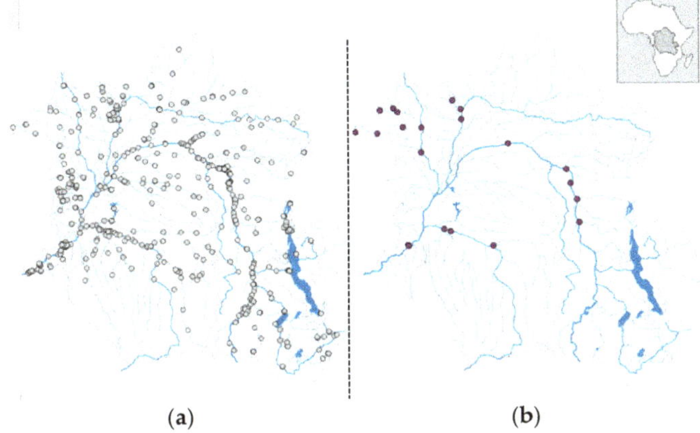

Figure 3. Hydrological network in Congo basin. (**a**) Past, before 1960 versus (**b**) Actual, after 1960 [7].

An inventory of existing data series can be found on the GRDC (Global Runoff Data Center) [21] and on the SIEREM website [6,20]. Thanks to BRLi (2016) [7] with the authorization of the CICOS, the Observatory Service SO-HYBAM [22] presents some daily and monthly hydrometric data today realized by the GIE-SCEVN and the RVF, on the five key stations with regular measurements from the beginning and from the second half of the past century, which are chosen for this study:

(1) The main station of Brazzaville-Kinshasa (BZV/KIN) on Congo River represents 97% of the entire Congo Basin. Measurements began in March 1902 in Kinshasa on its left bank, while measurements at Brazzaville on the other bank, opposite of Kinshasa, began in 1947.

Observed data on the Congo River, at the Kinshasa left bank station (1902–1983) are available from the GRDC and in the Mateba 22 report (1984) [23].

These data have been used to extend the water-stage data series from the Brazzaville station on the right bank, back to 1902 [14,24] to produce a single time series discharge of 117 years from 1902 to today, available on https://hybam.obs-mip.fr/ [22].

(2) On the mainstream of the Congo River (Figure 1), apart from the main gauging station of Brazzaville-Kinshasa (BZV/KIN), there exists only the Kisangani gauging station with a good daily discharge data record, for the period from 1951–2012 [7]. From upstream of the Bayoma falls (ex. Stanley Falls) at Kisangani, the river is known as Lualaba River.

(3) There are very little discharge data for left bank tributaries, although at many former gauging stations, water heights have been acquired during several decades. However, they have no known discharge time series associated with them, due mainly to the lack of gauging and rating curves. Fortunately, daily discharge from the Kasai River, the largest tributary of Congo, are available at its main Kutu Moke station for the period of 1948–2012 [7].

The main right bank hydrometric stations that are still active, and with a long time series of daily discharges are:

(4) Bangui on the Ubangi River (the second largest tributary to the Congo River, available from 1936) and,

(5) Ouesso on the Sangha River, available from 1948, both on https://hybam.obs-mip.fr/ [22].

A remarkable gap remains in understanding the hydro-climate processes in this region, thus leading to uncertainties and risks in decision-making for the major water resources development plans currently under discussion.

3.3. TSS, TDS, DOC Data

From 2005, with the continuous efforts of French and Congolese hydrologists, the SO-HYBAM (Observatory Service of larges rivers around the Atlantic intertropical ocean) was launched in the Congo Basin to preserve and continue acquisition of numerous varieties of data at the station of Brazzaville, such as daily discharge and monthly concentrations of total suspended solids (TSS), total dissolved solid (TDS), dissolved organic carbon (DOC), trace and rare earth elements, and also to ensure a free dissemination of data to the wider community [22].

This latest hydro-sediment-geo-chemical database at Brazzaville Station was completed with similar data coming from the previous PEGI/GBF (Programme Environnement Géosphère Intertropicale/opération Grands Bassins Fluviaux) Research program that monitored 10 stations on the right bank tributaries of the Congo River initially compiled and presented in the work of Laraque et al. (1995) [25]; Laraque et Orange (1996) [26]; Laraque et al. (1996 and 1996) [27,28], and Orange et al. (1996 and 1996) [29,30].

A non-exhaustive bibliographical summary of the work on the five categories of data mentioned above (Rainfall, flow and 'TSS, TDS, DOC') is presented in the next chapter.

4. State-of-the Art Hydroclimatological and Hydrosedimentological Studies of the Congo River Basin

4.1. Hydroclimatologial Studies

Over the entire Congo Basin, the annual average of rainfall for the 1940–99 period is 1500 mm year^{-1} [31,32]. The rainfall variability over the basin has been studied by many authors [18,33–38]. For example, Riou (1983) [39] studied the potential evapotranspiration in Central Africa. Samba Kimbata (1991) [40] quantified the water budget between 1951 and 1980 using monthly data, and Matsuyama et al. (1994) [41] studied the relationship between seasonal variations in the Congo Basin's water balance and the atmospheric vapor flow. From the Congo Basin's left bank data, Kazadi and Kaoru (1996) [42] proposed that the climatic variations over this basin can be related to solar activity. However, if the hydro-climatology of the basin presents similar cycles of 10–12 years, starting from the second half of the 20th century as shown by Laraque et al. (2013) [43], how can we explain their absence during the first half of this same century? Orange et al. (1997) [44] and Mahé et al. (2000, 2013) [45,46] linked the effects of the rainfall shortage since the 70s drought with the reduction of the groundwater reserves, leading to a decrease in the discharges in both humid and dry seasons. On the basis of the evolution of the regimes of major West and Central African rivers [47,48], some authors such as Olivry et al. (1993) [49], Mahé and Olivry (1999, 1995) [2,50] have noticed that the discharge reduction during the last decades is stronger in tropical Sudanian areas than in humid Equatorial areas, thus having an impact on the northern part of the Congo Basin.

Bricquet (1995) [51], categorized the main types of hydrograph that exist in the entire Congo Basin and proposed maps (Figure 4) to identify the hydrology contributions and times of water transit from upstream to downstream, according to different locations in the basin and during the different phases of the hydrological cycle of the Congo River at the Brazzaville-Kinshasa station. One of his maps shows the spatially contrasting values of specific flows, which can vary between extremes values of 0.5 (outlet of Lake Tanganyika) to 35 L s^{-1} km^{-2} (Batéké plateau).

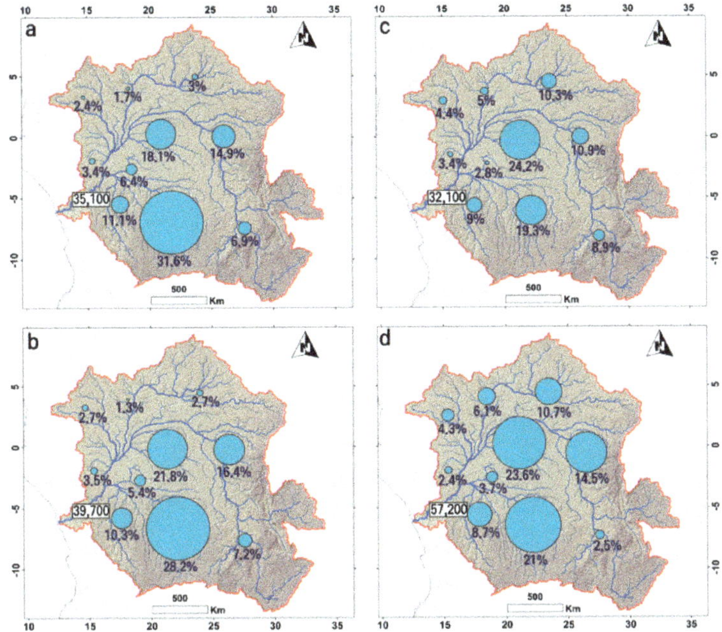

Figure 4. Participation of the different hydrological zones of the Congo Basin in the flow observed in Brazzaville/Kinshasa at different periods of the hydrological cycle (modified from Bricquet, 1995) [51] in March (**a**), May (**b**), August (**c**), and December (**d**). The area of the circles is proportional to the flow, and the percentage specifies the relative contribution of each zone to the flow in Brazzaville. For each zone, the flow retained takes into account its transfer time at the outlet. The units values are in $m^3 \ s^{-1}$.

In their study of the inter-tropical regions of Africa during the 1950–90s period, Bricquet et al. (1997) [52] highlights, that since the 1970s there has been a hydrological step change, which is thought to be a memory effect of the drought in the flow of the large river basins, reflecting a change in contribution of base flow to the flood hydrograph. Bricquet et al. (1997) [52] specify that this effect is more accentuated in the Sudano/Sahelian region. In the equatorial area, where the example studied was the Sangha River in Congo Basin, the secondary flood (named also spring flood) was more affected than the annual principal flood (or autumn flood). The authors hypothesize that there is a specific stream drought, which could be referred to as a "phreatic drought", in addition to the climatic drought.

Nguimalet and Orange (2019) [3] suggest that rainfall runoff ratios over the Ubangi Basin at Bangui do not vary in time and also comment on the impact of phreatic drought. They studied rainfall data from 1935 to 2015 over the basin and confirmed a single discontinuity in 1971 causing a decrease (<5%) in the rainfall data series (Figure 5). As Laraque et al. (2013) [42], they identified three hydrological discontinuities (1960, 1971, 1982), with the strongest decrease of 22% in 1982. At last, they underlined a probable change in the relationship between surface flows and groundwater flows, before and after 1970 and after 2000.

Moukandi N'kaya et al. (accepted and in press) [13] show that the high flow period of the 1960s is the major hydrological anomaly of the Congo River over its 116 years continuous record. They show that the hydropluviometric discontinuity in 1970 is common in most of the tributaries of the Congo River basin, accompanied by significant reductions in flows, depending on various factors (geographical location, vegetation cover, surface conditions and land use, etc.).

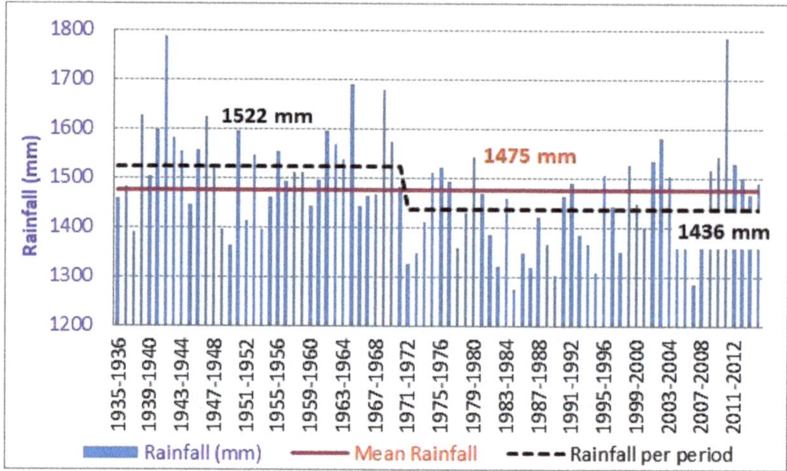

Figure 5. Discontinuities in the annual rainfall series for Ubangi basin, which comes from Nguimalet & Orange (2019) [3].

Although this synthesis prioritizes field data, it is worth mentioning the complementary studies of Lee et al. (2011, 2014) [53,54] focused on terrestrial water dynamics on the "Cuvette Centrale", using gravimetry data from the GRACE satellite and also using satellite radar altimetry. Becker et al. (2014) [55] presents the largest altimetry dataset of water levels ever published over the entire Congo Basin. This dataset, based on a total of 140 water level time series extracted using ENVISAT altimetry over the period of 2003 to 2009, allows illustration of the hydrological regimes of various tributaries of the Congo River. This analysis reveals nine distinct hydrological regions, a number similar to the ten zones of Bricquet (1995) [51], based on fields measurement. The proposed regionalization scheme is validated and therefore considered reliable for estimating monthly water level variations in the Congo Basin.

4.2. Hydrosedimentological and Biogeochemical Studies

Here, we consider TSS measurement and its extension to the other main complementary parameters like TDS, DOC, pH and EC (Electrical Conductivity).

Within the Congo Basin, the first solid (TSS) and dissolved transport (TDS + DOC) analyses were carried out during the first half of the 20th century at the main hydrological station of BZV/KIN. Table 3 summarizes their results in chronological order.

For BZV/KIN, this paper presents the mean results of the study of Moukandi N'Kaya et al. (2020) [56], about the comparison of the only two longer multiannual monthly chronicles of available data (1987–1993 versus 2006–2017) issued respectively from two French-Congolese projects, i.e., 8 years (1987–1993) of PEGI and 12 years of SO-HYBAM (during 2006–2017 period), the latter still operating. The parameters studied are TSS, TDS, DOC, pH and EC.

5. Recent Results

5.1. Spatially Interpolated Rainfall Data

The collected rainfall data were stored, analyzed, and spatialized using the inverse distance weighted method to obtain a rainfall value for every half a square degree (Figure 2b). Over the 60 year period 1940–1999, the inter-annual averaged rainfall on the Congo Basin ranges between 1000 and 2000 mm year^{-1}. Over one-third of the surface of the basin receives more than 1500 mm year^{-1} of rainfall, mainly in the area of the "Cuvette Centrale". The monthly rainfall regime (Figure 6) has changed slightly over time, as during the last 20 years of the time series, rainfall has decreased from February to April, while it has increased in July and August, and did not change much from September to December. This decrease of rainfall during the 'Spring' has provoked a decrease of the 'Spring' flood, which has also been observed in other equatorial rivers like the Ogooué River in Gabon and the Kouilou-Niari River in Congo [46,57], and the South Cameroon rivers [58].

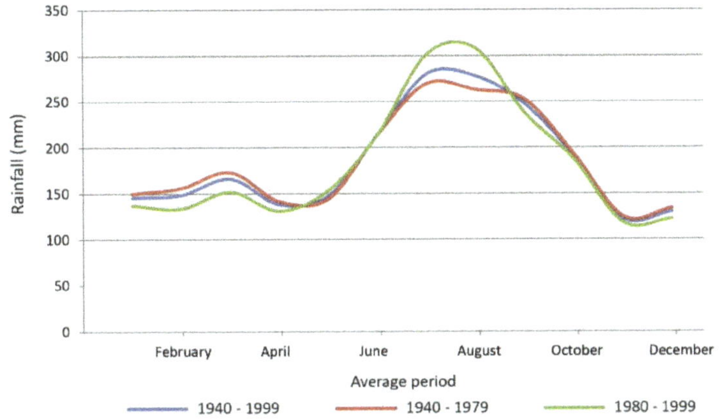

Figure 6. Mean monthly rainfall regime on the Congo River Basin over two different hydroclimatic periods (1940–1979 versus 1980–1999) and over the whole 1940–1999 period.

Two sets of gridded rainfall data have been used, from CRU and from SIEREM_HSM [32]; both time series are quite close, except for the 1940s where SIEREM is slightly higher, and the 1980s where the differences are a bit higher. These two periods correspond to low density of the observed stations, which can explain these differences when the interpolation exacerbates some smaller differences between raw data in each grid. However, a comparison with the Congo discharge time series shows that its variability closely follows the SIEREM rainfall time series. Therefore, for this study, we use the SIEREM gridded data, which are based on a larger number of observed stations.

5.2. Stationarity and Step Changes in Time Series and Wavelet Analysis of Rainfall Series

Statistical tests were performed on rainfall time series to detect segments, indicated by step changes, in the rainfall regimes [59]. Hubert's segmentation of regionalized annual rainfall (Figure 7) shows four steps in the time series, which occurred in 1970, 1983, 1987, and 1990. The main change in time series over the 60 years is found in 1983, which corresponds to the worst drought year in West and Central Africa in the 20th century. The time series presented in Figure 7 shows an increase of annual rainfall in 1987 and 1988, which seems exaggerated, as this change is not observed in the discharge time series, even if the discharge of 1988 shows a slightly higher value than the other years of the 1980s. This difference is mainly due to the lack of data over large parts of the Congo basin since the end of the 1980s.

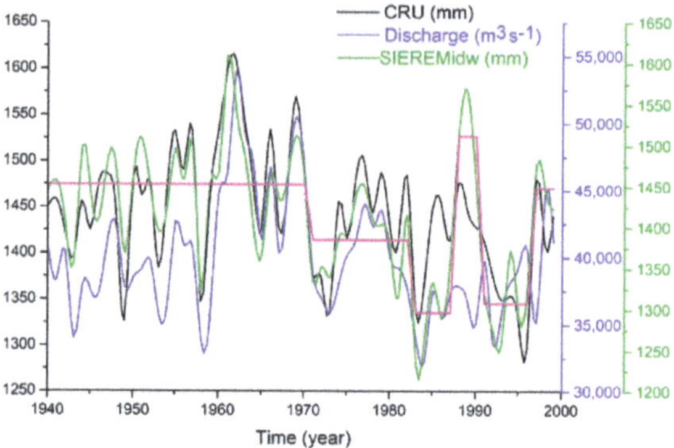

Figure 7. Regionalized annual rainfall over the whole Congo River Basin with phases of homogeneous rainfall (horizontal red lines). CRU (black) and SIEREM (green) gridded data set are compared, together with discharges time series (blue).

Following the work of Moron et al. (1995) [60], linking rainfall variability in sub-equatorial America and Africa with STT (Sea Surface Temperature), it is interesting to operate wavelet analysis on this rainfall time series, to detect periods with higher energy and possible associated features. The spectrum analysis (Figure 8) reveals a frequency shift that occurs by the mid-1970s, from the 4–8 year band to the 8–16 year band, and in 1983, when the 4–8 year band re-appears while the 8–16 year band is still observed. This increase of the energy since the mid-70s in the 8–16 years band also corresponds to a high level of correlation between annual rainfall and SST in the South Atlantic [61]. Both CRU and SIEREM databases give the same period for this shift, around the mid-70s, but there is a difference after this date, i.e., the CRU rainfall does not show a signal in the 8–16 band, which is the case for the SIEREM base, and moreover, the SIEREM base also shows a return of a signal in the 4–8 band from the mid-80s onward.

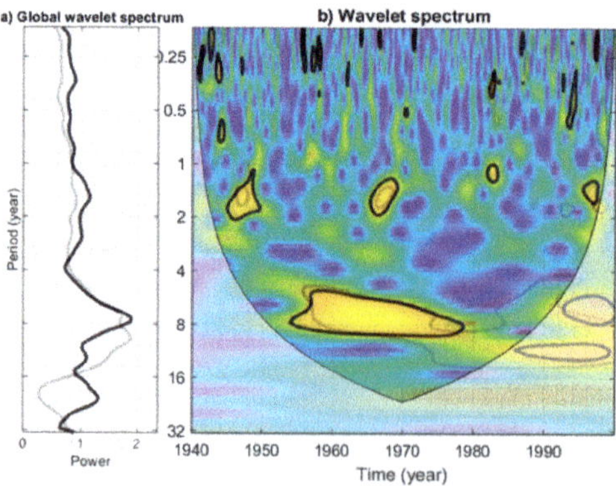

Figure 8. Graph overlay of (**a**) Global wavelet spectrum and (**b**) wavelet spectrum of the regionalized monthly rainfall over the Congo river basin (from CRU (dark line) and SIEREMv1 gridded rainfall) (gray line) (1940–1999).

5.3. Rainfall and Discharge Trends in the Congo River Basin

In Figure 9a,b, the graphs illustrate the synthesis of the study of Moukandi N'Kaya et al. (accepted and in press) [13] on the trends of multi-decadal rainfall records and the evolution of hydrological regimes during different homogeneous periods within the Congo Basin.

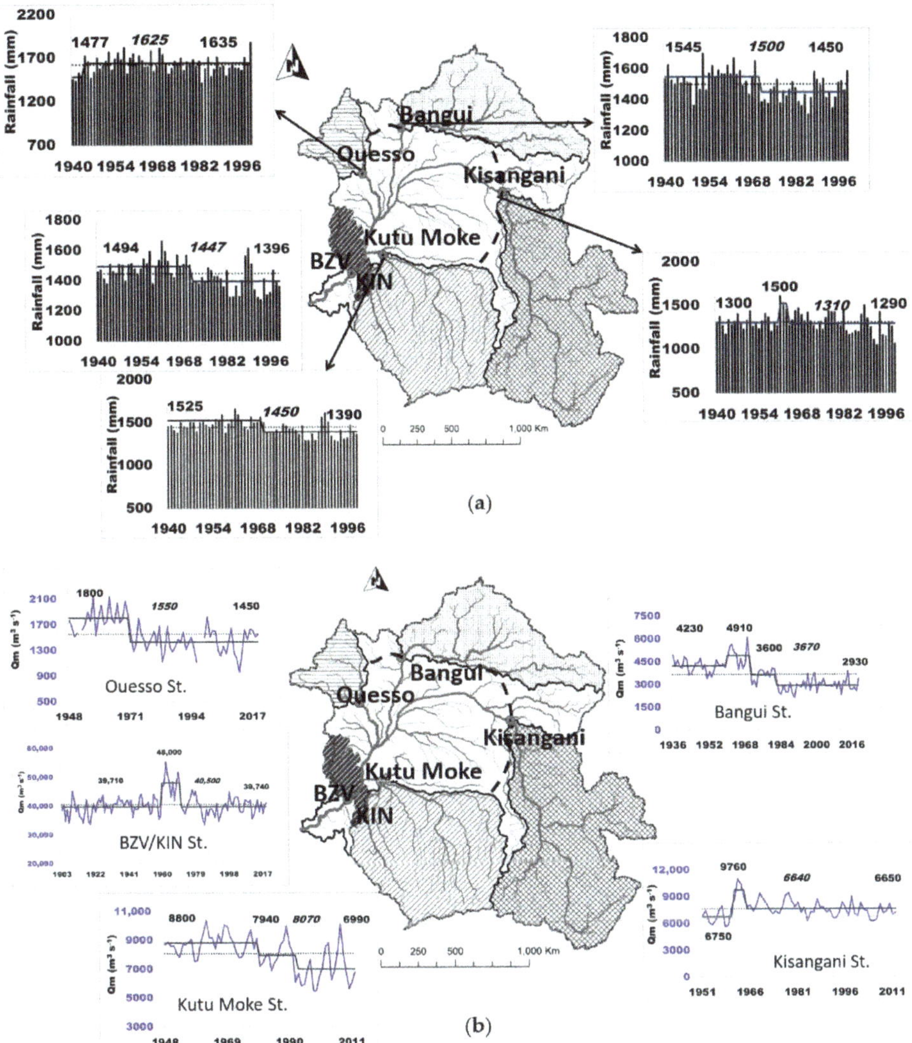

Figure 9. Graph discontinuities in the annual (**a**) rainfall and (**b**) hydrological chronicles of the five main drainage units studied from field data, including that of the entire Congo Basin. Legend: Values in italics are the interannual averages of the entirety of each series studied. (Rainfall series comes from www.hydrosciences.fr/sierem) [6] and Discharges series from https://hybam.obs-mip.fr/database [22].

The multi-decadal rainfall and hydrological records were analyzed by various statistical tests available in Khronostat software [62,63]. The tests included in the software are largely taken from the technical note of the World Meteorological Organization [64]. The description of these tests is presented in Laraque et al. (2013 and 2001) [42,65] and Moukandi N'Kaya et al. (accepted and in press) [13].

This preliminary study by catchment area will then make it possible to compare their hydro-pluviometric behaviors for the purpose of better understanding the hydrological responses within the Congo River Basin (CRB).

It can be concluded for the rainfall records between 1940 and 1999 (except for Ubangi basin with data from 1936 to 2018), that:

(i)   the rains have globally decreased on the Ubangi, Kasaï, and entire Congo Basin, with the main change around 1970. Nevertheless, for the Ubangi basin, Nguimalet et Orange (under review) [66] note a resumption of rainfall from 2007 onwards (Figure 5);

(ii)   the Lualaba basin shows high rainfall in the early 1960s and a decreasing rainfall trend around the 1980s;

(iii) in the Sangha basin, annual rainfall has not significantly changed;

(iv) for all the basins analyzed, the end of the 1980s and the beginning of the 1990s seem to have experienced relatively more rain, in a context of declining rainfall since 1970;

(v)   trends and changes in the rainfall can also be seen in the runoff record, sometimes with a time lag of a few years.

In Equatorial Africa, rainfall has returned to previous annual total and the runoff variability does not show any long-term trend, compared to the high runoff decrease observed in West African rivers. After the 1980s' drought years, the runoff of the Central African rivers returns to average values. The Congo River presents a certain hydrological inertia due to its vast surface area distributed on both sides of the Equator, more than 40% of which is covered by rainforest under the influence of the Atlantic and Indian Oceans and continental southern Africa.

The statistical "sequencing" of river flow records available (Figure 9b), shows one to three discontinuities, depending on the river analyzed (1970 for the equatorial Sangha River at Ouesso; 1960, 1971, 1983 for Ubangi at Bangui; 1960, 1964 for Lualaba at Kisangani; 1980 and 1990 for Kasaï at Kutu Moke and 1960, 1970 for Congo River at BZV/KIN gauging stations. Unfortunately, due to the lack of in situ data, one cannot estimate with accuracy, how river flows are changing on the left-bank tributaries since 2012. However, for the Kasaï basin, the later occurrence of hydrologic discontinuities in comparison with the rainfall ones can be explained mainly by a lag in groundwater flows.

The first analysis of the 1902–1996 record of annual discharge of the Congo River at Brazzaville and of the Ubangi River at Bangui made by Laraque et al. (2001) [65] identifies four homogeneous periods: the first period before 1960 showed no significant trend; 1960–70 experienced high flows and was followed by two periods of successive flow drops, in 1971–81 and 1982–95, the periods of lowest flows ever recorded.

The most recent analysis of the longest hydrological series of the Congo River at BZV/KIN and the Ubangi River at Bangui until 2019 (Figure 9b), shows that the wet decade of the 1960s with an interannual discharge of 48,000 m^3 s^{-1} separates two periods of equivalent mean flows for Congo River, 39,710 versus 39,740 m^3 s^{-1} [13].

Indeed, since 1996, measured values are close to the overall mean annual discharge of 40,500 m^3 s^{-1}, and a renewal of runoff is observed from 1990 onwards, after almost 15 years of deficit flow.

For the Ubangi River at Bangui, the same wet decade of the 1960s preceded a decline in flows in 1970, which in 1981 became more pronounced and remained stabilized until 2013; it seems to have a rising wave from this recent date.

In recent years, the presence of monthly extremes has been noted. For example, in 2011, the lowest levels for 65 years were observed in Brazzaville (with 23,550 m^3 s^{-1} in August) and in 2012, the Ubangi reached its lowest level for 100 years (with 207 m^3 s^{-1} in April), with new sandbanks appearing, which present fluvial navigation obstacles.

The end of 2019 was marked by an extreme rising stage on Ubangi River at Bangui, with a daily discharge of 12,400 m^3 s^{-1} in November, which correspond to a return period of a dozen years. This event seems to confirm a new humid period announced by the discontinuity in 2014 [3]. One month later, this flood reached Brazzaville-Kinshasa, where the maximum December discharge of the annual hydrologic cycle reached 70,000 m^3 s^{-1} for an mean monthly flow in this month of around 56,000 m^3 s^{-1}. The high flow events caused catastrophic flooding and affected thousands of people in the three capitals of Bangui, Brazzaville, and Kinshasa, in addition to the damage suffered by others localities along the rivers' courses.

These hydropluviometric recoveries are found in other regions of Africa like in the endoreic basin of Lake Chad [67] and in East Africa [68].

5.4. Hydrological Regimes of the Congo River and the Special Case of Ubangi River

Mahé et al. (2013) [46] highlighted that in West Africa and in a part of Central Africa, climate has changed since 1970, with the beginning of an intense and long dry period, but this change is less active since the 1980s in Central Africa. Nevertheless, a change in the intra-seasonal rainfall distribution (a slight rainfall decrease from March to May) most probably led to one of the biggest changes in the hydrological regimes in Africa, mainly visible in the Ogooue, Kouilou-Niari, and South Cameroonian rivers [58], but also visible partly in the Congo River. This reduction of the spring flood since the 1980s is therefore visible at the sub-continental scale. It might be associated with a change in global weather regime, linked to ITCZ (Intertropical Convergence Zone) changes in seasonal variability.

Figure 10 shows the hydrological regimes at the main gauging stations of the six main drainage units of the Congo Basin and of the whole Congo Basin at BZV/KIN.

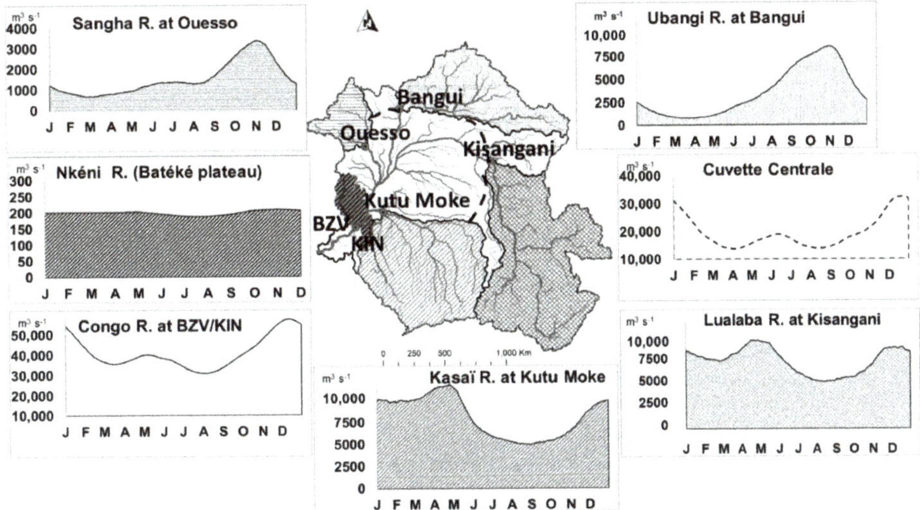

Figure 10. Graph Comparison of the hydrological regimes of the six main drainage units of the Congo Basin and of the whole Congo basin [13]. Legend: the dashed line encircles the 'Cuvette Centrale'.

The hydrological regime of the remote, swampy 'Cuvette Centrale', which is very difficult to measure in situ, has been deduced from the budget between the BZV/KIN gauging station and the other main stations (Ouesso, Bangui, Kisangani, Kutu Moke, …) with long hydrological records available, situated around its location. The results are in agreement with the reconstitution by hydropluviometric modelling made by BRLi (2016) [7]. The Congo River regime at BZV/KIN is the combination of the alternating inputs of its tributaries from both hemispheres. The local regime is strongly influenced and attenuated by the more regular equatorial regime of the Central Basin. Although its flows are much lower, the stable hydrological regime of the Batéké plateau (illustrated here with Nkéni R.) also mitigates low flows in the Congo River.

The evolution of hydrological regimes by periods of homogeneous flow for the studied drainage entities, according to their available flow records are presented in Moukandi N'kaya et al. (accepted and in press) [13]. The two longest records (Congo at BZV/KIN and Ubangi at Bangui) are now detailed.

The Congo River at BZV/KIN (Figure 11a) has a bimodal regime with its rising stage usually starting in October and ending in February. Afterwards, a large recession period begins and ends in

October. During this recession stage, a secondary smaller flood appears with a slightly rounded peak in May.

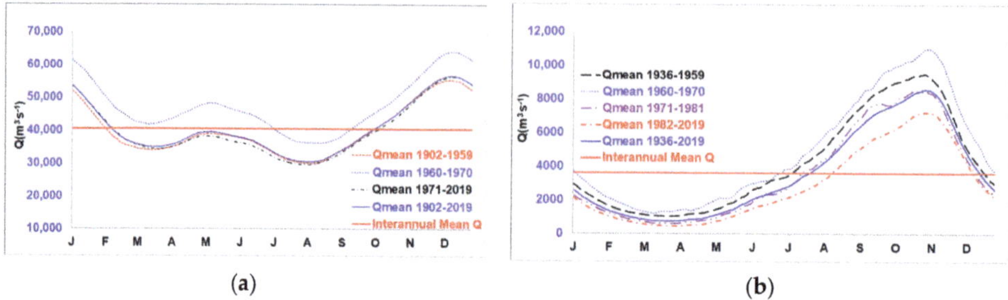

Figure 11. (**a**) Changes in mean daily hydrological regimes by homogeneous flow periods for Congo River at BZV/KIN, (**b**) and Ubangi at Bangui. The total periods of available daily Q are presented for each station.

The variations of hydrological regimes by periods of homogeneous flows of the Congo River in BZV/KIN (Figure 11a) does not show any clear trend except for the 1960s humid period, with runoff during all the hydrological cycle always higher than that of the two other homogenized periods. During the wet phase from 1960 to 1970, the secondary flood greatly exceeded the interannual mean discharge calculated since 1903, but since 1970, it has fallen below this.

The Ubangi River is monitored by the Bangui gauging station in the northern hemisphere and flow peaks are in end November-early-December. The falling stage occupies the rest of the year and lasts seven months.

In contrast to Congo River at BZV/KIN (Figure 11a), the variations of hydrological regimes by periods of homogeneous flows of the Ubangi River at Bangui (Figure 11b) shows clear trend, except for the 1960s humid period, with runoff during all the hydrological cycle always higher than from the others ones. The volume of water flowing out of the Ubangi almost halved between 1936 and 2018. Since 1982, the current period presents an important decrease of total runoff. This is due to the reduction of the magnitude and duration of the floods, but also due to a reduction of low flows. It is this latter period that presents the most altered hydrological regime for the Ubangi River.

As a complement to the study of Bricquet et al. (1997) [52], which showed the increase of its baseflow recession constant since the 1970s' break, Nguimalet and Orange (2019) [3] carried out some analyses of one hydropluviometric time series from 1935 to 2015 for the Ubangi River.

The average rainfall-runoff relationship did not change during the four hydroclimatic periods already identified by Laraque et al. (2013) [43], suggesting that the hydrological functioning of the Ubangi has not changed during the climate disruption. However, this is not the same pattern seen from its annual depletion coefficients, which continuously increased from 1935 to 2015 with an emphasis in 1970, but has declined since (Figure 12a). Nothing in the recent evolution of the annual rainfall helps explain this change of trend. The recent changes in the baseflow recession constant indicate that the support of baseflows by groundwater inflows is no longer assured. For this river, Nguimalet and Orange (2019) [3] showed the parallel responses of the changes of daily flood flows, as well as mean and low annual flows (Figure 12b). Nevertheless, these authors find that its mean daily flow in Bangui fell by 30% between the wet (1959–1970) and dry (1982–2013) periods. On the other hand, the mean daily low water flow fell by about 60%, which probably underlines a significant drop in the contribution of the water table of the Ubangi basin.

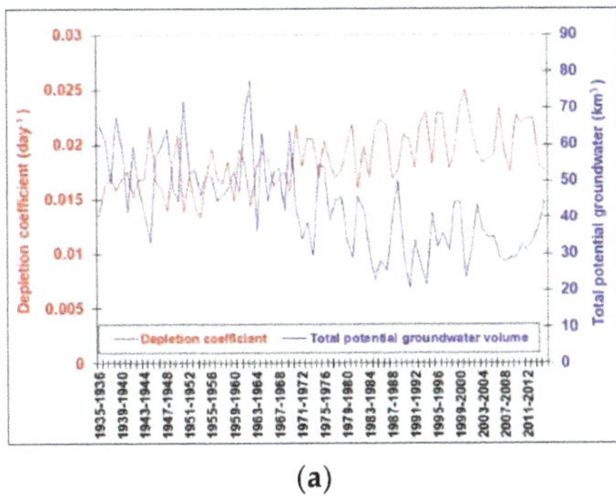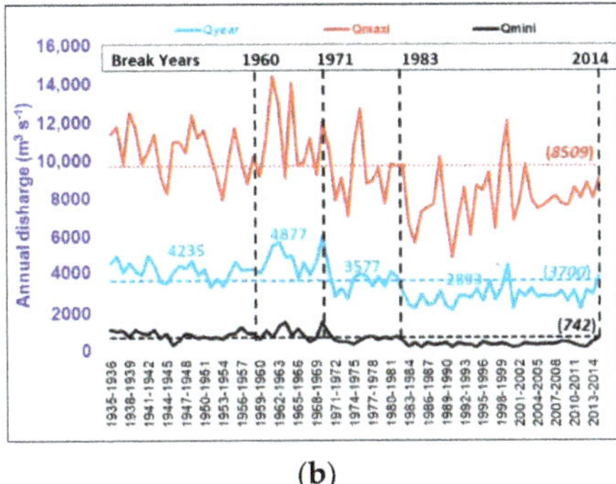

(a)   (b)

Figure 12. (**a**) Interannual evolution of the baseflow recession constant of the Ubangi River at Bangui and of the volume mobilized by the groundwater contribution of its catchment, from 1935 to 2015 hydrological cycles [3]. (**b**) Comparative evolution of annual (Qyear), maximum daily flood (Qmax), and minimum daily flood (Qmini) of the Ubangi at Bangui, from 1935 to 2015 [3].

5.5. Main Features, Concentrations, and Fluxes of the Congo River

On the Congo River, in the year 1980, African and French researchers continued the studies pioneered by Belgian scientists on TSS and TDS, by Van Mierlo (1926) [69] and Spronck (1941) [70] and then later on DOC by Eisma et al. (1978) [71]. Figure 13 presents their historical and recent results, also available in Table 3. These show that since 1967, a majority of authors, 12 versus 19, present TSS values with a relatively good convergence, between 20 and 30 mg L^{-1}. For TDS, it is similar, with 13 versus 16 authors presenting values between 30 and 40 mg L^{-1} since 1968. For DOC, all the authors found values between 8 and 13 mg L^{-1} for analysis carried out from 1982. For TSS and TDS, the measurement of pre-1968 data results from less accurate sampling protocols and laboratory analyses, with samples that do not regularly cover the entire hydrological cycle.

Moukandi et al. (2020) [56] show that in comparison with the PEGI data, SO-HYBAM data presents a slight increase (+7.5%) of interannual mean TSS concentrations, from 25.3 to 27.2 mg L^{-1}, a lower TDS concentration in the current period, i.e., −15% with a drop from 36.5 to 31.1 mg L^{-1} and a DOC concentration increase (+28%) from 9.9 to 12.8 mg L^{-1}. These variations appear to be essentially due to the increase in flows (nearly 4.5%), from 38,000 to 39,660 m^3 s^{-1} (inter-annual average), respectively for the two study periods. Indeed, generally the increase in discharge causes an increase of TSS by basin leaching and a decrease of TDS by dilution effect. For DOC, always in the absence of significant anthropogenic impacts in the Congo Basin, we suspect that its relatively high increase is due to the leaching of the vast 'Cuvette Centrale' covered by flooded forests and swamps rich in organic matter, with more water during the recent period (2006–2017).

We present in Table 2 the main statistics of Congo River water quality parameters collected by the SO-HYBAM project at the BZV/KIN gauging station.

The Congo River at the BZV/KIN section is characterized by a slightly acid mean pH value of 6.8, a mean specific electrical conductivity of 28.4 µS cm^{-1} at 25 °C, and a mean temperature of 27.7 °C. The total matter transported (TSS + TDS + DOC) is around 71 mg L^{-1}, with 38% TSS (27.2 mg L^{-1}) and 62% of Total Dissolved Matter (TDM = TDS + DOC = 43.8 mg L^{-1}). TDM comprises 29% of DOC (12.7 mg L^{-1}) and 71% of TDS (31.1 mg L^{-1}). The TDS comprises 20.2 mg L^{-1} of ionic load (cations + anions) and 10.9 mg L^{-1} of neutral mineral oxides, in which silica concentrations (10.5 mg L^{-1}) predominate. Consequently, the Congo River channel surface waters at Brazzaville contain relatively low concentrations of TSS and ionic load, but are rich in concentrations of SiO$_2$ and DOC with a mixed-bicarbonate geochemical facies.

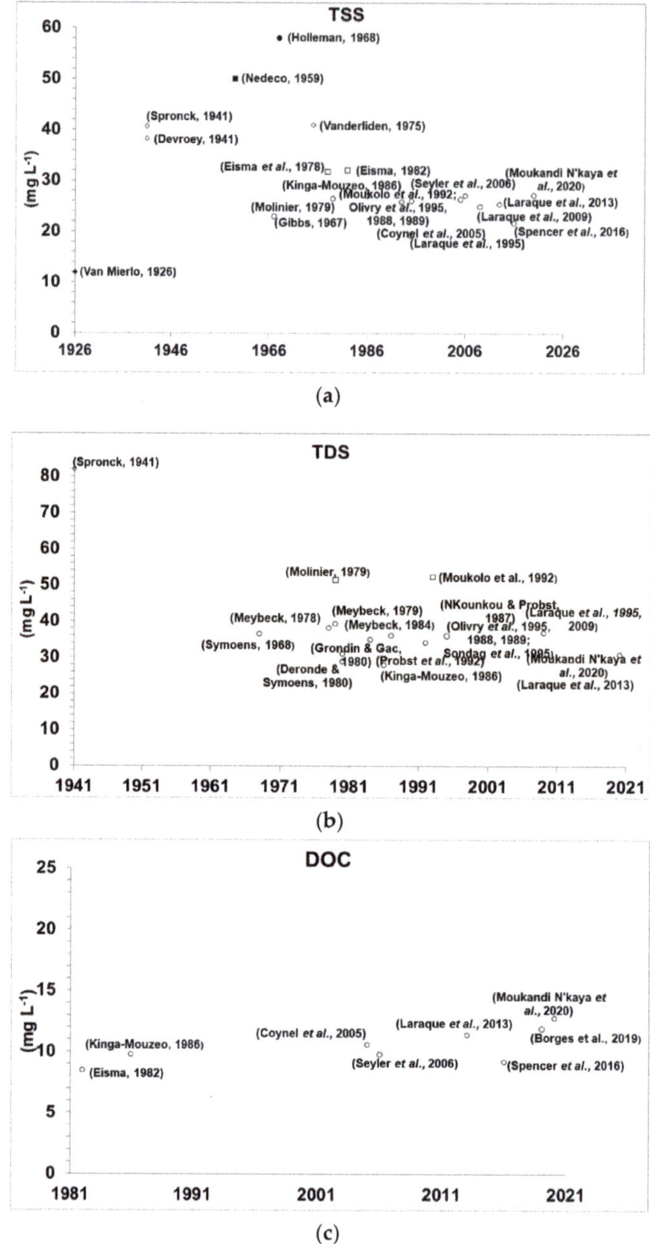

Figure 13. (a) TSS, (b) TDS, and (c) DOC concentrations versus time on the Congo River at BZV/KIN from different authors.

Unfortunately, POC (Particulate Organic Carbon) was not sampled during the SO-HYBAM project, but was studied by a few authors like Coynel et al. (2005) [72] and Spencer et al. (2012 and 2016) [73,74]. This last author carried out a monthly study for the period from September 2009 to November 2010 and found a mean POC concentration of 1.46 mg L^{-1} with extreme low and high values of 1.01 and

1.97 mg L^{-1}, respectively. For comparison purposes, from 1992 to 1993, Seyler et al. (2006) [75] found an almost constant POC on the Ubangi River at Bangui, around 1.4 mg L^{-1}. More recently, based on 15-day sampling over the hydrological cycle (2011–2012), Bouillon et al. (2014) [76] found a lower average of 0.9 mg L^{-1}.

Table 2. Interannual Yearly discharge (Qy) and main surface water quality parameters for the Congo River at BZV/KIN gauging station, for the 2006–2017 period.

Parameters & Units	Mean ± Std	Max	Min	Max/Min
Qy (m^3 s^{-1})	39,660 ± 8.70	61,330	22,710	2.7
Temperature (T °C)	27.7 ± 1.6	31.7	20.0	1.6
pH	6.8 ± 0.7	8.9	5.1	1.8
EC (μs cm^{-1} at 25 °C)	28.4 ± 4.98	36.0	20.0	1.8
Ionic load (mg L^{-1})	20.3 ± 4.1	27.6	11.6	2.4
SiO$_2$	10.5 ± 1.1	14.4	5.7	2.5
TSS (mg L^{-1})	27.2 ± 7.9	53.2	10.6	5.0
TDS (mg L^{-1})	31.1 ± 3.8	39.5	18.2	2.2
DOC (mg L^{-1})	12.7 ± 5.0	29.3	5.2	5.6
TOTAL (mg L^{-1})	71.0	-	-	-

TOTAL = TDS + DOC + TSS; Std: Standard deviation.

Table 3. Chronological compilation of TSS, TDS, and DOC concentrations since the beginning of the studies on Congo River quality at Brazzaville/Kinshasa gauging stations.

References	TSS	TDS	DOC
	(mg L^{-1})	(mg L^{-1})	(mg L^{-1})
(Van Mierlo, 1926) [69]	12		
(Spronck, 1941) [70]	40.7	82	
(Devroey, 1941) [77]	38.2		
(Nedeco, 1959) [78]	50		
(Gibbs, 1967) [79]	23		
(Symoens, 1968) [80]		36.6	
(Holeman, 1968) [81]	58		
(Vanderliden, 1975) [82]	41		
(Eisma et al., 1978) [71]	31.8		
(Meybeck, 1978) [83]		38.25	
(Meybeck, 1979) [84]		39.3	
(Molinier, 1979) [85]	26.5	51.4	
(Deronde & Symoens, 1980) [86]		28.93	
(Grondin & Gac, 1980) [87]		31	
(Eisma, 1982) [88]	32		8.5
(Meybeck, 1984) [89]		34.99	
(Kinga-Mouzeo, 1986) [90]	27	27.93	9.8
(NKounkou & Probst, 1987) [91]		36.05	
(Olivry et al., 1988) [92]	25.4	59	
(Olivry et al., 1989) [24]	25.4	59	
(Probst et al., 1992) [93]		34.18	
(Moukolo et al., 1992) [94]	25.9	52.3	
(Laraque et al., 1995) [25]	19.27	36.35	
(Olivry et al., 1995) [95]	26	36	
(Sondag et al., 1995) [96]		36	
(Coynel et al., 2005) [72]	26.3		10.6
(Seyler et al., 2006) [75]	27.1		9.8
(Laraque et al., 2009) [15]	24.98	36.89	
(Laraque et al., 2013) [97]	25.4	29.4	11.4
(Spencer et al., 2016) [74]	21.7		9.2
(Borges et al., 2019) [98]			11.9
(Moukandi N'kaya et al., 2020) [56]	27.13	30.92	12.78

Laraque et al. (2013) [97] explained the variations of TSS, TDS, and DOC, as follows: (i) the low monthly variation in TSS concentrations, a ratio of 5 (Table 2), during the hydrological year (Figure 14a) reflects the different timing and successive contributions of the different tributaries from one sub-basin to another, on each side of the Equator, within the overall Congo Basin; (ii) TDS concentration (Figure 14a) decreases during the rising stage of the annual flood. This decrease is due to the well-known TDS dilution effect during the rainy season. Conversely, TDS concentrations increase during the falling stage, due to the concentration effect; (iii) DOC concentrations show a ratio of monthly variation of 5.6 (Table 2) over the hydrological year. The highest DOC concentrations are generally observed during the main flood period (December–January). This logically comes from the leaching of the marshes and flooded forests in the 'Cuvette Centrale'.

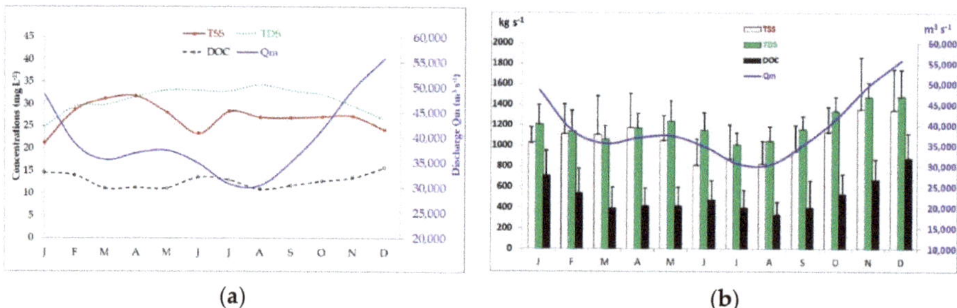

(a)   (b)

Figure 14. (**a**) Mean monthly discharge (Q), Total Suspended-Sediment (TSS), Total Dissolved Solid (TDS) and Dissolved Organic Carbon (DOC) concentrations and their (**b**) fluxes of the Congo River at Brazzaville (for the period 2006–2017) [56].

In conclusion, TSS, TDS, and DOC (without correction for atmospheric inputs) exhibit low concentrations and limited seasonal variation for the Congo at Brazzaville, which is not always synchronous with discharge variations (Figure 14a). However, due to their weak concentrations, TSS, TDS, and DOC monthly fluxes are mainly controlled by the magnitude of the discharge and show temporal variations more like those of the hydrograph (Figure 14b).

As discussed by Laraque et al. (2013) [97], within the Congo catchment, the TSS load is lower than the dissolved load, due to the ancient and peneplaned landscape, protected by a dense rainforest cover. Moreover, the higher contribution of dissolved matter is due to the dominating geochemical weathering (TDS) and the biochemical contribution (DOC concentrations) produced by the equatorial rain forest.

The "Cuvette Centrale" has a strong influence on the biogeochemical characteristics of the rivers crossing it, diluting the TSS and TDS, increasing organic matter concentrations, and lowering the pH due to the release of acids from the forest and swamps. Although this is a gradual process, differences were observed in Dissolved Organic Matter (DOM) production load. They are lower for semi-terra firma belt (Figure 15a) around the central flooded "Cuvette", illustrated respectively by the examples of the Likouala Mossaka River (Figure 15a), with DOM = 32 mg L^{-1} and pH = 5 to 6, and by the Likouala aux Herbes River (Figure 15b), with DOM = 80 mg L^{-1} and pH = 3.5 to 5 [15]. DOM corresponds to the difference between the weight of the dry residue at 105 °C and of the total dissolved solids (TDS = Σ(cations, anions, silica)) determined in laboratory.

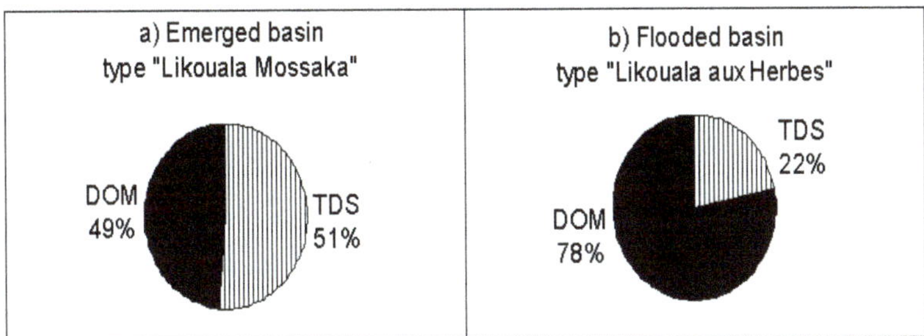

Figure 15. Distribution of dissolved matter (mg L^{-1}) in the emerged (**a**) and flooded (**b**) part of the 'Cuvette Centrale" [15].

These observations were confirmed by measurements along the river course of the Ubangi Rivers. During the scientific cruise of November 1992, at 350 km from its confluence with the Congo River, Figure 16 shows at the same time a progressive increase of TDM and a decrease of TDS. The difference is due to the increase of DOM when the river leaves a mixed zone of shrubby and woody savannah, before entering the dense forest covering the 'Cuvette Centrale'. This last system corresponds to a real DOM reservoir that releases almost 25 t km^{-2} year^{-1} of organic matter after slow maceration due to the soaking of the forest litter [15].

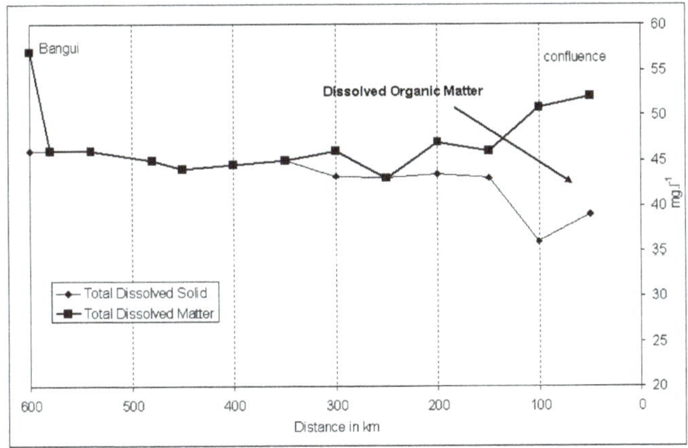

Figure 16. Evolution of dissolved matters during the Ubangi cruise in November 1992 [15].

For material fluxes, nearly 88 × 10^6 tonnes of material are exported annually from the Congo Basin to the Atlantic Ocean, composed of 33.6 × 10^6 tonnes of TSS, 38.1 × 10^6 tonnes of TDS and 16.2 × 10^6 tonnes of DOC for the study period (2006–2017). These values are in accordance with the first TSS fluxes given by Gibbs (1967) [79] and with those obtained during the period 1987–93 by Olivry et al. (1989, 1988, 1995) [24,92,95], Coynel et al. (2006) [72] and Laraque et al. (2009) [15], taking account of the discharge differences between these two periods.

In the case of the Congo, chemical weathering (10.6 t km^{-2} year^{-1} of TDS) slightly predominates physical weathering (9.3 t km^{-2} year^{-1} of TSS), followed by biological production (4.5 t km^{-2} year^{-1} of DOC).

Finally, thanks to these previous in situ studies and the use of remote sensing, Mushi et al. (2019) [99] was able to carry out a first assessment of soil erosion at the scale of the Congo River basin. They conclude that it is necessary to develop a program of high-frequency water and sediment collection at several strategic points in the basin, in order to understand its different transport dynamics.

5.6. Hydrological and Hydrogeochemical Anomalies Congo River Basin

The work of Laraque et al. (1998) [10] carried out on the right bank tributaries of the Congo river in Republic of Congo shows two close regions with very different hydrological regimes: the Batéké plateau and the 'Cuvette Centrale' (Figure 17), although they present similar annual rainfall amounts (1700 versus 1900 mm year^{-1}). The first region lies upon Tertiary sandstones formations of about 200 to 400 m in thickness, covered by savannah. The second one corresponds to the floodplain of the lowest sector of the wide Congo depression, according to Burgis and Symoens (1987) [100], and lies upon Quaternary alluvial deposits, covered with swamps and dense equatorial rain forests (Figure 1b).

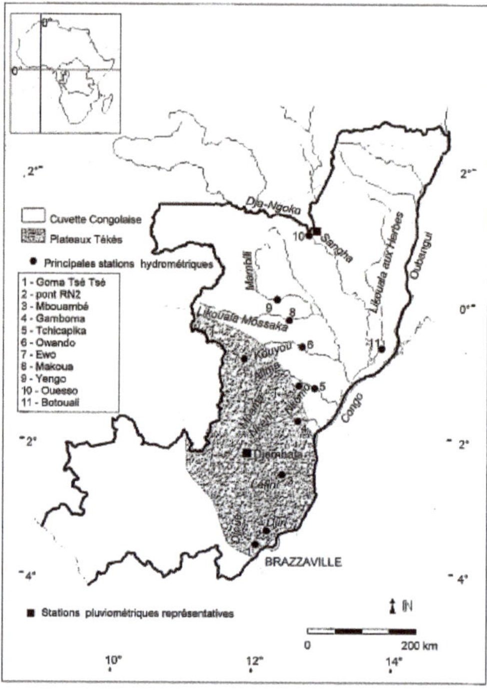

Figure 17. Congolese tributaries of the Congo River and smoothed contours of the Batéké plateau and of the "Cuvette Centrale' on the right bank of the Congo River [10]". Legend: 'Plateaux Tékés' = Batéké plateau'; 'Principales stations hydrométriques' = Main gauging stations; 'Stations pluviométriques représentatives' = Representative rainfall stations.

The hydrological regimes of the Batéké rivers are a representative example of a hydrologic paradox. Indeed, their flow regimes are almost independent of the regional rainfall regime because of the great infiltration capacity of the thick aquifer, which regulates their flows. This region holds some of the most regular rivers of the world, as illustrated by Figure 18a. In contrast, in the "Cuvette Centrale", the lower permeability of the soils, the interception of rainwaters, the evapotranspiration of the forest cover, and the direct evaporation on the floodplain areas as well as the swamps, lead to an important water deficit. The hydrological regime here is closer to the rainfall seasonality (Figure 18b).

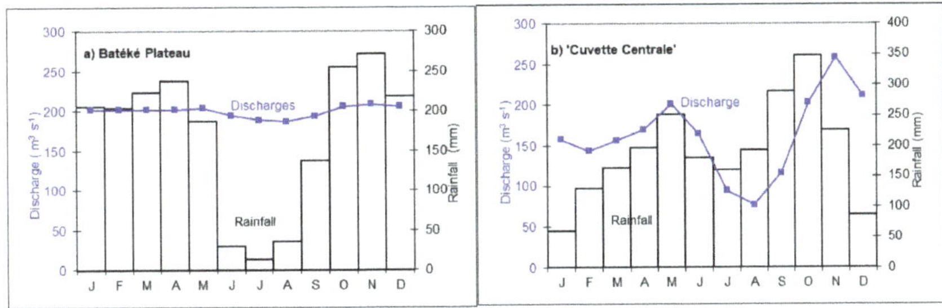

Figure 18. Mean monthly typical rainfall histogram and hydrological regimes of the (**a**) Batéké Plateau and the (**b**) "Cuvette Centrale" (adapted from Laraque et al., 2001) [65].

From a qualitative point of view, these two regions are also very distinct. On one side, after crossing an important sandy aquifer whose alterable minerals have been totally washed out, the Batéké rivers are 'clear waters' with remarkably few minerals, but relatively hypersiliceous. These waters, with dissolved inorganic matter ranging from 1 to 3 mg L^{-1} (not taking into account the dissolved silica), have similar TDS concentrations to rainfall waters (Figure 19a). These are among the most dilute surface waters of the world. On the other side, waters outflowing from the "Cuvette Centrale" stay for a long time under the forest cover and passes through the swamp. These are very rich in organic matter (up to 44% of particulate organic carbon) and very acid (pH can be lower than 4) and correspond to the typical 'black water rivers'. Their mineralization, although relatively low for rivers (11 to 30 mg L^{-1}, without dissolved silica), can be considered high when compared to the Batéké plateau waters (Figure 19b). These particularities point to the dominant roles of the geological formations and of the vegetation cover on the runoff of the concerned rivers, as well as on their water quality.

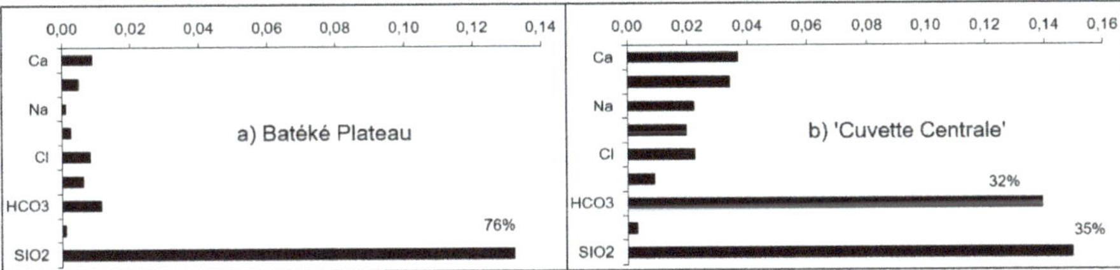

Figure 19. Composition of dissolved mineral phases (issue from mean concentration in mmol.l^{-1}, of (**a**) Batéké Plateau rivers versus (**b**) "Cuvette Centrale" rivers (adapted from Laraque et al., 1998) [10].

For the Batéké rivers, the interannual runoff coefficient (ratio between the runoff and the rainfall) ranges between 45–60%, versus 20–30% for the "Cuvette Centrale". The average seasonal variability of discharges (mean ratio between the highest and lowest monthly discharge for each studied year) ranges between 1.1 and 1.5 for the Batéké rivers, versus 2.5 and 5.5 for the "Cuvette Centrale" rivers. The specific discharges of the first region varies from 25 to 35 L s^{-1} km^{-2}, versus 10 to 15 L s^{-1} km^{-2} for the second region. These considerations allowed Laraque et al. (1998) [10] to propose an original hydrological classification of the Congolese right bank tributaries, based on these two hydrological parameters. Their variability can be explained by the physiographic characteristics of the drained basin (Figure 20). In this last figure, the Kouyou River presents an intermediate hydrological regime. Indeed, it originates in the Batéké Plateau with important annual runoff coefficients due to a strong and rapid infiltration in the aquifer, which will smooth the seasonal variability of its flows. It ends its course in the 'Cuvette Centrale' where, in the absence of a buffer aquifer and in the presence of a significant forest cover favoring evapotranspiration, the annual runoff coefficient becomes weaker and the seasonal variability of its flows, higher.

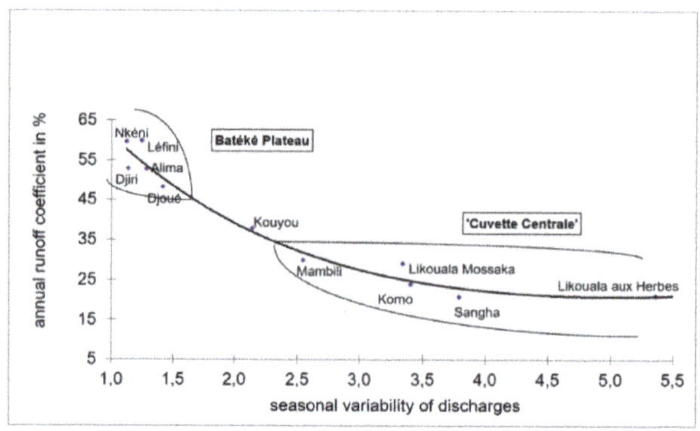

Figure 20. Hydrological classification of the Congolese tributaries of the Congo River [10].

Another anomaly in the "Cuvette Centrale", emphasized by Lee et al. (2011; 2014) [53,54] using remote sensing, is the difference between water levels in the river mainstem and the floodplain, which are lower by 0.5 to 3.0 m than the adjacent wetlands. Moreover, the mean seasonal variation of water levels on the mainstem of the Congo river, 3.5, 2.5, and 2.5 m, respectively at Kinshasa, Mbandaka, and Kisangani, repeated from decades of stage value studies by O'Loughlin et al. (2013) [101] and Betbeder et al. (2013) [102], are not matched with the seasonal variation of water levels in the adjacent wetlands of only 0.5 to 1.0 m, measured by Envisat radar altimetry [53,54]. This range of about 1 m of amplitude (Figure 21), presented in Alsdorf et al. (2016) [103], was also observed by in situ observations on the Télé Lake on the right bank wetland by Laraque et al. (1998) [104]. When taking into consideration, these studies conducted using remote sensing and ground observations on shallow lakes of the "Cuvette Centrale", such as the Mai Dombe and Tumba lakes [100,105] on the left bank and the Télé Lake on the right bank of Congo River, it appears that the vertical water exchange (rainfall versus PET) is dominant in the center of the Congo Basin. Laraque et al. (1998) [104], already stated that mainly meteorological contributions (~1650 mm year^{-1}) were largely compensated for (>80%) by evaporation (~1300 mm year^{-1}) in the wetland area of Télé Lake, where convective thunderstorms vertically recycle moisture. The "Cuvette Centrale" is filled mostly by rain (which accumulates in the depression) rather than by the overflowing of the river. In the low flow period, this place seems to drain slowly by gravity, through seepage, drainage, and desaturation of its spongy peat soils towards the mainstem river situated at a lower elevation.

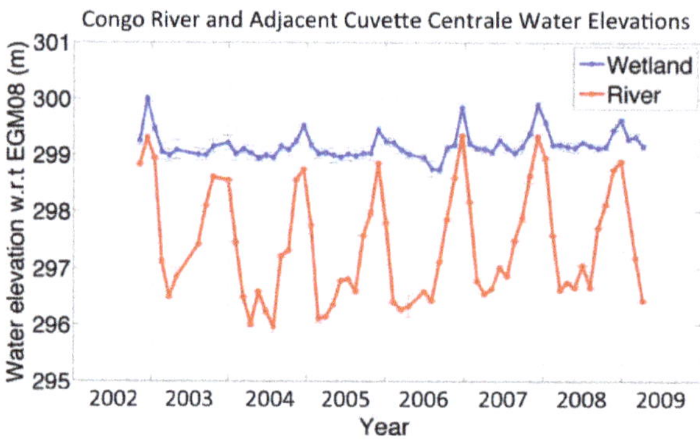

Figure 21. Water surface elevations for the Congo River and immediately adjacent wetlands of the Cuvette Centrale at 17.54° E, 0.74° S. Elevations are from the EnviSat radar altimeter, pass number 973. Elevations are with respect to the EGM08 geoid model [103].

Obviously, its water exchange with the river mainstem is different from those of other great rivers with vast lateral floodplains like the Amazon River, where most exchanges are assured by several interconnected channels [106–110], named locally as "furos".

6. Hydrological Modeling in the Congo River Basin

In this basin, one of the most complete rainfall-runoff modeling approaches which is well calibrated with all the in situ parameters available, is the work of BRLi (2016) [7], which uses the GR4J 'rainfall-ETP-Q daily' and Mike Hydro Basin models.

However, it is necessary to recall that hydrological modeling is used to address many water resource management issues, as models can be used with limited data, but generate sufficiently reliable information for management purposes. Tshimanga (under review) [111] provides a comprehensive review of hydrological model applications in the Congo Basin during the two last decades. This review reveals that there has been a great deal of effort from the past modeling studies to address both data issues and process understanding, and to establish models that represent the basin's hydrological processes. The author argues that many hydrological models' development during the past two decades in the basin have been initiated mainly in an experimental phase, testing the models' performance in a new and data scarce environment, and not for solving the real hydrological and water resources management issues of day-to-day life. For this reason, it has been difficult to use the models in support of policy decisions for river basin management and development. There are still a number of issues and uncertainties that should attract the attention of future modeling studies—right modeling for the right reason. The author's findings collectively reveal that researchers need to change their mindsets and to look for adequate or novel approaches to model hydrology and generate knowledge at appropriate scales of policy decision and management in the Congo Basin.

As mentioned above, the review of hydrological modeling studies in the Congo Basin shows that both data issues and processes understanding have been at the heart of model's development in the basin. With regard to data, and given a critical lack of in-situ data, there have been efforts made towards the use of global datasets, including global climate (both reanalysis and satellite-based) and physical basin properties datasets as well as satellite altimetry for prediction of water level and discharge. With regard to process understanding, a number of functions have been developed to improve existing model structures, most specifically to address the issues of modeling natural storages of wetlands and lakes that dominate the hydrology of the Congo Basin as well as the routing functions of the large river channels [112–115]. However, there have also been a number of problems that challenged the various modeling exercises. These problems relate to a lack of appropriate data as well as the lack of a thorough understanding of climate-hydrology processes, the lack of integration of this understanding in model structures and therefore, the lack of integrated and critical model assessment [111]. Currently, there is a number of studies that focus on model development to address some of these problems. Some of these studies include the use of the LISFLOOD-FP model to inform the wetland functions of the GW-Pitman model [116]; the use of SWAT (Soil and Water Assessment Tool) to highlight the attenuation functions of the 'Cuvette Centrale' and its hydrological budget [117]; the application of the MGB model to develop rating curves and derive river discharge in real time based on satellite altimetry [118]; and the development of a hydraulic and bathymetric HEC-RAS 2D channel model, which allowed Carr et al. (under review) [119] to create high-definition bathymetric maps of a complex and anastomosing section along a 120-km reach of the Congo River, downstream of the Ubangi confluence (Figure 22). This HEC-RAS model is a key step in the simulation of the river hydrodynamics necessary to facilitate and secure navigation on the main river "highway" in the basin [120].

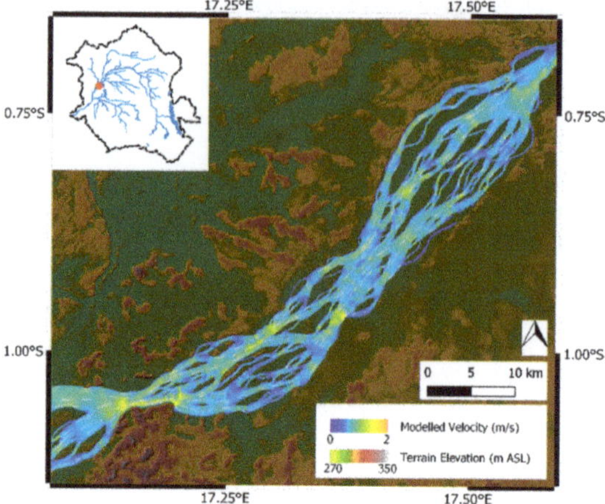

Figure 22. Water surface output plot of flow velocity downstream of the Ubangi confluence, from the Hec-Ras 2D channel model, described by Carr et al. (under review) [119].

Based on old navigational charts, made sometime between 1917 and 1931 and recent remote sensing images, these last authors were able to quantify river planform change on the same place over the last century.

On a longer timescale, a study by Molliex et al. (2019) [121] focused on a reconstitution of the water and sediment flows during the 155 ka BP, thanks to the cross-analysis of the evolutions of various proxies (geochemical and isotopic on the sedimentary cores of the submarine canyon of the Congo oceanic outlet) and by using the HydroTrend model calibrated in particular for the hydrological data of the last 116 years. Figure 23 presents the results focused on the last 23 ka. This study gives an idea of the long-term variations in the Congo's flows, according to the glacial and interglacial cycles that have affected our planet.

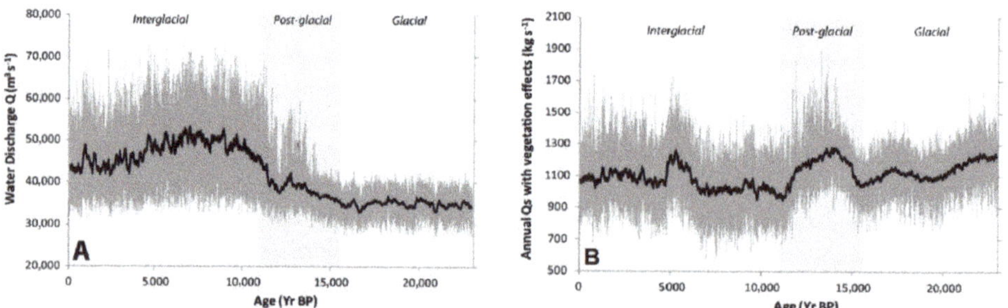

Figure 23. Water and suspended sediment simulation results focused over the last 23 ka. Gray curves represent annual data, while black curves are running means over 100 years. Three climatic periods (interglacial, post-glacial, and glacial) are individualized by the light gray/white bands in background. (**A**) Water discharge. (**B**) Mean suspended sediment load taking into account vegetation changes [121].

7. Human Use, Integrated Management, and the Future

7.1. Human Use Exploitation

Some works such as Shem and Dickinson (2006) [1] relate to the alteration of the water cycle under the effects of deforestation in the Congo Basin, but until now, the most part of the Congo River basin hydrology is still natural and relatively little affected by a population, whose standard of life is generally based on subsistence activities. Upstream of the Brazzaville-Kinshasa capitals, there are only a few hydropower dams on relatively small rivers like Lobaye, Léfini, Nzoro. There is no bridge over the course of the main river upstream or even between the two capitals, which are the nearest in the world, separated by only 3 km of river width. There are very few large cities upstream and no major industrial activities. However, there are some impacts from mining activities, especially for gold and diamonds near Kisangani city and copper near the town of Lubumbashi in the Precambrian massifs, which border the wide central sedimentary basin. There is also some wood exploitation, but deforestation is limited for example in comparison with other intertropical catchments like the Amazon basin. A large part of the Congo Basin is difficult to reach because of the lack of transport infrastructure, and large insecure areas, especially in the North-East. With the absence of a substantial road network, the fluvial navigation, even though precarious, is often the only way for the population to circulate and exchange goods, especially between states. Until the end of the 20th century, the wood was transported by floating rafts descending the river to Brazzaville to then be transported to the Congo delta by one of the few railways line, the "Congo-Ocean", from Brazzaville.

Under these conditions, the human impact is still very weak and probably much lower than that of climate changes.

We have already noted the longer duration of the recession stage, which reduces the fluvial navigation, especially on the Ubangi tributary (Figure 11b), where the runoff deficit is responsible for the degradation of the navigation conditions since 1975 [122]. This causes increases in sandbanks, which need huge dredging operations. The length of the annual interruption period of the navigation of large boats on the Ubangi river has increased from 9 days (period 1935–1977) to 84 days (period 1978–1991), and then to 108 days (period 1983–1991) according to the ATC, ACCF, SCEVN (2000) [123] report, and has been more than 200 days since 2002. The same report indicates a critical situation for the Sangha River, a right bank tributary on which the navigation was interrupted for 30 days per year for the 2002–2004 period, when it was only 2 days a year during the years 1954–1977.

More recently, the Congo Basin Atlas of CICOS (2016) [124] pointed out that the severe low water levels that have affected the Ubangi and Sangha rivers for several years have caused navigation to stop for 4 to 6 months a year.

Fluctuations in discharges and TSS fluxes need to be monitored for the PANAV project of regional navigation and for the next steps of the construction of the "Grand" Inga Dam on the Congo River in RDC, downstream of Brazzaville and Kinshasa. This project could produce twice the electricity generating capacity as the Three Gorges Dam [125].

7.2. Need for a Coherent Policy for Future Protection and Management

Despite its size and its importance and in the context of current climate change, the Congo River is poorly known and monitored, since the almost abandonment of the hydrological network after 1960 for left bank tributaries and 1993 for right bank tributaries.

As a consequence, the countries of the Congo Basin need coordination and organization to restore/create functional hydrological services, which have to reactivate operational hydrological stations and collaborate in the management of shared databases. The hydrological and hydrosedimentological information will serve in particular to improve the fluvial navigation, which is crucial for the economy of the neighboring states.

For these reasons, CICOS was created in 1999, and has objectives to promote inland interstate waterway transport and also to coordinate programs of regional importance like the Congo-HYCOS for the Hydrological Cycle Observation System on the Congo Basin, under the WMO (World Meteorological Organization) framework, AMESD for the African Monitoring of the Environment for Sustainable Development, etc., to estimate and better manage the water resources of this vast catchment and to strengthen capacity building of the NHS, especially with the use of the remote sensing. CICOS has the mandate of the six riparian countries, occupying more than 80% of the basin, to promote the integrated management of the water resources of the Congo Basin. This institution is also responsible for coordination of the hydrological monitoring of the basin and could take advantage of the results of this publication in order to further improve its scheme for the development and management of water resources and update its management tools, especially its Hydrological Information System and Water Resources Allocation tools, with the help of other national institutions of each country.

8. Conclusions

As the longest river in Africa after the Nile, and the second-largest global river in discharge after the Amazon, the Congo represents half of the African water input to the Atlantic Ocean. The understanding of its hydro-meteorological functioning, as well as that of its solid and dissolved matter flows is essential for water, material, and energy transfer modeling on a global scale and also for the sustainable implementation of the world's largest hydroelectric dam, the 'Grand Inga', which has been in the project planning stage for the last few decades.

Nguimalet and Orange (2019) [3] show that it is on the Ubangi river that the drought is the strongest, especially in this last decade, where the maximum flow rates are the most affected.

It raises an increasing problem for inland navigation, as the annual duration of which is reduced due to the longer annual period of low flows. This is crucial because it is the key method of exchanging goods and persons between the nearby countries in the absence of any good road network.

These considerations also call into question the usefulness of the TRANSAQUA project for the construction of a channel between the northeastern Ubangi tributary of the Congo and the Logone-Chari Rivers to offset water deficits in Lake Chad and highlight the possible impacts of such a huge project [126]. The polemic on this pharaonic project is amplified by the return of the backfilling of Lake Chad, which has historically experienced several natural water level fluctuations [67]. It is a wave-like phenomenon of irregular frequency. In addition, the use and management of the resources of this ecosystem is complex because the issues are political, environmental, economic, and security-related [127]. Therefore, it is necessary to take into account the socio-environmental disruption, which the TRANSAQUA project could lead to, as this is one of world's last fairly 'natural' large inter-tropical hydro-eco-systems present in central Africa.

Within the Congo Basin, the wet decade of the 1960s was the most striking anomaly in the river's hydraulic behavior over the last 118 years, which separates two symmetrical periods (1902–1959 versus 1970–2017), with equivalent mean flows around 39,700 m^3 s^{-1} for the Congo River at BZV/KIN. As the results of the 2006–2017 period of the recent SO-HYBAM monthly survey correspond to this same mean interannual discharge, it can also be considered representative of these two periods, which cover 104 years. The recent budget over the last 12 years of the SO-HYBAM survey shows specific discharge of 10.8 L s^{-1} km^{-2}, with specific fluxes of 9.3×10^6 t km^{-2} $year^{-1}$ (i.e., 33.6×10^6 t $year^{-1}$) for TSS; 10.6 t km^{-2} $year^{-1}$ (i.e., 38.1×10^6 t $year^{-1}$) for TDS; 4.5 t km^{-2} $year^{-1}$ (i.e., 16.2×10^6 t $year^{-1}$) for DOC. These fluxes need to be constantly documented on a regular basis and at a more regional scale by taking into account technological improvements like remote sensing in data collection and possible changes due to climate and human impacts.

These results are consistent with those of past periods, illustrating the significant inertia of this basin by: (i) being located on both sides of the equator line, (ii) the virtue of its very large area which may mitigate further possible changes, and (iii) highlighting the buffering role of the vast forest cover and of the huge 'sponge' formed by the 'Cuvette Centrale', which helps to sustain flows. Even if the drought has been particularly severe in the northern part of its basin, this paper underlines that its hydrological functioning was not very impacted due to its large untouched 'cuvette', which is the opposite of the Amazon rainforest where the impact of deforestation on the basin and on the flows of the Amazon River had begun to be suspected by Callède et al. (2008) [128], who indicated a 5% increase in the mean annual discharge over the 22 years from 1981 to 2003 and a number of strong floods five time higher during this period than before. Since then, the situation has continued to worsen.

Of course, short-term observations (one century) on large basins such as the Congo, should not obscure the relativity of the hydrological standards that have been established on the concepts of flow stationarity. This concept shows that our temporal myopia is moreover questioned [129,130] and must therefore be revised, as shown by the study of highly reactive sub-basins such as the Ubangi, which has evidenced accentuated hydropluviometric regime changes since 1970.

To face its future challenges like IWRM, the Congo Basin urgently needs the emergence of a true regional hydrological awareness to ensure the sustainable and balanced development of one of the last and rare regions of the world, which is still relatively pristine.

Author Contributions: Conceptualization, A.L., and G.D.M.N.; methodology, A.L., G.D.M.N., and D.O.; software, A.L., G.D.M.N., and J.M.T.; validation, A.L., D.O., R.T., and J.M.T.; formal analysis, A.L., G.D.M.N., R.T., and J.M.T.; investigation, A.L., G.D.M.N., D.O., R.T., J.M.T., G.M., C.R.N., M.A.T., and S.Y.; resources, A.L., G.D.M.N., D.O., R.T., J.M.T., G.M., C.R.N., M.A.T., S.Y. and G.G.; data curation, A.L. and G.D.M.N.; writing—original draft preparation, A.L.; writing—review and editing, all authors.; visualization, all authors; supervision, A.L. All authors have read and agreed to the published version of the manuscript.

Funding: This research received no external funding.

Acknowledgments: "On the eve of my retirement, and to use an African expression, I dedicate this summary to my ORSTOM "Dads", Jean Claude Olivry and Bernard Pouyaud, who entrusted me with the responsibility of the last hydrology laboratory at the ORSTOM Centre in Brazzaville. On the Congolese side, my "Dad" was Paul Bassema, an intrepid hydrologist technician, capable of solving everything in the bush. I also dedicate this paper to Bienvenu Maziezoula, endowed with the wisdom of a great Chief with whom I shared this direction. I cannot forget the generations of ORSTOM hydrologists who have succeeded one another in Congo and CAR since the 1940s to set up and manage their hydrological networks under epic and very difficult conditions. This synthesis has been made possible thanks to data from international projects such as PIRAT, PEGI/GBF & SO-HYBAM, (https://hybam.obs-mip.fr/), SIEREM (www.hydrosciences.fr/sierem), and the ECHOBACO/AUF project, thanks to the international and inter-institutional collaboration and cooperation between various institutions such as BRLi, CICOS, the Directorate of Meteorology and Hydrology of CAR, GET, GIE-SCEVN, HSM, IRD (e.g., ORSTOM), RVF, UMNG, and UNIKIN. Special thanks go to L. Ayissou, A. Carr, S. Délichère, P. Datok, G. Mokando, A. Pandi, N. Rouché, A. Yambélé, and S. Yepez for their various collaborations. Finally, I would also like to warmly thank Doug Alsdorf, specialist in remote sensing at the Ohio State University, for his initiative to relaunch research in the Congo Basin. With the AGU Chapman Conference in Washington in 2018 and this special issue on the Congo Basin, which he is leading, he has been keen to involve the nationals of this basin, so that they can be actors in their own destiny".

Conflicts of Interest: The authors declare no conflict of interest.

References

1. Shem, O.W.; Dickinson, R.E. How the Congo basin deforestation and the equatorial monsoonal circulation influences the regional hydrological cycle. In Proceedings of the 86th Annual American Meteorological Society Meeting, Atlanta, GA, USA, 29 January–2 February 2006; Available online: http://www.ametsoc.org/ (accessed on 4 June 2020).

2. Mahé, G.; Olivry, J.-C. Assessment of freshwater yields to the ocean along the intertropical Atlantic coast of Africa. *Comptes Rendus de l'Académie des Sci. Ser. IIAEarth Planet. Sci.* **1999**, *328*, 621–626.

3. Nguimalet, C.R.; Orange, D. Caractérisation de la baisse hydrologique actuelle de la rivière Oubangui à Bangui, République Centrafricaine. *La Houille Blanche* **2019**, *1*, 78–84. [CrossRef]

4. Runge, J. The Congo River, Central Africa. In *Large Rivers: Geomorphology and Management*; Gupta, A., Ed.; Wiley and Sons: London, UK, 2007; pp. 293–309.

5. De Wasseige, C.; Marshall, M.; Mahé, G.; Laraque, A. Interactions between climate characteristics and forests. In *The Forests of the Congo Basin-Forest and Climate Change*; De Wasseige, C., Tadoum, M., Atyi, E.A., Doumenge, C., Eds.; Weyrich: Neufchâteau, Belgium, 2015; Volume 3, pp. 53–64.

6. SIEREM (Système d'Informations Environnementales sur les Ressources en Eau et leur Modélisation), mai. Available online: www.hydrosciences.fr/sierem (accessed on 20 May 2020).

7. BRLi (Bas Rhône Languedoc Ingénierie). *Développement et Mise en Place de L'outil de Modélisation et D'allocation des Ressources en eau du Bassin du Congo. Rapport Technique de Construction et Calage du Modèle. Rapport d'étude Pour le Compte de la CICOS*; Commission Internationale du bassin Congo-Oubangui-Sangha: Kinshasa, Democratic Republic of the Congo, 2016; p. 210.

8. Runge, J.; Nguimalet, C.R. Physiogeographic features of the Ubangi catchment and environmental trends reflected in discharge and floods at Bangui 1911–1999, Central African Republic. *Geomorphology* **2005**, *70*, 311–324. [CrossRef]

9. Bwangoy, J.-R.B.; Hansen, M.C.; Roy, D.P.; de Grandi, G.; Justice, C.O. Wetland mapping in the Congo Basin using optical and radar remotely sensed data and derived topographical indices. *Remote Sens. Environ.* **2010**, *114*, 73–86. [CrossRef]

10. Laraque, A.; Mietton, M.; Olivry, J.C.; Pandi, A. Impact of lithological and vegetal covers on flow discharge and water quality of Congolese tributaries from the Congo River. *Rev. Des Sci. De L Eau* **1998**, *11*, 209–224.

11. USGS. Africa Land Cover Characteristics database Version 2.0. 2000. Available online: https://www.usgs.gov/media/files/africa-land-cover-characteristics-data-base-version-20 (accessed on 5 May 2020).

12. Farr, T.G.; Rosen, P.A.; Caro, E.; Crippen, R.; Duren, R.; Hensley, S.; Kobrick, M.; Paller, M.; Rodriguez, E.; Roth, L.; et al. The Shuttle Radar Topography Mission. *Rev. Geophys.* **2007**, *45*. [CrossRef]

13. Moukandi N'kaya, G.D.; Laraque, A.; Paturel, J.M.; Gulemvuga, G.; Mahé, G.; Tshimanga Muamba, R. A new look at hydrology in the Congo Basin, based on the study of multi-decadal chronicles. In *Congo Basin-Hydrology, Climate, and Biogeochemistry: A Foundation for the Future*; Alsdorf, D., Tshimanga Muamba, R., Moukandi N'kaya, G.D., Eds.; AGU, John Wiley & Sons Inc.: Malden, MA, USA, 2021; Unpublished work.

14. Laraque, A.; Maziezoula, B. *Banque de Données Hydrologiques des Affluents Congolais du Fleuve Congo-Zaïre et Informations Physiographiques-programme PEGI/GBF, Volet Congo-UR22/DEC*; ORSTOM-Laboratoire d'hydrologie: Montpellier, France, 1995; p. 250.

15. Laraque, A.; Bricquet, J.-P.; Pandi, A.; Olivry, J.-C. A review of material transport by the Congo River and its tributaries. *Hydrol. Process.* **2009**, *23*, 3216–3224. [CrossRef]

16. Orange, D.; Sigha-Nkamdjou, L.; Mettin, J.-L.; Malibangar, D.; Debondji, D.; Feizouré, C. *Evolution Menuselle de la Lame D'eau Précipitée sur le Bassin de L'Oubangui Depuis 1972. Rapport ORSTOM, Bangui*; Géographie, La Faculté des Lettres et Sciences Humaines, Université de Bangui: Bangui, Central African Republic, 1994; p. 132.

17. Callède, J.; Boulvert, J.; Thiébaux, J.P. *Le Bassin de l'Oubangui*; IRD: Marseille, France, 2009; ISBN 978-2-7099-1680-6. Available online: https://horizon.documentation.ird.fr/exl-doc/pleins_textes/divers16-07/010050233.pdf (accessed on 16 December 2019).

18. Mahé, G.; Rouché, N.; Dieulin, C.; Boyer, J.F.; Ibrahim, B.; Crès, A.; Servat, E.; Valton, C.; Paturel, J.E. *Carte des Pluies Annuelles en Afrique/Annual Rainfall Map of Africa*; IRD: Bondy, France, 2012.

19. Bultot, F. Atlas climatique du bassin congolais-les composantes du bilan de l'eau-t. 1971. Available online: https://donum.uliege.be/handle/2268.1/2165 (accessed on 5 May 2020).

20. Boyer, J.F.; Dieulin, C.; Rouché, N.; Crès, A.; Servat, E.; Paturel, J.E.; Mahé, G. SIEREM: An environmental information system for water resources. In *Water Resource Variability: Processes, Analyses and Impacts*; IAHS Publications: La Havana, Cuba, 2006; Volume 308, pp. 19–25.

21. UNH/GRDC Composite Runoff Fiels. Available online: http://www.grdc.sr.unh.edu (accessed on 20 May 2020).

22. SO-HYBAM Amazon Basin Water Resources Observation Service Website. Available online: https://hybam.obs-mip.fr/ (accessed on 25 February 2020).

23. *Mateba 22 Report, Observations Limnimétriques: Kinshasa-Matadi-Boma, Inventaire 1903–1983 Navigabilité du Bief Maritime du Fleuve Zaïre*; Laboratoire de Recherches Hydrauliques: Chatelet, Belgium, 1984; 50p.

24. Olivry, J.-C.; Bricquet, J.-P.; Thiébaux, J.-P. Bilan annuel et variations saisonnières des flux particulaires du Congo à Brazzaville et de l'Oubangui à Bangui. *La Houille Blanche.* **1989**, *3*, 311–316. [CrossRef]

25. Laraque, A.; Bricquet, J.-P.; Olivry, J.-C.; Berthelot, M. Transports solides et dissous du fleuve Congo (Bilan de six années d'observations). In *Grands Bassins Fluviaux Périatlantiques: Congo, Niger, Amazone-Actes du Colloque PEGI-INSU-CNRS-ORSTOM*; Olivry, J.-C., Boulègue, J., Eds.; ORSTOM Editions: Paris, France, 1995; pp. 133–145.

26. Laraque, A.; Orange, D. *Banque de Données Hydrochimiques Des Eaux de Surface d'Afrique Centrale (Congo et Oubangui) de 1987 à 1994-Programme PEGI/GBF*; Volet Afrique Centrale-Laboratoire d'hydrologie, ORSTOM: Montpellier, France, 1996; p. 145.

27. Laraque, A.; Orange, D.; Gries, S. *Typologie Géochimique des Eaux de Surface Congolaises du Fleuve Congo-Statistique Descriptive par Population Géographique (SYSTAT) et Analyses Qualitatives et Temporelles par Station (EXCEL)-Tome 1*; Lab. Hydrologie, ORSTOM: Montpellier, France, 1996; p. 204. Available online: https://horizon.documentation.ird.fr/exl-doc/pleins_textes/divers16-09/010051468.pdf (accessed on 20 May 2020).

28. Laraque, A.; Orange, D.; Gries, S. *Typologie Géochimique des Eaux de Surface Congolaises du Fleuve Congo-Systématique des Eaux par Station (SYSEAU). Tome 2*; Lab. Hydrologie, ORSTOM: Montpellier, France, 1996; p. 193. Available online: https://horizon.documentation.ird.fr/exl-doc/pleins_textes/divers16-09/010051468.pdf (accessed on 20 May 2020).

29. Orange, D.; Laraque, A.; Gries, S. *Typologie Géochimique des Eaux de Surface de L'Oubangui-Statistique Descriptive par Population Géographique (SYSTAT) et Analyses Qualitatives et Temporelles par Station (EXCEL). Tome 1*; Lab. Hydrologie, ORSTOM: Montpellier, France, 1996; p. 240. [CrossRef]

30. Orange, D.; Laraque, A.; Gries, S. *Typologie Géochimique Des Eaux de Surface de L'Oubangui-Systématique Des Eaux Par Station (SYSEAU). Tome 2*; Hydrologie, ORSTOM: Montpellier, France, 1996; p. 136. [CrossRef]

31. Tshimanga, R.M.; Tshitenge, J.M.; Kabuya, P.; Alsdorf, D.; Mahe, G.; Kibukusa, G.; Lukanda, V. A regional perceptive of flood forecasting and disaster management systems for the Congo River basin. In *Flood Forecasting: An International Perspective*; Adams, T.E., Pagano, T.C., Eds.; Academic Press: Cambridge, MA, USA; Elsevier: Amsterdam, The Netherlands, 2016; pp. 87–124.

32. Dieulin, C.; Mahe, G.; Paturel, J.E.; Ejjiyar, S.; Tramblay, Y.; Rouche, N.; El Mansouri, B. A new 60-year 1940–1999 monthly gridded rainfall data set for Africa. *Water* **2019**, *11*, 387. [CrossRef]

33. Lempicka, M. *Bilan Hydrique du Bassin du Fleuve Zaïre. I: Ecoulement du Bassin 1950–1959*; Office National de la Recherche et du Développement: Kinshasa, Democratic Republic of the Congo, 1971.

34. Anthony, C.; Hirst, A.C.; Hastenrath, S. Diagnostics of hydrometeorological anomalies in the Zaïre (Congo) basin. *J. R. Met. Soc.* **1983**, *109*, 881–892.

35. Janicot, S. Spatio-temporal variability of West African rainfall; part I: Regionalizations and typing. *J. Clim.* **1992**, *5*, 489–511. [CrossRef]

36. Mahé, G. Modulation annuelle et fluctuations interannuelles des précipitations sur le bassin-versant du Congo. In *Grands Bassins Fluviaux Périatlantiques: Congo, Niger, Amazone-Actes du colloque PEGI-INSU-CNRS-ORSTOM*; Olivry, J.-C., Boulègue, J., Eds.; ORSTOM: Paris, France, 1995; pp. 13–26.

37. Bigot, S.; Moron, V. Synchronism between temporal discontinuities in African rainfall and sea-surface temperatures. In *Climate Dynamics*; Tellus: Stockholm, Sweden, 1997; p. 26.

38. Samba, G.; Nganga, D.; Mpounza, M. Rainfall and temperature variations over Congo Brazzaville between 1950 and 1998. *Theor. Appl. Climatol.* **2008**, *91*, 85–97. [CrossRef]

39. Riou, C. Experimental study of potential evapotranspiration (PET) in Central Africa. *J. Hydrol.* **1983**, *72*, 88. [CrossRef]

40. Samba Kimbata, M.-J. Précipitations et bilans de l'eau dans le bassin forestier du Congo et ses marges. Ph.D. Thesis, Université de Bourgogne, Dijon, France, 1991.

41. Matsuyama, H.; Oki, T.; Shinoda, M.; Masuda, K. The seasonnal change of the Water Budget in the Congo River Basin. *J. Meteorol. Soc. Jpn.* **1994**, *72*, 281–299. [CrossRef]

42. Kazadi, S.N.; Kaoru, F. Interannual and long term climate variability over the Zaïre River Basin during the last thirty years. *J. Geophys. Res. Atmos.* **1996**, *101*, 351–360. [CrossRef]

43. Laraque, A.; Bellanger, M.; Adele, G.; Guebanda, S.; Gulemvuga, G.; Pandi, A.; Paturel, J.E.; Robert, A.; Tathy, J.P.; Yambélé, A. Recent evolutions of Congo, Ubangi and Sangha rivers flows. *Geo-Eco-Trop* **2013**, *37*, 93–99.

44. Orange, D.; Wesselink, A.; Mahé, G.; Feizouré, C.T. The effects of climate changes on river baseflow and aquifer storage in Central Africa. In *Sustainability of Water Resources under Increasing Uncertainty*; Rosbjerg, D., Boutayeb, N., Gustard, A., Kundzewicz, Z., Rasmussen, P., Eds.; IAHS: Wallingford, UK, 1997; Volume 240, pp. 113–123. ISBN 1-901502)05-8.

45. Mahé, G.; Olivry, J.-C.; Dessouassi, R.; Orange, D.; Bamba, F.; Servat, E. Relations eaux de surface–eaux souterraines d'une rivière tropicale au Mali. *C. R. Acad. Sci.* **2000**, *330*, 689–692. [CrossRef]

46. Mahé, G.; Lienou, G.; Descroix, L.; Bamba, F.; Paturel, J.E.; Laraque, A.; Meddi, M.; Moukolo, N.; Hbaieb, H.; Adeaga, O.; et al. The rivers of Africa: Witness of climate change and human impact on the environment. *Hydrol. Process.* **2013**. [CrossRef]

47. Sircoulon, J. Variations des débits des cours d'eau et des niveaux des lacs en Afrique de l'Ouest depuis le 20ième siècle. In *The Influence of Climate Change and Climatic Variability on the Hydrologic Regime and Water Ressources*; Solomon, S., Beran, M., Hogg, W., Eds.; IAHS: Waterloo, ON, Canada, 1987; Volume 168, pp. 13–25.

48. Wesselink, A.; Orange, D.; Feizouré, C.; Randriamiarisoa. Les régimes hydroclimatiques et hydrologiques d'un bassin versant de type tropical humide: L'Oubangui (République Centrafricaine). In *L'hydrologie Tropicale: Géosciences et Outil Pour le Développement: Mélanges à la Mémoire de Jean Rodier*; Chevallier, P., Pouyaud, B., Eds.; IAHS: Wallingford, UK, 1996; Volume 238, pp. 179–194.

49. Olivry, J.-C.; Bricquet, J.-P.; Mahé, G. Vers un appauvrissement durable des ressources en eau de l'Afrique humide. In *Hydrology of Warm Humid Regions. (Proc. Yokohama Symp., July 1993)*; Gladwell, J.S., Ed.; IAHS: Paris, France, 1993; Volume 216, pp. 67–78.

50. Mahé, G.; Olivry, J.-C. Variations des précipitations et des écoulements en Afrique de l'Ouest et centrale de 1951–1989. *Sci. et Chang. planétaires/Sécheresse* **1995**, *6*, 109–117.

51. Bricquet, J.-P. Les écoulements du Congo à Brazzaville et la spatialisation des apports. In *Grands Bassins Fluviaux Périatlantiques: Congo, Niger, Amazone-Actes du colloque PEGI-INSU-CNRS-ORSTOM*; Olivry, J.-C., Boulègue, J., Eds.; ORSTOM: Paris, France, 1995; pp. 27–38.

52. Bricquet, J.-P.; Bamba, F.; Mahe, G.; Toure, M.; Olivry, J.C. Evolution récente des ressources en eau de l'Afrique atlantique. *Rev. des Sci. de l'Eau* **1997**, *3*, 321–337.

53. Lee, H.; Beighley, R.E.; Alsdorf, D.; Jung, H.C.; Shum, C.K.; Duan, J.; Guo, J.; Yamazaki, D.; Andreadis, K. Characterization of terrestrial water dynamics in the Congo Basin using GRACE and satellite radar altimetry. *Remote Sens. Environ.* **2011**, *115*, 3530–3538. [CrossRef]

54. Lee, H.; Jung, H.-C.; Yuan, T.; Beighley, R.E.; Duan, J. Controls of terrestrial water storage changes over the central Congo Basin determine by integrating PalSAR scanSAR, EnviSat altimetry, and GRACE data. In Proceedings of the Chapman Conference, Washington, DC, USA, 25 September 2018; Volume 2014.

55. Becker, M.; Santos da Silva, J.; Calmant, S.; Robinet, V.; Linguet, L.; Seyler, F. Water Level Fluctuations in the Congo Basin Derived from ENVISAT Satellite Altimetry. *Remote Sens.* **2014**, *6*, 9340–9358. [CrossRef]

56. Moukandi N'kaya, G.D.; Orange, D.; Bayonne Padou, S.; Datok, P.; Laraque, A. Temporal Variability of Sediments, Dissolved Solids and Dissolved Organic Matter Fluxes in the Congo River at Brazzaville/Kinshasa. *Geosciences* **2020**, *10*, 341. [CrossRef]

57. Mahé, G.; Lerique, J.; Olivry, J.-C. L'Ogooué au Gabon. Reconstitution des débits manquants et mise en évidence de variations climatiques à l'équateur. *Hydrol. Continent.* **1990**, *5*, 105–124.

58. Liénou, G.; Mahé, G.; Paturel, J.E.; Servat, E.; Sighomnou, D.; Sigha-Nkamdjou, L.; Ekodeck, G.E.; Dezetter, A.; Dieulin, C. Impact de la variabilité climatique en zone équatoriale: Exemple de modification de cycle hydrologique des rivières du sud-Cameroun. *Hydrol. Sci. J. Sci. Hydrol.* **2008**, *53*, 789–801. [CrossRef]

59. Hubert, P. The segmentation procedure as a tool for discrete modeling of hydro-meteorological regimes. *Stoch. Environ. Res. Risk Assess.* **2000**, *14*, 297–304. [CrossRef]

60. Moron, V.; Bigot, S.; Roucou, P. Rainfall variability in subequatorial America and Africa and relationships with the main sea-surface temperatures modes (1851–1990). *Int. J. Clim.* **1995**, *15*, 1297–1322. [CrossRef]

61. Tshitenge Mbuebue, J.-M.; Lukanda Mwamba, V.; Tshimanga Muamba, R.; Javaux, M.; Mahe, G. Wavelet Analysis on the Variability and the Teleconnectivity of the Rainfall of the Congo Basin for 1940–1999. In Proceedings of the Conférence Internationale sur l'hydrologie des grands bassins Fluviaux d'Afrique, Hammamet, Tunisia, 26–30 October 2015.

62. Lubes Niel, H.; Masson, J.M.; Paturel, J.E.; Servat, E. Variabilité climatique et statistiques. Etude par simulation de la puissance et de la robustesse de quelques tests utilisés pour vérifier l'homogénéité de chroniques. *Rev. des Sci. de l'Eau* **1998**, *11*, 383–408. [CrossRef]

63. KHRONOSTAT. *Logiciel D'analyse Statistique de Séries Chronologiques*; IRD Ex: Paris, France; Available online: http://www.hydrosciences.org/spip.php?article239 (accessed on 14 October 1999). (In French)

64. WMO. *World Meteorological Organization, Climatic Change. Report of a Working Group of the Commission for Climatology*; WMO 195, TP 100, Tech. note n°79; WMO: Geneva, Switzerland, 1966.

65. Laraque, A.; Mahé, G.; Orange, D.; Marieu, B. Spatiotemporal Variations in hydrological regimes within Central Africa during the XXth century. *J. Hydrol.* **2001**, *245*, 104–117. [CrossRef]

66. Nguimalet, C.R.; Orange, D. Evolution annuelle de l'indice pluviométrique sur le bassin de l'Oubangui à Mobaye (1938–2015). In *Congo Basin-Hydrology, Climate, and Biogeochemistry: A Foundation for the Future*; Alsdorf, D., Tshimanga, M.R., Moukandi N'kaya, G.D., Eds.; AGU, John Wiley & Sons Inc.: Malden, MA, USA, 2021; Unpublished work.

67. Pham-Duc, B.; Sylvestre, F.; Papa, F.; Frappart, F.; Bouchez, C.; Crétaux, J.-F. The Lake Chad hydrology under current climate change. *Nat. Sci. Rep.* **2020**, *10*, 5498. [CrossRef]

68. Nicholson, S. Climate and climatic variability of rainfall over eastern Africa. *Rev. Geoph.* **2017**. [CrossRef]

69. van Mierlo, J.G. Le mécanisme des alluvions du Congo. *Annls. Ass. Ingrs. Éc Gand.* **1926**, *5*, 349–354.

70. Spronck, R. Mesures hydrographiques dans la région divagante de bief maritime du fleuve Congo. *Mém. Inst. R. Colon. Sect. Sci. Technol.* **1941**, *3*, 56.

71. Eisma, D.; Kalf, J.; Vandergaast, S.J. Suspended Matter in the Zaire Estuary and the Adjacent Atlantic Ocean. *Neth. J. Sea Res.* **1978**, *12*, 382–406. [CrossRef]

72. Coynel, A.; Seyler, P.; Etcheber, H.; Meybeck, M.; Orange, D. Spatial and seasonal dynamics of total suspended sediment and organic carbon species in the Congo River. *Glob. Biogeochem. Cycles* **2005**, *19*, GB4019. [CrossRef]

73. Spencer, R.G.M.; Hernes, P.J.; Aufdenkampe, A.K.; Baker, A.; Gulliver, P.; Stubbins, A.; Aiken, G.R.; Dyda, R.Y.; Butler, K.D.; Mwamba, V.L.; et al. An initial investigation into the organic matter biogeochemistry of the Congo River. *Geochim. Cosmochim. Acta* **2012**, *84*, 614–627. [CrossRef]

74. Spencer, R.G.M.; Hernes, P.J.; Dinga, B.; Wabakanghanzi, J.N.; Drake, T.W.; Six, J. Origins, seasonality, and fluxes of organic matter in the Congo River. *Glob. Biogeochem. Cycles* **2016**, *30*. [CrossRef]

75. Seyler, P.; Coynel, A.; Moreira-Turcq, P.; Etcheber, H.; Colas, C.; Orange, D.; Bricquet, J.-P.; Laraque, A.; Guyot, J.L.; Meybeck, M. Organic carbon transported by the equatorial rivers: Example of Zaire-Congo and Amazon Rivers. In *Soil Erosion and Carbon Dynamics*; Roose, E.J., Lal, R., Feller, C., Barthes, B., Stewart, B.A., Eds.; CRC Press: Boca Raton, FL, USA, 2006; Volume 15, pp. 255–274.

76. Bouillon, S.; Yambélé, A.; Gillikin, D.P.; Teodoru, C.; Darchambeau, F.; Lambert, T.; Borges, A.V. Contrasting biogeochemical characteristics of the Oubangui River and tributaries (Congo River basin). *Sci. Rep.* **2014**, *4*, 5402. [CrossRef]

77. Devroey, E. *Le Bassin Hydrographique Congolais*; Coll Mémoires in-8° de la Section des Sciences Techniques Institut Royal Colonial Belge: Bruxelles, Belgium, 1941; p. 172.

78. Nedeco (Netherlands Engineering Company). *River Studies, Niger and Benué*; North Holland Publishing Co.: Amsterdam, The Netherlands, 1959.

79. Gibbs, R.J. Amazon river-environmental factors that control Iits dissolved and suspended load. *Science* **1967**, *156*, 1203–1232. [CrossRef]

80. Symoens, J.J. La minéralisation des eaux naturelles. Résultats scientifiques. *Explor. Hydrobiol. Bassin du Lac Bangwelo et du Luapula* **1968**, *2*, 1–199.

81. Holleman, J.N. The sediment yield of major rivers of the world. *Water Resour. Res.* **1968**, *4*, 737–747. [CrossRef]

82. Vanderlinden, M.J.H. Reactions between Acids and Leaf Literi; Premier colloque Int. Biodegradation et humification. Ph.D. Thesis, Univ. Nancy, Nancy, France, 1975.

83. Meybeck, M. Note on e1emental contents of the Zaïre River. *Neth. J. Sea Res.* **1978**, *12*, 293–295. [CrossRef]

84. Meybeck, M. Concentrations des eaux fluviales en éléments majeurs et apports en solution aux océans. *Revue de Géologie Dyn. Géogr. Phys.* **1979**, *21*, 215–246.

85. Molinier, M. Note sur les débits et la qualité des eaux du Congo à Brazzaville. *Cah. ORSTOM Série Hydrol.* **1979**, *16*, 55–66.

86. Deronde, L.; Symoens, J.J. L'exportation des éléments dominants du bassin du fleuve Zaïre: Une réévaluation. *Ann. Limnol.* **1980**, *16*, 183–188. [CrossRef]

87. Grondin, J.L.; Gac, J.Y. Apports de matières aux océans: Bilan des six principaux fleuves d'Afrique. 1980; unpublished work.

88. Eisma, D. Suspended matter as a carrier for pollutants in estuaries and the sea. In *Marine Environmental Pollution, 2. Mining and Dumping*; Geyer, R.A., Ed.; Elsevier: Amsterdam, The Netherlands, 1982; Volume 27, pp. 281–295.

89. Meybeck, M. Carbon, nitrogen and phosphorus transport by world rivers. *Am. J. Sci.* **1982**, *282*, 401–450. [CrossRef]

90. Kinga-Mouzeo. Transport Particulaire Actuel du Fleuve Congo et de Quelques Affluents; Enregistrement Quaternaire dans L'éventail Détritique Profond (Sédimentologie, Minéralogie et Géochimie). Ph.D. Thesis, Université de Perpignan, Perpignan, France, 1986; p. 251.

91. Nkounkou, R.; Probst, J.-L. Hydrology and geochemistry of the Congo River system. *SCOPE/UNEP Sonderband* **1987**, *64*, 483–508.

92. Olivry, J.-C.; Bricquet, J.-P.; Thiébaux, J.-P.; Sigha-Nkamdjou, L. Transport de matière sur les grands fleuves des régions intertropicales: Les premiers résultats des mesures de flux particulaires sur le bassin du fleuve Congo. In *Sediment Budgets*; AISH 174: Porto-Alegre, Brazil, 1988; pp. 509–521.

93. Probst, J.-L.; Nkounkou, R.; Krempp, G.; Bricquet, J.-P.; Thiebaux, J.-P.; Olivry, J.-C. Dissolved major elements exported by the Congo and the Ubangi rivers during the period 1987–1989. *J. Hydrol.* **1992**, *135*, 237–257. [CrossRef]

94. Moukolo, N.; Laraque, A.; Olivry, J.-C.; Bricquet, J.-P. Transport en solution et en suspension par le fleuve Congo (Zaïre) et ses principaux affluents de la rive droite. *Hydrol. Sci. J.* **1993**, *38*, 133–145. [CrossRef]

95. Olivry, J.-C.; Bricquet, J.-P.; Laraque, A.; Guyot, J.-L.; Bourges, J.; Roche, M.A. Flux liquides, dissous et particulaires de deux grands bassins intertropicaux: Le Congo à Brazzaville et le Rio Madeira à Villabella. In *Grands Bassins Fluviaux Périatlantiques: Congo, Niger, Amazone-Actes du Colloque PEGI-INSU-CNRS-ORSTOM*; Olivry, J.-C., Boulègue, J., Eds.; ORSTOM: Paris, France, 1995; pp. 345–355.

96. Sondag, F.; Laraque, A.; Riandey, C. Chimie des eaux du fleuve Congo à Brazzaville et de l'Oubangi à Bangui (années 1988 à 1992). In *Grands Bassins Fluviaux Périatlantiques: Congo, Niger, Amazone-Actes du colloque PEGI-INSU-CNRS-ORSTOM*; Olivry, J.-C., Boulègue, J., Eds.; ORSTOM: Paris, France, 1995; pp. 121–131.

97. Laraque, A.; Castellanos, B.; Steiger, J.; Lopez, J.-L.; Pandi, A.; Rodriguez, M.; Rosales, J.; Adèle, G.; Perez, J.; Lagane, C. A comparison of the suspended and dissolved matter dynamics of two large inter-tropical rivers draining into the Atlantic Ocean: The Congo and the Orinoco. *Hydrol. Process.* **2013**. [CrossRef]

98. Borges, A.V.; Darchambeau, F.; Lambert, T.; Morana, C.; Allen, G.H.; Tambwe, E.; Toengaho Sembaito, A.; Mambo, T.; Nlandu Wabakhangazi, J.; Descy, J.-P.; et al. Variations in dissolved greenhouse gases (CO_2, CH_4, N_2O) in the Congo River network overwhelmingly driven by fluvial-wetland connectivity. *Biogeoscience* **2019**, *16*, 3801–3834. [CrossRef]

99. Mushi, C.A.; Ndomba, P.M.; Trigg, M.A.; Tshimanga, R.M.; Mtalo, F. Assessment of basin-scale soil erosion within the Congo River Basin: A review. *Catena* **2019**, *178*, 64–76. [CrossRef]

100. Burgis, M.J.; Symoens, J.J. *African Wetlands and Shallow Water Bodies-Zones Humides et lacs peu Profonds d'Afrique*; ORSTOM: Paris, France, 1987; p. 650.

101. O'Loughlin, F.; Trigg, M.A.; Schumann, G.J.-P.; Bates, P.D. Hydraulic characterization of the middle reach of the Congo River. *Water Res. Resear.* **2013**, *49*, 5059–5070. [CrossRef]

102. Betbeder, J.; Gond, V.; Frappart, F.; Baghdadi, N.N.; Briant, G.; Bartholomé, E. Mapping of central Africa forested wetlands using remote sensing. *IEEE J. Sel. Top. Appl. Earth Obs. Remote Sens.* **2013**. [CrossRef]

103. Alsdorf, D.; Beighley, D.; Laraque, A.; Lee, H.; Tshimanga, R.; O'Loughlin, F.; Mahé, G.; Dinga, B.; Moukandi N'kaya, G.; Spencer, R.G.M. Opportunities for hydrologic research in the Congo Basin. *Rev. Geophys.* **2016**, *54*. [CrossRef]

104. Laraque, A.; Pouyaud, B.; Rocchia, R.; Robin, R.; Chaffaut, I.; Moutsambote, J.M.; Maziezoula, B.; Censier, C.; Albouy, Y.; Elenga, H.; et al. Origin and function of a closed depression in equatorial humid zones: The lake Tele in the north Congo. *J. Hydrol.* **1998**, *207*, 236–253. [CrossRef]

105. Marlier, G. Limnology of the Congo and Amazon Rivers. In *Tropical Forest Ecosystems in Africa and South America: A Comparative Review*; Meggers, B.J., Ayensu, E.S., Duckworth, W.D., Eds.; Smithsonian Institution Press: Washington, DC, USA, 1973; pp. 223–238.

106. Alsdorf, D.; Bates, P.; Melack, J.; Wilson, M.; Dunne, T. The spatial and temporal complexity of the Amazon flood measured from space. *Geophys. Res. Lett.* **2007**, *34*, L08402. [CrossRef]

107. Rudorff, C.M.; Melack, J.M.; Bates, P.D. Flooding dynamics on the lower Amazon floodplain: 2 Seasonal and interannual\hydrological variability. *Water Resour. Res.* **2014**, *50*, 635–649. [CrossRef]

108. Bonnet, M.-P.; Barroux, G.; Seyler, P.; Pecly, G.; Moreira-Turcq, P.; Lagane, C.; Cochoneau, G.; Viers, J.; Seyler, F.; Guyot, J.-L. Seasonal links between the Amazon corridor and its flood plain: The case of the varzea of Curuai. In Dynamics and Biogeochemistry of River Corridors and Wetlands. In Proceedings of the Symposiums Held During the Seventh IAHS Scientific Assembly, Foz do Iguaçu, Brazil, 3–9 April 2005; p. 294.

109. Martinez, J.M.; Le Toan, T. Mapping of flood dynamics and vegetation spatial distribution in the Amazon floodplain using multitemporal SAR data. *Remote Sens. Environ.* **2007**, *108*, 209–223. [CrossRef]

110. Trigg, M.A.; Bates, P.D.; Wilson, M.D.; Schumann, G.; Baugh, C. Floodplain channel morphology and networks of the middle Amazon River. *Water Resour. Res.* **2012**, *48*. [CrossRef]

111. Tshimanga, R.M. Two decades of hydrologic modeling and predictions in the Congo River Basin: Progress and prospect for future investigations. In *Congo Basin-Hydrology, Climate, and Biogeochemistry: A Foundation for the Future*; Alsdorf, D., Tshimanga Muamba, R., Moukandi N'kaya, G.D., Eds.; AGU, John Wiley & Sons Inc.: Malden, MA, USA, 2021; Unpublished work.

112. Beighley, R.E.; Ray, R.L.; He, Y.; Lee, H.; Schaller, L.; Andreadis, K.M.; Durand, M.; Alsdorf, D.; Shum, C.K. Comparing satellite derived precipitation datasets using the Hillslope River Routing (HRR) model in the Congo River Basin. *Hydrol. Process.* **2011**, *25*, 3216–3229. [CrossRef]

113. Tshimanga, R.M.; Hughes, D.A. Basin scale performance of a semi distributed rainfall runoff model for hydrological predictions and water resources assessment of large rivers: The Congo River. *Water Resour. Res.* **2014**. [CrossRef]

114. Aloysius, N.; Saiers, J. Simulated hydrologic response to projected changes in precipitation. *Hydrol. Earth Syst. Sci.* **2017**, *21*, 4115–4130. [CrossRef]

115. Munzimi, Y. Satellite-Derived Rainfall Estimates (TRMM Products) Used for Hydrological Predictions of the Congo River Flow: Overview and Preliminary Results. START Report. 2008. Available online: http://start.org/alumni-spotlight/yolande-munzimi.html (accessed on 12 January 2020).

116. Kabuya, P.M.; Hughes, D.A.; Tshimanga, R.M.; Trigg, M.A. Understanding factors influencing the wetland parameters of a monthly rainfall-runoff model in the Upper Congo River Basin. In Proceedings of the EGU Conference Presentation, EGU2020-642, (in the Format Sharing Geoscience Online). 4–8 May 2020.

117. Datok, P.; Fabre, C.; Sauvage, S.; Moukandi N'kaya, G.D.; Paris, A.; Dos Santos, V.; Laraque, A.; Sánchez Pérez, J.M. Investigating the role of the central cuvette of the Congo River in the hydrology of the basin. In *Congo Basin-Hydrology, Climate, and Biogeochemistry: A Foundation for the Future*; Alsdorf, D., Tshimanga Muamba, R., Moukandi N'kaya, G.D., Eds.; AGU, John Wiley & Sons Inc.: Malden, MA, USA, 2021; Unpublished work.

118. Paris, A.; Calmant, S.; Gosset, M.; Fleischmann, A.; Conchy, T.; Garambois, P.A.; Bricquet, J.-P.; Papa, F.; Tshimanga, R.; Moukandi N'kaya, G.D.; et al. Monitoring hydrological variables from remote sensing and modelling in the Congo River basin. In *Congo Basin-Hydrology, Climate, and Biogeochemistry: A Foundation for the Future*; Alsdorf, D., Tshimanga Muamba, R., Moukandi N'kaya, G.D., Eds.; AGU, John Wiley & Sons Inc.: Malden, MA, USA, 2021; Unpublished work.

119. Carr, A.; Trigg, M.A.; Tshimanga, R.; Smith, M.W.; Borman, D.J.; Bates, P.D. Modelling the Bathymetry and Hydraulics of the Congo River Multichannel Mainstem using Spatially Limited In-situ Data. In *Congo Basin-Hydrology, Climate, and Biogeochemistry: A Foundation for the Future*; Alsdorf, D., Tshimanga Muamba, R., Moukandi N'kaya, G.D., Eds.; AGU, John Wiley & Sons Inc.: Malden, MA, USA, 2021; Unpublished work.

120. Trigg, M.; Carr, A.B.; Smith, M.W.; Tshimanga, R. Measuring Geomorphological Change on the Congo River Using Century Old Navigation Charts. Geomorphological Change on the Congo River. In *Congo Basin-Hydrology, Climate, and Biogeochemistry: A Foundation for the Future*; Alsdorf, D., Tshimanga Muamba, R., Moukandi N'kaya, G.D., Eds.; AGU, John Wiley & Sons Inc.: Malden, MA, USA, 2021; Unpublished work.

121. Molliex, S.; Kettner, A.J.; Laurent, D.; Droz, L.; Marsset, T.; Laraque, A.; Rabineau, M.; Moukandi N'Kaya, G.D. Simulating sediment supply from the Congo watershed over the last 155 ka. *Q. Sci. Rev.* **2019**, *203*, 38–55. [CrossRef]

122. Pandi, A.; Ibiassi, G.; Tondo, B.; Ladel, J.; Laraque, A. Impact de la variabilité des écoulements sur la navigabilité de l'Oubangui, un affluent du fleuve Congo. In Proceedings of the International Conference, (WCCCAA: Water issues/challenges and Climate Change Adaptation in Africa), Bangui, Central African Republic, 24–28 October 2012; pp. 38–39.

123. ATC; ACCF; SCEVN. *Renseignements Sur Les Conditions de Navigabilité du Fleuve Congo et Rivières du Bassin Congo-Oubangui-Sangha*; Rapport Technique: Brazzaville, République du Congo, 2000; p. 69.

124. CICOS. *Atlas du Bassin du Congo*; CICOS: San Jose, CA, USA, 2016; p. 95.

125. Wachter, S.J. Giant Dam Projects Aim to Transform African Power Supplies. *The New York Times*. 19 June 2007. Available online: https://www.nytimes.com/2007/06/19/business/worldbusiness/19iht-rnrghydro.1.6204822.html (accessed on 25 February 2020).

126. Umolu, J.C. Macro perspectives for Nigeria's water resources planning. Discussion of Ubangi-Lake Chad diversion schemes. In Proceedings of the First Biennial National Hydrology Symposium, Maiduguri, Nigeria, 26–28 November 1990; pp. 218–262.

127. Kiari Fougou, H.; Lemoalle, J. Variabilité du lac Tchad: Quelle gestion hydraulique pour préserver les ressources naturelles? In *Congo Basin-Hydrology, Climate, and Biogeochemistry: A Foundation for the Future*; Alsdorf, D., Tshimanga Muamba, R., Moukandi N'kaya, G.D., Eds.; AGU, John Wiley & Sons Inc.: Malden, MA, USA, 2021; Unpublished work.

128. Callède, J.; Ronchail, J.; Guyot, J.L.; Oliveira, E. Amazonian deforestation: Its influence on the Amazon discharge at Óbidos (Brasil). *Rev. Sci. L Eau* **2008**, *21*, 59–72.

129. Milly, P.C.D.; Betancourt, J.; Falkenmark, M.; Hirsch, R.M.; Kundzewicz, Z.W.; Lettenmaier, D.P.; Stouffer, R.J. Stationarity is dead: Whither water management? *Science* **2008**, *319*, 573–574. [CrossRef]

130. Milly, P.C.D.; Betancourt, J.; Falkenmark, M.; Hirsch, R.M.; Kundzewicz, Z.W.; Lettenmaier, D.P.; Stouffer, R.J.; Dettinger, M.D.; Krysanova, V. On critiques of 'Stationarity is Dead: Whither water management?'. *Water Res. Resear.* **2015**, *51*, 7785–7789. [CrossRef]

Article

Water and Sediment Budget of Casiquiare Channel Linking Orinoco and Amazon Catchments, Venezuela †

Alain Laraque [1], Jose Luis Lopez [2], Santiago Yepez [3,*] and Paul Georgescu [4]

[1]   Géosciences Environnement Toulouse, Université de Toulouse, CNES, CNRS, IRD, UPS, 31400 Toulouse, France; alain.laraque@ird.fr

[2]   Instituto de Mecánica de Fluidos, Universidad Central de Venezuela—UCV, Caracas 1050, Venezuela; lopezjoseluis7@gmail.com

[3]   Departamento Manejo de Bosques y Medio Ambiente, Facultad de Ciencias Forestales, Universidad de Concepción—UdeC, Concepción 407374, Chile

[4]   Centro de Mecánica de Fluidos y Aplicaciones—CEMFA, Universidad Simón Bolívar—USB, Edificio Fluidos y Operaciones Unitarias, Caracas 89000, Venezuela; pgeorg@cantv.net

*   Correspondence: syepez@udec.cl

†   This paper was written in tribute to our colleague and friend Paul Georgescu, who passed away shortly before this publication. He supported the development of this work and followed it step by step. As first author, I must thank Paul for inviting me to accompany him in this fantastic adventure in the Casiquiare, carried out in September 2000, in one of the least known and unexplored regions of the planet. Rest in peace, after an exciting existence as an explorer and adventurous scientist, navigating thousands of kilometers of waterways throughout South America.

Received: 4 September 2019; Accepted: 27 September 2019; Published: 3 October 2019

Abstract: The Casiquiare River is a natural channel that connects two of the greatest rivers in the world, the Orinoco and the Amazon in the South American continent. The aim of this paper is to present a review and synthesis of the hydrological and sedimentological knowledge of the Casiquiare River, including the first hydro-sedimentary balance of the Casiquiare fluvial system conducted 9–12 September 2000 at the bifurcation and mouth during the expedition 'Humboldt-Amazonia 2000'. Bathymetric flow discharge and physico-chemical measurements were made at the inlet and outlet of the Casiquiare Channel. The main conclusions of this study indicate that Casiquiare is taking a significant proportion of flow (20% to 30%) from the Upper Orinoco basin to the Amazon basin. Throughout its 356 km-course, this chameleon channel undergoes significant morphological, hydrological, and bio-geochemical variations between the inlet and outlet, whose most visible witnesses are the increase in its width (3 to 4 times), flow (7 to 9 times), and its change in water color (white to black water), under the influence of tributaries coming from vast forest plains.

Keywords: Casiquiare; Orinoco; Amazon; bifurcation; hydro-sedimentary budget

---

1. Introduction

The Casiquiare River is a geographical rarity and a true hydrological anomaly. It is a natural channel that connects two of the greatest rivers in the world, the Orinoco and the Amazon rivers (Figure 1A). Located on the border between Venezuela and Brazil, the Upper Orinoco deviates part of its waters to the Amazon basin. This is the largest example of river capture in the world. Thus, the same source feeds two different marine estuaries on the Atlantic coast of the South American continent, which makes it unique, taking in consideration the magnitude of the flows transferred from the Orinoco River (more than 320 $m^3 \cdot s^{-1}$), and the distance between the mouths of the Orinoco and Amazon rivers (about 1500 km).

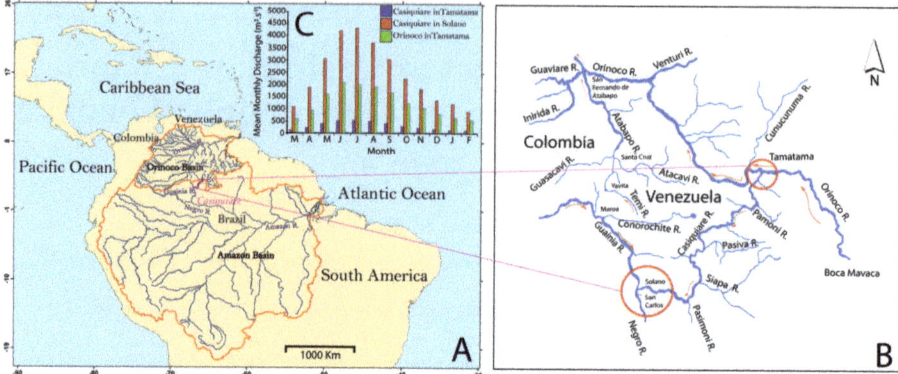

Figure 1. (**A**) Location map of the Casiquiare natural channel, connecting the Orinoco and Amazon basins. (**B**) Close view of the fluvial system showing main tributaries of the Casiquiare Channel. (**C**) Mean hydrographs of the Casiquiare channel (in Tamatama and Solano) and the Orinoco River in Tamatama, using discharge data from the period 1970–1986.

The Casiquiare Channel is a very isolated watercourse (accessible only by river or air), located in an environment virtually devoid of riparian population but subject to some degree of insecurity due to illegal mining activities and guerrilla and paramilitary incursions. For these reasons hydrological measurements have been scarce, the latest being in September 2000 [1] and in December 2001 [2]. Figure 1 presents a close view of the fluvial system. The Upper Orinoco bifurcates downstream at the Tamatama village and runs for about 356 km before meeting Guainia River, which from this point on becomes the Negro River. The main tributaries of the Casiquiare River are all concentrated along its left bank (Pamoni, Pasiba, Siapa, and Pasimoni rivers) (Figure 1B). Figure 1C presents the mean hydrographs of Casiquiare in Tamatama and Solano, as well as the Orinoco in Tamatama, covering the same period of common discharge data between 1970 and 1986 (from The Global Runoff Data Centre—GRDC website [3]).

This paper aims to present: (a) a review and synthesis of the hydrological and sedimentological knowledge of this unique river; and (b) the first and unpublished hydro-sedimentary balance of the Casiquiare fluvial system at the bifurcation and at the mouth, based on the measurements made by an expedition that took place in September 2000 with the participation of Laraque and Georgescu, two of the authors of this paper [1]. The paper also presents a brief history of the discovery of the Casiquiare Channel and examines various theories concerning its origin. The problem of tracing the boundaries of the Casiquiare basin is also addressed.

2. Discovery of the Casiquiare Channel

Spanish conquistadors heard about the existence of the Casiquiare Channel from native people during the seventeenth century. The first account of this hydrologic connection is generally attributed to Acuña in the year 1641 [4]. Sanson Fer, cartographer of the King of Spain, presented a communication between the two rivers on their maps in 1707. Some documents in 1726 refer to trips of Portuguese traders on this channel [5]. Historical sources are also found in Humboldt's narrative [6]. The Jesuit priest Manuel Román, superior of the Spanish missions, registered the existence of a connection between the Orinoco and the Amazon rivers in 1744. The first scientific description of the Casiquiare Channel is attributed to La Condamine [7], but firmly denied at the Academy of Sciences in Paris by Philippe Buarche, the royal geographer, who described Casiquiare as a geographical monstrosity. The waterway was not officially referred to by the name of the Casiquiare Channel until 1772 in the so-called "Plan Geográfico de Santa Fe de Bogota, Nuevo Reino de Granada" signed by the cartographer Joseph Aparicio Moreta. However, this is not totally accepted by the scientific community until 1800, when Alexander von Humboldt and Aimé Bonpland landed in Venezuela. They navigated the Casiquiare Channel on 10 May 1800, and closed the controversy over the existence (or not) of a communication between the Amazon and the Orinoco [6]. From this moment, the existence of the Casiquiare Channel has been definitively accepted by the modern science community. Figure 2 shows a map presented by Humboldt indicating the course of the river channel between the Orinoco and Negro rivers [8]. Most recently in 1943, during World War II, the U.S. Army Corps of Engineers carried out the first hydrologic study for the navigability in this channel in order to develop a secure route (away from the German submarines navigating along the South American coast off the mouth of the Amazon) for transportation of latex between Brazil, Colombia, and Venezuela, to be used for war purposes [9].

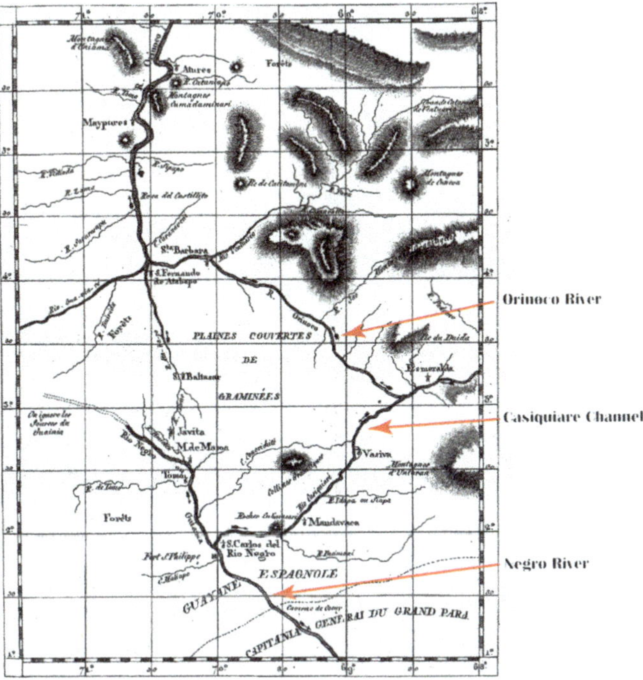

Figure 2. Map presented by Humboldt in 1812 showing the course of the Casiquiare Channel between the bifurcation of the Orinoco and the confluence with the Negro River. Used by permission of Société de Géographie, from Georgescu et al. [10].

3. General Description

The description and geographical location of the "hydrological singularity" of the Casiquiare Channel is mentioned in several articles on the hydrology of the Orinoco and Amazon basins [4,11,12]. However, scientific knowledge of the Casiquiare Channel is generally limited in time [13–15], subjective, incomplete, approximate, and often erroneous because it results largely from summary observations, without the support of physical measurements. No hydro-sedimentary balance of a complete hydrological cycle has yet been carried out.

Figures 3 and 4 show the morphological differences between the Orinoco bifurcation which coincides with the beginning of the Casiquiare Channel and the confluence of the latter with the Guainia River. This bifurcation begins abruptly and perpendicularly to the drainage axis of the Orinoco River (Figure 1B) just downstream of the locality of Tamatama. The Casiquiare Channel turns southwest to meet the Guainia River from the northwest at the end of its stream to form the Negro

River, just upstream of San Carlos locality. Further downstream, the Negro River meets Rio Solimões to form the Amazon River near the town of Manaus in Brazil. The coordinates of the bifurcation at its origin, in the Upper Orinoco in Venezuela are N. 03° 08′ 25.26″ and W. 65° 52′ 48.66″, and the coordinates of its mouth, also in Venezuela are N. 02° 00′ 15.18″ and W. 67° 06′ 52.44″.

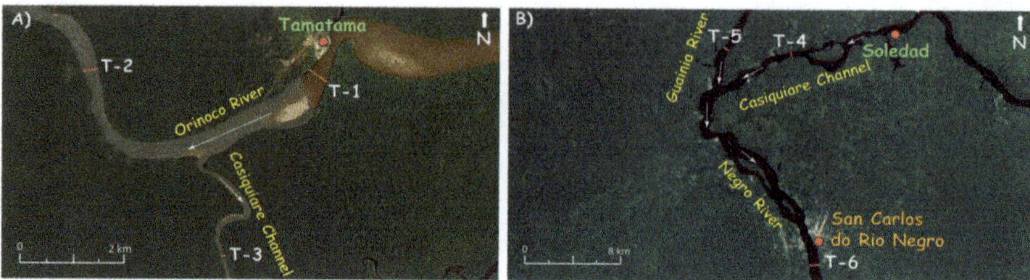

Figure 3. Satellite views of the Casiquiare Channel: (**A**) at the bifurcation, and (**B**) at the confluence, showing location of transects T1 to T6 (images from Google Earth).

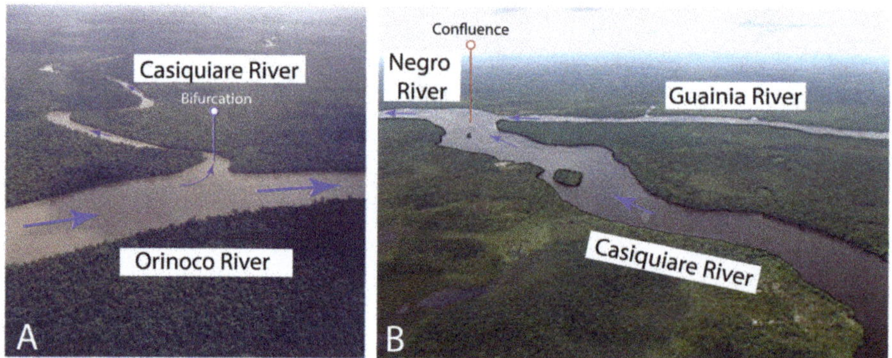

Figure 4. Aerial photos of the bifurcation (**A**) and confluence (**B**) taken from a helicopter by A. Laraque on September 10, 2000.

The Casiquiare Channel, bordered by gallery forests, is 356.4 km long and its width varies from 46 to 610 m. Minimum water depths are about 0.30 m upstream and 1.2 m downstream. Between its two extremities, the elevation difference is about 21 m, with an average slope of 6 cm·km^{-1} [9]. But this hydraulic gradient can vary from 7.82 to 8.55 cm·km^{-1} between low and high water. The surrounding plain has a slope of 10.3 cm·km^{-1}.

On the Casiquiare left bank, the contribution of the Pamoni, Pasiba, Siapa, and Pasimoni rivers come from the Sierras Curupira, Tapirapeco, Imeri, and Neblina located in southern Venezuela, with peak elevations between 2000 and 3000 m.a.s.l., the highest of which reaches 3014 m (Pico da Neblina in Brazil). The land drained by the left bank of the Casiquiare and by the Upper Orinoco is composed of undifferentiated Proterozoic rocks [16]. This mountain range with high rainfall rates, about 4000 mm·yr^{-1}, constitutes a real water reservoir that feeds the left bank of the Casiquiare Channel. This northwest-facing runoff reflects the slight slope in the same direction, which drains the plains surrounding the Guyanese shield between 150 and 200 m.a.s.l., where some steep residual landforms are observed at about 500 m.a.s.l. [14].

4. Origin and Evolution of the Casiquiare Channel

Geologists agree that the capture of the Orinoco by the Casiquiare River is more recent than the formation of the Andean range which has profoundly modified the direction of circulation of the Amazon and Orinoco rivers. According to ichthyological observations, Goldstein, [17] attributes the birth of the Casiquiare Channel to the capture of a secondary channel of the Orinoco River on the Pleistocene-Holocene limit, about 10,000 years ago.

Stern [18] was the first one to describe the genesis of the Casiquiare. Vareschi [19] mentioned an incomplete capture by overflow and regressive erosion produced by a northern affluent of the Negro River. This is one of the largest examples of natural hydrological capture in the world. The mechanism of capture, the current situation, and future trends are illustrated in Figure 5 [18,19]. At remote times the Casiquiare did not exist. The Orinoco River passed in two channels around a fluvial island denoted by I (Figure 5-1). The Upper Casiquiare used to be a branch of the Orinoco River. The confluence of the Cunucunuma River (CU) with the Orinoco (O) was located at the north branch (S) of the island. The lower reach of the south branch was Caño Seco (CS), which used to flow around Quiritari island (Q) before entering the main channel (O) of the Orinoco River. At some distance to the south, the Pamoni River, a perennial river, was the headwaters of the river today called Casiquiare, which flowed into the Negro River and the Amazon basin. The approximate drainage division between the Orinoco and Amazon basins used to be the dashed line denoted by D (Figure 5-1).

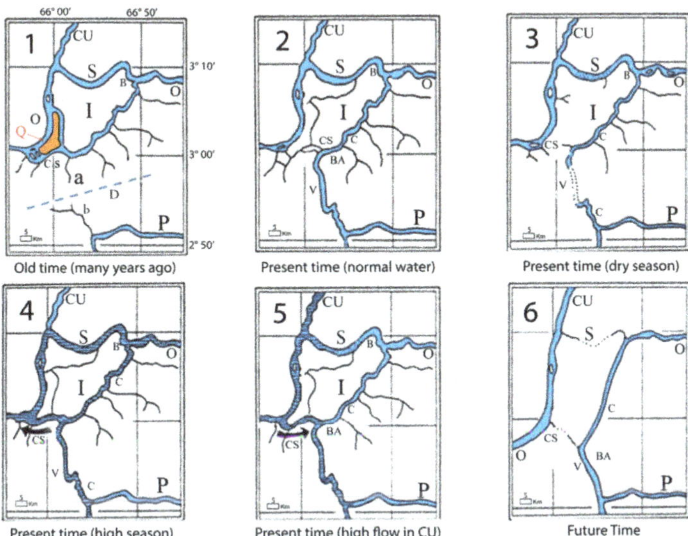

Figure 5. Sketch-maps of the Casiquiare fluvial system showing past and future trends in the evolution of the channel, according to Stern [18] as illustrated by Vareschi [19]. O = Orinoco; C = Casiquiare; CU = Cunucunuma River; B = bifurcation; I = fluvial island; CS = Caño Seco; BA = Buenos Aires sector; V = Venados (abandoned town); P = Pamoni River; S = north channel of the island; Q = Quiritari Island (in orange); D = water divide; a = small channel; b = small channel.

Probably, during high flows, the south branch of the Orinoco River began flowing to the Pamoni River through channel (a) and channel (b). The alluvial material composing the drainage divide (D) was eroded and a new channel was formed, which was the genesis of the Casiquiare River by avulsion processes. The capture of the Orinoco waters was helped by the existence of a terrain slope larger toward the south than toward the west. Caño Seco was abandoned and became a dry channel, which flows only during the high flow season. Near an abandoned town, called Venados site (V), the Casiquiare is almost dry during the low season, making navigation difficult (Figure 5-3). This observation was made during the Oriampla Expedition to the Orinoco in January 1980 [10]. At present time, during a combination of high flows in the Cunucunuma River and low flows in the Orinoco, the flow in Caño Seco (CS) could revert as it is shown in Figure 5-5.

Stern [18] reported that the flow discharge, flow depth and width in the upstream reach of the Casiquiare Channel had increased in the last 50 years, according to stories from old pilots, navigators, and local inhabitants. This is reinforced by the accumulation of sediment and point bars in the Orinoco River downstream from the point of bifurcation. Stern suggested that in a future time the Casiquiare will totally capture the Orinoco flows and that the headwaters of the Orinoco would be the headwaters of the Cunucunuma River (Figure 1B). Although a few authors like [20] questioned the Stern theory of the Casiquiare formation, Vareschi [19] supports the opinion of Stern based on field observation of two abandoned channels connecting the Orinoco with the Casiquiare, noting that their channel widths were significantly smaller than the Casiquiare width at the present time.

Using a geomorphometric analysis of the region with Digital Elevation Models, Hypsometry, Isobases and Swath Profiles, Grohmann et al. [21] explain the capture of the Orinoco River and the formation of Casiquiare canal by tectonic controls.

A recent study by Stokes at al. [22] used measurements from the 1943 U.S. Army Corps of Engineers and hydraulic calculations to show that the Rio Casiquiare is actively capturing the Upper Orinoco River, driven by a combination of in-channel sedimentation in the Orinoco and seasonal inundation of the drainage divide that favors the steeper Casiquiare River. They consider that the ongoing reorganization of the river system, is under tectonic controls and also conclude that the Casiquiare River course will eventually become the dominant channel, which confirms Stern's pioneering studies [18,23].

To conclude this paragraph, it seems necessary to clarify that the location of the initial capture of the Orinoco is located at the level of the south branch and downstream of the river island "I" in Figure 5-1–5, and not at the bifurcation of the Casiquiare, which is located upstream of the same river island and downstream of Tamatama, as is too often mentioned in the literature. Indeed, at this point in past time, it was a simple division of the Orinoco River into two branches that bypassed the river island "I". The critical area of the initial avulsion that diverted the south branch of the Orinoco is located at Caño Seco (Figure 5-1–5). Subsequently, in the event of a continuation of this river capture, the northern branch of the Orinoco, which bypasses the river island, will dry up and so will the Caño Seco. At that time the capture will be total and the upper Orinoco River will be completely diverted towards the Amazon basin (Figure 5-6).

5. Data and Methodology

5.1. Sampling and Measurements

During the September 2000 expedition, bathymetric, discharge and physico-chemical profile measurements were made along six transects denoted by T1 to T6 (Figures 3 and 6), using a small boat provided with an echo-probe (Eagle Strata 128) coupled with a GPS (GARMIN, Etrex model; Ap = +/− 1m), a turbidimeter (AQUALITYC; Ap = +/− 0.01 NTU), a thermometer, a portable pH meter (WTW pH 320; +/− 0.01), and a conductivity meter (WTW LF 318; Ap = +/− 0.1 μS·cm^{-1}). The physico-chemical measurements were made at three points on the vertical (surface, middle, and bottom) at the center of the river. Total suspended sediment (TSS) was also collected with a 500 mL sampler. For discharge measurement, first, a bathymetric survey is carried out using the echo sounder and GPS to obtain the flow area at each respective section of the river. The flow velocity is determined using the boat (canoe) as a drifting float on three longitudinal lines distributed at 25%, 50%, and 75% of the bank distance of the section. GPS provides us the drift speed of the canoe, and therefore the surface velocity. Thus, the average velocity of the flow section is calculated by applying a correction coefficient between 0.8 and 0.95 [24], generally accepted for this type of flat watercourse. Water discharge was determined as the product of the velocity and the area, and material flows (in kg·s^{-1}) were obtained by multiplying the average values of the measured TSS concentrations (in mg·L^{-1}) with the discharge (in m^3·s^{-1}).

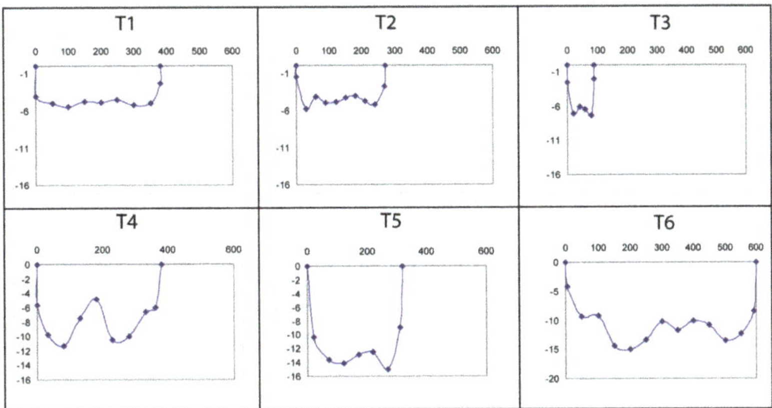

Figure 6. Bathymetric profiles obtained at 6 different sections in the bifurcation and in the confluence of the Casiquiare Channel. T1: Orinoco upstream of bifurcation; T2: Orinoco downstream of bifurcation; T3: Casiquiare at inlet; T4: Guainia River before confluence; T5: Casiquiare at outlet; T6: Negro River at San Carlos. Depths and widths are given in meters. See Figure 3 for location of cross sections collected in the 2000 expedition.

5.2. Analytical Techniques

Water samples were filtered with filters of acetate (cellulose) with a porosity of 0.45 μm for the total suspended sediment (TSS). After being dried, these fractions were weighed to a precision of 0.1 mg (Sartorius scale). In the laboratory, 100 mL of filtrates were heated in a steamer at 105 °C to obtain the dry residue, which corresponds to the sum of total dissolved solids (TDS) and the total dissolved organic matter (DOM).

5.3. Morphometric Analysis

A morphometric analysis was carried out to accurately check the contributing area of each basin, including a description of the elevation and slope changes both of the Upper Orinoco basin and the Casiquiare Canal basin. For this, we used the SRTM (Shuttle Radar Topography Mission)—Digital Elevation Model (DEM) of the Nasa at 1 arc of a second, with approximately 30 m [25] in combination with the Fluvial Corridor Toolbox software [26]. Thus, a synoptic diagram was created for each basin. From this set of digital cartographic data in three coordinates (XYZ) or DEM, "elevation and slope tool" (in Fluvial Corridor Toolbox) enables the extraction of topographical data such as the elevation and the slope of a set of polylines (e.g., a hydrographic network). It is based on the add surface information (3D Analyst Tools) and can be applied over a linear network. The "Z-mean" elevation was calculated over the entire segment. Finally, the "slope" field is the ratio between the upstream and downstream difference of elevation and the segment length. The drained area at the points is also important information for the dynamics of the river system. Therefore, the "watershed tool" was also used to determine the drainage area along a continuum. It is based on the extraction of data included in a flow accumulation raster, provided by the user and derived from a DEM [26].

6. Results

The hydrologic data (maximum depth, width of section, and surface stream velocity) and physical-chemical parameters (T°, electrical conductivity, pH, turbidity, and suspended sediment concentration) are presented in Table 1 for each section in consideration. Figure 6 presents the bathymetric profiles at the three cross sections of the Orinoco–Casiquiare's bifurcation and at the three cross sections of its confluence with the Guainia River to form the Negro River. Comparisons of these data and calculations of hydro-sedimentary budgets at the three measuring points of the bifurcation and confluence of the Casiquiare Channel provide a better understanding of its hydro-geodynamic functioning.

Table 1. Hydrologic and physico-chemical data obtained from measurements at the bifurcation and confluence of the Casiquiare Channel in September 2000. Q = flow discharge; V = surface flow velocity; W max = maximum channel width; H max = maximum depth; Temp. = temperature; EC = electrical conductivity; Turb = turbidity; TSS = total suspended sediment concentration.

Transect Number	Date	Station	River	Q	V	W Max	Depth of Sampling	H Max	Temp	EC	Turb	pH	TSS	TSS Fluxes
				$m^3 \cdot s^{-1}$	$m \cdot s^{-1}$	m	m	m	°C	$\mu S \cdot cm^{-1}$	NTU		$mg \cdot L^{-1}$	$kg \cdot s^{-1}$
	9-09-00	Puerto Ayacucho	Orinoco		1.94		0		27.3	7.0	26.0	5.00	49.20	
T1	10-09-00	Tamatama	Orinoco	2323	1.28	380	0	5.20	26.0	14.0	22.0	5.74	34.40	79.69
	"	"	"				−2		"	14.0	25.0	5.78	34.20	
	"	"	"				−4		"	13.0	30.0	5.84		
T2	10-09-00	Downstream bifurcation	Orinoco	1343	1.11	270	0	5.80	26.5	15.0	24.0	5.70	32.00	42.99
T3	10-09-00	Bifurcation	Casiquiare	720	1.39	88	0	7.30	26.0	14.0	18.0	5.87	43.60	31.66
	"	"	"				−3				30.0		44.40	
	"	"	"				−6				34.0			
T4	11-09-00	Mouth	Casiquiare	5438	1.39	320	0	14.10	26.7	7.0	7.9	4.20	8.00	48.94
	"	"	"				−6					7.0		10.00	
	"	"	"				−12					7.9		9.00	
T5	11-09-00	Confluence	Guainia	2573	0.83	380	0	11.30	26.8	14.0	0.4	3.73	0.20	0.51
T6	11-09-00	San Carlos	Negro	8034	1.19	600	0	14.30	26.9	9.0	1.5	3.20	5.60	44.99
T7	12-09-00	San Gabriel	Negro	6878			0		28.0	10.0	0.9	3.48	3.40	

6.1. Morphology Features and Hydro-Sedimentary Budget at the Bifurcation

While the Orinoco River has a width of 380 m and 270 m, respectively, at T1 and T2 (see Figure 6), before and after the bifurcation, the width of the Casiquiare Channel is smaller (88 m) and has a greater depth (7.30 m) than the Orinoco River (5.20 m). Flow velocity increases slightly from 1.28 m·s^{-1} in the Orinoco River (upstream) to 1.39 m·s^{-1} at the entrance of the Casiquiare (Table 1). The section at the entrance of the Casiquiare corresponds to about 28.5% of the section of the Orinoco before the bifurcation (Figure 6). The water balance is nearly 90% complete. Indeed, the sum of the flows at the Orinoco downstream of the bifurcation (1343 m^3·s^{-1}) and the one at the entrance to the Casiquiare (720 m^3·s^{-1}) is 2063 m^3·s^{-1} instead of 2324 m^3·s^{-1} for the Orinoco upstream of the bifurcation. It is possible that this inaccuracy is due to the flow measurement protocol. At the time of the study, the Casiquiare diverts about 30% of the flow of the Upper Orinoco River.

According to the annual averages calculated over the period 1970–1975, the bifurcation with a discharge of 320 m^3·s^{-1} diverts 25% of the flow from the Upper Orinoco [11]. Ron et al. [27] report that the minimum and maximum flows at the entrance to the Casiquiare, 127 and 680 m^3·s^{-1} respectively, maintain the same proportion between 22% and 25% of the flows in the Upper Orinoco River. These two rivers obviously have the same hydrological behavior with flow differences by a factor of 5 between rising and falling stages.

With regard to the physico-chemical data, we obtained pH, electrical conductivity, and turbidity values of the Orinoco water, respectively in the order of 5.80, 14 µS·cm^{-1} at 25 °C and 25 NTU for TSS concentrations of 35 mg·L^{-1} (Table 1). Note a higher value in TSS at the entrance to the Casiquiare (44 mg·L^{-1}) probably due to the increase in velocities, which results in a resuspension of particulate matter. In terms of sediment fluxes, the bifurcation balance closed to nearly 94%. Indeed, the sum of the sediment flows of the Orinoco downstream of bifurcation and the one at the entrance of Casiquiare is 75 kg·s^{-1}, very close to 80 kg·s^{-1} for the Orinoco upstream of the bifurcation. A report of the field work made in December 2001 [2] mentioned TSS concentrations around 20–25 mg·L^{-1}, composed of 20% or 5 mg·L^{-1} of sand (> 62 µm) and 80% or 15–20 mg·L^{-1} of fine TSS (between 62 and 0.45 µm).

6.2. Morphology Features and Hydro-Sedimentary Budget at the Confluence

At the confluence with the Guainia, the Casiquiare Channel presents widths, maximum depths, flow velocities, and discharges of 320 m, 14.10 m, 1.39 m·s^{-1}, and 5439 m^3·s^{-1}, respectively, compared to 380 m, 11.3 m, 0.83 m·s^{-1}, and 2573 m^3·s^{-1} for the Guainia River, and 600 m, 14.30 m, 1.19 m·s^{-1}, and 8034 m^3·s^{-1} for the Negro River just after the confluence (Table 1).

It should be noted that the measurements carried out in September 1970 yielded similar results [28]. At the bifurcation they reported 2430 m^3·s^{-1} for the Orinoco and 646 m^3·s^{-1} at the entrance of the Casiquiare. At the outlet, they measured 5440 m^3·s^{-1} at the Casiquiare and 3336 m^3·s^{-1} at Guainia. The mean flow discharge is about 2800 m^3·s^{-1} according to water levels recorded during the period 1970–1975 [2]. Using sixteen years of available discharge data during the period 1970-1986 (from the GRDC website), the hydrographs in Figure 1C were constructed, which illustrates comparatively the monthly variations of: (i) the Orinoco River discharge in Tamatama, (ii) the Casiquiare in its mouth downstream of Tamatama and (iii) in its mouth at the Solano hydrometric station. These last two hydrographs provide information on the extent of the contributions of the tributaries of the left bank of the Casiquiare.

At their confluence, Casiquiare and Guainia have morphologically similar sections (Figure 6). The flow of Casiquiare is two times higher than the one of the Guainia, mainly due to its 1.67 times higher flow velocity compared to that of the Guainia. The water balance is nearly 100% complete. Indeed, the sum of the flows measured at the outlet of the Casiquiare and at Guainia is 8011 m^3·s^{-1}, compared to 8034 m^3·s^{-1} for the newly formed Negro River (see T6 in Figure 3B).

With regard to physical-chemical data, the pH, electrical conductivity, turbidity, and TSS concentrations of the three confluence points are respectively: 3.73; 14 µS·cm^{-1}; 0.4 NTU; 0.2 mg·L^{-1}

for Guainia, 4.2; 7 μS·cm^{-1}; 7.6 NTU; 9 mg·L^{-1} for Casiquiare and 3.9; 9 μS·cm^{-1}; 1.5 NTU; 5.6 mg·L^{-1} for Negro, values similar to those found by Vegas et al. [29].

The electrical conductivity of Guainia River is twice as high as Casiquiare, probably due to a high presence of organic anions from humic and fulvic forest acids, which make them more acidic. Indeed, the black waters of the Guainia River, typical of "black water rivers", have just crossed a huge forest plain and are therefore much depleted of TSS compared to those of the Casiquiare, part of which comes from the loaded TSS waters of the Upper Orinoco.

The balance of sedimentary flows at the confluence is nearly 91% complete. Indeed, the sum of the flows from the Casiquiare and Guainia is 49.5 kg·s^{-1} instead of 45 kg·s^{-1}, for Negro River in San Carlos station, just downstream of the confluence (see T6 in Figure 3B). Given the approximations on the flow calculations obtained from the adopted summary and simplified protocol, both from the point of view of flow measurement and from that of the calculations of the average concentrations of TSS, it is reasonable to assume that the law of conservation of mass is respected, both at the bifurcation and at the confluence.

6.3. Global Budget and Functioning of the Casiquiare Hydraulic Complex

The flow area of the channel at the outlet is 7.5 times larger than the one at the inlet, due to maximum depths and widths which are, respectively, 2 and 3.6 times greater. However, the flow velocity remains constant along the channel (1.39 m·s^{-1}). These flow velocities are higher than those of the Orinoco upstream and downstream of the bifurcation (1.28 and 1.11 m·s^{-1}), and that those of the Guainia (0.83 m·s^{-1}) and of the Negro River (1.19 m·s^{-1}).

These higher Casiquiare flow velocities would come from both its elevation difference and the hydrological thrust from about 100 tributaries, mainly on the left bank, which input a total flow of 4720 m^3·s^{-1}, i.e., 6.6 times higher than those at the entrance to the channel.

The flow discharge at the outlet is 7.5 larger than at the entrance, but its sedimentary fluxes are only 1.5 times higher, due to TSS concentrations 4.9 times lower at the end of the channel. The travel time of the waters of the Casiquiare, from its entry to its exit, is about 3 days during the rising stage of the hydrograph. Eight thresholds define its course and make it difficult to navigate in the falling stage, when many rocks are exposed at the surface of the water.

The values of turbidity, electrical conductivity, and pH decreased from 24 to 7.6 NTU; 14 to 7 μS·cm^{-1} and 5.87 to 4.2, respectively, from the inlet to the outlet. This longitudinal variability on pH and EC had already been highlighted by Edwards and Thornes [14]. Thus, the white waters at the entrance of the Casiquiare, which correspond to those of the Upper Orinoco, loaded with TSS, acquire, during their travel, more acidic pHs, turbidity, and lower conductivity, due to lateral inputs that have drained the forest cover and swampy plains on its left bank. In these ecosystems, there is little erosion due to lack of relief and slow runoff allows water to soak up dissolved organic matter. A progressive dilution of the waters of the Casiquiare then gives it a facies more and more similar to that of the "black rivers".

Figure 7 presents a schematic summary of the main parameters measured in the fluvial system. It represents the first model of the hydro-sedimentary behavior of the Casiquiare Channel with a distribution of morpho-hydrodynamic parameters in the three sections of its bifurcation and the three sections of its confluence.

Lateral sediment fluxes have been obtained from mass conservation principles. The TSS flux is equal to 17 kg·s^{-1} and the corresponding mean concentration is 3.6 mg·L^{-1} (Figure 7C). Similarly, the electrical conductivity, pH, and turbidity of lateral inputs are equal to 7.15 μS·cm^{-1}, 3.42, and 4.4 NTU (Figure 7B). These values are typical of the "black water" rivers in the Amazon basin.

A relationship between the TSS (mg·L^{-1}) and the turbidity (NTU) was obtained from the measurements in the Casiquiare Channel hydrological complex (Figure 8):

$$TSS = 1.4975 \times Turb + 0.6206 \tag{1}$$

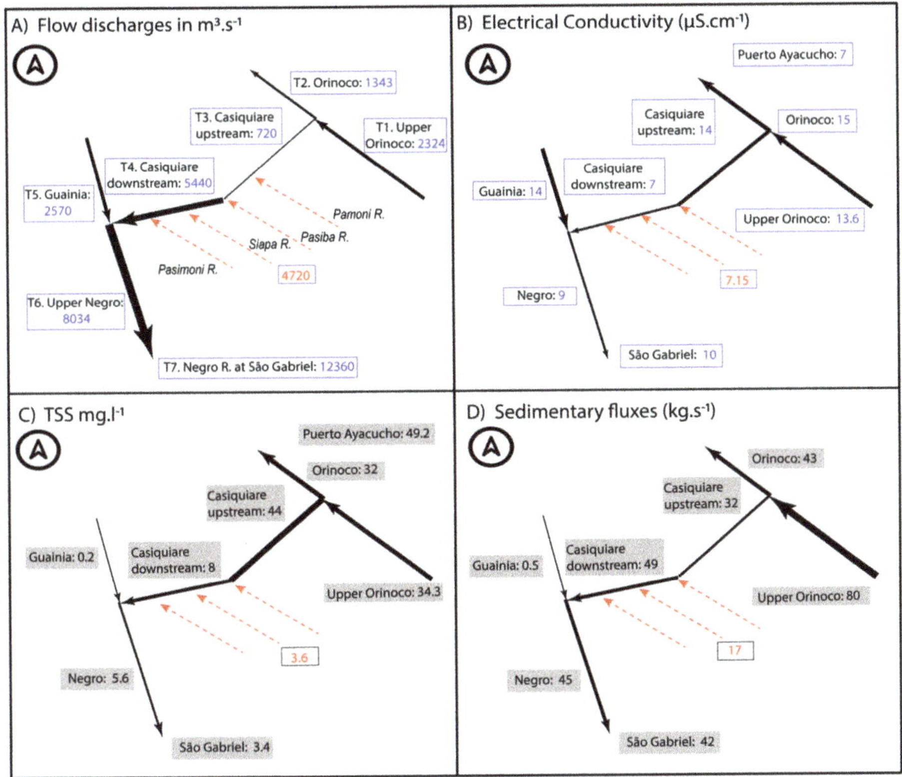

Figure 7. Schematic diagrams of the main morpho-dynamic and physico-chemical parameters measured in the Casiquiare hydrological complex from 9 to 12 September 2000. Numbers inside rectangles indicate flow discharges (Chart **A**), electrical conductivity (Chart **B**), suspended sediment concentrations (Chart **C**) and sediment fluxes (Chart **D**).

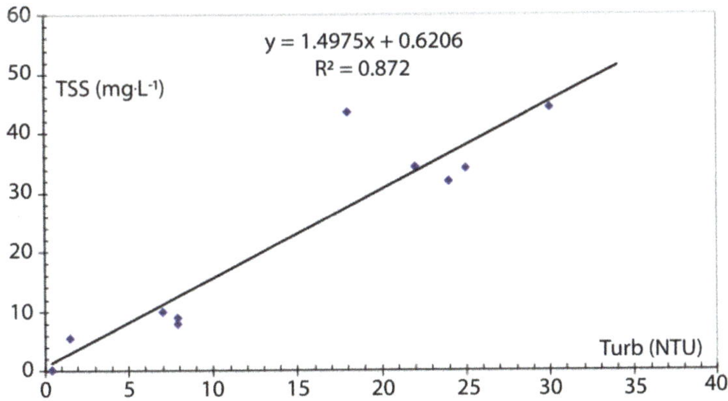

Figure 8. Relation between TSS (mg·L^{-1}) and turbidity (NTU) for the Casiquiare Channel.

It is worth noting that the interannual discharge of the Casiquiare bifurcation (320 m^3·s^{-1}) corresponds to only 0.85% of the interannual discharge of the Orinoco at its delta, estimated at 37,600 m^3·s^{-1} [30–32]. In its mouth, the Amazon River presents an interannual discharge of 209,000 m^3·s^{-1} [33], of which 0.16% comes from the Upper basin of the Orinoco, and not 1.50% as reported by some authors such as Palacios [34], who takes into account the Casiquiare outlet discharge, instead of the inlet discharge at its bifurcation.

7. The Problem of Tracing the Boundaries of the Watershed

Tracing the boundaries between the watersheds of the Casiquiare and Upper Orinoco is an insoluble problem with the concepts of classical hydrology, since depending on the season, 20% to 30% of the waters of the Orinoco above the bifurcation flows into the Casiquiare. Thus, the Upper part of the Orinoco watershed feeds two basins and it is a kind of diffuse area which is not well defined.

At its bifurcation and confluence with the Negro River, the problem arises because fuzzy boundaries cross the marshy plains characterized by shallow lagoons and flooded forests under which superficial two-way laminar-flow runs according to the seasons and to the water levels of the rivers they connect. In addition to this, there are small secondary canals connecting the Orinoco to the Casiquiare and the latter with the Guainia River, mentioned by Georgescu [8], Carreteiro [35], and CAF [36]. These features are found in the depressions and vast plains of major river basins, such as the Congo [37]. Another peculiarity of the Casiquiare watershed is that it is limited almost exclusively to the north-west by its own river course. Indeed, the tributaries of its right bank are small streams, very few in number, whose drained areas are negligible compared to that of the whole basin.

Lewis and Weibezahn [38], Lewis and Saunders [39], Paolini [40], and Brown et al. [41] present acceptable boundaries between the two largest basins on the continent, although the ambiguity of the surface area of the Upper Orinoco River that feeds both this river and the Casiquiare is not mentioned.

According to Ron et al. [27], the area sensu stricto (s.s.) of the Casiquiare watershed, i.e., the area whose rainwater flows directly and solely towards the Casiquiare Channel, is about 33,000 km^2, with an average flow rate at its outlet of 2790 m$^3 \cdot$s^{-1} [36] or 2100 m$^3 \cdot$s^{-1} [42]. According to the same author, the area of the Upper Orinoco River above the bifurcation is 37,870 km^2. However, in this study a morphometric analysis using the Fluvial Corridor Toolbox and the DEM SRTM at 30 m showed that the area of the Upper Orinoco River Basin presents a contributing area above the bifurcation of 40,110 km^2, coinciding quite well with the 40,000 km^2 estimate made by Stokes et al. [22]. Likewise, for the Casiquiare Channel basin a contributing area of 42,810 km^2 was estimated (Figure 9). Since it supplies both the Orinoco downstream of the bifurcation and the Casiquiare (in the order of 25% of its drainage water), it can be deduced that a total area of 82,920 km^2 contributes to the Casiquiare water supply.

For this reason, the specific flow rate of the Casiquiare basin cannot be calculated, because its catchment is composed of two parts. The first one (42,810 km^2) flows directly to the Casiquiare Channel (it is the sensu stricto (s.s.) watershed of the Casiquiare Channel) and the second one (40,110 km^2) flows at the same time to the Orinoco and Casiquiare. That means that the Casiquiare basin is delimited by both a typical watershed line (surrounding an area of 42,810 km^2) and by an unusual watershed area of 40,110 km^2 (which corresponds to the Upper Orinoco catchment) shared by the Orinoco and Amazon basins (Figure 9). The addition of these two contours delimits the surface area of the basin sensu lato (s.l.) of the Casiquiare, including the famous shared surface of the Upper Orinoco which does not only flow to the Casiquiare.

Figure 9 also compares the calculated slopes with the DEM SRTM (pixel ~30m) for the Casiquiare Channel with respect to the slope calculated locally by USACE [9], small differences can be observed (8.5 versus 6 cm·km^{-1}). These small differences can be explained as a result of uncertainties in local sampling (elevations using barometric measurements surveyed by the USACE), as well as by the low spatial resolution of the DEM used (in the current study). While terrain elevation models are very helpful for direct quantification of slope in channels, the accuracy of these analyses is a function of the spatial resolution of the DEM, the data acquisition system, the smoothing algorithms, as well as the methodology used for channel profile extraction. Currently, with the use of DEMs derived from Light Detection and Ranging (LIDAR) data, it is possible to obtain accurate slope estimates over relatively short distances. However, considerable shortening of channels and excessively steep slope estimates have also been documented when using a coarse data system such as ASTER GDEM 30 m or similar [43]. When evaluating the profiles of both channels a very subtle slope break can be observed that effectively shows the occurrence of the bifurcation in the main channel from the Upper Orinoco to the Casiquiare Channel. However, in order to better understand this hydrological phenomenon in the future it will be crucial to acquire and analyze new field measurements with the most recent technologies.

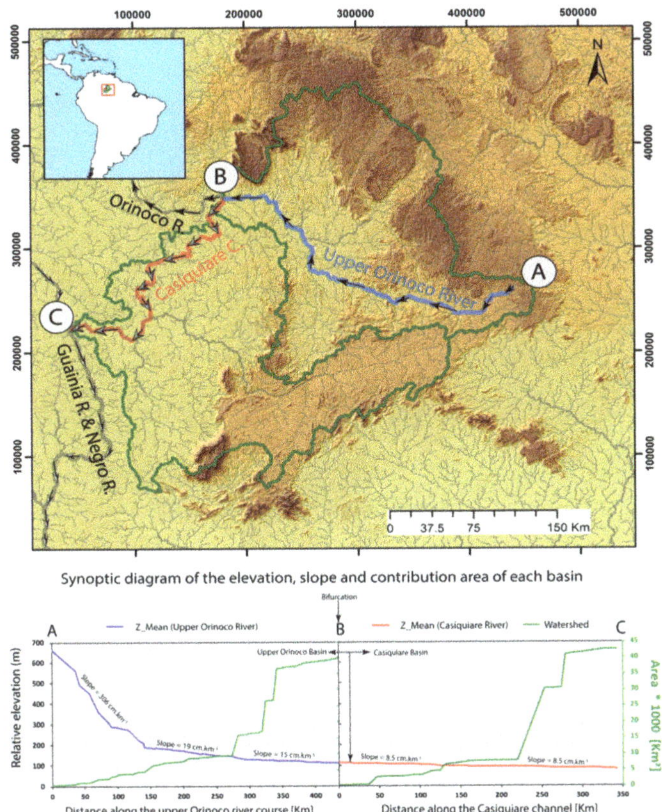

Figure 9. Boundaries of the Casiquiare s.s. catchment and Upper Orinoco catchment (green lines). The blue line delineates the Upper Orinoco River and the red line the Casiquiare Channel with a total length of ~350 km. A synoptic diagram highlights the contributing area in each basin, as well as the elevation and slope of each of the two rivers.

Depending on the season, 20 to 30% of the water from the Upper Orinoco upstream of Tamatama flows into the Casiquiare, which means that the same percentage of rainfall on the part of the catchment area upstream of the bifurcation is discharged through this channel, while the rest (70–80%) flows into the Orinoco River. Thus, it can be assumed that the same portion of a watershed feeds two rivers unevenly in the proportions previously stated.

This natural situation is unique in the world in terms of: (i) its location at the headwaters of a watershed, (ii) the areas of the sharing basin, (iii) discharge involved, (iv) for flows in only one direction, (v) the length of the channel, and (vi) the distance between the mouth of the Amazon and Orinoco rivers that are fed by the Upper Orinoco River.

8. An Ecological Connectivity Between the Orinoco and Amazon Basins

The Casiquiare links two basins characterized by different geomorphologic and geochemical conditions. The upper basin of the Orinoco River is a typical mountain relief of the tropics, covered with dense vegetation, containing mostly clear-water streams with slightly acid pH and moderate concentrations of dissolved organic and inorganic substances [44]. On the other hand, the upper basin of the Negro River, at lower elevations and covered in tropical forests, with areas of savanna occurring in upland regions, is dominated by 'blackwater streams' with high transparency but strong staining by tannins and other organic compounds leached from vegetation, extremely low pH, negligible amounts of solutes, and substrates of fine quartz sand [42].

Above the Casiquiare bifurcation, the Orinoco is a white-water stream with a common light brown color, so the Casiquiare begins its course with the same water conditions. However, during its ~350 km long-course, along its left bank, the Casiquiare receives the flows of many predominant black-water streams, like the Pamoni, Pasiba, Siapa, and Pasimoni rivers, carrying clear-acid-swamp water due to the high content of dissolved-organic humic acids. Thus, the Casiquiare suffers a gradual transition in the color of the river until its discharge into the Negro River, representing a hydrochemical gradient between white waters at its origin and black waters at its mouth.

The Casiquiare influences the migration of aquatic organisms between the Orinoco and Amazon basins and seems to function as an interbasin dispersal corridor for fish, but the effectiveness of this connection is mitigated by the strong physical-chemical and ecological gradient that spans its length. Winemiller and Willis [45] concluded that the degree to which the river serves as a dispersal corridor or barrier is variable and depends on the physiological and ecological tolerances of individual species.

9. Other River Captures

Very few natural hydrological captures exist worldwide that present some similarities with the Casiquiare. Without being exhaustive, it can be mentioned the Echimamish River in Canada, which connects the Hayes and Nelson Rivers flowing into Hudson Bay [46]. In this small connection that crosses a marshy region, the weak currents in the Echimamish River are reversed according to the water levels of the two connected rivers. This is the case for most of the world's bifurcations, many of which are intermittent and others are inside the same river basin. Another analogue ongoing river capture is the Selinda spillway of Southern Africa, which diverts water from the Okavango towards Kwando and Zambezi rivers, probably by overflow during the rising stage of Okavango River [47].

It seems that the only other current bifurcation similar to the Casiquiare is the Tarendö-elf canal in Sweden, which diverts 80 $m^3.s^{-1}$ of the flows of the Torne River towards the Kalix River, along a winding path of 52 km in a marshy terrain. Its origin would also come from an overflow of the Torne River, which would be the cause of its partial capture. Two hundred km downstream of this junction, the Torne and Kalix rivers flow into the Bay of Bothnia in the Baltic sea, with only 20 km distance between its rivers mouths [48]. However, the hydrological singularity of the Casiquiare is unique in its size. It is the only permanent and one-way bifurcation that extends for more than 300 km long with flow discharges varying between several hundred and a few thousand $m^3{\cdot}s^{-1}$, it also connect two neighboring basins with very distant deltas separated about 1500 km apart.

10. Conclusions

The Casiquiare Channel is a bifurcation flowing all year in the same direction from the Orinoco basin to the Amazon basin. Throughout its course, this natural channel undergoes significant morphological, hydrological, and bio-geochemical variations, whose most visible witnesses are the increase in width (3 to 4 times), flow (7 to 9 times), and its change in water color (white to black), under the influence of tributaries coming from vast forest plains. For this reason, it can be called a chameleon channel.

Studies made by other researchers [2,9] were interested more in the hydrology and hydraulics of the system, than in the sediment balance. Water and sediment budget is important in the Casiquiare system because of future plans for establishing a South American river integration system through the Orinoco, Casiquiare, Negro, Amazon, Paraguay, Parana, and Río de la Plata Rivers [36]. In this paper, a hydro-sedimentary balance of the Casiquiare fluvial system has been presented based on field data collected in September 2000.

At the beginning of the falling stage of the hydrograph, the flow at the bifurcation inlet channel corresponds to about 13% of those of the outlet flows. The latter, in the order of 5500 m^3·s^{-1} or an annual average of 2800 m^3·s^{-1}, are respectively 7 to 9 times higher than those at the inlet. The relatively high average slope (6 cm·km^{-1}) of the channel leads to relatively high velocities of up to 1.4 m·s^{-1} during rising stage. The natural slope and these velocities produce a river channel without many meanders and its operating dynamics should be more one of erosion-transport rather than sedimentation.

The geochemical facies of the waters range from "white water" (pH = 5.87) loaded in suspended sediment in the order of 44 mg·L^{-1}, coming from the relief belt that delimits the Upper Orinoco basin with that of the Amazon, to a facies characteristic of the "black rivers", more acidic (pH = 4.20) and less loaded by sediment (8 mg·L^{-1}) and dissolved mineral matter. This is due to significant dilution by numerous inputs from left bank tributaries crossing vast forest plains. Additionally, the electric conductivity changed from 14 to 7 µS·cm^{-1}. The longitudinal physical-chemical gradients clearly illustrate the evolution of this chameleon channel where the outlet waters are mixed by lateral inflows (tributaries), which are respectively 6 to 8 times higher than those at the inlet, depending on whether one is in rising stage or whether one considers the average module.

The surface of the Casiquiare catchment area sensu lato (s.l.) is about 82,920 km^2, and approximately half of it feeds two rivers at the same time, the Orinoco and Casiquiare. Thus it can be considered that only about 1/4 of the drainage area of the Upper Orinoco (40,110 km^2) is flowing to the Casiquiare. It is an unusual watershed area shared between the Orinoco and Amazon basins. In total, 0.9% of the Orinoco's average annual flow at its ocean mouth is diverted by the Casiquiare Channel to the Amazon. Through this channel, the Upper Orinoco River contributes 0.16% of the Amazon's average annual flow at its mouth.

Author Contributions: Conceptualization: A.L., J.L.L. and P.G.; Methodology: A.L., J.L.L. and S.Y.; Software: S.Y.; Validation: A.L. and P.G.; Formal analysis: A.L., J.L.L. and S.Y.; Investigation: A.L., J.L.L. and S.Y.; Data curation: A.L., S.Y. and P.G.; Writing—original draft preparation: A.L., J.L.L., P.G. and S.Y.; Writing—review and editing: A.L., J.L.L. and S.Y.; Project administration: A.L. and P.G.; Funding acquisition: P.G.

Funding: This research received no external funding.

Acknowledgments: This study has been carried out under the sponsorship of the Simón Bolívar University (USB, Venezuela) and the University of Brasilia (UnB, Brazil) with the assistance of the Venezuelan and Brazilian armed forces and the Institute of Research for Development (IRD, France), during the international fluvial expedition 'Humboldt-Amazonia 2000'.

Conflicts of Interest: The authors declare no conflict of interest.

References

1. Laraque, A.; Castellanos, B.; Steiger, J.; Lopez, J.L.; Pandi, A.; Rodriguez, M.; Rosales, J.; Adèle, G.; Perez, J.; Lagane, C. A comparison of the suspended and dissolved matter dynamics of two large inter-tropical rivers draining into the Atlantic Ocean: The Congo and the Orinoco. *Hydrol. Process.* **2013**, *27*, 2153–2170. [CrossRef]

2. Pérez, H.D. *Mediciones Hidrológicas en el Sector Alto Orinoco—Caño Casiquiare*; MARNR: Caracas, Venezuela, 2002.

3. The Global Runoff Data Centre GRDC—The World-Wide Repository of River Discharge Data and Associated Metadata. Available online: https://www.bafg.de/GRDC/EN/01_GRDC/grdcnode.html (accessed on 15 March 2019).

4. Sternberg O'Reilly, H. *The Amazon River of Brazil*; Geographische Zeitschrift, Beihefte Erdk. Wissen 40; Franz Steiner: Wiesbaden, Germany, 1975; p. XII+ 74.

5. Goodland, R.; Irwin, H. *A Selva Amazônica: Do Inferno Verde ao Deserto Vermelho*; Universidade de Sao Paulo: São Paulo, Brazil, 1976.

6. Von Humboldt, A. *Voyage Dans Les Régions Equinoxiales du Nouveau Continent*; Tome II; N. Maze: Paris, France, 1819; Available online: http://gallica.bnf.fr/ark:/12148/bpt6k61299w (accessed on 14 March 2018).

7. La Condamine, C.M. *Relation Abrégé D'un Voyage Fait Dans L'amérique Méridionale*; Académie des Sciences: Paris, France, 1981.

8. Georgescu, C.P. *Los Ríos de la Integración Suramericana*; USB: Caracas, Venezuela, 1984.

9. USACE (U.S. Army Corps of Engineers). Orinoco-Casiquiare-Negro Waterway. In *Venezuela-Colombia-Brasil*; USACE: Washington, DC, USA, 1943.

10. Georgescu, P.; Georgescu, C.P.; Laraque, A. *Le Casiquiare ... un Fleuve qui Relie Deux Fleuves! Bulletin de la Société de Géographie*; Société de Géographie: Paris, France, 2008.

11. Nordin, C.F.; Meade, R.H. The Amazon and the Orinoco. In *McGraw-Hill Yearbook of Science & Technology*; McGraw-Hill Inc.: New York, NY, USA, 1986; pp. 385–390.

12. Meade, R.H.; Koehnken, L. Distribution of the river dolphin, tonina "Inia geoffrensis", in the Orinoco River Basin of Venezuela and Colombia. *Interciencia* **1991**, *16*, 300–312.

13. Gessner, F. Ensayo de una comparación química entre el Río Amazonas, el Río Negro y el Río Orinoco. *Acta Científica Venez.* **1960**, *11*, 63–64.

14. Edwards, A.M.C.; Thornes, J.B. Observations on the dissolved solids of the Casiquiare and Upper Orinoco, April-June. *Amazoniana* **1968**, *2*, 245–256.

15. Edmond, J.; Palmer, M.; Measures, C.; Grant, B.; Stallard, R. The fluvial geochemistry and denudation rate of the Guayana Shield in Venezuela, Colombia and Brazil. *Geochim. Cosmochim. Acta* **1995**, *59*, 3301–3325. [CrossRef]

16. Sidder, G.B.; Mendoza, S.V. *Geology of the Venezuelan Guayana Shield and Its Relation to the Entire Guayana Shield*; USGS Open-File Rept; US Geological Survey: Reston, VA, USA, 1991; pp. 91–141.

17. Goldstein, R.J. *Angelfish. Barron's*; B.E.S. Publishing: Hauppauge, NY, USA, 2001; 96p.

18. Stern, K.M. Der Casiquiare-Kanal, einst und jetzt. *Amazoniana* **1970**, *2*, 401–416.

19. Vareschi, V. La Bifurcación del Orinoco: Observaciones Hidrográficas Ecológicas de la Expedición Conmemorativa de Humboldt del año 1958. *Acta Científica Venez.* **1963**, *14*, 9.

20. Besler, H. Beobachtungen zum Casiquiare-problem (Remarques sur le problème du Casiquiare). *Erdkunde* **1995**, *49*, 152–156. [CrossRef]

21. Grohmann, C.H. Tectonic Controls on the Capture of the Orinoco River and Formation of the Casiquiare Canal, Venezuela. *Geol. Soc. Am. Abstr. Programs* **2012**, *44*, 188.

22. Stokes, M.F.; Goldberg, S.L.; Taylor, J. Ongoing river capture in the Amazon. *Geophys. Res. Lett.* **2018**. [CrossRef]

23. Stern, K.M. La Génesis del Casiquiare. *Acta Científica Venez.* **1954**, *5*, 52–53.

24. Subramanya, K. *Flow in Open Channels*, 3rd ed.; Tata McGraw-Hill Education: New York, NY, USA, 1982.

25. USGS. *Shuttle Radar Topography Mission, 1 Arc Second Scene SRTM_u03_n008e004, Unfilled Unfinished 2.0, Global Land Cover Facility*; University of Maryland: College Park, MD, USA, 2000.

26. Roux, C.; Alber, A.; Bertrand, M.; Vaudor, L.; Piégay, H. "Fluvial Corridor": A new ArcGIS toolbox package for multiscale riverscape exploration. *Geomorphology* **2015**, *242*, 29–37. [CrossRef]

27. Ron, E.; León, R.; Mejía, A.; Colmenares, G. *Informe de Las Condiciones Hidráulicas del Caño Casiquiare*; Revista El Agua: Caracas, Venezuela, 1982.

28. MOP. *Mediciones en Ríos Grandes*; Ministerio de Obras Públicas, División de Hidrología: Caracas, Venezuela, 1972.

29. Vegas, V.; Paolini, J.; Herrera, R. A physico-chemical survey of black water rivers from the Orinoco and the Amazon basins in Venezuela. *Arch. Hydrobiol.* **1988**, *111*, 491–506.

30. Cordoba, J.R. Caracterización del Funcionamiento Hidrológico e Hidráulico-Fluvial del Delta del Rio Orinoco. In *PDVSA-Informe Desarrollo Armónico de Oriente DAO*; FUNINDES, USB: Caracas, Venezuela, 1999.

31. Silva, G. La cuenca del rio Orinoco: Visión hidrográfica y balance hídrico. The Orinoco River Basin: Hydrographic view and its hydrological balance. *Rev. Geogr. Venez.* **2005**, *46*, 75–108.

32. Laraque, A.; Pouyaud, B.; Rocchia, R.; Robin, R.; Chaffaut, I.; Moutsambote, J.M.; Maziezoula, B.; Censier, C.; Albouy, Y.; Elenga, H.; et al. Origin and function of a closed depression in equatorial humid zones: The lake Tele in the north Congo. *J. Hydrol.* **1998**, *207*, 236–253. [CrossRef]

33. Molinier, M.; Guyot, J.L.; Oliveira, E.; Guimarães, V. *Les Régimes Hydrologiques de L'amazone et de Ses Affluents*; IAHS: Paris, France, 1996.

34. Palacios, M.O. *El Orinoco, Tercer río Del Mundo*; Fundación Avensa y Fundación Terramar: Caracas, Venezuela, 1998.

35. Carreteiro, R.P. *A Navegação na Amazônia*; Editora Calderaro Ltda.: Manaus, Brazil, 1987; pp. 62–65.

36. CAF. *Los Ríos Nos Unen. Integración Fluvial Suramericana*; Corporación Andina de Fomento: Caracas, Venezuela, 1998.

37. Laraque, A.; Georgescu, P.; Hernández, D.P. El canal del Casiquiare: Una singularidad hidrológica única en el mundo. In Proceedings of the XXII Congreso Latinoamericano de Hidráulica, Ciudad Guayana, Venezuela, 9–12 October 2006.

38. Lewis, W.M.; Weibezahn, F.H. The chemistry and phytoplankton of the Orinoco and Caroni rivers, Venezuela. *Arch. Hydrobiol.* **1981**, *91*, 521–528.

39. Lewis, W.M.; Saunders, J.F. Concentration and transport of dissolved and suspended substances in the Orinoco River. *Biogeochemistry* **1989**, *7*, 203–240. [CrossRef]

40. Paolini, J. Organic carbon in the Orinoco River (Venezuela). *Verh. Internat Verein. Limnol.* **1991**, *24*, 2077–2079. [CrossRef]

41. Brown, E.; Edmond, J.; Raisbeck, G.; Bourlès, D.; Yiou, F.; Measures, C. Beryllium isotope geochemistry in tropical river basins. *Geochim. Cosmochim. Acta* **1992**, *57*, 1607–1624. [CrossRef]

42. Winemiller, K.; Lopez-Fernandez, H.; Taphorn, D.C.; Nico, L.G.; Duque, A.B. Fish assemblages of the Casiquiare River, a corridor and zoogeographical filter for dispersal between the Orinoco and Amazon basins. *J. Biogeogr.* **2008**, *35*, 1551–1563. [CrossRef]

43. Carretier, S.; Tolorza, V.; Regard, V.; Aguilar, G.; Bermúdez, M.A.; Martinod, J.; Guyot, J.L.; Hérail, G.; Riquelme, R. Review of erosion dynamics along the major NS climatic gradient in Chile and perspectives. *Geomorphology* **2018**, *300*, 45–68. [CrossRef]

44. Weibezahn, F.H.; Alvarez, H.; Lewis, W.M., Jr. (Eds.) *The Orinoco River as an Ecosystem*; EDELCA: Caracas, Venezuela, 1990.

45. Winemiller, K.; Willis, S.C. The Vaupes Arch and Casiquiare Canal: Barriers and Passages. In *Historical Biogeography of Neotropical Freshwater Fishes*; Albert, J.S., Reis, R., Eds.; The Regents of the University of California: Oakland, CA, USA, 2011.

46. McLeod, S.L. *The Evolution of the Echimamish River: Northern Manitoba*; FGS—Electronic Theses & Dissertations (Public): Winnipeg, MB, Canada, 1975; Available online: http://hdl.handle.net/1993/6165 (accessed on 7 August 2019).

47. Gumbricht, T.; McCarthy, T.S.; Merry, C.L. The topography of the Okavango Delta, Botswana, and its tectonic and sedimentological implications. *S. Afr. J. Geol.* **2001**, *104*, 243–264. [CrossRef]

48. Siergieiev, D. Impact of Hydropower Regulation on River Water Geochemistry and Hyporheic Exchange. Ph.D. Thesis, Luleå Tekniska Universitet, Luleå, Sweden, 2013.

Article

Inverse Estuaries in West Africa: Evidence of the Rainfall Recovery?

Luc Descroix [1,2,*], Yancouba Sané [2,3], Mamadou Thior [2,3], Sylvie-Paméla Manga [2,3,4], Boubacar Demba Ba [2,3], Joseph Mingou [2,3], Victor Mendy [3,5], Saloum Coly [3,5], Arame Dièye [2,3], Alexandre Badiane [1,2,3], Marie-Jeanne Senghor [2], Ange-Bouramanding Diedhiou [2], Djiby Sow [2,3], Yasmin Bouaita [2], Safietou Soumaré [2,6,7], Awa Diop [2,8], Bakary Faty [9], Bamol Ali Sow [3,5], Eric Machu [5,10], Jean-Pierre Montoroi [11], Julien Andrieu [2,7] and Jean-Pierre Vandervaere [12]

[1] IRD UMR PALOC MNHN/IRD/Sorbonne-Université, 75231 Paris, France; a.badiane785@zig.univ.sn
[2] LMI PATEO, UGB, BP 234 Saint Louis, Senegal; saneyancouba@gmail.com (Y.S.); thioryaz@yahoo.fr (M.T.); s.manga4555@zig.univ.sn (S.-P.M.); badembaba@gmail.com (B.D.B.); josephmingou30@gmail.com (J.M.); dieyearame91@gmail.com (A.D.); mjsenghor@gmail.com (M.-J.S.); bouramanding@gmail.com (A.-B.D.); sowsowdjiby@gmail.com (D.S.); yasmin.bouaita@gmail.com (Y.B.); safisoumare1@gmail.com (S.S.); a.diop754@zig.univ.sn (A.D.); Julien.ANDRIEU@univ-cotedazur.fr (J.A.)
[3] UASZ Université Assane Seck de Ziguinchor, BP 523 Ziguinchor, Senegal; victormendy97@yahoo.fr (V.M.); colysaloum@gmail.com (S.C.); bsow@univ-zig.sn (B.A.S.)
[4] Université de Lorraine, UFR des Sciences Humaines et Sociales, 54015 Nancy, France
[5] LMI ECLAIRS, UASZ, ENS, BP 5036 Dakar, Senegal; eric.machu@ird.fr
[6] LaSTEE, Ecole Polytechnique de Thiès, DPA 10 Thiès, Senegal
[7] ESPACE Lab, Université Côte d'Azur, UFR Espaces & Cultures Campus, 06204 Nice, France
[8] Université Versailles St Quentin en Yvelines, UFR Sciences Sociales, 78280 Guyancourt, France
[9] DGPRE, Direction de la Gestion et la Planification des Ressources en Eau, section 2, 20000 Diamniadio, Senegal; bakaryfaty@gmail.com
[10] IRD/Laboratoire d'Océanographie Physique et Spatiale (LOPS), IUEM, Univ. Brest, CNRS, IRD, Ifremer, 29280 Plouzané, France
[11] IRD/IEES; Institut d'écologie et des sciences de l'environnement, 93143 Bondy, France; jean-pierre.montoroi@ird.fr
[12] IGE/Université Grenoble Alpes; Institut des Géosciences et Environnement, 38058 Grenoble, France; jean-pierre.vandervaere@univ-grenoble-alpes.fr
*   Correspondence: luc.descroix@ird.fr

Received: 26 January 2020; Accepted: 24 February 2020; Published: 28 February 2020

Abstract: In West Africa, as in many other estuaries, enormous volumes of marine water are entering the continent. Fresh water discharge is very low, and it is commonly strongly linked to rainfall level. Some of these estuaries are inverse estuaries. During the Great Sahelian Drought (1968–1993), their hyperhaline feature was exacerbated. This paper aims to describe the evolution of the two main West African inverse estuaries, those of the Saloum River and the Casamance River, since the end of the drought. Water salinity measurements were carried out over three to five years according to the sites in order to document this evolution and to compare data with the historical ones collected during the long dry period at the end of 20th century. The results show that in both estuaries, the mean water salinity values have markedly decreased since the end of the drought. However, the Saloum estuary remains a totally inverse estuary, while for the Casamance River, the estuarine turbidity maximum (ETM) is the location of the salinity maximum, and it moves according to the seasons from a location 1–10 km downwards from the upstream estuary entry, during the dry season, to a location 40–70 km downwards from this point, during the rainy season. These observations fit with the functioning of the mangrove, the West African mangrove being among the few in the world that are markedly increasing since the beginning of the 1990s and the end of the dry period, as mangrove growth is favored by the relative salinity reduction. Finally, one of the inverse estuary behavior factors is the low fresh water incoming from the continent. The small area of the Casamance and Saloum basins

(20,150 and 26,500 km^2 respectively) is to be compared with the basins of their two main neighbor basins, the Gambia River and the Senegal River, which provide significant fresh water discharge to their estuary.

Keywords: water salinity; inverse estuaries; West Africa; drought; mangrove

1. Problematics, State of the Art

West African Sahelian and Sudanian areas commonly have very flat coast plains. Therefore, rivers have long estuaries in front of their low hydraulicity. The Gambia River estuary, 450 km long, is considered the second longest in the world after that of the Amazon River, although its basin has an area 100 times and a discharge 1000 times smaller. In West Africa, only great basins provide enough fresh water to reach pushing saline waters seasonally out of the river mouths. In this constraining context, it is very easy for marine salt water to enter profoundly in the continent, and it is also easy for the tide to influence deeply the river water level within the continent. An inverse estuary is an estuary in which freshwater input is less than the losses due to evaporation; such estuaries contain hyperhaline water (e.g., Laguna Madra in Texas; [1]). Inverse estuaries were also defined by Pritchard [2] as the ones where salinity increased with distance to the mouth. In an inverse estuary, sea water enters the estuary from downstream to upstream to compensate for the losses due to evaporation, carrying salt, which raises concentration [3].

Some of these estuaries have very low fresh water income and are located in areas with very high evaporation. This leads to an increase in water density and a hypersaline density downwelling. Therefore, a surficial water flow is noticed upwards, and another flow is noticed near the bottom downwards [4]. In the case of the Saloum River, the increase is observed to be in a roughly linear fashion with distance to the sea [5].

These inverse estuaries are commonly found in semiarid or arid areas, such as southern Australia [6], northern Australia [7], some areas of Texas [1], Baja California (NW Mexico) [8,9], and in the Sahel [10], among others. Such hypersaline estuaries are relatively rare on the Earth. Other examples are the Red Sea and the Persian Gulf in the northern Indian Ocean [11], Shark Bay and Exmouth Gulf in Western Australia [12], and the hypersaline Coorong estuary/lagoon in South Australia [13,14]. The characteristics of Spencer Gulf, South Australia, are that evaporation exceeds precipitation all year round and that the spring–neap tidal cycle is greatly exaggerated [6].

In West Africa, as in the other estuaries, enormous volumes of marine water are entering the continent [15]. Fresh water discharge is commonly strongly linked to rainfall level [5]. During the Great Sahelian Drought (1968–1993), some main rivers completely dried up; this was the case of the Niger River in Niamey in May 1985 [16]. In extreme cases, "a negative water budget has even more drastic effects on smaller rivers: discharge becomes negative, (and) seawater may invade the estuary which becomes hyperhaline" [17]. As an example, "the Casamance River estuary, in a dry year, and during the dry season, can be changed into an evaporation basin, concentrating marine salts coming from [the] Atlantic Ocean and becoming a threat to fluvio-marine areas soils" [17].

Southward Dakar Senegambian estuaries are subject to [this] "unusual hydrodynamical regime caused by weak or absent run-off" [18]. Such a process has been occurring in two coastal "rivers" of Senegal, the Casamance and the Saloum Rivers, which are both actually tide-influenced "inverse estuaries" [5]. Therefore, the West African Sudano-Sahelian coast is one of the regions in the world where inverse estuaries are observed [19].

West Africa has shown a great interdecadal rainfall variability. Few data before 1950 allow highlighting two dry periods in 1910–1915 and 1940–1944. After the Second World War, the number of rain gauges increased significantly, and the following evolution can be described [20–22]:

- a very rainy period from 1950 to 1967
- a long and very dry period from 1968 to 1993 (whole West Africa) and from 1968 to 1998 in Senegambia and Mauritania
- a rainfall recovered period (1994 to 2018 in WA, 1999 to 2018 in Senegambia), which was characterized by a rainfall annual amount close to the 1918–2017 average and an intensification of rainfall: an increase in rainfall intensity [23] and in the number of extreme rainfall events [24] are observed.

The evolution in coastal Senegal is similar to the regional one. Figure 1 shows the location of Senegambia (Figure 1a) and the location of rain gauge stations (Figure 1b). Figure 2 gives the rainfall evolution at different spatiotemporal scales. Figure 2a shows the evolution at three Senegalese coastal stations from 1917 to 2017, and Figure 2b shows the one in the inverse estuaries areas, southward from Dakar, from 1950 to 2017. Figure 2c,d give the SPI (Standard Precipitation Index) respectively for Senegambia and for the Casamance River basin. This confirms also the conclusions of Faye and Sané (2015; [25]), who observed the end of the long dry period in 1996 for the Casamance River Basin.

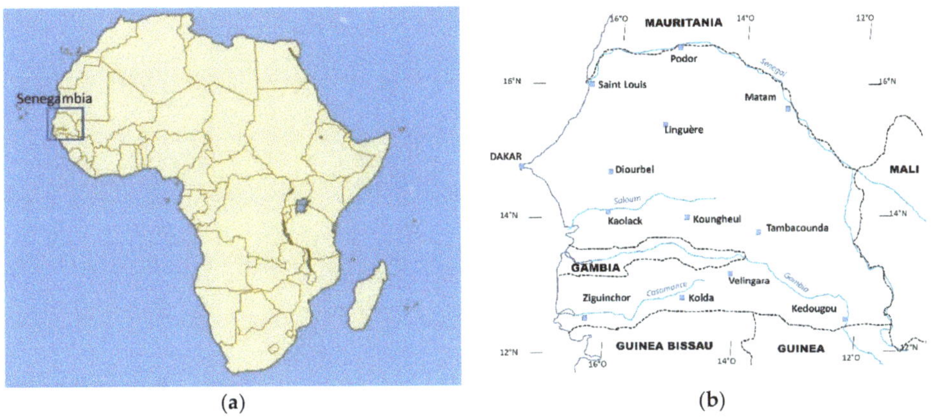

(**a**)   (**b**)

Figure 1. Location of Senegambia (**a**) and location of main rivers, boundaries, and rain gauge stations (**b**).

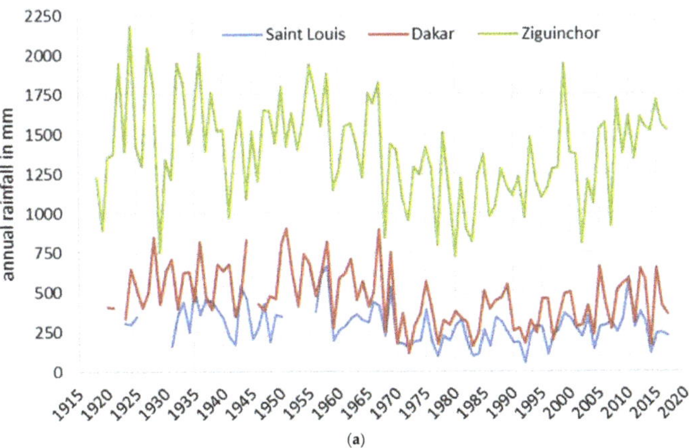

Figure 2. *Cont.*

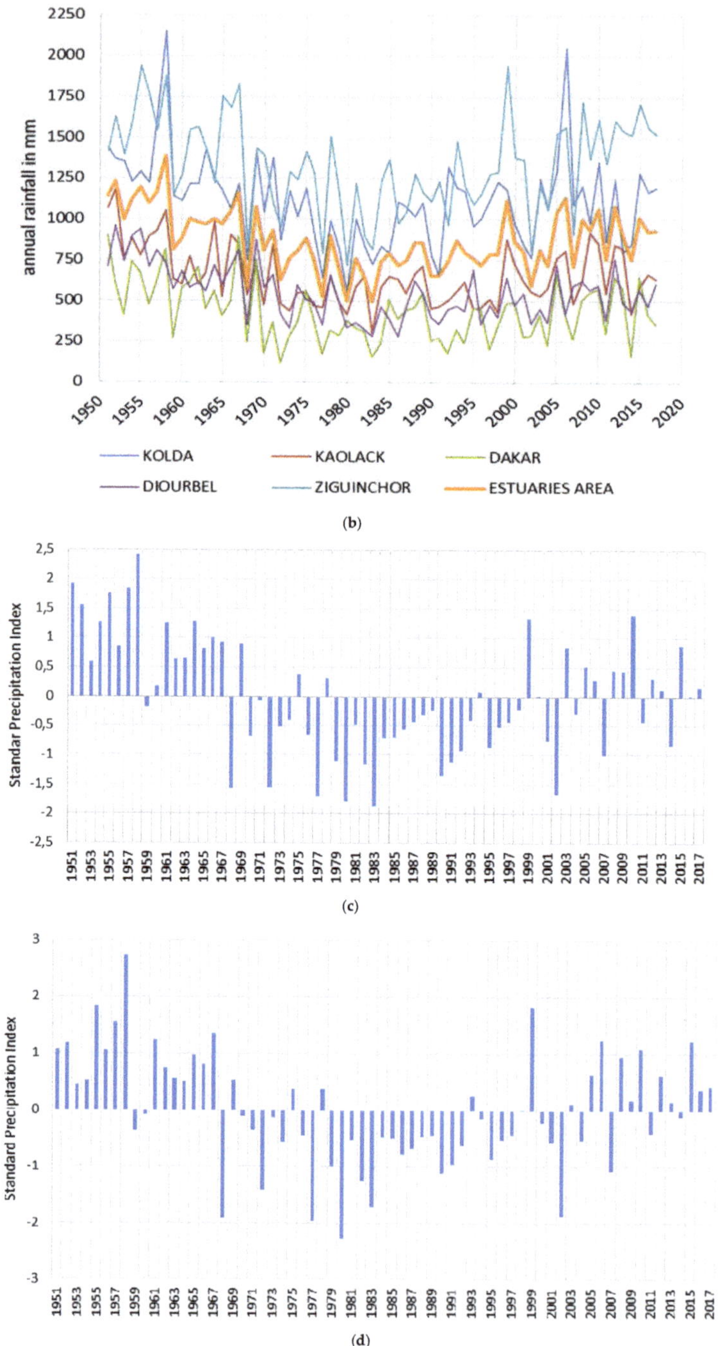

Figure 2. (**a**) Evolution of annual rainfall from 1917 to 2017 in three coastal Senegalese stations; (**b**) evolution of annual rainfall in the inverse estuary area; Rainfall Index Evolution in Senegambia (SPI, Standard Precipitation Index) (**c**), and in Casamance River basin: Ziguinchor, Kolda, Velingara (SPI Standard Precipitation Index) (**d**).

Teleconnections with the ITCZ (Inter Tropical Convergence Zone) fluctuations and their impact on flooding were analyzed by Mahé et al. (2013) [26].

When a river bed is close to the sea level, sea water may ingress during the low-water seasons. If the freshwater inflow is less than the loss through evaporation, salinity becomes higher than that in the sea. During the drought (1968–1998) Binet et al. 1995 [27] wrote, "This happens in the Senegal and the Casamance, particularly during the "current" drought".

Before 1960, marine water intrusions into the Senegal River during the dry season were negligible; but, in 1969 they reached 150 km upstream, and in 1978, they reached even 200 km [28]. Before the construction of the Diama (1986) and Manantali (1988) dams, high salinity values were observed especially during the dry period (1968–1986), particularly in the later years when salinity was concentrated due to a succession of very dry years in the second paxorysm of the Great Drought (1982–1986). Values above 35‰ were almost each year noticed at the end of the dry season (June and July) 80 km upstream from the mouth by Saos et al. (1984) [29]. These authors evidenced the inverse functioning of the estuary. Such behavior is widespread in the other Senegambian estuaries (Sine-Saloum, Gambia, and Casamance Rivers); however, it was strictly exceptional and of very short duration in the Senegal estuary [28]. These authors observed that the recovery of normal functioning is quite long (27 days in 1981 and 38 days in 1982).

Similar to the Casamance and Saloum estuaries, the semidiurnal microtidal Somone River estuary (80 km southeast of Dakar), where the maximum tidal range is about two meters, is characterized by an inverse salinity gradient [30,31]. Fluvial flows in the Somone estuary are null. Therefore, the salinity gradient is inverse; the only fresh water incomes observed during the 2007–2010 period are provided by rainfall and groundwater [31].

Hydrochemical analysis confirms that the Casamance River and Saloum River are "inverse estuaries", in which the water is salted during most of the year (at least 9 months) and hypersaline at the end of dry season [32].

Some estuaries within dry tropical areas (e.g., in Australia) show a combination of estuarian "normal" and "inverse" modes [4]. Inverse functioning of the Saloum River estuary, where salinity can reach 130 g/L [33], strongly affects fish and all marine species populations.

To conclude, Baran (1994) [34] summarized the context explaining that "Gambia river has a normal estuary, i.e., with decreasing salinity upwards. The Casamance estuary is inverse (rising salinity upwards) during the dry season and normal during the rainy season, while the Sine-Saloum is a "ria" where fluvial flows are null and the estuary is always inverse".

This study proposes to describe the current behavior of two West African inverse estuaries and to compare it with that observed during the Sahelian dry period of the end of the 20th century.

2. Methodology

In order to document the current functioning of the Saloum and the Casamance estuaries (see location in Figure 3), two measurement devices were implemented:

- a twice a year direct measuring campaign through the Saloum and the Casamance estuaries since the end of 2016:

This measuring campaign included a network of measurement sites in both Saloum and Casamance estuaries, salinity measures with a refractometer PCE© 0100 (67250 Soultz-sous-forêt, France) and, for values lower than 20 mS/cm, a conductimeter HANNA© HI 98130 (Woonsocket, RI, the USA), which also gives temperature and pH. This campaign was carried out at the end of the dry season (may) and at the end of the rainy season (November);

- a settled ensemble of five multi-sensor devices localized in the Casamance River estuary only, since Jan 2014;

These devices are CTD sensors (Conductivity, Temperature, Depth) model Decagon© CTD 10 (Pullman, WA, the USA), each one was coupled with a Decagon© EM50 data logger.

According to Noblet (2012) [35], salinity is calculated as follows:

$$Sa = (0.72 \times \sigma - 3.06) \times (1 + 0.02\,(T - 25)) \tag{1}$$

where:

- S_a is salinity in psu,
- σ is conductivity (mS/cm).
- T is temperature in °C.

Equation (2) was validated with the measured values with both a refractometer and field conductimeter. Since the bolons' water temperature always ranged between 21 and 28 °C, the deviation between measured and calculated values was low, rarely exceeding 2.5% (the highest observed difference was 4.8%).

Table 1 summarizes the data collected thanks to the implemented instruments.

Table 1. Data collected with Conductivity, Temperature, Depth (CTD) sensors around the Casamance estuary.

Site *	Number of Measurements	Number of Averaged Points	% Missing Data	% Corrected Data
Karabane	129,500	804	50	10
Ziguinchor	127,400	873	50	12
Goudomp	178,200	1653	20	10
Baila	214,700	2228	5	
Niambalang	168,000	1547	15	8

* see location in Figure 3.

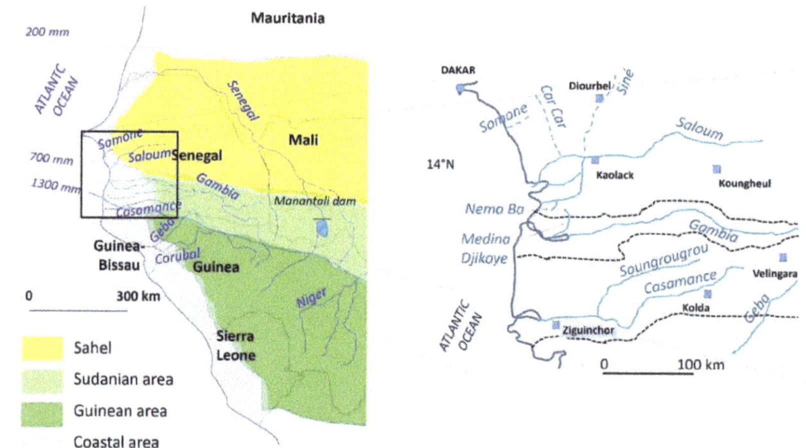

Figure 3. Location of cited West African estuaries. At **left**: geoclimatical areas (some annual rainfall amounts are indicated) and the main rivers network; at **right**: zoom on the study area at the local scale (dotted lines are the countries boundaries).

CTD data were regularly calibrated with that of the conductimeters (up to 20 mS/cm) and that of the refractometer (from 5 g/L, i.e., approximately 3 mS/cm, up to 100 mS/cm).

- Refractometers were calibrated with distillated water at the beginning and at the end of each measurement fraction of the day.

\-   Conductimeters were calibrated with standard dilution products supplied by the provider at the beginning and at the end of each measurement fraction of day in order to ensure the quality of measured data.

3. Results

Figure 4a (Saloum estuary) and 4b (Casamance estuary) give the location of the observations and implementation devices.

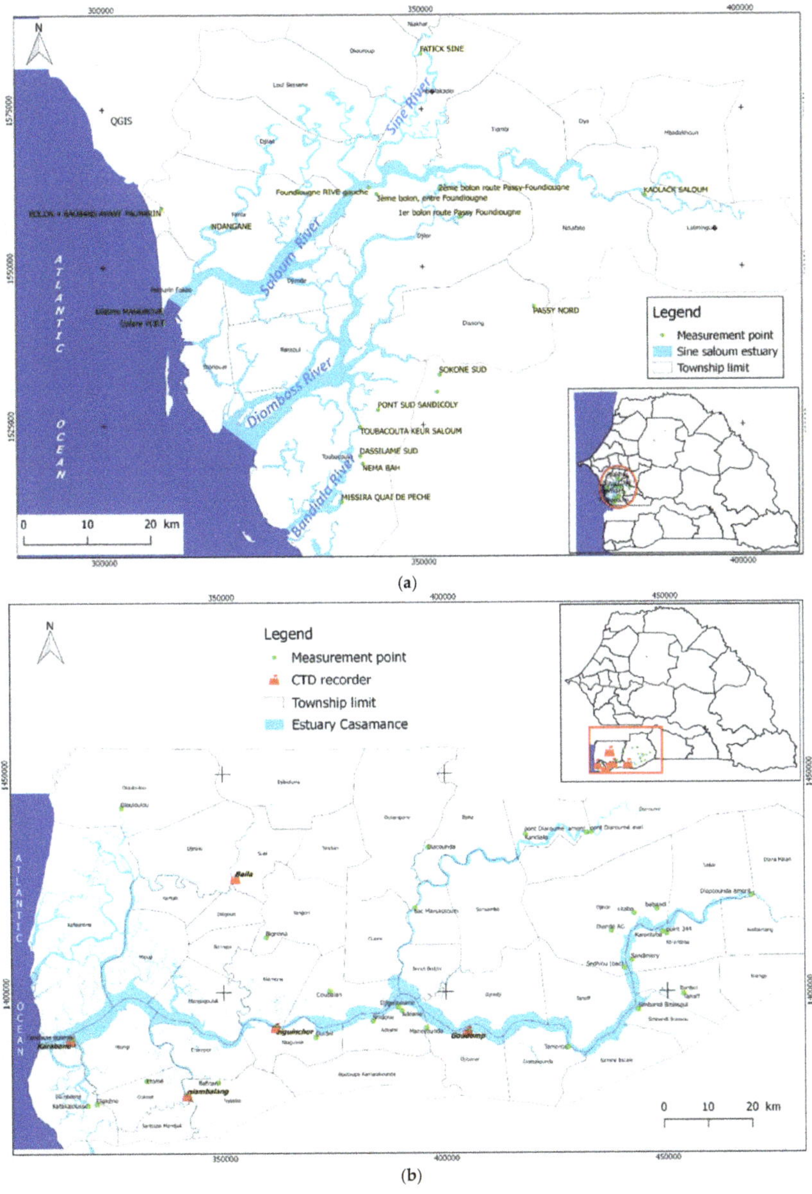

Figure 4. Location of devices and measurement points: (**a**) Saloum; (**b**) Casamance.

Despite the different periods of observation in each one of the settled devices, some general observations can be made about the current behavior of the salinity in the estuaries.

3.1. Casamance Estuary

Figure 5a–e show the seasonal evolution of water salinity in five points of the Casamance estuary from 2013 to 2019. The locations of the five salinity measurement stations are shown in Figure 4b.

- The salinity annual variation increases from the mouth (located 2 km downstream from Karabane station) to the upstream part of the estuary.

    ○ this variation increases in the main reach of the Casamance river, from Karabane (Figure 5a) to Ziguinchor (70 km from the mouth; Figure 5b) and then to Goudomp (125 km from the mouth, Figure 5c)

    ○ it also increases from the main branch to secondary branches, the "tributaries" coming from the North, at Baila on the "Baila bolon" (Figure 5d; *bolon* is the mandinka name given to the saline rivers of the mangroves in West Africa), and from the South, at Niambalang on the Kamobeul bolon (Figure 5e);

- The mean salinity values remain close to those of the sea (slightly above, at 40 g/L instead of 35 g/L) in the estuary (at least until Goudomp), as well as in the south branch of Kamobeul Bolon; it is significantly higher in the northern branch of the Baila Bolon (55 g/L).

Figure 6a–f shows the spatial variation of salinity at two stages of the year in the Casamance estuary. Figure 6a,c,e gives the salinity at the end of the rainy season in 2016, 2017, and 2018, respectively; Figure 6b,d,f gives the salinity at the end of the dry season, in 2017, 2018, and 2019, respectively. The following characteristics are observed:

- There is along all the year a fresh water income at the upstream entry of the main branch of the Casamance estuary (at Diopcounda Bridge); the same observation is made at the upstream origin of its main tributary, the Soungrougrou (Diaroumé Bridge); however, fresh water discharge is significantly lower in this river;
- There is always an estuarine turbidity maximum (ETM, [36,37]) in the upper part of both the Soungrougrou and the Casamance;

    ○ This area moves upstream during the dry season and it reaches the highest salinity values of the main reach of the Casamance (70 g/L in the Casamance, 100 g/L in the Soungrougrou);

    ○ It moves downstream during the rainy season, pushed by the fresh water discharge coming from the (small) basin of the Casamance and Soungrougrou rivers. The salinity values decrease during this period;

    ○ Downstream of this moving peak, salinity decreases all year long; then, Casamance river has an inverse estuary, however, its upper part has a normal functioning during a few kilometers in the dry season and over some tens of kilometers in rainy season;

    ○ In the main branch (Casamance), a second salinity peak is observed during some seasons at the confluence with the Soungrougrou river, due to the upper salinity values of the latter;

    ○ As observed in Figure 4a–f, salinity values are lower in the rainy season and higher in the dry season in the tributary bolons than in the main reach of the Casamance river estuary.

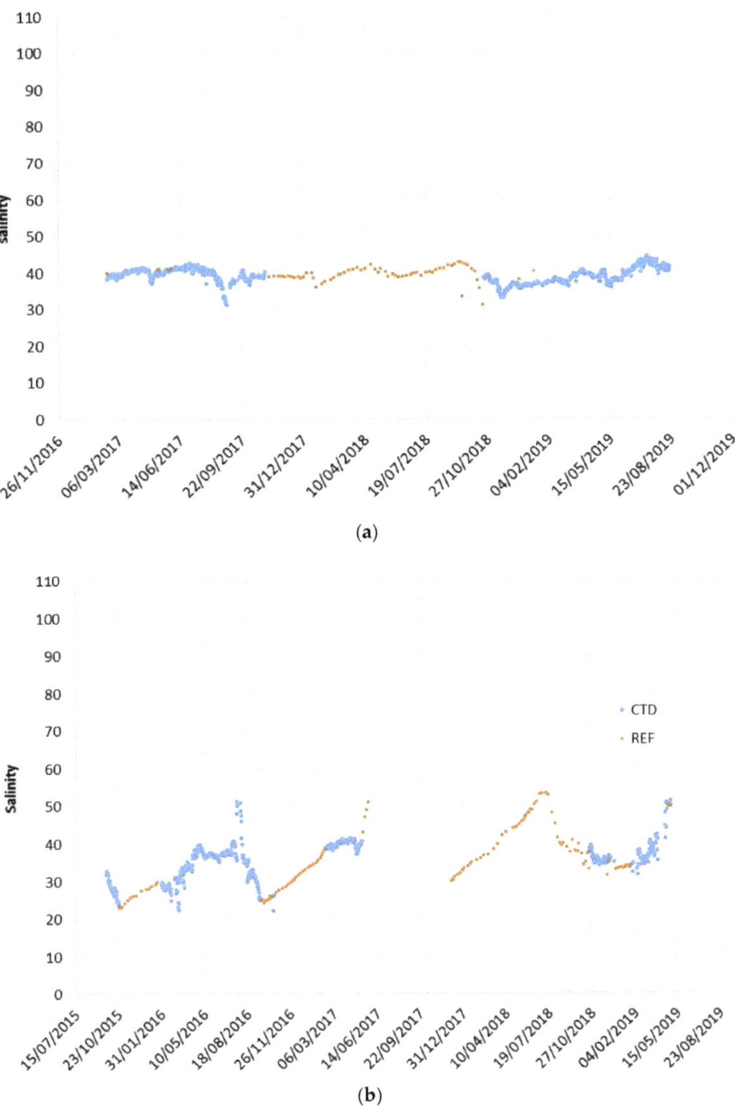

(**a**)

(**b**)

Figure 5. *Cont.*

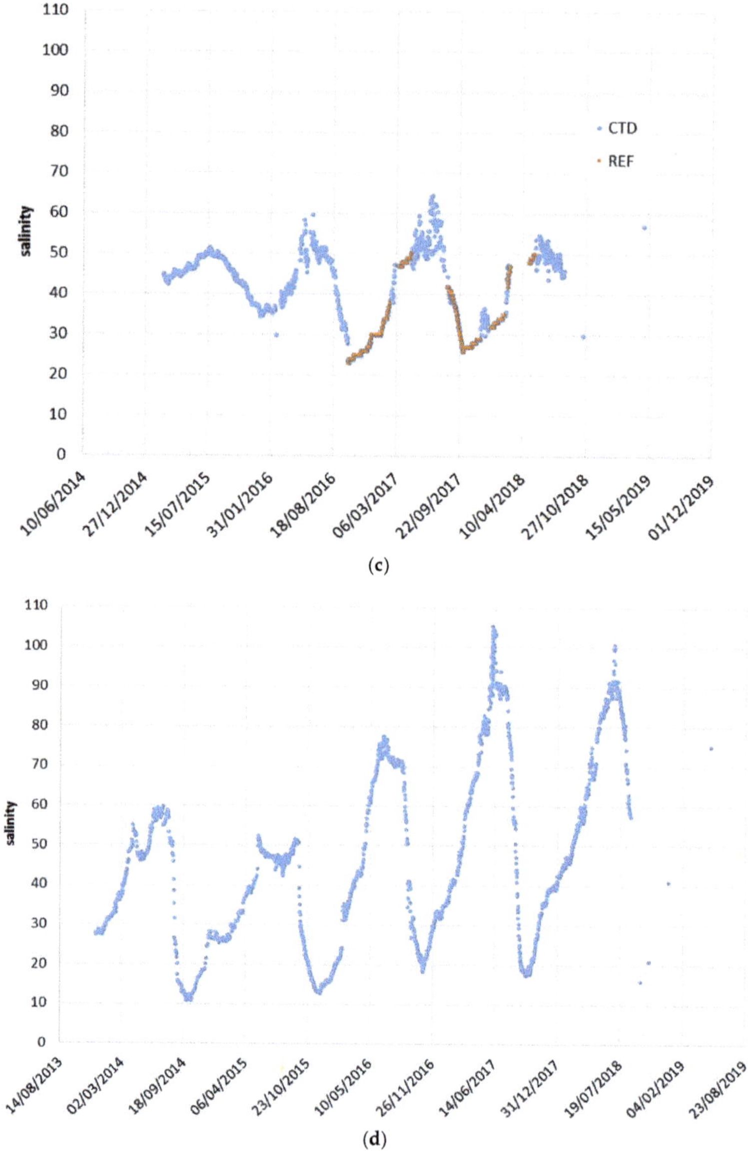

Figure 5. *Cont.*

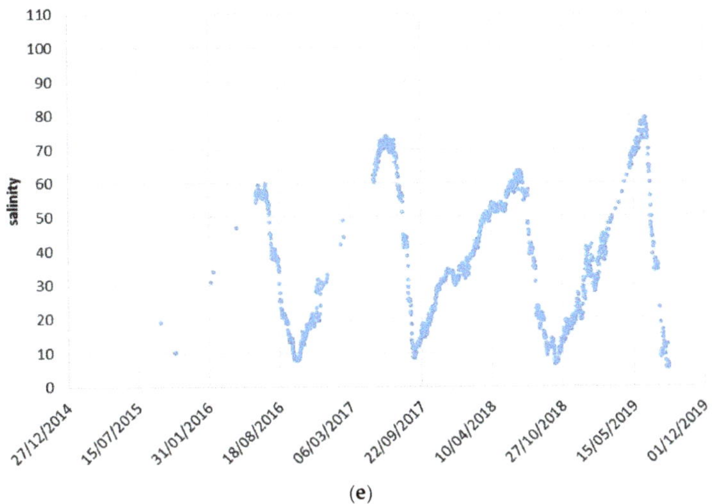

(**e**)

Figure 5. Evolution of salinity (g.L^{-1}) at the recording CTD stations: (**a**) Karabane (February 2017–August 2019); (**b**) Ziguinchor (September 2015–August 2019); (**c**) Goudoump (February 2015–October 2018); (**d**) Baila (December 2013–August 2018); (**e**) Niambalang (June 2016–August 2019). Red points are reconstructed data based on tens of water salinity measurement made by a refractometer.

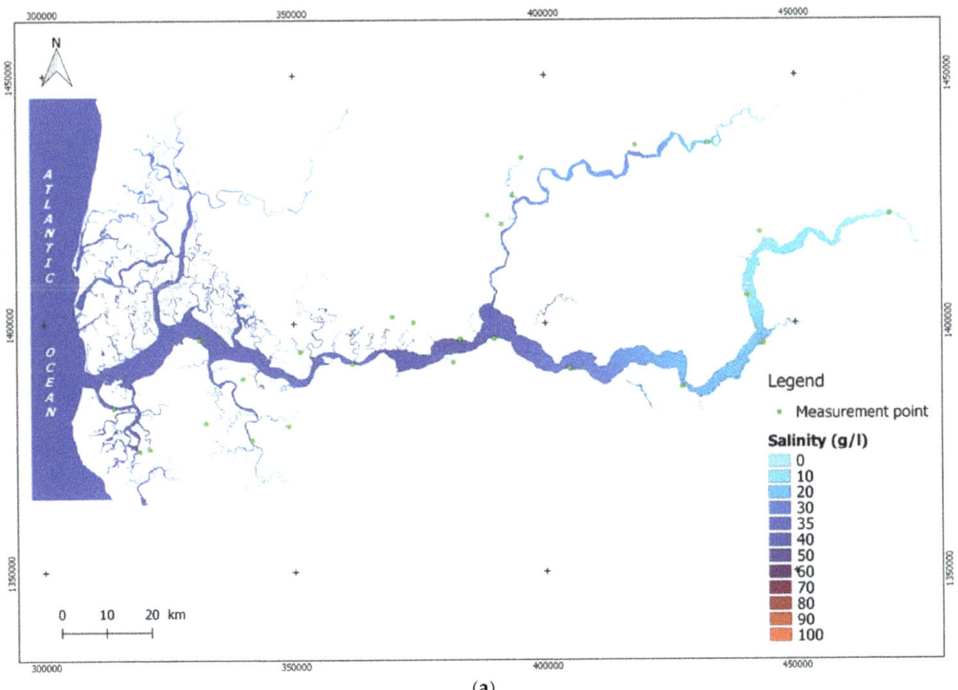

(**a**)

Figure 6. *Cont.*

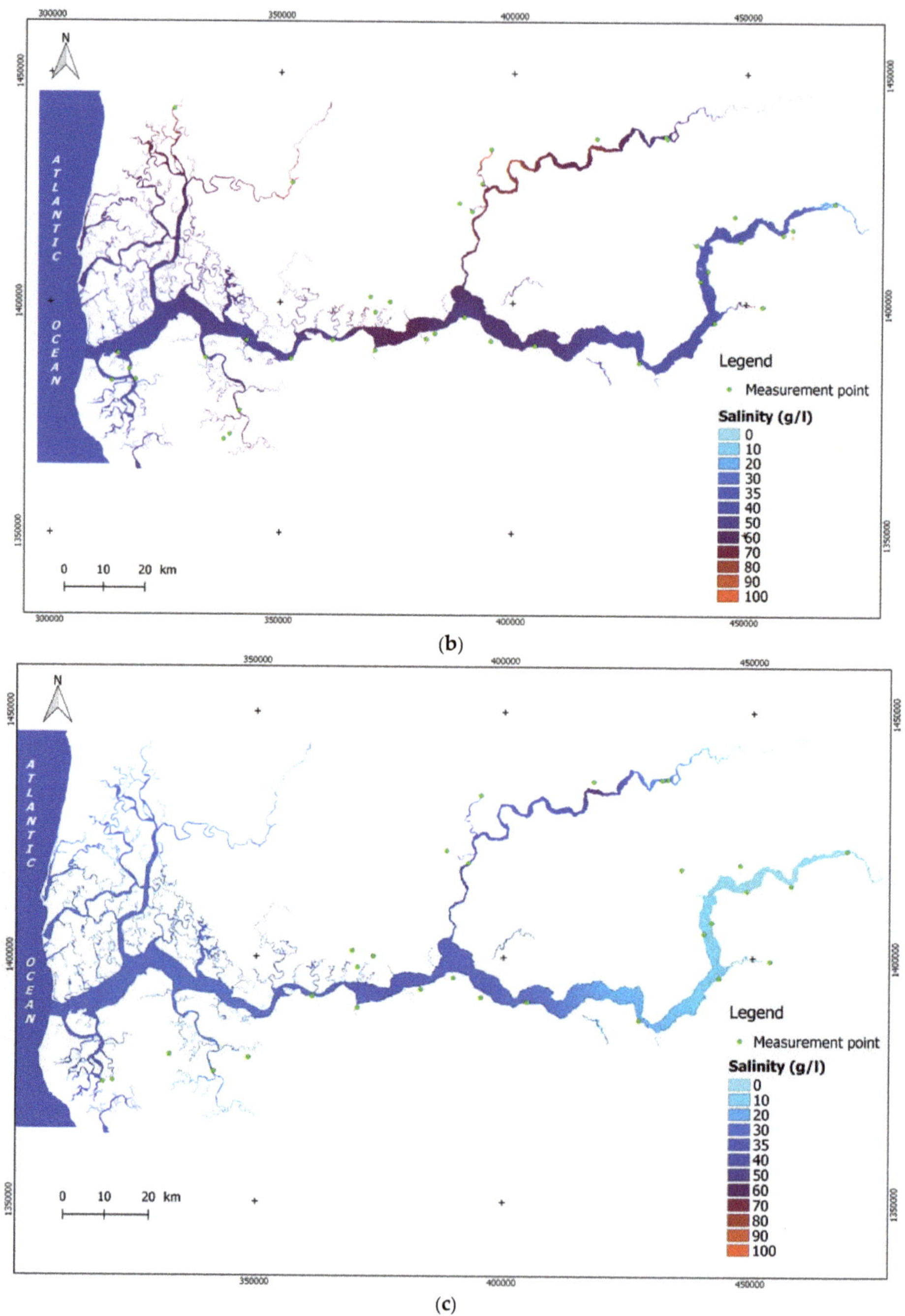

Figure 6. *Cont.*

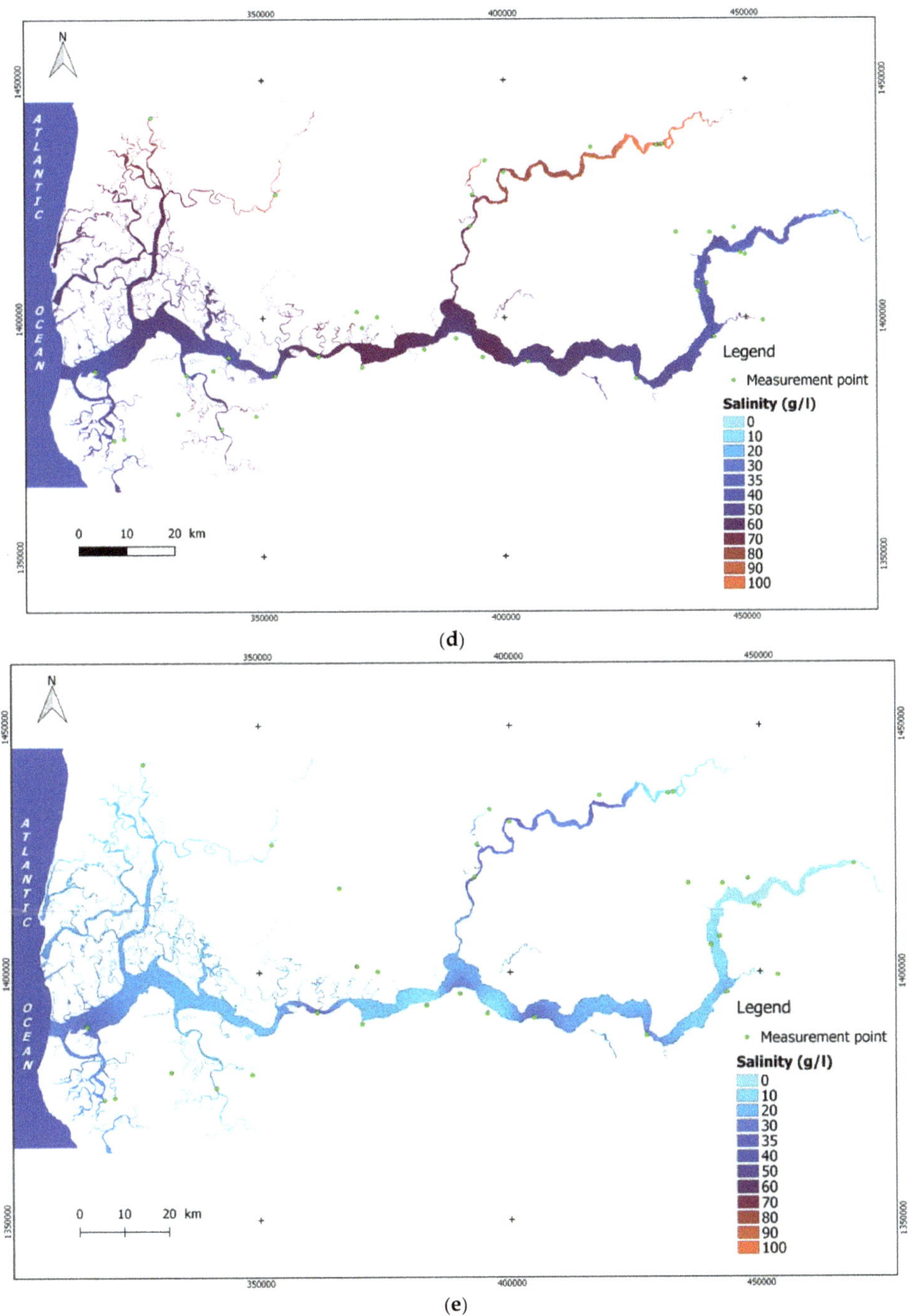

(**d**)

(**e**)

Figure 6. *Cont.*

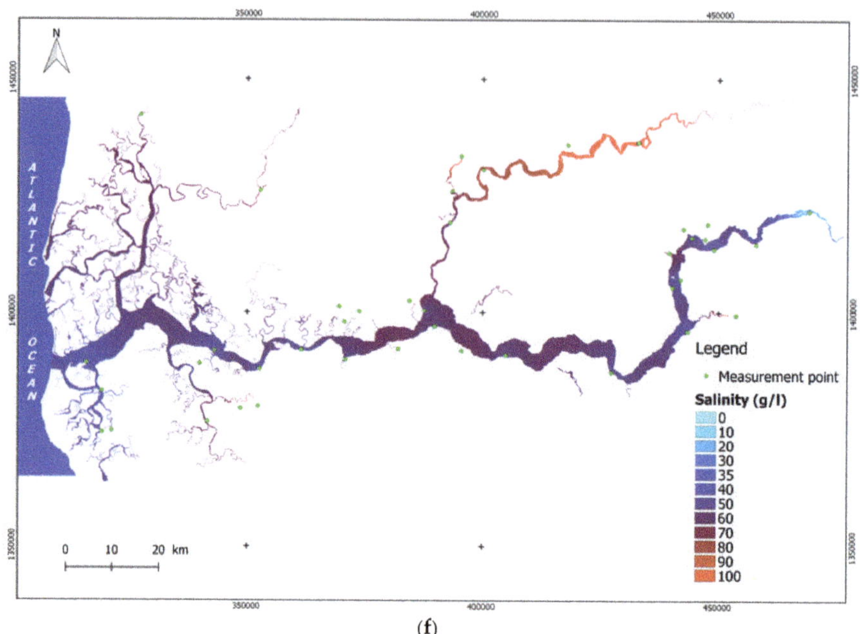

(f)

Figure 6. Casamance estuary salinity evolution: (**a,c,e**) at the end of the rainy season. (**a**) December 2016. (**c**) December 2017. (**e**). October 2018. (**b,d,f**), at the end of the dry season: (**b**) May 2017. (**d**) May 2018. (**f**) April 2019.

3.2. Saloum Estuary

Figure 7a–f documents the salinity of the Saloum estuary at the end of the rainy season (Figure 7a,c,e for 2016, 2017, and 2018, respectively) and at the end of the dry season (Figure 7b,d,f for 2017, 2018, and 2019 respectively). The functioning of this estuary is simpler than that of the Casamance.

- The fresh water discharge in the rainy season is quasi null and thus completely negligible;
- Values are lower during the rainy season due to lower evaporation, rain fallen within the wide estuary zone, and the sum of many small inputs by surficial runoff and small bolons;
- The estuary has an inverse behavior all year long;
- The salinity always increases upwards; the maximal values are always measured completely upstream, at Kaolack bridge in the Saloum and at Fatick Bridge in the Sine river;
- The salinity is higher in the north branch of Saloum estuary than that in the mid one (Diomboss) and overall than that in the southern one (Bandiala) (see location Figure 4a);
- The Bandiala bolon is provided in fresh water by the Nema Bah river, which is a small permanent fresh water river; water comes from the abundant water table of the southern Saloum plateau.

The temporal interannual variability is discussed in the last part of the discussion.

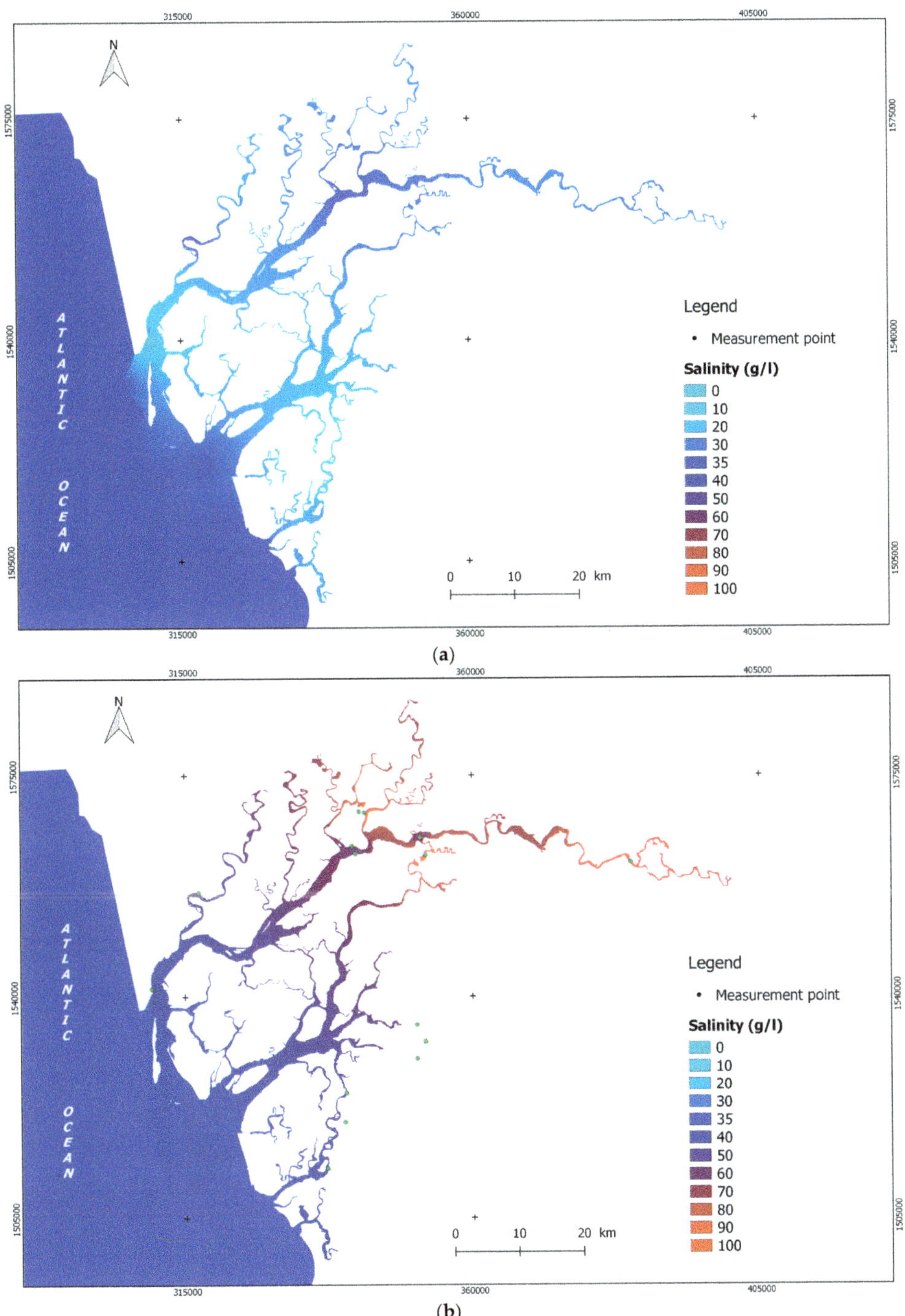

Figure 7. *Cont.*

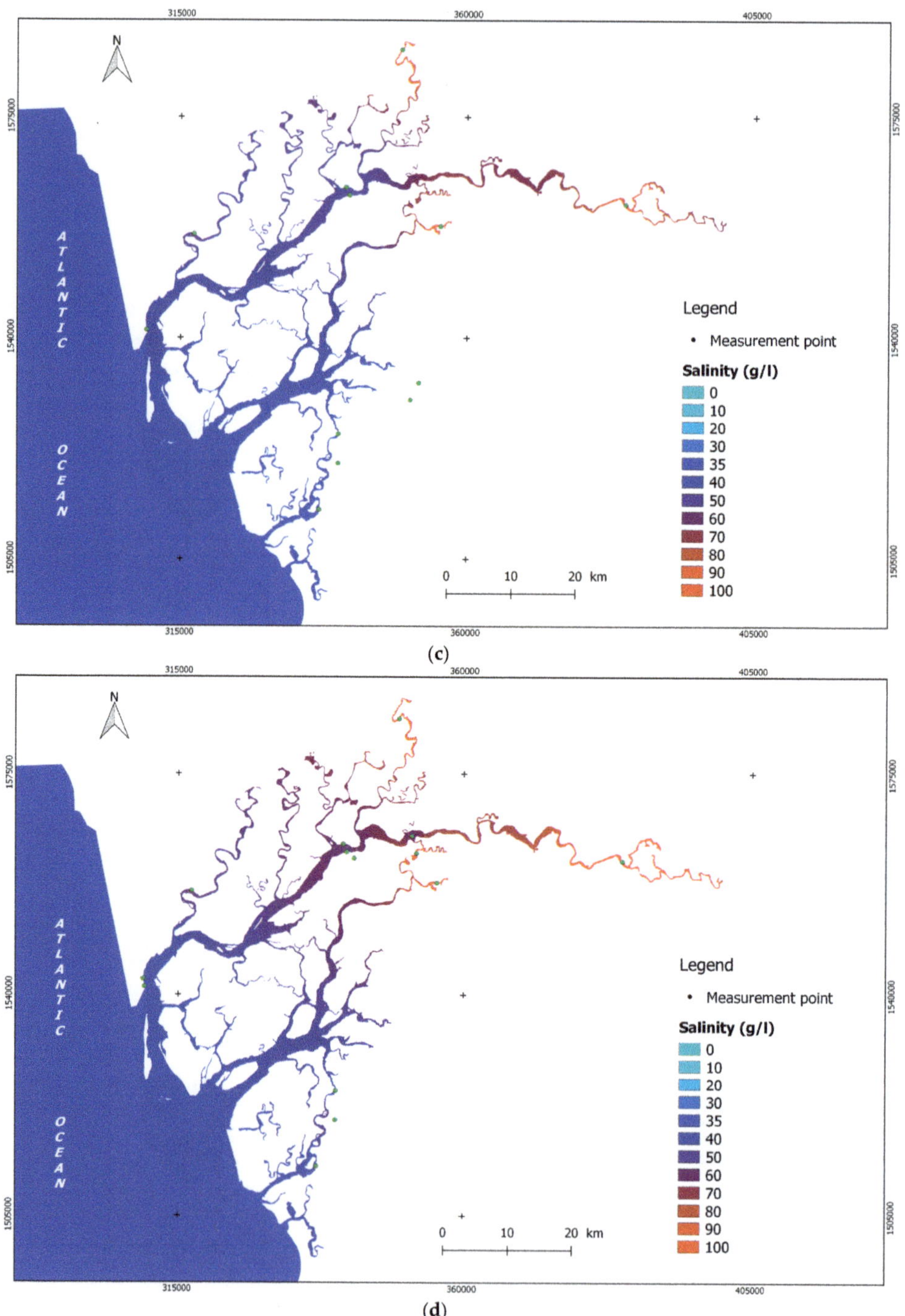

Figure 7. *Cont.*

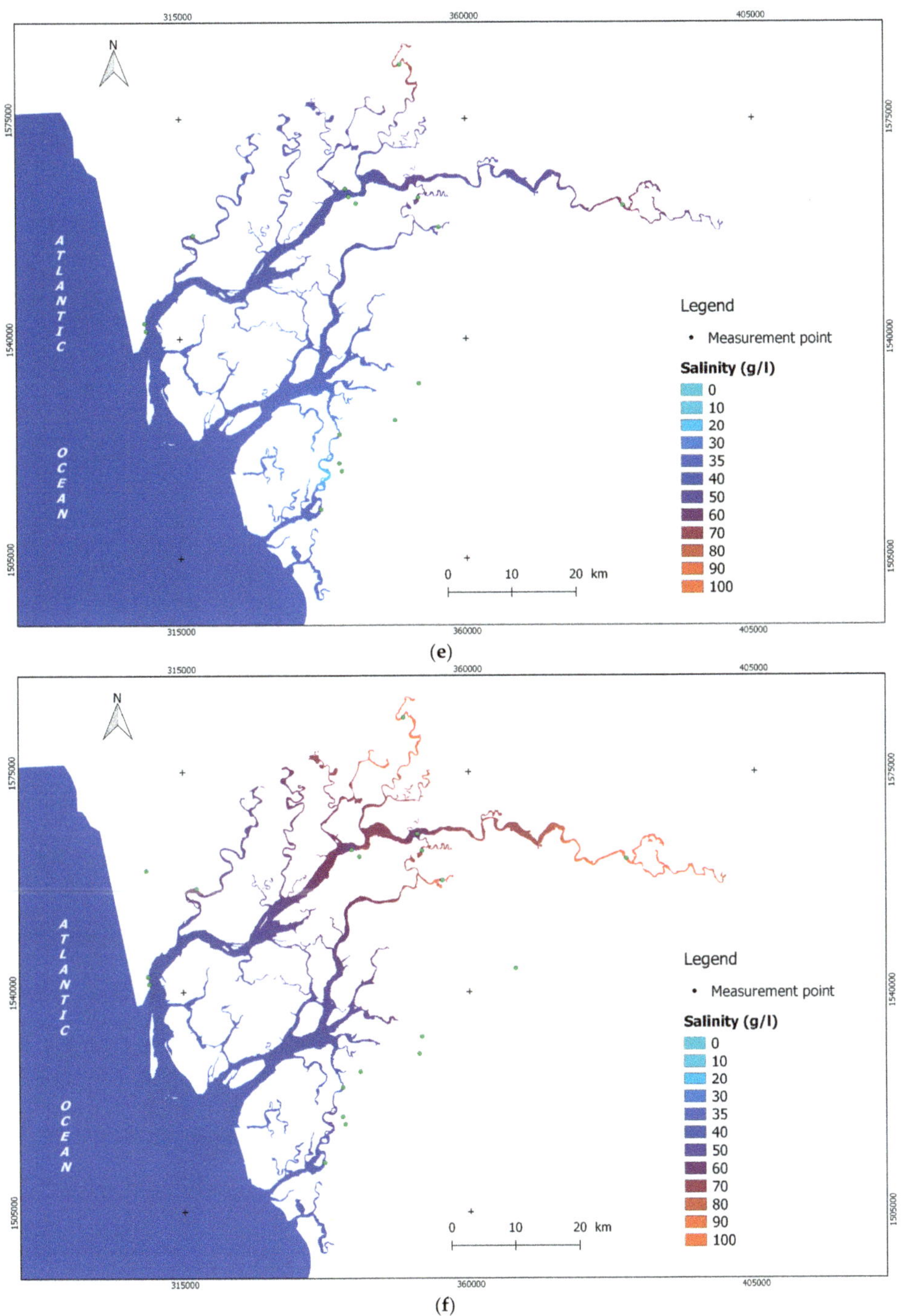

Figure 7. Saloum estuary salinity evolution: (**a**,**c**,**e**), at the end of the rainy season. (**a**) September 2016. (**c**) Jan 2018. (**e**) November 2018. (**b**,**d**,**f**), at the end of the dry season: (**b**) June 2017. (**d**) May 2018. (**f**) April 2019.

4. Discussion

4.1. A Comparison with Historical Data

One of the main references of past salinization (during the dry period 1968–1993) is the documentation about the Baila Bolon [3,38,39]. These authors highlighted the hypersalinization of the Baila bolon at its peak in 1984, as well as within the whole Senegambian valley [29]. In the Baila bolon, the mean salinity value was 96.8 g/L in 1983–1984; in the following years, salinity decreased until reaching 45.4 g/L in 1988–1989. Figure 8 shows the evolution of salinity at Baila (Baila bolon) during two periods. Salinization is not irreversible, and the rainfall recovery after 1988 allowed a quasi-complete desalinization at the end of the rainy season at the upstream sites of the estuary. The groundwaters have been recharged, and they flow through the salted rivers. They recharge the aquifers, rising water tables and allowing the conservation of all its fresh water [3].

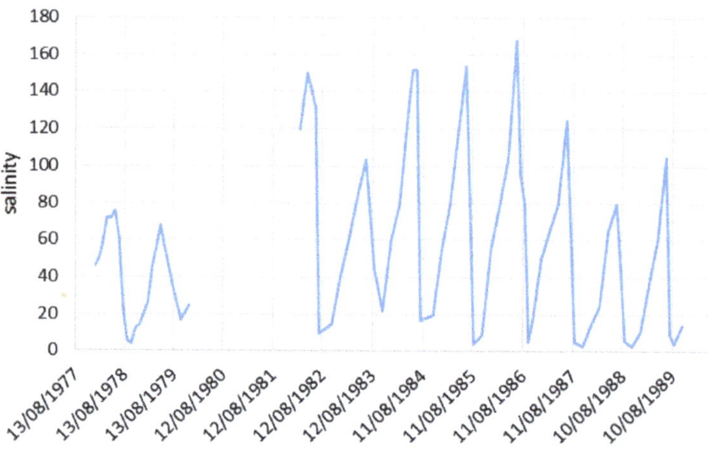

Figure 8. Salinity (g/L) measured at Baila during the Sahelian « great drought: 1978–1979 [32] and 1982–1989 [3].

The salinity at Baila station during the current period (Figure 5d), although presenting high interannual variability, is closer to the level observed during the 1978–1979 and 1988–1989 periods [3,32] than that measured during the 1982–1987 period [3]. The latter corresponds to the driest period of the Great Drought, the end of the second peak of dry years, which therefore is the period when the cumulated rainfall deficit was the highest. The 1978–1979 years are the period with a temporal rainfall recovery between the two "peaks" of droughts (1972–1974 and 1982–1984); the 1988–1989 period is also considered as the beginning of the rainfall recovery after the drought.

Table 2 shows the evolution of salinity at different points of the Casamance estuary, during two periods:

-   1978–1979 (in [32])
-   2016–2019 (our measurements)

Table 2. Salinity evolution at several places in Casamance estuary in 1977–1979 (Marius, 1985, p. 48, [32]) and in 2015–2019 periods (see location in Figure 4b).

Date Salinity in g.L^{-1} Place	January 1978	May 1978	November 1978	May 1979	November 2015	May 2016	November 2016	May 2017	November 2017	May 2018	November 2018	May 2019
Baila	46	76	6	68	8	60	18	82	18	88	16	75
Pointe St Georges	44	48	30	56	/	/	/	41	35	40	31	43
Etomé	41	93	4	91	3	88	5	99	15	90	8	100
Guidel	53	89	0	88	/	50	16	65	32	65	17	55
Marsassoum	42	68	20	66	/	/	35	65	45	65	35	62
Kandialo	16	60	0	52	/	/	31	75	59	87	43	101
Sedhiou	8	30	5	33	/	/	10	35	9	38	10	45
Diopcounda	1	14	0	15	/	/	4	21	2	25	0	11
Diaroumé	30	60	13	63	/	/	20	56	27	120	10	126

The comparison of the two periods allow noticing some differences:

- The minimal salinity values are more pronounced during the first period; however, this is partially due to the fact that in the second period, only two measurements per year are made;
- Salinity increases between the first and the second period in all the upper valleys (Baila Bolon at Baila, Soungrougrou at Diaroumé and Kandialo, and Casamance at Diopcounda);
- It decreases only at the Guidel station;
- It remains approximately equal in the mid basins (Sedhiou in the Casamance, Marsassoum in the Soungrougrou) and the lower valley (Etomé and Pointe St Georges).

This suggests that in spite of the rainfall recovery, the salinity did not decrease completely after the second "peak" of the drought, during the 1980s.

Figure 9 allows the same conclusion about the Saloum estuary; values of salinity are higher in the 2016–2019 period than during the 1980–1982 one. It is likely due to the strong increase in salinity caused by the second peak of the drought (1983–1985); the rainfall recovery since the 1990s was not sufficient to complete a true desalinization.

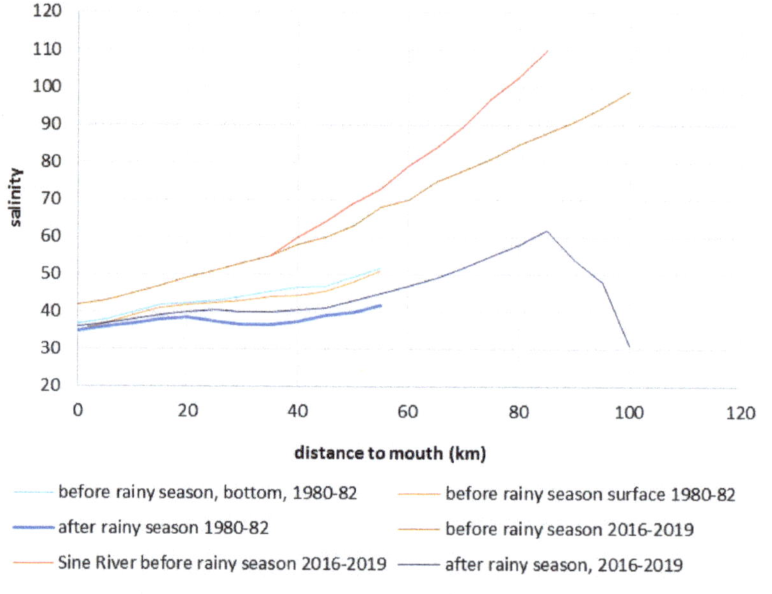

Figure 9. Salinity (g/L) vs. distance to mouth in the Saloum estuary, after [15].

Debenay and Pagès (1987) [40] defined the Casamance estuary, and they noticed five areas to be distinguished from downstream to upstream:

- From 0 to 50 km: a marine domain, with salinity, tides, and behavior close to those of the sea;
- From 50 to 85 km: an intermediary area, with increasing salinity;
- From 85 to 175 km: an hyperhaline area where salinity can reach 100 g/L;
- From 175 to 225 km: an alternative domain where salinity can vary from 0 to 100 g/L between rainy and dry season and during a few weeks only;
- Upstream from 225 km: fresh water with low discharge of the continental area.

The rainfall recovery in Casamance [25] and in the whole Senegambia [21] did not modify this general functioning; however, it caused a significant decrease in salinity in both estuaries of the Saloum and Casamance rivers.

4.2. An Integrative Indicator: The Mangrove

The mangrove forest is located at the boundary between the ocean and continent. The global mangrove area is declining mainly due to human activities [41,42]. Inversely, it is worth noticing that in West Africa, after a decline period during the Great Drought of 1968–1993, the mangrove area was significantly rising [43–45]; it almost reached its ancient extension level within the Saloum estuary and it even exceeds it within the Casamance estuary. It is expanding quickly [44,46] due to rainfall recovery and overall due to sea level rise. An improvement of mangrove governance and reforestation (unfortunately, mainly proceeded with Rhizophora in places where Avicennia was most indicated) locally could have contributed to this extension (only 4% of the increase in the mangrove area is due to reforestation, with the other 96% being spontaneous in the Saloum estuary [46]; respectively, these figures are 7% and 93% in the Casamance estuary [45]).

The mangrove is very sensitive to salinity levels; it can resist a few weeks or months to very high or very low salinity; however, if these peaks are repeated each year or if salinity values remain very high for more than 7 or 8 months a year, mangrove would eventually die.

Gilman et al. (2008) [47] describe the role of sedimentation on the mangrove stability and the ways that sedimentation may impact mangrove resilience (such as sediment accretion and erosion, biotic contributions, belowground primary production, autocompaction, fluctuations in water table levels, and pore water storage) but conclude that there is no correlation between sedimentation rates and sea level rise (SLR).

Osland et al. (2018) [48] show how the accelerated SLR could favor sedimentation and biotic contributions and thus the extension upland or upriver of the mangrove forest. Mangrove forests in arid and semiarid climates are known to be particularly vulnerable to changes in rainfall and freshwater availability [48]; the SLR probably accelerated the desalinization of the West African bolons, and this is probably the main explaining factor of the current mangrove expansion.

As an example, and contrary to the conclusion of Dièye et al. (2013) [49], the natural opening of the "fleche de Sangomar" sand spit in 1987, at the northern side of the Saloum estuary, allowed the entry of important volumes of marine water (35 g/L) in an hyperhaline estuary; then, it provoked a reduction of the salinity in the inland estuary. The mangrove regeneration was firstly observed near Djiffère in 1992 (Mamadou Sow, personal communication), where the opening of the "fleche de Sangomar" allowed a strong relative desalinization and the mangrove recovery. Mangrove decline in the lower Saloum estuary after 1987 and the end of the drought (rainfall annual total amount began increasing after 1985) observed by Dieye et al. (2013) [49] must be due to factors other than salinization.

SLR seems to be influencing the mangrove extension by leading to a relative desalinization of inverse estuaries water and by helping greater volumes of sea water entering in estuaries with much higher salt rates. The rainfall recovery since the end of the 1980s made the Casamance estuary show normal behavior during an increasing proportion of the year; this recovery did not allow at the time the Saloum River to have periods with normal functioning. However, there is no evidence that Saloum was not yet, before the drought, a completely inverse estuary at least since the African Humid Period, before 4000 BP.

The current expansion of the mangrove is the result of the significant relative desalinization of the inverse West African estuaries.

4.3. Low Discharges Explaining the Inverse Estuaries

Saloum and Casamance rivers have very low discharge values due to geology (sedimentary originated mainly sandy soils), the topography is very flat and, in addition, in Casamance, vegetation is very dense and rice cropping is organized to store most of the rainwater during the rainy season. Table 3 indicates the runoff coefficient values to be compared with that of surrounding basins.

Table 3. Mean rainfall, discharge, and runoff coefficient (KE) is the inverse estuaries, compared with their surrounding main basins.

River Basin	Area km^2	Annual Rain m^3	Annual Discharge m^3	% Total	Rain Depth mm	Mean Discharge m^3 s^{-1}	KE %
Casamance at Diana Malari [1]	4710	52,987 × 10^6	158,962 × 10^3	7.1	1125	5.08	3
Soungrougrou [1]	4480	51,609 × 10^6	51,610 × 10^3	2.3	1152	1.65	1
Mid Casamance basin [1]	4150	54,158 × 10^6	216,630 × 10^3	9.7	1305	6.93	4
low Casamance basin Right bank [1]	4323	56,631 × 10^6	283,156 × 10^3	12.7	1310	9.05	5
low Casamance basin Left bank [1]	1560	23,587 × 10^6	235,872 × 10^3	10.6	1512	7.54	10
water body [2]	927	12,857 × 10^6	1,285,749 × 10^3	57.6	1387	41.1	100
CASAMANCE BASIN	20,150	251,875 × 10^6	2,231,978 × 10^3	100	1250	71.4	8.9
Nema Bah [3]	50	318 × 10^6	2365 × 10^3	0.31	637	0.075	8.2
Medina Djikoye [3]	300	2145 × 10^6	10,373 × 10^3	1.38	716	0.33	6
Car Car at Tataguine [4]	1950	11,349 × 10^6	5674 × 10^3	0.75	582	0.18	0.5
Sine at Fatick [4]	3600	21,348 × 10^6	12,809 × 10^3	1.7	593	0.41	0.6
Saloum at Kaolack [4]	9502	61,180 × 10^6	91,770 × 10^3	12.19	644	2.93	1.5
Lower basin [4]	10,710	74,984 × 10^6	299,936 × 10^3	39.85	700	9.59	4
Water Body [2]	388	3298 × 10^6	329,800 × 10^3	43.82	850	10.55	100
SALOUM BASIN	26,500	174,625 × 10^6	752,727 × 10^3	100	659	24.07	4.3
Senegal [5]	337,000	2,527,500 × 10^6	20,183,040 × 10^3		751	640	8.0
Gambia [5]	60,000	612,000 × 10^6	8,675,400 × 10^3		1020	275	14.2
Geba [6]	12,440	164,208 × 10^6	1,970,496 × 10^3		1320	62.5	12
Corubal [2]	26,000	445,640 × 10^6	10,249,200 × 10^3		1714	325	23

[1] Dacosta, 1989 [50]; [2] calculated from GIS and rainfall data, + Guinea Bissau hydrological service for Geba River; [3] Mendy, 2010 [51]; [4] estimated from few existing data and double mass comparison with similar or surrounding basins; [5] [22]; [6] Sambou, 2019 [52]. See location of basins in Figure 3.

Clearly, runoff coefficients are lower in the Saloum and Casamance basins compared with other basins that have a comparable total rainfall amount. This is one of the main explaining factors of their inverse functioning.

The Casamance River basin has a runoff coefficient just slightly higher than that of the Senegal River basin, although it receives 55% more rainwater than the latter. The runoff coefficient of the Saloum River basin is half of that of the Casamance River. The Geba River basin also has a low runoff coefficient compared with that of the other basins with similar rainfall amount.

These three basins have little runoff due to geological and topographical factors. In the two first cases, runoff is significantly enhanced by the rain water fallen directly in the estuary; this constitutes 44% of the total discharge of freshwater in the Saloum basin and 58% of that for the Casamance basin.

5. Conclusions

These observations about salinity spatiotemporal evolution carried out within two West African inverse estuaries allow us to attempt a few statements:

Firstly, about the functioning of estuaries, we can confirm that:

- The Saloum River estuary has a total inverse behavior, with salinity increasing upwards;
- The Casamance estuary has a spatially partial inverse functioning, with a point of maximum salinity migrating from 20–30 km of the upstream end estuary at the end of the dry season to 50–80 km downstream from this point at the end of the rainy season.
- Therefore, about the spatial variability, we observed:
- decreasing salinity downwards from the peak of the ETM in the Casamance estuary and from the upstream entry of the estuary at Kaolack in the Saloum river;
- Increasing salinity seasonal variability in the tributary bolongs;
- A similar behavior in the bolongs than in the Casamance upper estuary.

Finally, the salinity seasonal variability increases with the distance to the ocean.

A relative desalinization is attested by the comparison of measurement realized since 2013. Besides, the progressive post-drought rainfall recovery led to a decreasing bolongs water salinity but answering with a 10–15-year delay. The mangrove spectacular recovery is a good indicator of the progressive reduction of the bolongs' water salinity.

However, the low discharge and runoff coefficient values in the continental part of the basins is one of the main and persistent explaining factors of this inverse functioning.

Author Contributions: Conceptualization, Y.S., M.T., S.-P.M., M.-J.S. and J.A.; Data curation, J.M., A.-B.D., D.S., Y.B., S.S. and A.D.(Awa Diop); Formal analysis, J.M., V.M., A.D. (Arame Dièye), A.B., A.-B.D. and B.F.; Funding acquisition, Y.S., M.T., S.-P.M., B.D.B. and J.M.; Investigation, Y.S., M.T., S.-P.M., B.D.B., J.M., S.C., A.D. (Arame Dièye), M.-J.S., D.S., S.S., A.D. (Awa Diop), B.F., B.A.S. and J.-P.V.; Methodology, V.M., S.C., A.D. (Arame Dièye), A.B., M.-J.S., A.-B.D., S.S., A.D. (Awa Diop) and B.F.; Project administration, B.A.S.; Resources, B.D.B., V.M. and S.C.; Software, A.B.; Supervision, E.M. and J.-P.M.; Validation, J.A. and J.-P.V.; Visualization, D.S. and Y.B.; Writing—original draft, L.D. Y.S. and M.T.; Writing—review and editing, L.D. S.-P.M., B.A.S., E.M., J.-P.M., J.A. and J.-P.V. All authors have read and agreed to the published version of the manuscript.

Funding: This research was funded by AFD (French Development Agency) (convention AFD CZZ 1894 01K).

Acknowledgments: The authors acknowledge the AFD (Agence Française de Développement) funded projet "Développement durable des zones littorales (Sénégal, Guinée Bissau, Guinée): vers une gouvernance citoyenne des territoires—(Sustainable development of coastal areas (Senegal, Guinea-Bissau, Guinea): towards a citizen governance of the territories (convention AFD CZZ 1894 01K), as well as the Grdr Migrations Citoyenneté Développement NGO for the project management.

Conflicts of Interest: The authors declare no conflict of interest.

References

1. Paturej, E. Estuaries, Types, Role and Impact on Human Life. 2008. Available online: https://www.researchgate.net/publication/228488312 (accessed on 11 September 2019).

2. Pritchard, D.W. What is an estuary: Physical viewpoint. In *Estuaries*; Lauff, G.H., Ed.; American Association for the Advancement of Science: Washington, DC, USA, 1967; Volume 83, pp. 3–5.

3. Saos, J.-L.; Thiebaux, J.-P. *Evolution de la salinité en Basse Casamance: Exemple du marigot de BAILA*; Equipe Pluri Disciplinaire d'Etude des Ecosystèmes Côtiers; Rapport Final; UNESCO Sea Sciences Division, PNUD, UCAD, UNESCO Breda Ed (the NL); Etude des Estuaires du Sénégal: Casamance, Sénégal, 1991.

4. Alvera-Azcarate, A. 2011. Available online: http://modb.oce.ulg.ac.be/mediawiki/upload/Aida/OCEA0011/3_ESTUARiES.pdf (accessed on 11 September 2019).

5. Pagès, J.; Citeau, J. Rainfall and salinity of a Sahelian estuary between 1927 and 1987. *J. Hydrol.* **1989**, *113*, 325–341. [CrossRef]

6. Nunes Vaz, R.A.; Lennon, G.W.; Bowers, D.G. Physical behaviour of a large, negative or inverse estuary. *Cont. Shelf Res.* **1990**, *10*, 277–304. [CrossRef]

7. Tomczak, M. *Examples of Estu Aries and Their Classification*2000. Available online: https://incois.gov.in/Tutor/ShelfCoast/chapter13.html (accessed on 11 September 2019).

8. Lavin, M.F.; Godinez, V.M.; Alvarez, L.G. Inverse-estuarine features of the upper gulf of California. *Estuar. Coast. Shelf Sci.* **1998**, *47*, 769–795. [CrossRef]

9. De Gutierrez Velasco, G.; Winant, C.D. Wind- and Density-driven circulation in a well-mixed inverse estuary. *J. Phys. Oceanogr.* **2004**, *34*, 1103–1116. [CrossRef]

10. Villanueva, M.C. Contrasting tropical estuarine ecosystem functioning and stability: A comparative study. *Estuar. Coast. Shelf Sci. March* **2015**, *155*, 89–103. [CrossRef]

11. Kampf, J.; Sadrinasab, M. The circulation of the Persian Gulf: A numerical study. *Ocean Sci.* **2006**, *2*, 1–15. [CrossRef]

12. Edyvane, K.S. Australia, coastal ecology. In *Encyclopedia of Coastal Science*; Schwartz, M., Ed.; Springer: Dordrecht, The Netherlands, 2005; pp. 96–109.

13. Kampf, J.; Bell, D. The Murray/Coorong estuary: Meeting of the waters. In *Estuaries of the World*; Wolanski, E., Ed.; Springer: Dordrecht, The Netherlands, 2013; pp. 31–47.

14. Kampf, J. South Australia's Large Inverse Estuaries: On the Road to Ruin. In *Estuaries of Australia in 2050 and Beyond, Estuaries of the World*; Wolanski, E., Ed.; Springer Science & Business Media: Dordrecht, The Netherlands, 2014.

15. Barusseau, J.-P.; Diop, E.S.; Giresse, P.; Monteillet, J.; Saos, J.-L. Conséquences sédimentologiques de l'évolution climatique fini-Holocène (10^2–10^3 ans) dans le delta du Saloum (Sénégal). *Océanographie Trop.* **1986**, *21*, 69–98.

16. Billon, B. Le Niger à Niamey: Décrue et étiage 1985. *Cah. l'ORSTOM Série Hydrol.* **1985**, *21*, 3–22.

17. Boivin, P.; Le Brusq, J.-Y. Désertification et salinisation des terres au Sénégal. Problèmes et remèdes. In Proceedings of the National Seminar on desertification, Dakar, Republic of Senegal, 22–26 April 1985.

18. Barusseau, J.-P.; Diop, E.S.; Saos, J.-L. Evidence of dynamics reversal in tropical estuaries, geomorphological and sedimentological consequences (Salum and Casamance Rivers, Senegal). *Sedimentology* **1985**, *32*, 543–552. [CrossRef]

19. Thior, M.; Sy, A.A.; Diedhiou, S.O.; Cissé, I.; Gomis, J.S.; Descroix, L. Low Casamance inverse estuary: Impacts on water quality and agrosystems in islander area/Estuaire inverse de basse Casamance: Impacts sur la qualité de l'eau et des agrosystèmes en milieu insulaire. *Environ. Water Sci. Public Health Territ. Intell. J.* **2019**, *3*, 192–197.

20. Nicholson, S.E. The West African sahel: A review of recent studies on the rainfall regime and its interannual variability. *ISRN Meteorol.* **2013**, *2013*, 453521. [CrossRef]

21. Descroix, L.; Diongue Niang, A.; Panthou, G.; Bodian, A.; Sané, T.; Dacosta, H.; Malam Abdou, M.; Vandervaere, J.-P.; Quantin, G. Evolution Récente de la Mousson en Afrique de l'Ouest à Travers Deux Fenêtres (Sénégambie et Bassin du Niger Moyen). *Climatologie* **2015**, *12*, 25–43.

22. Descroix, L.; Guichard, F.; Grippa, M.; Lambert, L.A.; Panthou, G.; Gal, L.; Dardel, C.; Quantin, G.; Kergoat, L.; Bouaïta, Y.; et al. Evolution of surface hydrology in the Sahelo-Sudanian stripe: An updated synthesis. *Water* **2018**, *10*, 748. [CrossRef]

23. Panthou, G.; Lebel, T.; Vischel, T.; Quantin, G.; Sané, Y.; Ba, A.; Ndiaye, O.; Diongue-Niang, A.; Diop Kane, M. Rainfall intensification in tropical semi-arid regions: The Sahelian case. *Environ. Res. Lett.* **2018**, *13*, 064013. [CrossRef]

24. Panthou, G.; Vischel, T.; Lebel, T. Recent trends in the regime of extreme rainfall in the Central Sahel. *Int. J. Climatol.* **2014**, *34*, 3998–4006. [CrossRef]

25. Faye, C.; Sané, T. Le changement climatique dans le bassin-versant de la Casamance: Évolution et tendances du climat, impacts sur les ressources en eau et stratégies d'adaptation. In Eaux et Sociétés Face au Changement Climatique Dans le Bassin de la Casamance. Actes de l'Atelier Scientifique et Lancement de l'initiative Casamance: Un Réseau Scientifique au Service du Développement en Casamance. *Ziguinchor Sénégal* **2015**, *15–17*, 280.

26. Mahé, G.; Lienou, G.; Descroix, L.; Bamba, L.; Paturel, J.-E.; Laraque, A.; Meddi, M.; Habaieb, M.; Adeaga, O.; Dieulin, C.; et al. The rivers of Africa: Witness of climate change and human impact on the environment. *Hydrol. Process.* **2013**, *27*, 2105–2114. [CrossRef]

27. Binet, D.; Le Reste, L.; Diouf, P.S. The influence of runoff and fluvial outflow on the ecosystems and living resources of West African coastal waters. In *Effects of Riverine Inputs on Coastal Ecosystems and Fisheries Resources*; Fisheries Technical Paper No.349; FAO—Food and Agriculture Organization of the United Nations: Rome, Italy, 1995; Volume 349, pp. 89–118. ISBN 978-9251036341.

28. Gac, J.-Y.; Kane, A. Le fleuve Sénégal: 2- flux continentaux de matières dissoutes à l'embouchure. *Sci. Géol. Bull.* **1986**, *39*, 151–172.

29. Saos, J.-L.; Kane, A.; Carn, M.; Gac, J.-Y. Persistance de la sécheresse au Sahel: Invasion marine exceptionnelle dans la vallée du fleuve Sénégal. In Proceedings of the 10th meeting of Earth Sciences, Bordeaux, France; 1984; p. 499.

30. Diop, E.S. La Côte Ouest Africaine du Saloum (Sénégal) à la Mellacorée (République de Guinée). Ph.D. Thesis, Université Louis Pasteur, Strasbourg, Paris, France, 1990.

31. Sakho, I. Evolution et Fonctionnement Hydro-Sédimentaire de la Lagune de la Somone, Petite Côte, Sénégal. Ph.D. Thesis, Université de Rouen, France and Université Cheikh Anta Diop de Dakar, Dakar, Sénégal, 2011.

32. Marius, C. Mangroves du Sénégal et de la Gambie. Ecologie, Pédologie, Géochimie, Mise en Valeur et Aménagement. Ph.D. Thesis, Louis Pasteur Strasbourg University, Strasbourg, Paris, 1985; 335p.

33. Diouf-Goudiaby, K. Influence de la salinité sur les déplacements et la croissance des juvéniles d'un poisson ubiquiste, Sarotherodon Melanotheron (Teleosteen, Cichlidae), dans les estuaires ouest-africains. Ph.D. Thèse, Université Montpellier 2, Montpellier, France, 2006; 176p.

34. Baran, E. Comparaison des ichtyofaunes estuariennes du Sénégal, de Gambie et de Guinée. In *Dynamiques et Usages de la Mangrove Dans les Pays des Rivieres du Sud (du Sénégal à la Sierra Leone) M-C*; Salem, C., Ed.; ORSTOM Editions, Paris-Bondy: Paris, France, 1994; pp. 84–92.

35. Noblet., J.-P. 2012. Available online: http://www.jf-noblet.fr/spe2012/2-eau/acti1-sel (accessed on 23 February 2020).

36. Toublanc, F.; Brenon, I.; Coulombier, T. Formation and structure of the turbidity maximum in the macrotidal Charente estuary (France): Influence of fluvial and tidal forcing. *Estuar. Coast. Shelf Sci.* **2016**, *169*, 1–14. [CrossRef]

37. Duy Vinh, V.; Ouillon, S.; Van Uu, D. Estuarine turbidity maxima and variations of aggregate parameters in the cam-nam trieu estuary, north vietnam, in early wet season. *Water* **2018**, *10*, 68. [CrossRef]

38. Olivry, J.-C.; Dacosta, H. *Le Marigot de Baila: Bilan des Apports Hydriques et Evolution de la Salinit*; Orstom: Dakar, Senegal, 1984; 150p. Available online: http://horizon.documentation.ird.fr/exl-doc/pleins_textes/divers11-07/17933.pdf#search=%22olivry%20dacosta%201984%20baila%22 (accessed on 21 September 2019).

39. Olivry, J.-C. Les conséquences durables de la sécheresse actuelle sur l'écoulement du fleuve Sénégal et l'hypersalinisation de la Basse-Casamance. In *Proceedings of the Influence of Climate Change and Climatic Variability on the Hydrologie Regime and Water Resources. In Proceedings of the Vancouver Symposium, Vancouver, BC, Canada, 9–22 August 1987).*; IAHS Publisher: Wallingford, UK, 1987.

40. Debenay, J.-P.; Pagès, J. Foraminifères et thécamoebiens de l'estuaire hyperhalin du fleuve Casamance (Sénégal). *Revue d'Hydrobiologie Trop.* **1987**, *20*, 233–256.

41. Valiela, I.; Bowen, J.L.; York, J.K. Mangrove forests: One of the world's threatened major tropical environments. *Bioscience* **2001**, *51*, 807–815. [CrossRef]

42. FAO. Forest Resources Assessment working paper n°63. Forest Resources Division. In *Status and Trends in Mangrove Area Extent Worldwide*; Wilkie, M.L., Fortuna, S., Eds.; FAO: Rome, Italy, 2003. Available online: http://www.fao.org/docrep/007/j1533e/j1533e00.htm (accessed on 27 February 2020).

43. Andrieu, J. Dynamiques des Paysages Dans les Régions Septentrionales des Rivieres-du-Sud (Sénégal, Gambie, Guinée-Bissau). Ph.D. Thèse, Université dé Paris, Paris, France, 2008; 524.

44. Conchedda, G.; Durieux, L.; Mayaux, P. An object-based method for mapping and change analysis in mangrove ecosystems. *ISPRS J. Photogramm. Remote Sens.* **2008**, *63*, 578–589. [CrossRef]

45. Andrieu, J. L'évolution de la Mangrove (1979–2019) du Saloum au Gêba, par Télédétection. In Proceedings of the Communication to the International Colloque Vulnerability of Societies and Environment of Coastal and Estuarine West Africa Held on in Ziguinchor (Senegal), Dakar, Senegal, 19–22 November 2019.

46. Andrieu, J.; Lombard, F.; Fall, A.; Thior, M.; Ba, B.D.; Diémé, B.E.A. Botanical field-study and remote sensing to describe mangrove resilience in the Saloum Delta (Senegal) after 30 years of degradation narrative. *For. Ecol. Manag.* **2020**, *461*, 117963. [CrossRef]

47. Gilman, E.L.; Ellison, J.; Duke, N.C.; Field, C. Threats to mangroves from climate change and adaptation options: A review. *Aquat. Bot.* **2008**, *89*, 237–250. [CrossRef]

48. Osland, M.J.; Feher, L.C.; Lopez-Portillo, J.; Daya, R.H.; Suman, D.O.; Guzman Menendez, J.M.; Rivera-Monroy, V.H. Mangrove forests in a rapidly changing world: Global change impacts andconservation opportunities along the Gulf of Mexico coast. *Estuar. Coast. Shelf Sci.* **2018**, *214*, 120–140. [CrossRef]

49. Dièye, E.B.; Diaw, A.T.; Sané, T.; Ndour, N. Dynamique de la mangrove de l'estuaire du Saloum (Sénégal) entre 1972 et 2010. Dynamics of the Saloum estuary mangrove (Senegal) from 1972 to 2010. Cybergeo. *Eur. J. Geogr. Environ. Nat. Environ. Nat. Paysage* **2013**. [CrossRef]

50. Dacosta, H. Précipitations et Ecoulement Sur le Bassin de la Casamance. Ph.D. Thesis, University Cheikh Anta Diop, Dakar, Senegal, 1989; 283p.

51. Mendy, A. Ressources en Eau des Bassins Versants de la Nema et de Medina Djikoye. Ph.D. Thesis, University Cheikh Anta Diop, Dakar, Senegal, 2010; 335p.

52. Sambou, S. Contribution à la Connaissance des Ressources en Eau du Bassin Versant du Fleuve Kayanga/Gêba (Guinée, Sénégal, Guinée-Bissau). Ph.D. Thesis, University Cheikh Anta Diop, Dakar, Senegal, 2019.

MDPI

St. Alban-Anlage 66

4052 Basel

Switzerland

Tel. +41 61 683 77 34

Fax +41 61 302 89 18

www.mdpi.com

Water Editorial Office

E-mail: water@mdpi.com

www.mdpi.com/journal/water